THE BIOLOGY OF THE NOCARDIAE

THE BIOLOGY OF THE NOCARDIAE

Edited by

M. GOODFELLOW

Department of Microbiology, University of Newcastle-upon-Tyne, England

G. H. BROWNELL

Department of Cell and Molecular Biology, Medical College of Georgia, U.S.A.

and

J. A. SERRANO

Centro de Microscopia Electronica, Universidad de los Andes, Venezuela

1976

ACADEMIC PRESS

LONDON · NEW YORK · SAN FRANCISCO

A Subsidiary of Harcourt Brace Jovanovich Publishers

ACADEMIC PRESS INC. (LONDON) LTD.
24/28 Oval Road,
London NW1 7DX

United States Edition published by
ACADEMIC PRESS INC.
111 Fifth Avenue
New York, New York 10003

Library of Congress Catalog Card Number: 76-016970
ISBN: 0-12-289650-5

PRINTED IN GREAT BRITAIN BY
J. W. ARROWSMITH LTD., BRISTOL

Contributors

J. N. ADAMS, *Department of Microbiology, School of Medicine, University of South Dakota, Vermillion, South Dakota 57069, U.S.A.*

F. ANTOINE-PORTAELS, *Prince Leopold Institute of Tropical Medicine, B-200 Antwerp, Belgium*

B. L. BEAMAN, *Department of Medical Microbiology, University of California School of Medicine, Davis, California 95616, U.S.A.*

H. BICKENBACH, *Institute of Hygiene, University of Cologne, Cologne, Federal Republic of Germany*

L. F. BOJALIL, *Departmento de Ecología Humana, Facultad de Medicina, U.N.A.M., México 20, D. F., Mexico*

CLAUDETTE BORDET, *Laboratoire de Biochimie Microbienne, Université de Lyon 1, 43 boulevard du 11 novembre 1918, 69621 Villeurbanne, France*

I. J. BOUSFIELD, *National Collection of Industrial Bacteria, Torry Research Station, Aberdeen AB9 8DG, Scotland*

S. G. BRADLEY. *Department of Microbiology, Virginia Commonwealth University, Richmond, Virginia 23298, U.S.A.*

G. H. BROWNELL, *Department of Cell and Molecular Biology, Medical College of Georgia, Augusta, Georgia 30902, U.S.A.*

MAGDALENA F. CONTRERAS, *Departmento de Ecología Humana, Facultad de Medicina, U.N.A.M., México 20, D.F., Mexico*

T. CROSS, *Postgraduate School of Studies in Biological Sciences, University of Bradford, Bradford BD7 1DP, England*

M. GOODFELLOW, *Department of Microbiology, The Medical School, The University, Newcastle-upon-Tyne, NE1 7RU, England*

A. GONZÁLEZ-OCHOA, *Instituto de Salubridad y Enfermedades Tropicales, México 17, D.F., Mexico*

RUTH E. GORDON, *Waksman Institute of Microbiology, Rutgers University, The State University of New Jersey, New Brunswick, New Jersey, U.S.A.*

J. LACEY, *Department of Plant Pathology, Rothamsted Experimental Station, Harpenden, England*

MARY P. LECHEVALIER, *Waksman Institute of Microbiology, Rutgers University, The State University of New Jersey, New Brunswick, New Jersey 08903, U.S.A.*

A. LIND, *Institute of Medical Microbiology, University of Göteborg, Göteborg, Sweden*

M. MAGNUSSON, *Tuberculin Department, Statens Seruminstitut, DK 2300 Copenhagen, S. Denmark*

G. MICHEL, *Laboratoire de Biochemie Microbienne, Université de Lyon 1, 43 boulevard du 11 novembre 1918, 69621 Villeurbanne, France*

D. E. MINNIKIN, *Department of Organic Chemistry, School of Chemistry, The University, Newcastle-upon-Tyne, NE1 7RU, England*

E. N. MISHUSTIN, *Institute of Microbiology, USSR Academy of Sciences, Moscow, U.S.S.R.*

M. MORDARSKI, *Ludwik Hirszfeld Institute of Immunology and Experimental Therapy, Polish Academy of Sciences, Wrocław, Chulubinskiego ul. 4, Poland*

L. ORTIZ-ORTIZ, *Departmento de Ecología Humana, Facultad de Medicina, U.N.A.M., México 20, D.F., Mexico*

P. V. PATEL, *Department of Biochemistry, The University of Hull, Hull, HU6 7RX, England*

H. PRAUSER, *Zentralinstitut für Mikrobiologie und experimentelle Therapie, Forschungszentrum für Molekularbiologie und Medizin, Akademie der Wissenschaften der DDR, Jena, German Democratic Republic*

C. RATLEDGE, *Department of Biochemistry, The University of Hull, Hull HU6 7RX, England*

M. RIDELL, *Institute of Medical Microbiology, University of Göteborg, Göteborg, Sweden*

T. J. ROWBOTHAM, *Postgraduate School of Studies in Biological Sciences, University of Bradford, Bradford BD7 1DP, England*

K. P. SCHAAL, *Institute of Hygiene, University of Cologne, Cologne, Federal Republic of Germany*

A. A. SERRANO, *Centro de Microscopia Electronica, Universidad de los Andes, Merida, Venezuela*

J. A. SERRANO, *Centro de Microscopia Electronica, Universidad de los Andes, Merida, Venezuela*

G. P. SHARPLES, *Department of Botany, University of Liverpool, Liverpool, England.*

P. H. A. SNEATH, *Medical Research Council Microbial Systematics Unit, University of Leicester, Leicester, England*

I. TARNOK, *Department for Enzymatical Microbiology, Research Institute Borstel, Institute for Experimental Biology and Medicine, 2061 Borstel, Federal Republic of Germany*

E. Z. TEPPER, *Institute of Microbiology, USSR Academy of Sciences, Moscow, U.S.S.R.*

S. T. WILLIAMS, *Department of Botany, University of Liverpool, Liverpool, England*

Preface

This volume is presented to microbiologists interested in actinomycetes from both the pure and applied aspects. At present it is unfortunately necessary to point out that scientific progress has rarely drawn a sharp distinction between the pure and applied sides, and the pioneering work of Professor S. A. Waksman on actinomycetes is eloquent testimony to this. This is the first book to be devoted wholly to the biology of the nocardiae for, until recently, these organisms have often been dismissed as an appendage of their famous neighbours the streptomycetes. In fact, it is only through the patient and thorough studies of workers like Professors R. E. Gordon, F. Mariat and N. M. McClung that interest in nocardiae remained alive, and in many ways this volume is a logical outcome of their researches.

Each chapter has been written by a specialist active in his field, so the book as a whole draws together a unique collection of information on a seriously neglected group of micro-organisms which have many fascinating and unusual features. The nocardiae are of interest from a biochemical, ecological, genetical, industrial, medical, taxonomic and veterinary point of view and all of these facets are covered. The various experts on nocardiae rarely venture beyond their own specialist interests but it is hoped that the present volume will encourage them to do so. The nocardiae deserve recognition in their own right and we hope that our efforts will help to stimulate further interest in them.

We would like to express our thanks to the chapter authors for their co-operation. Many colleagues have assisted us in a variety of ways and we extend our thanks to them also. In particular, we would like to thank some of our enthusiastic younger colleagues: Grace Alderson, Jim Clarke, David and Hazel Collins, David Dawson, Anne and Terry Hudson and Valerie Orchard, and also Professor C. A. Green, Professor P. H. A. Sneath and Dr D. E. Minnikin for their constant help, encouragement and support.

Throughout, we have been greatly encouraged and assisted by Academic Press.

M. Goodfellow
G. H. Brownell
J. A. Serrano

June 1976

Contents

3. A Taxonomist's Obligation

R. E. GORDON

4. An Evaluation of Numerical Taxonomic Techniques in the Taxonomy of *Nocardia* and Allied Taxa

P. H. A. SNEATH

5. The Micromorphology and Fine Structure of Nocardioform Organisms

S. T. WILLIAMS, G. P. SHARPLES, J. A. SERRANO, A. A. SERRANO and J. LACEY

6. Cell Walls of Nocardiae

G. MICHEL and CLAUDETTE BORDET

7. Lipid Composition in the Classification and Identification of Nocardiae and Related Taxa

D. E. Minnikin and M. Goodfellow

8. Serological Relationships Between *Nocardia, Mycobacterium, Corynebacterium* and the "*rhodochrous*" taxon

A. Lind and M. Ridell

9. Sensitin tests as an aid in the taxonomy of *Nocardia* and its pathogenicity

M. Magnusson

13. The Ecology of Nocardioform Actinomycetes

T. Cross, T. J. Rowbotham, E. N. Mishustin, E. Z. Tepper, F. Antoine-Portaels, K. P. Schaal and H. Bickenbach

14. Lipid-Soluble, Iron-Binding Compounds in *Nocardia* and Related Organisms

C. Ratledge and P. V. Patel

15. Possible Mechanisms of Nocardial Pathogenesis

B. L. Beaman

16. Delayed Hypersensitivity to *Nocardia* Antigens

L. ORTIZ-ORTIZ, MAGDALENA F. CONTRERAS and L. F. BOJALIL

17. Nocardiae and Chemotherapy

A. GONZÁLEZ-OCHOA

18. Metabolism in Nocardiae and Related Bacteria

I. TARNOK

1. The Taxonomy of the Genus *Nocardia*: Some light at the end of the tunnel?

MARY P. LECHEVALIER

Waksman Institute of Microbiology, Rutgers University, State University of New Jersey, New Brunswick, New Jersey 08903, U.S.A.

Contents

1. Introduction

Nocardia is the oldest name now in current use for an aerobic actinomycete genus.* It is a taxon that has had a long and turbulent history over the years, becoming a sort of microbiological duffel

* It is if one disregards the Russian school's continued use of *Actinomyces* for *Streptomyces*.

bag stuffed fuller and fuller of heterogeneous strains. This was as much a consequence of there being no obvious taxonomic slot for them elsewhere as the lack of a clearcut definition of the genus itself. Further, the history of the proposals and counterproposals in the nomenclature of nocardiae, has and still does resemble an immense free-for-all in a game of nomenclatural musical chairs.

The current eighth edition of Bergey's Manual (McClung, 1974) recognizes 31 species of *Nocardia* and lists 40 others under *Species incertae sedis.* The list of genera to which these organisms were originally or subsequently assigned, reads, in fact, a little like a "Who's-Who" or even "Who's-That" of the microbial world:

Actinobacillus, Actinococcus, Actinomadura, Actinomyces, Anthracillus, Asteroides, Bacillus, Bacterium, Carterii, Cladothrix, Cohnistreptothrix, Corynebacterium, Discomyces, Flavobacterium, Indiellopsis, Microbacterium, Micrococcus, Mycobacterium, Oerskovia, Oospora, Pionnothrix, Proactinomyces, Serratia, Streptomyces and *Streptothrix* (Lessel, 1960).

This review is not intended to be exhaustive. The author will discuss only the high and low points of the history of the genus that she feels to have been important in giving it some sort of shape, however grotesque.

2. Early Definitions and Concepts

The genus *Nocardia* was originally described by Trevisan (1889) to accommodate five species. These included *N. farcinica* (Nocard, 1888, "bacille du farcin"), which, because it was listed first, was subsequently assumed to be the type of the genus, and *N. foersteri*, the first actinomycete ever described (Cohn, 1875). Trevisan described his new genus as "*Cladothrix* without a sheath" showing "false branching." In this, he reflected the uncertainty that Cohn himself felt concerning the nature of the branching of his strain, and that Nocard's description of his farcy bacillus did nothing to dissipate: "This ramification (of the filaments at the edge of the colony) is more apparent than real; the careful examination of young colonies shows that the development of the filaments takes place by elongation; when the bacillus has reached twice its initial length, it segments and the newly formed units bend, most often at a right angle to the older segment which continues to grow in a straight line, in sum, it is a question of false branching analogous to that which one sees among the *Cladothrix*."

As a matter of fact, although belittled by later writers, these early authors were not entirely wrong. Besides true branching, which both Cohn's and Nocard's drawings also show in an unmistakable fashion, many organisms now classified as *Nocardia*, not to mention *Actinomyces*, *Mycobacterium* and *Corynebacterium*, show a kind of "false-branching," albeit without a sheath. If one reserves "false branching" for the sheathed bacteria, then "pseudo-ramiform growth" might be an appropriate substitute for this phrase when it is applied to actinomycetes. As can be seen in the many drawings and descriptions of earlier workers (Ørskov, 1923, 1938; Jensen, 1931; Gray & Thornton, 1928; McCarter & Hastings, 1935; Topping, 1937) as well as more recent ones (Bisset and Moore, 1949; McClung, 1949; Adams & McClung, 1962; Gordon & Mihm, 1962*a*) such pseudo-ramiform growth can arise by different means. These include the lateral germination of segments *in situ* as well as the kind of "trompe l'oeil" growth such as Nocard described for *N. farcinica* where one fragment shifts to lie across or at an angle to the main growing filament.

From the time Cohn first described his *Streptothrix foersteri*, and even after Trevisan's proposal of *Nocardia*, other authors such as Winter (1879), Macé (1888), Migula (1895) and Engler (1907) continued to place the actinomycetes with the false-branching bacteria (Breed & Conn, 1919).

Trevisan also included in his new genus, as *N. actinomyces*, the *Actinomyces bovis* of Harz (1879) which causes actinomycosis of cattle. Trevisan cannot be faulted for grouping what subsequently became to be accepted as an anaerobic species with the strictly aerobic *N. farcinica* since Harz did not actually succeed in growing his actinomycete and its oxygen requirements were unknown at the time Trevisan published. What is remarkable, however, is that later authors such as Breed & Conn (1919) and even Ørskov (1923) and Jensen (1931) saw no contradiction in including both *N. farcinica* and *A. bovis* in the same genus. Lieske (1921), although making some very interesting morphological and physiological studies, could conceive of no better course than to return all the actinomycetes he examined back into *Actinomyces* regardless of differences in spore formation, relation to free oxygen and in physiological characters. He even stated that, because of their variability, the definition of species was impossible in the actinomycetes, a sentiment that many later actinomycetologists have privately echoed in their gloomier moments.

The first to recognize the relationship of nocardiae and other actinomycetes to mycobacteria and corynebacteria was Lachner-Sandoval (1898) who placed them all together in the family *Actinomycetes.*

Wright (1905) proposed to restrict the use of *Actinomyces* to the anaerobic actinomycetes growing above room temperature and producing no spores, leaving *Nocardia* for the rest. He was quite sure that actinomycoses, which he defined as inflammatory processes the lesions of which contain the characteristic granules or "drusen," were caused only by anaerobes and that no "*Nocardia*" had ever been proven to form lesions containing such structures. He proposed "nocardiosis" as the term to be used for infections caused by members of that genus.

Pinoy (1913), after having examined many different cases of actinomycosis in both animals and man, came to the conclusion that there were indeed different types of actinomycetes which could cause actinomycoses, including those which were aerobic, anaerobic and facultative. Although these resembled each other on superficial microscopic examination, the anaerobes were seen to differ from the aerobes by the lack of formation of arthrospores and by their slow growth. Pinoy proposed *Nocardia* for the aerobes and *Cohnistreptothrix* for the anaerobes.

Another proposal was that of Chalmers & Christopherson (1916) whose key for pathogenic actinomycetes divided the group into three genera, *Actinomyces, Cohnistreptothrix* and *Nocardia,* depending on the color of the granules contained in the pus and the requirements for free oxygen.

A very large proportion of the published work on actinomycetes up until 1919 was concerned with strains of medical importance. In that year, Waksman, having studied a number of saprophytes and parasites both morphologically and by means of a quite sophisticated group of physiological tests, did not find sufficient differences between the two groups to warrant placing them in different genera.

Langeron (1922) refined Pinoy's definition of *Nocardia* and *Cohnistreptothrix*, the latter being described as facultative anaerobes. *Nocardia* (facultative aerobes), section *Parasitica*, was divided into three sub-sections, *Majores, Minores* and *Breviores*, depending on ease of cultivation, growth on potato, formation of aerial mycelium, branching, acid-fastness, odor production and production of diastases. Like all such systems, it was extremely dependent on the conditions of culture and subject to the vagaries of subjective interpretation.

On the basis of an extensive microscopic study of previously described strains as well as recent isolates of his own, Ørskov (1923) proposed to divide the actinomycetes into three large groups. The first, *Cohnistreptothrix*, was defined as containing strains having an undividing substrate mycelium and an aerial mycelium which was almost always present, that was thicker than the substrate mycelium when viewed with oil immersion. Spores were formed on the aerial hyphae by constriction rather than by septation and were more resistant to desiccation and heat than the vegetative hyphae. In liquid cultures, this group of organisms was slow to show surface growth. Among the cultures he placed here were "*Actinomyces*" *madurae, Nocardia dassonvillei* as well as many strains that would now be classified as streptomycetes. Ørskov's choice of generic name for this group was unfortunate since his definition was quite different from Pinoy's original, and only served to enrich the stew of confusion.

Ørskov's second group, *Actinomyces*, contained organisms which he felt were morphologically identical but which he nevertheless divided further on the basis of presence or absence of aerial mycelium. Group IIa formed aerial hyphae which did not differ in size from the substrate hyphae when examined under oil. Both vegetative and aerial mycelium ultimately segmented by means of septa, such segments not being more resistant to heat or desiccation than the vegetative hyphae. This group yielded early surface growth in liquid culture. *Nocardia asteroides* and *N. rubra* were included here. Group IIb was like IIa except that no aerial mycelium was produced and "angular growth" (his term for pseudo-ramiform growth) like that described by Nocard for *N. farcinica*, would always be observed during the growth cycle. Ørskov strongly underlined the close morphological relationship that Group IIb had with the coryne- and mycobacteria. He included *N. farcinica, N. polychromogenes* and *A. bovis* in IIb.

One of Ørskov's real contributions to actinomycetology as a whole and to the study of *Nocardia* in particular, was his emphasis on studying cultures in their undisturbed state growing on agar media, rather than by the use of the classical stained slide. Even today, the literature is probably full of descriptions of "*Arthrobacter* sp." that would turn out to be actinomycetes belonging to Group IIb if properly examined in undisturbed culture.

Jensen (1931) was unhappy with Ørskov's choice of genus for Group II and proposed, instead, the genus *Proactinomyces.* He included it with *Mycobacterium* and *Corynebacterium* in the family *Proactinomycetaceae* of Lehmann & Neumann (1927). Jensen stated that the "constant formation of an initial mycelium" distinguished the proactinomycetes from the corynebacteria and mycobacteria and that the lack of spore formation separated them from Group I types. In later years (1953) he was to underline their unfortunate capacity to vary in such morphological characteristics.

Puntoni & Leonardi (1940) removed the species *asteroides* from *Actinomyces* since they felt that this genus should be restricted to organisms causing the actinomycosis of the lymphatics of cows. They placed it instead in a new genus *Asteroides* which was to contain the causative agents of pulmonary "pseudoactinomycosis." This proposal was typical of the tunnel vision that afflicted medical actinomycetologists even at that late date. Although saprophytic actinomycetes continued to be studied along with parasites by generalists such as Waksman, Lieske, Ørskov, Jensen and Erikson, the medically-oriented microbiologists continued to act as if there were no other actinomycetes but those that cause disease.

Krassil'nikov (1938) recognized Jensen's *Proactinomyces* as a useful and valid genus. Like Jensen, he and Tausson also underlined these organisms' capacity to vary in morphological characteristics (1938).

Baldacci (1939) insisted that *Proactinomyces* be limited to aerobes. In consequence of the uncertain identity of *Actinomyces bovis,* he resurrected Haass' *Actinobacterium* for the anaerobes and assigned *Cohnistreptothrix* to the microaerophilic group.

The splitting of hairs from a morphological point of view continued unabated. Umbreit (1939) divided the proactinomycetes into two types. The α type was characterized by instability of the mycelium, softness of the colonies and turbid growth in liquid media. The β type had a long stable mycelium, a hard colony and growth in liquid media left the broth clear. He remarked, however, on the occurrence of intermediate types which force the investigator to "resort to other characteristics."

While Waksman & Henrici (1943) argued that *Actinomyces* should be reserved for the anaerobic organisms causing "lumpy jaw" of cattle, they recognized the importance of Ørskov's separation of the actinomycetes into two major groups on the basis of the presence or lack of fragmentation of vegetative mycelium. They correctly pointed out that *Nocardia* had priority over *Proactinomyces* for the aerobic fragmenting actinomycetes of Ørskov's Group II and proposed the new genus *Streptomyces* to accommodate his Group I.

Their classification was followed in the 1948 edition of Bergey's Manual:

I. Mycelium rudimentary or absent, no spores formed. Acid-fast.
Family Mycobacteriaceae.
Genus: *Mycobacterium*

II. True mycelium produced.
A. Vegetative mycelium divides by segmentation into bacillary or coccoid elements. Some species partially acid-fast.
Family Actinomycetaceae
I. Obligate aerobic. The colonies are bacteria-like in nature, smooth, rough or folded, of a soft to a dough-like consistency, sometimes compact and leathery in young stages. Most forms do not produce any aerial mycelium; a few produce a limited mycelium, the branches of which also break up into oidiospores or segmentation spores. Some species are partially acid-fast.
Nocardia
II. Anaerobic or microaerophilic, parasitic; non-acid-fast, non-proteolytic and non-diastatic.
Actinomyces
B. Vegetation mycelium normally remains undivided.
Family Streptomycetaceae

Despite Waksman & Henrici's (1943) argument that *Nocardia* had priority and Jensen's (1953) subsequent recognition of this, the Russian school continued (and continues) to use *Proactinomyces* when referring to nocardiae. The Russians were not the only ones who did not care for this American system. Prévot in France (1946) divided his new order "*Actinobacteriales*" into the Gram-negative *Spherophoraceae* and the Gram-positive *Actinomycetaceae*. The latter family contained *Actinomyces* (=*Streptomyces* and other assorted actinomycetes), *Proactinomyces*, *Corynebacterium*, *Actinobacterium* (=anaerobic *Actinomyces*), *Bifidobacterium* and *Erysipelothrix*. Mycobacteria were segregated into a new mono-generic order "*Mycobacteriales*" containing all acid-fast or partially acid-fast forms. *Proactinomyces* were defined as "small rods or filaments frequently swollen and sometimes branched, generally forming very little mycelium at the beginning of growth. Short rod or coccoid forms in old cultures; no conidia. Non-motile, non-sporulating, aerobic, Gram-positive. Type species: *P. agrestis*." Prévot later (1961) came to recognize *Nocardia* as having priority over *Proactinomyces*.

Bisset & Moore (1949) tackled the problem of the aerobic coryneform organisms by proposing a new genus, *Jensenia*. This was classed in the family *Actinomycetaceae* along with the anaerobic to microaerophilic *Actinomyces*. The two genera were described as having "cells several times as long as broad. Individual bacillus unicellular." The *Mycobacteriaceae* (*Mycobacterium*, *Nocardia* and *Corynebacterium*) were distinguished from the *Actinomycetaceae* on the basis that the breadth of their cells were approximately the same as their length. The individual cell was "multicellular" (sic). It was not an easy system to follow.

But then, neither was that in Bergey's Manual. It may have been an improvement over previous systems, but as Jensen (1953) was soon to point out, "Although this classification looks simple enough, it conceals many difficulties that become obvious when it is tried to identify or merely assign generic position to an unknown organism with the general properties of the proactinomycetes." He also remarked that, "Standing between the mycobacteria and corynebacteria and the mould-like *Streptomyces* and linking these extremes through a gradual sequence of intermediate types, the proactinomycetes represent a . . . heterogeneous collection . . . ." He pointed out that (1) some *Nocardia* may be less mycelial than some mycobacteria, (2) that some mycobacteria form rudimentary aerial hyphae and (3) that acid-fastness was very dependent on the growth conditions and on the strain examined (see also Erikson, 1949). With the fervor of a carpenter driving nails into a coffin, he reminded his listeners that there are strains which have aerial spores typical of *Streptomyces* but a vegetative mycelium which fragments easily like a *Nocardia.* He also discussed the problems of classifying mutants and variants which no longer show the morphological features of the parents.

In summing up he remarked, "... we may say that while the typical representatives of *Corynebacterium, Mycobacterium* and *Nocardia* are easy enough to recognize on the basis of morphology, staining reactions and type of carbon metabolism, the genera are linked by boundary forms of which the position is difficult or impossible to define, because the distinctive features overlap, and each feature represents a spectrum that varies through many degrees."

Despite the fact that such objections were increasingly heard, the actinomycetologists had no new criteria to substitute for the old. Hence, in the seventh edition of Bergey's Manual (1957) we see that Waksman retained his key to the genus *Nocardia* exactly as it appeared in the previous edition. In his long, detailed description of the genus, however, he added a long footnote to the effect that although nocardiae were defined as nonmotile, some could be motile. In this description he also continued to make certain statements such as (nocardiae form) "Slender filaments or rods, frequently swollen and occasionally branched" followed subsequently by the statement that "they form true mycelium; hyphae branch abundantly ...." Or, "Typical nocardias never form an aerial mycelium, but there are cultures whose colonies are covered with a thin coating of short aerial hyphae which break up into cylindrical oidiospores." The exceptions seemed as numerous as the rules.

In 1957, Gordon & Mihm were still valiantly trying to apply criteria including formation of spores and colony margin morphology to the task of separating *Streptomyces, Nocardia* and *Mycobacterium.* They could distinguish two distinct groups of morphological characteristics which they felt quite adequately separated *Streptomyces* and *Mycobacterium* but *Nocardia* strains were found in both groups. However, they decided to assign the species *rhodochrous* to the genus *Mycobacterium* because some of the strains showed colonies with entire margins. They felt that *Nocardia* should be restricted to species whose colonies showed only filamentous margins. Arai & Kuroda (1959), however, reported that some of their strains of nocardiae, namely *N. asteroides* and *N. brasiliensis,* showed non-filamentous margins from the beginning of growth.

Schneidau & Shaffer (1957) maintained that a chain of spores was not sufficient to distinguish between *Nocardia* and *Streptomyces.* Gordon & Mihm themselves subsequently (1958) published electron micrographs showing that the spores of certain strains of *N. asteroides* could not be differentiated from those of *Streptomyces.* Bradley (1959) found that on the basis of formation of spores on the aerial mycelium alone, the separation of the two genera was impossible.

Arai & Kuroda (1959) also pointed out that the statement in the seventh edition of Bergey's that "typical *Nocardia* never form an aerial mycelium" might be confusing since many nocardiae formed aerial hyphae visible both macroscopically and microscopically varying from short shoots to extensive mats. The difficulties inherent in another statement in the 1957 Bergey's that nocardiae form "mycelium that breaks up into regularly cylindrical short cells then into coccoid cells" was pointed out by the report of Nikitina & Kalakoutskii (1971) that certain streptomycetes gave rise to stable nocardioform variants on fructose-containing media. Other factors, both physical and chemical, can influence such fragmentation (Williams *et al.*, 1974; see chapter 5). Mariat (1965) stated that since morphology varied even within the same strain of *Nocardia,* and was heavily dependent on temperature, pH, light, medium and the like, it was not useful in nocardial classification.

Thus, the usefulness of morphology in the differentiation of nocardiae from mycobacteria, corynebacteria and streptomycetes was being peppered from all sides; better criteria were obviously needed.

3. New Directions; Some Light

(a) *Cell wall chemistry*

Help was not long in coming. Cummins & Harris (1956*a,b*), in their pioneering use of cell wall composition in the classification of actinomycetes, had a catalytic action on the whole field. They reported on the cell wall constituents of strains of *Corynebacterium,*

Actinomyces, Streptomyces, Mycobacterium and *Nocardia.* All the walls contained alanine, glutamic acid, glucosamine and an unknown hexosamine. In addition, *Actinomyces* (anaerobes) contained lysine and galactose, *Streptomyces* and one non-acid-fast *Nocardia* (*gardneri*), L-diaminopimelic acid and glycine and the corynebacteria, acid-fast nocardiae and mycobacteria, *meso*-diaminopimelic acid and arabinose. The unknown hexosamine was later found to be 3-0-lactyl glucosamine by Strange (1956) who named it "muramic acid."

Earlier, Avery & Blank (1954) had concluded that the lack of cellulose and chitin in strains of *Actinomyces, Micromonospora, Nocardia* and *Streptomyces* clearly separated them from the fungi.

Other workers (Romano & Sohler, 1956) reported that, based on the examination of a small number of strains, nocardial and streptomycete cell walls could be distinguished on the basis of lysozyme-sensitivity and sugar constituents. All their nocardiae were lysozyme-insensitive and contained arabinose and galactose; streptomycetes were lysozyme-sensitive and contained no pentose. Unfortunately, later work showed that not all streptomycetes were lysozyme-sensitive (Sohler, Romano & Nickerson, 1958) and that not all nocardiae are lysozyme-resistant (Lechevalier & Lechevalier, 1974). Other more extensive studies (Cummins & Harris, 1958; Becker, Lechevalier & Lechevalier, 1965) pointed to the fact that not all strains then classed as *Nocardia* contained arabinose.

In these and later surveys (Davis & Baird-Parker, 1959; Cummins, 1965; Lechevalier, Lechevalier & Becker, 1966; Baboolal, 1969; Lechevalier, M. P. & Lechevalier, H., 1970) it was found that:

(1) Nocardiae of the *asteroides* and *farcinica* types, *N. brasiliensis, N. caviae* and *N. vaccinii* had the same type of cell wall as mycobacteria, corynebacteria of the *diphtheriae* type, leptotrichiae and members of the taxa *Micropolyspora, Thermomonospora viridis* (now *Saccharomonospora*), *Pseudonocardia* and *Bacterionema.* This cell wall type was termed Type IV by Lechevalier and co-workers (Table I).

(2) *Nocardia madurae, N. pelletieri, N. flexuosa* and *N. dassonvillei* had a Type III cell wall.

(3) *Nocardia turbata* was found to have a Type VI cell wall plus galactose.

Table 1
Major constituents of cell walls to actinomycetes

Cell wall type	Typical of*	Major distinguishing constituents†
I	*Streptomyces*	L-DAP, glycine
II	*Micromonospora*	*Meso*-DAP, glycine; hydroxy DAP may also be present
III	*Actinomadura*	*Meso*-DAP
IV	*Nocardia*	*Meso*-DAP, arabinose, galactose
	Mycobacterium	
	Corynebacterium (*diphtheriae*-type)	
	rhodochrous Group	
V	*Actinomyces israelii*	Lysine, ornithine
VI	*Rothia*	Lysine, aspartic acid
	Actinomyces bovis	
VI+Gal	*Oerskovia*	Lysine, aspartic acid, galactose
VII	*Agromyces*	DAB, glycine
—	*Mycoplana*	*Meso*-DAP, numerous amino acids

* Not all genera of each given wall type are listed.
† All cell wall preparations contain major amounts of alanine, glutamic acid, glucosamine and muramic acid.
DAP = 2,6-diaminopimelic acid; DAB = 2,4-diaminobutyric acid.

(4) Streptomycetes could be distinguished from all nocardiae by their content of the L isomer of diaminopimelic acid.

An interesting variation in the composition of the cell walls of certain actinomycete genera appears to hold hope for use as a criterion in differentiating nocardiae and mycobacteria from corynebacteria. Adam *et al.* (1969) and Azuma *et al.* (1970) reported that certain mycobacteria but not *Corynebacterium fermentans* contained *N*-glycolyl muramic acid instead of the usual *N*-acetylated variety. Subsequently, Vacheron *et al.* (1972) and Bordet *et al.* (1972) reported that strains of *N. asteroides, N. caviae* and *N. kirovani* (the last belongs to the *rhodochrous* group) contain this compound but that *N. piracicabensis* and *N. mediterranei* do not. The last two strains contain no nocardic acids (see (d) *Lipids*) whereas the first three do. *N*-glycolyl muramic acid is not unique to the nocardiae and mycobacteria; it has also been reported from *Micromonospora* (Vilkas, Massot & Zissman, 1970).

(b) *Whole cell composition*

It was also found that, without preparing pure walls, the cell wall types I through IV could conveniently be determined indirectly by the analysis of whole cell hydrolysates for the isomer of diaminopimelic acid and the type of sugars present (Becker *et al.*, 1964; Mordarski, 1966; Lechevalier, 1968; Lechevalier & Gerber, 1970) (Table 2). Furthermore, by whole cell analysis, *N. dassonvillei* could be distinguished from *N. madurae* and *N. pelletieri* by the lack of 3-0-methyl D-galactose (madurose) in the whole cells. It should be emphasized that madurose is found only in whole cells. It is lacking in cell wall preparations and is probably produced as an extracellular polysaccharide which is removed during the cell wall purification. Whole cell analyses have been used successfully by others (Pridham & Lyons, 1969; Berd, 1973; Staneck & Roberts, 1974) to help classify nocardiae.

Table 2

Whole cell sugar and DAP patterns of actinomycetes with cell wall types I to IV

Cell wall type	Whole cell constituents Diagnostic components		Whole cell sugar pattern
	DAP isomer	Sugars	
I	L	None	—
II	*Meso*	Xylose, arabinose	D
III	*Meso*	Madurose*	B
		None	C
IV	*Meso*	Arabinose, galactose	A

* Madurose = 3-0-methyl-D-galactose. Sugar present in whole cells only.

(c) *Phage sensitivity*

Actinomycetes, including nocardiae, are sensitive to phage. Although phage sensitivity has been used to delineate the actinomycete genus, *Nocardioides* (Prauser, 1974), there is conflicting evidence on its utility in nocardial taxonomy (see chapter 10).

Prauser (1967) and Prauser & Falta (1968) found that nocardiae fell into three sensitivity groups: (1) strains having a cell wall of Type IV, (2) strains with a cell wall of Type VI and (3) strains with a cell wall of Type III, thus confirming the divisions in *Nocardia* made previously on the basis of cell wall composition. Prauser emphasized that a given phage did not necessarily attack every strain having the same wall type, simply that its range was limited to organisms having the same wall composition.

In contrast, Bradley and his coworkers found phages that cross the boundaries of cell wall composition. Bradley & Anderson (1958) and Jones & Bradley (1961) found *Streptomyces* phage which attacked nocardiae, including *N. paraguayensis, N. madurae, N. brasiliensis, N. asteroides* and *Jensenia canicruria.* Bradley, Anderson & Jones (1961) also found *Nocardia* phage which attacked streptomycetes.

(d) *Lipids*

Although cell wall composition distinguished nocardiae from streptomycetes and many other taxa, it did not permit one to differentiate nocardiae from mycobacteria and corynebacteria of the *diphtheriae* type. Fortunately, lipid composition turned out to be of assistance. Members of the genera *Nocardia, Mycobacterium* and *Corynebacterium* with a cell wall composition of Type IV produce characteristic long chain, α-branched, β-hydroxy fatty acids collectively called mycolic acids (Asselineau, 1966) (Fig. 1).

These mycolic acids may be distinguished from each other by pyrolytic gas chromatography (Etémadi, 1964). In this reaction, the heat of the injector causes the molecule to break down into two fragments, a "mero" aldehyde and a fatty ester (Fig. 2). The latter is recognized by its retention time by comparison to known compounds. In the mycobacterial mycolic acids (eumycolates), the fatty ester released contains 22–26 carbon atoms; in nocardiae and corynebacteria, it contains 12–18 carbon atoms. In the nocardiae, the meroaldehyde fragment is of large molecular weight and remains on the column; in corynebacteria, the meroaldehyde is smaller and is eluted. A rapid method for obtaining pure mycolates has been reported (Lechevalier, Lechevalier & Horan, 1973). An examination of the mycolic acids of strains of corynebacteria, mycobacteria and nocardiae showed that the method was effective in separating members of the three genera (Lechevalier, Horan & Lechevalier, 1971).

During the course of these and other (Lanéelle, Asselineau & Castelnuovo, 1965; Asselineau, Lanéelle & Chamoseau, 1969; see

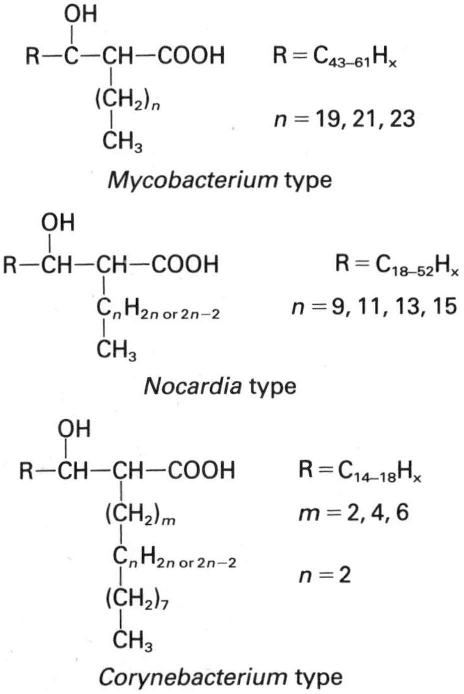

$$\begin{array}{l} \text{R}-\underset{|}{\overset{\text{OH}}{\overset{|}{\text{C}}}}-\text{CH}-\text{COOH} \qquad R = C_{43-61}H_x \\ \qquad\qquad (CH_2)_n \\ \qquad\qquad CH_3 \qquad\qquad n = 19, 21, 23 \end{array}$$

Mycobacterium type

$$\begin{array}{l} \text{R}-\overset{\text{OH}}{\overset{|}{\text{CH}}}-\text{CH}-\text{COOH} \qquad R = C_{18-52}H_x \\ \qquad\qquad C_nH_{2n \text{ or } 2n-2} \qquad n = 9, 11, 13, 15 \\ \qquad\qquad CH_3 \end{array}$$

Nocardia type

$$\begin{array}{l} \text{R}-\overset{\text{OH}}{\overset{|}{\text{CH}}}-\text{CH}-\text{COOH} \qquad R = C_{14-18}H_x \\ \qquad\qquad (CH_2)_m \qquad m = 2, 4, 6 \\ \qquad\qquad C_nH_{2n \text{ or } 2n-2} \qquad n = 2 \\ \qquad\qquad (CH_2)_7 \\ \qquad\qquad CH_3 \end{array}$$

Corynebacterium type

Fig. 1. Mycolic acids.

chapter 7) studies it was found the strains of the *rhodochrous* group contained nocardomycolates and that *N. farcinica* strains from African cases of farcy were, by this criterion, really mycobacteria since they contained eumycolates. The two presumably identical strains of Nocard's original isolate turned out to be different; one, NCTC 4524, contained eumycolates and the other, ATCC 3318,

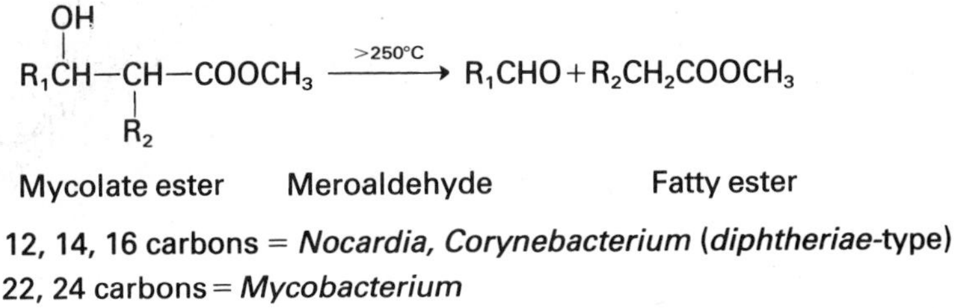

$$R_1\overset{\text{OH}}{\overset{|}{\text{CH}}}-\underset{R_2}{\underset{|}{\text{CH}}}-COOCH_3 \xrightarrow{>250^\circ C} R_1CHO + R_2CH_2COOCH_3$$

Mycolate ester — Meroaldehyde — Fatty ester

R_2 = 10, 12, 14, 16 carbons = *Nocardia, Corynebacterium* (*diphtheriae*-type)
R_2 = 20, 22, 24 carbons = *Mycobacterium*

Fig. 2. Pyrolysis of mycolic acids.

contained nocardomycolates. As a consequence, Lechevalier, Horan & Lechevalier (1971) suggested that *N. farcinica* be considered a *nomen dubium.*

Another method using mycolic acids is that of Kanetsuna & Bartoli (1972) which relies on solubility and quantity of these compounds to distinguish *Nocardia* from *Mycobacterium.* Corynebacteria have not been tested by this technique.

Nocardiae may be separated from mycobacteria by the detection of LCNAs ("lipids characteristic of *Nocardia*") (Mordarska & Mordarski, 1969; Mordarska, Mordarski & Goodfellow, 1972; Minnikin, Alshamaony & Goodfellow, 1974) through extraction of whole dry cells and analytical thin-layer chromatography. LCNA's have recently been shown to be mycolic acids (Goodfellow *et al.*, 1973).

Acid-fastness, one of the characters on which many classifications of nocardiae and mycobacteria are based, was once thought to be related to the presence of mycolic acids (Asselineau, 1966). This theory is no longer held; however, it has been found that the compound(s) responsible for acid-fastness in nocardiae are extractable with pyridine, whereas those in mycobacteria are not (Beaman & Burnside, 1973).

The fatty acids produced by nocardiae are like those of mycobacteria but differ from those produced by streptomycetes. The latter produce mainly branched-chain fatty acids of the *iso* and *anteiso* series whereas nocardiae and mycobacteria produce mostly normal and mono-unsaturated types. (Bordet & Michel, 1963; Ballio & Barcellona, 1968; Campbell & Naworal, 1969; Guinand, Vacheron & Michel, 1970; Farshtchi & McClung, 1970; Vestal & Perry, 1971; Thoen, Karlson & Ellefson, 1971*a,b*).

(e) *Nocobactins*

The complex, lipid-soluble, iron-transporting factors known as "mycobactins" were initially discovered when Twort & Ingram (1912) found growth factors for Johne's bacillus (*Mycobacterium paratuberculosis*) in extracts of various other mycobacteria. Related to the water-soluble sideramines, the structure of many of the mycobactins is now known (Snow, 1970). Analogous compounds are also produced by corynebacteria and nocardiae, and in the latter may be species-specific, different ones being found in

N. asteroides, N. brasiliensis, N. caviae and *N. uniformis* (Patel & Ratledge, 1973; Ratledge & Snow, 1974; see chapter 14). They are absent however, from the *rhodochrous* strains that have been examined to date, although small amounts have been detected in the *rhodochrous*-related *Gordona* group (Ratledge & Patel, 1974).

(f) *Serology*

As early as 1953, using serological techniques, González-Ochoa and Vasquez-Hoyos argued that *N. madurae* and *N. pelletieri* differed from *N. asteroides, N. brasiliensis* and *Actinomyces bovis* and from *Streptomyces somaliensis.* Cummins (1962, 1965) found, not unexpectedly, that the Type IV cell walls of corynebacteria, mycobacteria and nocardiae cross-reacted when measured by both agglutination and immunofluorescence. He also found that prior solvent extraction of the cells increased their reactivity. Presumably the removal of the outer lipid layers permitted access to the antigens of the cell wall.

Earlier workers had found that acid-fast actinomycetes including *N. asteroides* and *N. farcinica* had antigens in common with both mycobacteria (Fritzsche, 1908) and corynebacteria (Goyal, 1937).

Schneidau (1956) and Schneidau & Schaffer (1957) using the whole cells of various strains of *Nocardia* and *Mycobacterium* and slide agglutination, found that their *N. asteroides, N. polychromogenes, N. corallina* and *N. opaca* strains could not be distinguished from each other serologically; only the *N. brasiliensis* strains were distinct. These all reacted with mycobacteria to some degree. In contrast, strains of *N. caviae, N. gardneri, N. leishmanii, N. madurae, N. pelletieri* and *N. rangoonensis* did not. One wonders, since all the species of the second group except *caviae* have been found to have cell walls other than Type IV, whether their *caviae* strains were properly identified. They also, on this basis, placed the *rhodochrous* strains closer to *Nocardia* and "*N.*" *intracellulare* close to *Mycobacterium.*

Kwapinski (1963) found that, depending on the cell fraction used, both cross-reactivity and genus specificity could be demonstrated in *Nocardia.* Using gel diffusion, Castelnuovo *et al.* (1968) reported some antigens in common between nocardiae and mycobacteria but that the common antigen between *N. asteroides* and *N. brasiliensis* strains and mycobacteria was not the same as the one between *rhodochrous* strains and mycobacteria. All the strains of *N. asteroides* and *N. brasiliensis* had one antigen in common; *brasiliensis-rhodochrous* reactions showed two distinct groups among the *rhodochrous.* She and her colleagues (1968) later reported an antigen common to all *Streptomyces, Nocardia* and *Mycobacterium* strains examined.

Using fluorescent antibody, Al-Doory (1965–1966) showed appreciable cross-staining of *N. asteroides* with other strains of this species as well as with *N. brasiliensis, N. corallina, N. caviae, N. convoluta, N. gibsonii, N. intracellularis, N. lutea, N. opaca, N. rubra, N. transvalensis, N. vaccinii, C. diphtheriae*, and, surprisingly, *S. somaliensis* and *N. madurae.* In contrast, he got no cross-reactions with *N. alba-schoen, N. blackwellii, N. coeliaca, N. eppingeri, N. farcinica, N. paraguayensis, N. polychromogenes, N. pretoriana, N. rangoonensis, N. convoluta, N. turbata* or *N. violaceus.* In general, his division of reactivity follows pretty well what one might expect from knowledge of the cell wall composition of the strains. One cannot help but remark that poor identification of strains is often the first rock on which much otherwise good work has foundered. Too many immunologists, chemists and the like have published good data on poor strains.

An extensive review (Kwapinski, 1969) of the cross-reactivities of the cell cytoplasms, cell walls and extracellular antigens of nocardiae with other actinomycetes points out the extreme complexity of the interrelationships among actinomycete taxa as portrayed by some serological analyses.

Takeya, Nakayama & Muraoka (1970) reported that guinea pigs sensitized with a *rhodochrous* strain showed no specific reaction with tuberculins purified from *N. polychromogenes, N. globerula* or *N. asteroides.* Using extracellular antigens, Pier & Fichtner (1971), reported certain antigens which were unique to *N. asteroides* and others which occurred in all strains of *N. asteroides, N. brasiliensis, N. caviae* and *N. rubra.* Cross-reactivity was found between *Streptomyces* and certain nocardiae. Ridell (1974*b*) has shown that, by immunodiffusion, *rhodochrous* strains appear to be equally related to mycobacteria and nocardiae.

By indirect fluorescent antibody technique and using solvent-extracted cells as antigens, Schaal (1974) obtained genus-specific antisera for *Nocardia* (*asteroides, brasiliensis, caviae, rubra*), *Actinomadura, Streptomyces, Actinomyces, Arachnia, Bacterionema, Mycobacterium* and *Corynebacterium.*

(g) *DNA composition and homologies*

GC%

The GC% (the guanine-cytosine content of purified DNA) values can only tell the taxonomist when he is dealing with unrelated strains; there is no assurance that because two organisms have the same GC% value, they are necessarily related.

Yamaguchi (1967) found that of 17 genera examined representing the four major cell wall types of aerobic actinomycetes, all but strains having a Type IV cell wall showed a GC% greater than 70%. The four strains representing the Type IV genera *Nocardia, Mycobacterium* and *Micropolyspora,* varied from 67·5 to 69·0. He was able to cluster all cell wall types but 1 if GC% and DNA homology relative to *S. griseus* DNA were plotted against each other.

Tewfik & Bradley (1967) reported that members of the *rhodochrous* group had GC%'s falling into two groups, one with a GC% of 62–64 and a second at 68–69%. Those with the higher percentages showed the greater homology with reference DNA from *S. venezuelae.* The GC%'s of five *Nocardia* strains were found by Tarnok, Röhrscheidt & Bönicke (1967) to be somewhat lower (53·6–61·2) than those reported by others. Yamada & Komagata (1970) reported that corynebacteria of cell wall Type IV had GC%'s ranging from 52·4 to 68·5. The corynebacteria which belong to the *rhodochrous* group ranged from 64·6–69·5. Bousfield (1972) reported a range of GC% for similar corynebacterial species to range from 55–61 with strains of the *rhodochrous* group ranging from 61–66.

Homology

The degree of relatedness of different strains may be estimated by reassociation of their denatured DNA's since the degree of renaturation depends on the number of nucleotide sequences shared. This homology is principally useful in evaluating specific relationships.

Using *S. venezuelae* mycelial DNA as reference, Enquist & Bradley (1968) found that DNA from other streptomycetes ranged in degree of homology from 27–88%, members of the *rhodochrous* group from 23–39% and mycobacteria 1–2%. This group later reported that they could not find a clearcut relationship between members of the *rhodochrous* group and members of the genera, *Nocardia, Mycobacterium* and *Streptomyces* (Bradley, Brownell & Clark, 1973). Bradley & Huitron (1973) continuing their work with the *rhodochrous* group, were also able to show that certain strains of *N. opaca* annealed extensively with *N. erythropolis* DNA whereas that from an *N. globerula* did not. The *opaca-erythropolis* relationship was confirmed by genetic recombination.

(h) *Infrared spectra of cells*

An attempt to differentiate between mycobacteria, nocardiae and streptomycetes on the basis of infrared spectra of whole killed cells was reported by Arai, Kuroda & Koyama (1963). By keeping growth conditions constant, they were able to derive typical spectra for each broad group; however, some strains which clearly belonged to *Nocardia* gave mycobacterial spectra and others had a streptomycete profile. Prauser on the contrary reported that he had not been successful in using infrared in taxonomy (1967–1968).

(i) *Antibiotic sensitivity*

In differentiating among certain genera having a Type IV cell wall, Goodfellow & Orchard (1974) found antibiotic sensitivity to be useful. The *rhodochrous* group alone among three species of *Nocardia* (*asteroides, brasiliensis* and *caviae*) and strains of *Gordona* and *Saccharopolyspora* ("Bagasse") was sensitive to 50 mcg/ml of rifampicin. Strains of *Saccharopolyspora* alone were insensitive to 100 mcg/ml of gentamicin. The nocardiae could be distinguished from *Gordona* and *rhodochrous* strains by their resistance to 100 mcg/ml of cephaloridine.

(j) *Morphology*

What had happened to morphology during this time? For the most part, apparently, nocardiologists' attention had turned elsewhere. Uesaka (1969) found that the breakdown of the filaments of both streptomycetes and nocardiae occurred by means of both fragmentation and autolysis. Jičínská (1973) reported that a coryneform culture having no sign of filamentation or branching gave rise on irradiation by u.v. to a nocardoid mutant which produced a well-branched mycelium. McClung (1954) was unable to relate nocardial morphological type with carbon compound utilization.

Morphology did however, prove useful in setting apart certain taxa, which, like nocardiae, had cell wall compositions of Type IV (see chapter 5). These included *Micropolyspora* (Lechevalier, Solotorovsky & McDurmont, 1961) *Pseudonocardia* (Henssen, 1957), *Saccharomonospora* (Nonamura & Ohara, 1971) and *Saccharopolyspora* (Lacey & Goodfellow, 1974).

(k) *Physiology*

The unquestioned champions in the use of physiology in the taxonomy of nocardiae are R. E. Gordon and her colleagues. Gordon's approach was a kind of early numerical taxonomy without computers. She sought out tests which might satisfactorily define given groups of organisms, then used the oldest legitimate name borne by a strain within each group to label the group as a whole (Gordon & Mihm, 1957, 1959, 1962*a,b*; Gordon, 1966*a,b*, 1967; Gordon & Horan, 1968; Gordon *et al.*, 1974). In this way she delineated, in physiological terms, the following taxa: *N. asteroides*, *N. brasiliensis*, *N. caviae*, the *rhodochrous* group (as "*M.*" *rhodochrous*), *N. madurae*, *N. pelletieri* and *N. autotrophica*. Although the numerical taxonomists do not always agree with her groupings (*vide infra*), there is no question that chemotaxonomists, serologists and numerical taxonomists owe her a debt of gratitude for having made available well-defined groups of nocardiae for their studies.

In other laboratories (Wayne, Juarez & Nichols, 1958), physiological studies showed that all mycobacteria but "*M.*" *rhodochrous* gave a positive aryl sulfatase tests whereas only certain nocardiae and no streptomycetes or corynebacteria did so. Mariat (1963) reported that 96% of his *N. asteroides* and 100% of his *N. brasiliensis* and West African *N. farcinica* strains were urease-positive; *N. farcinica* from Tchad, *N. madurae*, *N. pelletieri* and *Streptomyces somaliensis* strains were negative. Kwapinski & Horsman reported that it was possible to distinguish among various nocardial species on the basis of growth in three defined media (1973).

(l) *Numerical taxonomy*

In 1957, Sneath first articulated the use of Adansonian principles and computers in bacterial taxonomy. In this method, organisms are examined, largely by automated methods, for a large number of characters, usually morphological and physiological, and the similarity or dissimilarity of the organisms determined by computer. Characteristics are usually not weighted, negative matches (when both organisms of a pair give negative reactions) may or may not be included in the computations and in general, it is considered better to have more tests of dubious applicability than fewer very reliable ones. Organisms are then quantitatively grouped into clusters or phenons, their similarity to other strains being used to make taxonomic conclusions.

Unfortunately, such large numbers of tests do not necessarily assure one of criteria which are easily useful in classification unless certain of the determined characters are subsequently weighted. This may lead to groupings which serve to define a species or sub-species more precisely, but they have failed completely until now, to formulate a clearcut definition of any genus. In other terms, distinctive supra-species criteria are not derived easily. This is because (1) it is very difficult to extract and articulate criteria of generic importance from these data, (2) it is rare that a character in a cluster or several clusters be shown by 100% of the strains, and (3) no one is quite sure how similar clusters need to be to warrant being placed in the same genus.

Not surprisingly, the groupings delineated by one numerical taxonomic study usually do not correspond in every respect to those made during another and numerical taxonomic "relationships" may not stand up when one uses other criteria such as DNA homology or cellular lipids (Stanier, 1970; Sneath, 1970, 1974).

In the first attempt to use this new concept in solving the problems of nocardial taxonomy, Jones & Bradley (1964) using 50 tests on 20 strains, concluded that certain *rhodochrous* strains were mycobacteria and others nocardiae. *Nocardia farcinica* was considered to belong to the mycobacteria and *N. turbata* was tentatively placed close to *Cellulomonas*.

Cerbón (1967) subjected 68 strains of nocardiae to numerical analysis. He found that *N. asteroides* was heterogeneous, *N. brasiliensis* tightly homogeneous and that a cluster including

N. corallina, *N. globerula*, *N. pretoriana*, *N. rhodnii* and *N. phenotolerans* could be distinguished. In the studies by various groups that were to follow, a broad consensus was reached on certain findings:

(1) the taxon *N. asteroides* as defined by Gordon & Mihm (1962*a*) is heterogeneous (2, 4 or 5 clusters depending on the study),
(2) *N. asteroides* strains clustered closely with *N. brasiliensis* and *N. caviae* strains,
(3) the *rhodochrous* taxon is heterogeneous (2, 6 or 9 clusters),
(4) the *rhodochrous* taxon is distinct from nocardiae and mycobacteria,
(5) *N. madurae* (and *N. pelletieri*, when tested, clustered relatively well but at some distance from the above.

Tsukamura (1969, 1974), on the basis of 100 or so tests on as many strains, Bradley (1971) testing 60 strains for 60 characters, including DNA homology, Goodfellow (1971) after running 283 strains representing 31 species for 241 characters and Kurup & Schmitt (1973) (157 strains and 117 characters) all endorsed these conclusions. Goodfellow also found *N. turbata* and *N. dassonvillei* strains clustered away from the other groups and from each other. Tacquet *et al.* (1971) having studied only strains of the *rhodochrous* group, agreed with (3) and (4).

In 1973, by weighting four morphological and 23 physiological characters culled from his studies, Goodfellow published tables to aid the taxonomist in classifying his unknown nocardioform strains to genus (see Tables 3 and 4). This is certainly a step in the right direction and a courageous one, since in setting up a target at which to aim, he is inviting flak from all sides. Nonetheless, he has definitely performed a useful service and when all the evidence, cytochemical, serological, numerical has been put together and hardened in this way under fire, along with newer criteria still to come, we shall, undoubtedly, be able to formalize a definition of the genus *Nocardia*.

4. New Lamps for Old

By analogy to the story of Aladdin, one sometimes wonders if new genera for old is really such a bargain. Certain of the newer

Table 3

Differentiation of Mycobacterium, Nocardia, Corynebacterium *and allied taxa using morphological, biochemical and fermentation characters*

Taxon	Mycelium	Rod-coccus cycle	Aerial hyphae	Weakly acid fast	Growth at 10°C	O-F test	Allantoin	Casein	Starch	Hypoxanthine	Tyrosine	Urea	Lysozyme sensitive	Penicillin sensitive
Mycobacterium	−	−	−	+	−	0	−	−	−	−	−	±	+	−
Corynebacterium	−	−	−	−	−	F/−	−	−	−	−	−	−	ND	+
Nocardia	+	−	+	+	±	0	+	±	−	±	±	+	−	±
'M.' rhodochrous	±	−	−	±	+	0	±	−	−	−	±	±	+	+
Actinomadura	+	−	+	−	−	0	−	+	±	+	+	±	+	+
Arthrobacter	−	+	−	−	−	0	−	+	+	±	+	−	ND	+
Oerskovia	+	−	−	−	+	F	+	+	+	−	−	+	+	+
Rothia	+	−	−	−	−	F	−	−	−	−	−	−	−	+

± = variable. ND = not determined.

Table 4

Differentiation of Mycobacterium, Nocardia *and allied taxa using carbon utilization tests*

Taxon	Cellobiose	Glycogen	Lactose	Melezitose	Sorbitol	Sucrose	Adipic acid	*p*-hydroxybenzoic acid	Sebacic acid	Na-*n*-butyrate	Na-(*d*, 1)-malate	Na succinate	Testosterone
Mycobacterium	±	−	−	−	±	±	−	−	±	+	±	−	±
Nocardia	−	−	−	−	−	±	±	−	±	+	+	+	+
'M.' rhodochrous	−	−	−	−	+	+	+	+	+	+	+	+	+
Actinomadura	−	±	−	−	−	−	−	−	±	+	+	+	−
Arthrobacter	+	−	±	+	+	+	−	±	−	+	+	±	−
Oerskovia	+	+	+	+	+	+	−	−	−	−	−	−	−

(Tables 3 and 4 are reproduced from Goodfellow, M. (1973). *Ann. Soc. belge Méd. trop.* **53,** 287.)

proposals for eliminating the distortions of the nocardial profile have been useful; others have a less certain value.

Because of the inhomogeneity of the cell wall and whole cell compositions of the various organisms assigned to the genus *Nocardia* (q.v.) and because both cell wall and whole cell composition were found to be stable characteristics (Šuput, Lechevalier & Lechevalier, 1967), it was proposed to limit the genus *Nocardia* to those organisms that have compositions like that of *N. farcinica*, the type species (Lechevalier, Lechevalier & Gerber, 1971). This proposal has been adopted in the current Bergey's Manual (McClung, 1974).

(a) Rothia

Toward this end, *Nocardia salivae* and *N. dentocariosa* were transferred by Georg & Brown (1967) to the new genus, *Rothia.* They based their decision on the cell wall composition of the strains (rothiae contain no diaminopimelic acid, arabinose or galactose), their fermentative rather than oxidative attack on sugars, on their more exacting nutritional requirements, lack of acid-fastness, morphology (lack of aerial mycelium and formation of conidia) and physiology (nitrite reduction). The morphological criteria were, once again, the weakest.

(b) Actinomadura *and* Oerskovia

The genus *Actinomadura* was proposed (Lechevalier, H. A. & Lechevalier, M. P. 1970) to accommodate the nocardial species having a cell wall of Type III; namely, *N. madurae, N. pelletieri* and *N. dassonvillei.* A second genus, *Oerskovia* was also later proposed (Prauser, Lechevalier & Lechevalier, 1970) for the motile "nocardiae" of the *turbata* group which had a cell wall of Type VI. The removal of the species *madurae,* and *pelletieri* from *Nocardia* had already been endorsed by González-Ochoa & Sandoval (1951) on the basis of morphology and physiology. The creation of all three genera, *Rothia, Actinomadura* and *Oerskovia,* were subsequently supported by phenetic data (Goodfellow, 1971, 1973). *Actinomadura* and *Oerskovia* were endorsed on the basis of phage typing (Prauser, 1967; Prauser & Falta, 1968).

(c) Gordona

Certain strains from both clinical sources and soil which are related to the *rhodochrous* group were described under the new genus name, *Gordona* (Tsukamura, 1971). Gordonae show rough pink or reddish colonies having no mycelium or branching. Members of the new taxon were differentiated from *Nocardia* and *Mycobacterium* on the basis of acid-fastness, fragmenting mycelium, nitrate reduction, arylsulfatase, acid from mannose, utilization of sucrose as a carbon source and trimethylene diamine as a nitrogen and carbon source. Gordonae were differentiated from the *rhodochrous* group by one test: acid from mannose. As already pointed out, degree of acid-fastness and fragmentation of the mycelium are difficult criteria to use. The other distinguishing characteristics of *Gordona* might be more worthy of a new species than a new genus. Tsukamura (1974) has recently proposed to change the name *Gordona* to *Rhodococcus*. Although this change is in the historical tradition of nocardial nomenclature, it probably pleases no one but Ruth Gordon.

(d) Proactinomyces

Proactinomyces was first proposed by Jensen in 1931 as a genus name for Ørskov's Group II organisms, the morphological equivalent to what are now referred to as nocardioform organisms. Although declared illegitimate because of the priority of Trevisan's *Nocardia*, the name is still in use by Russian actinomycetologists. Bradley & Bond (1974) have recently proposed to resurrect *Proactinomyces* as the genus name for the orphaned *rhodochrous* group, with *P. corallinus* as the type. This proposal has several limitations: (1) just as it is very difficult to define the *rhodochrous* group, so is it difficult to make a clearcut description of *Proactinomyces*; (2) Bradley & Bond's description of properties of the genus contains some errors* and reflects some of the difficulties of translating phenetic data into generic descriptions; and (3) although Jensen did not

* Statement: "(*Proactinomyces*) do not produce aerial mycelium." Some do: Gordon (1966*a*); Goodfellow (1971); Lechevalier, M. P. (unpublished observations). Statement: "form acid from . . . fructose, maltose, sucrose . . .." Some do not: Goodfellow (1971); Lechevalier & Lechevalier (1974). Statement: "utilize *n*-butyrate, malate, propionate, succinate, acetate." Not all do: Goodfellow (1971). Statement: "are sensitive to lysozyme." Not all are: Goodfellow (1971); Lechevalier & Lechvalier (1974). Statement: "Grow well at 10°C." Not all do: Goodfellow (1971).

designate a type species of *Proactinomyces*, Prévot, in 1946, designated *P. agrestis* as type, and it is not at all clear that this proposal is illegitimate in view of the present uncertainty concerning the status of all *rhodochrous*-related taxa. Thus, there may be a question of priority for the type species. Goodfellow and his colleagues (1974) on the other hand, after further extensive numerical taxonomical studies which included for the first time, both serological and chemotaxonomic data, concluded that further studies must be made before the generic status of the *rhodochrous* group can be decided.

5. A Question of a Type Species of *Nocardia*

Nocard's original description of the "bacille de farcin" (1888) which he isolated from material sent to him from Guadeloupe, may have been excellent for the time, but gives us very little help in trying to resolve some of the problems that confront us today.

The differences in the mycolic acids of the two presumably identical strains of Nocard's original isolate, ATCC 3318 and NCTC 4524, have already been previously discussed (see (d) *Lipids*). In addition, a complex glycolipid, mycoside C′, very similar to others found in mycobacteria, has been reported to be produced by certain African *N. farcinica* strains. These compounds have not been reported from nocardiae (Lanéelle, Asselineau & Chamoiseau, 1971).

DNA of strains of *N. farcinica* (origins of the cultures were not given) reassociated extensively with the DNA of *Mycobacterium stercoides* but to a slightly lesser extent with DNA from *M. bovis* BCG than did most other mycobacteria included for comparison (Bradley, 1972). Further work (Bradley, 1973), showed that ATCC 3318 (his 330) was more closely related to other nocardiae than to mycobacteria confirming the results from mycolic acid analysis. However, the DNA of 3318 reassociated to a greater, albeit modest, degree (38%) with *M. smegmatis* DNA than the DNA's of other nocardiae tested.

Using delayed-type skin reactions of guinea pigs to extracellular antigens of certain nocardiae, Magnusson & Mariat (1968) grouped together a number of recent African isolates from bovine farcy with ATCC 3318. Three other *N. farcinica* strains, one from

Dakar and others whose origin is not given, formed two other groups. No mycobacteria were included in these tests.

Immunodiffusion studies showed that strains presently designated *N. farcinica* fall into two groups: one, centering around NCTC 4524 and reacting like mycobacteria, and the other centering around ATCC 3318 and closely related to *N. asteroides* (Ridell, 1974*a*).

Phage sensitivity studies (Anderson & Bradley, 1961) showed that *N. farcinica* strains were attacked only by mycobacterial phage and not nocardiophage. No other nocardiae except one *rhodochrous* strain were sensitive to the mycobacterial phage. The nocardiophages were not active on the mycobacteria or the *N. farcinica* and were variably active on other nocardiae. However, subsequent work (Manion *et al.*, 1964) showed that a *N. brasiliensis* phage which had been modified by growth on alternative hosts, attacked *N. farcinica.*

It thus appears that considerable uncertainty attends the final disposition of the type species, *N. farcinica.* Nocard's original description speaks unmistakably of colonies "à surface mamelonnée, terne et comme poussiéreuse . . ." ("with a mammillate surface, matte, and as if powdery") a clear implication that his strain formed aerial mycelium visible to the naked eye. By the time Ørskov (1923) examined the strain, it had lost this capacity, but this is, of course, not an uncommon occurrence in laboratory strains of actinomycetes. However, no *N. farcinica* strains I have examined, regardless of source, whether recently isolated or old, growing on media such as Nocard used or on any other, produced macroscopically visible aerial mycelium. F. Mariat of the Pasteur Institute in Paris is absolutely certain no strain of Nocard's culture is extant in France. Unfortunately, farcy no longer exists in Guadeloupe, so no new isolates may be made from that region. The strains now isolated from African farcy appear by many criteria to be more closely related to mycobacteria than nocardiae. As a consequence, Chamoiseau (1973) has proposed to call them *Mycobacterium farcinogenes.* It is entirely possible that Nocard's strain differed from these altogether; however, we shall probably never know.

Thus, in view of the uncertainty which surrounds *N. farcinica,* it would seem reasonable that the Judicial Commission reverse its fiat making *N. farcinica* the type species of the genus (Trevisan did not specify a type) and, appoint *N. asteroides* in its stead, a course already recommended by Gordon many years ago (Gordon & Mihm, 1962*a*).

6. Towards a Definition of the Genus

On looking over the more recent history of the genus *Nocardia*, one has the impression that nocardiologists are still doing a bit of stumbling around in the tunnel trying to find that light at the end. There has been a vast improvement compared to the period prior to 1950, with some new aids in finding our way, and the search for better methods in nocardial taxonomy going ahead on many fronts and in many countries. We may yet make it to a "solid" definition of the genus.

Until then a "semisolid" definition of *Nocardia* is proposed as follows: *Actinomycetes containing nocardomycolic acids, having Type IV cell walls and Type A whole cell sugar patterns and not capable of being sharply characterized by morphological criteria.*

7. References

Adam, A., Petit, J. F., Wietzerbin-Falzspan, J., Sinay, P., Thomas, D. W. & Lederer, E. (1969). L'acide N-glycolyl muramique, constituent des parois de *Mycobacterium smegmatis*: Identification par spectrometrie de masse. *FEBS Letters* **4,** 87.

Adams, J. M. & McClung, N. M. (1962). Comparison of the developmental cycles of some members of the genus *Nocardia. J. Bact.* **84,** 206.

Al-Doory, Y. (1965–1966). Fluorescent antibody studies with *Nocardia asteroides.* I. *Sabouraudia* **4,** 135.

Anderson, D. L. & Bradley, S. G. (1961). Susceptibility of nocardiae and mycobacteria to actinophage. *Antimicrob. Ag. Chemother.* 898.

Arai, T. & Kuroda, S. (1959). Comparative studies of certain strains of *Nocardia. Symposium on Taxonomy of Actinomycetes, Tokyo,* p. 22.

Arai, T., Kuroda, S. & Koyama, Y. (1963). Infrared absorption spectra of whole cells of *Nocardia* and related organisms. *J. gen. appl. Microbiol. Tokyo* **9,** 119.

Asselineau, J. (1966). *The Bacterial Lipids.* p. 122, Paris: Hermann.

Asselineau, J., Lanéelle, M. A. & Chamoiseau, G. (1969). De l'étiologie du farcin de zébus tchadiens: nocardiose ou mycobactériose? II. Composition lipidique. *Revue Élev. Méd. vét. Pays trop.* **22,** 205.

Avery, R. J. & Blank, F. (1954). On the chemical composition of the cell walls of the *Actinomycetales* and its relation to their systematic position. *Can. J. Microbiol.* **1,** 140.

Azuma, I., Thomas, D. W., Adam, A., Ghuysen, J.-M., Bonaly, R., Petit, J.-F. & Lederer, E. (1970). Occurrence of *N*-glycolylmuramic acid in bacterial cell walls. *Biochim. biophys. Acta* **208,** 444.

Baboolal, R. (1969). Cell wall analysis of oral filamentous bacteria. *J. gen. Microbiol.* **58,** 217.

Baldacci, E. (1939). Introduzione allo studio degli Attinomiceti. *Mycopathologia* **II,** 84.

Ballio, A. & Barcellona, S. (1968). Relations chimiques et immunologiques chez les *Actinomycétales.* 1. Les acides gras de 43 souches d'actinomycetès aerobies. *Annls Inst. Pasteur, Paris* **114,** 121.

Beaman, B. L. & Burnside, J. (1973). Pyridine extraction of nocardial acid fastness. *Appl. Microbiol.* **26,** 426.

Becker, B., Lechevalier, M. P., Gordon, R. E. & Lechevalier, H. A. (1964). Rapid differentiation between *Nocardia* and *Streptomyces* by paper chromatography of whole cell hydrolysates. *Appl. Microbiol.* **12,** 421.

Becker, B., Lechevalier, M. P. & Lechevalier, H. (1965). Chemical composition of cell-wall preparations from strains of various form-genera of aerobic actinomycetes. *Appl. Microbiol.* **13,** 236.

Berd, D. (1973). Laboratory identification of clinically important aerobic actinomycetes. *Appl. Microbiol.* **25,** 665.

Bisset, K. A. & Moore, F. W. (1949). The relationship of certain branched bacterial genera. *J. gen. Microbiol.* **3,** 387.

Bordet, C., Karahjoli, M., Gateau, O. & Michel, G. (1972). Cell walls of nocardiae and related actinomycetes: Identification of the genus *Nocardia* by cell wall analysis. *Int. J. Syst. Bact.* **22,** 251.

Bordet, C. & Michel, G. (1963). Etude des acides gras isolés de plusieurs espèces de Nocardia. *Biochim. biophys. Acta* **70,** 613.

Bousfield, I. J. (1972). A taxonomic study of some coryneform bacteria. *J. gen. Microbiol.* **71,** 441.

Bradley, S. G. (1959). Sporulation by some strains of nocardiae and streptomycetes. *Appl. Microbiol.* **7,** 89.

Bradley, S. G. (1971). Criteria for definition of *Mycobacterium, Nocardia* and the rhodochrous complex. *Adv. Fronts Pl. Sci.* **28,** 349.

Bradley, S. G. (1972). Reassociation of deoxyribonucleic acid from selected mycobacteria with that from *Mycobacterium bovis* and *Mycobacterium farcinica. Am. Rev. resp. Dis.* **106,** 122.

Bradley, S. G. (1973). Relationships among mycobacteria and nocardiae based on deoxyribonucleic acid reassociation. *J. Bact.* **113,** 645.

Bradley, S. G. & Anderson, D. L. (1958). Taxonomic implication of actinophage host-range. *Science, N.Y.* **128,** 413.

Bradley, S. G., Anderson, D. L. & Jones, L. A. (1961). Phylogeny of actinomycetes as revealed by susceptibility to actinophage. *Devs ind. Microbiol.* **2**, 223.

Bradley, S. G. & Bond, J. S. (1974). Taxonomic criteria for mycobacteria and nocardia. *Adv. Appl. Microbiol.* **18,** 131.

Bradley, S. G., Brownell, G. H. & Clark, J. (1973). Genetic homologies among nocardiae and other actinomycetes. *Can. J. Microbiol.* **19,** 1007.

Bradley, S. G. & Huitron, M. E. (1973). Genetic homologies among nocardiae. *Devs ind. Microbiol.* **14,** 189.

Breed, R. S. & Conn, H. J. (1919). The nomenclature of the *Actinomycetaceae. J. Bact.* **4,** 585.

Campbell, I. M. & Naworal, J. (1969). Composition of the saturated and mono-unsaturated fatty acids of *Mycobacterium phlei. J. Lipid Res.* **10,** 593.

Castelnuovo, G., Bellezza, G., Giuliani, H. I. & Asselineau, J. (1968). Relations chimiques et immunologiques chez les *Actinomycétalès. Annls Inst. Pasteur, Paris* **114,** 139.

Cerbón, J. (1967). Taxonomic analysis of *Nocardia. Revta lat.-am. Microbiol.* **9,** 65.

Chalmers, A. J. & Christopherson, J. B. (1916). A sudanese actinomycosis. *Ann. trop. Med. Parasit.* **10,** 223.

Chamoiseau, G. (1973). *Mycobacterium farcinogenes* agent causal du farcin du boeuf en Afrique. *Ann. Microbiol.* (*Inst. Pasteur*) **124A,** 215.

Cohn, F. (1875). Untersuchungen über Bacterian. II. *Beitr. Biol. Pfl.* **1,** 141.

Cummins, C. S. (1962). The chemical composition and antigenic structure of cell walls of *Corynebacterium, Mycobacterium, Nocardia, Actinomyces* and *Arthrobacter. J. gen. Microbiol.* **28,** 35.

Cummins, C. S. (1965). Chemical and antigenic studies on cell walls of mycobacteria, corynebacteria and nocardias. *Am. Rev. resp. Dis.* **92,** 63.

Cummins, C. S. & Harris, H. (1956*a*). The chemical composition of the cell wall in some gram-positive bacteria and its possible value as a taxonomic character. *J. gen. Microbiol.* **14,** 583.

Cummins, C. S. & Harris, H. (1956*b*). A comparison of cell wall composition in *Nocardia, Actinomyces, Mycobacterium* and *Propionibacterium. J. gen. Microbiol.* **15,** ix.

Cummins, C. S. & Harris, H. (1958). Studies on the cell wall composition and taxonomy of actinomycetales and related groups. *J. gen. Microbiol.* **18,** 173.

Davis, G. H. G. & Baird-Parker, A. C. (1959). The classification of certain filamentous bacteria with respect to their chemical composition. *J. gen. Microbiol.* **21,** 612.

Enquist, L. W. & Bradley, S. G. (1968). DNA homologies among actinomycetes. *Bact. Proc.* **1968,** 20.

Erikson, D. (1949). Differentiation of the vegetative and sporogeneous phases of actinomycetes. 4. The partially acid-fast proactinomycetes. *J. gen. Microbiol.* **3,** 361.

Etémadi, A. H. (1964). Techniques microanalytiques d'etude de structure d'esters α-ramifiés, β-hydroxylés. Chromatographie en phase vapeur et spectrométrie de masse. *Bull. Soc. chim. Fr.* **1964,** 1537.

Farshtchi, D. & McClung, N. M. (1970). Effect of substrate on fatty acid production in *Nocardia asteroides. Can. J. Microbiol.* **16,** 214.

Fritzsche, E. (1908). Experimentelle untersuchungen über biologische Beziehungen des Tuberkelbazillus zu einigen anderen saurefesten Mikroorganismen und Actinomyzeten. *Arch. Hyg. Bakt.* **65,** 181.

Georg, L. K. & Brown, J. M. (1967). *Rothia,* gen. nov. an aerobic genus of the family *Actinomycetaceae. Int. J. Syst. Bact.* **17,** 79.

González-Ochoa, A. & Sandoval, M. A. (1951). Caracteristicas de los actinomycetos más communes. *Congreso Cientifico de la Universidad autonóma de Mexico.*

González-Ochoa, A. & Vasquez-Hoyos, A. (1953). Relationes serologicas de los principales actinomycetes patogenos. *Revta Inst. Salubr. Enferm. trop.* **13,** 177.

Goodfellow, M. (1971). Numerical taxonomy of some nocardioform bacteria. *J. gen. Microbiol.* **69,** 33.

Goodfellow, M. (1973). Characterization of *Mycobacterium, Nocardia, Corynebacterium* and related taxa. *Ann. Soc. belge Méd. trop.* **53,** 287.

Goodfellow, M., Lind, A., Mordarska, H., Pattyn, S. & Tsukamura, M. (1974). A co-operative numerical analysis of cultures considered to belong to the '*rhodochrous*' taxon. *J. gen. Microbiol.* **85,** 291.

Goodfellow, M., Minnikin, D. E., Patel, P. V. & Mordarska, H. (1973). Free nocardomycolic acids in the classification of nocardias and strains of the "rhodochrous" complex. *J. gen. Microbiol.* **74,** 185.

Goodfellow, M. & Orchard, V. A. (1974). Antibiotic sensitivity of some nocardioform bacteria and its value as a criterion for taxonomy. *J. gen. Microbiol.* **83,** 375.

Gordon, R. E. (1966*a*). Some strains in search of a genus—*Corynebacterium, Mycobacterium, Nocardia* or what? *J. gen. Microbiol.* **43,** 329.

Gordon, R. E. (1966*b*). Some criteria for the recognition of *Nocardia madurae* (Vincent) Blanchard. *J. gen. Microbiol.* **45,** 355.

Gordon, R. E. (1967). The Taxonomy of Soil Bacteria. In *Ecology of Soil Bacteria.* Eds T. R. G. Gray & B. Parkinson. p. 293. Liverpool: Liverpool University Press.

Gordon, R. E., Barnett, D. A., Handerhan, J. E. & Pang, C. H. (1974). *Nocardia coeliaca, Nocardia autotrophica* and the nocardin strain. *Int. J. Syst. Bact.* **24,** 54.

Gordon, R. E. & Horan, A. H. (1968). *Nocardia dassonvillei,* a macroscopic replica of *Streptomyces griseus. J. gen. Microbiol.* **50,** 235.

Gordon, R. E. & Mihm, J. M. (1957). A comparative study of some strains received as nocardiae. *J. Bact.* **73,** 15.

Gordon, R. E. & Mihm, J. M. (1958). Sporulation by two strains of *Nocardia asteroides. J. Bact.* **75,** 239.

Gordon, R. E. & Mihm, J. M. (1959). A comparison of *Nocardia asteroides* and *Nocardia brasiliensis. J. gen. Microbiol.* **20,** 129.

Gordon, R. E. & Mihm, J. M. (1962*a*). The type species of the genus *Nocardia. J. gen. Microbiol.* **27,** 1.

Gordon, R. E. & Mihm, J. M. (1962*b*). Identification of *Nocardia caviae* (Erikson) *nov. comb. Ann. N.Y. Acad. Sci.* **98,** 628.

Goyal, R. K. (1937). Étude microbiologique, expérimentale et immunologique de quelques streptothricées. *Annls Inst. Pasteur, Paris* **59,** 94.

Gray, P. H. H. & Thornton, H. G. (1928). Soil bacteria that decompose certain aromatic compounds. *Zentbl. Bakt Parasitkde Abt. II.* **73,** 74.

Guinand, M., Vacheron, M. J. & Michel, G. (1970). Structure des parois cellularies des *Nocardia.* I. Isolement et composition des parois de *Nocardia kirovani. FEBS Letters* **6,** 37.

Harz, C. O. (1879). *Actinomyces bovis*, ein neuer Schimmel in den Geweben des Rindes. *Z. Tiermed.* **5,** 125.

Henssen, A. (1957). Beiträge zur Morphologie und Systematik der thermophilen Actinomyceten. *Arch. Mikrobiol.* **26,** 373.

Jensen, H. L. (1931). Contributions to our knowledge of the actinomycetales. II. *Proc. Linn. Soc. N.S.W.* **56,** 345.

Jensen, H. L. (1953). The genus *Nocardia* (or *Proactinomyces*) and its separation from other actinomycetales, with some reflections on the phylogeny of the actinomycetes. In *Actinomycetales, Morphology, Biology and Systematics. Symposium. VI. Intern. Congs. Microbiol., Roma,* p. 69.

Jičínská, E. (1973). Nocardia-like mutants of a soil coryneform bacterium. *Arch. Mikrobiol.* **89,** 269.

Jones, L. A. & Bradley, S. G. (1961). Susceptibility of actinomycete type species to actinophages. *Bact. Proc.* **1961,** 74.

Jones, L. A. & Bradley, S. G. (1964). Phenetic classification of actinomycetes. *Devs ind. Microbiol.* **5,** 267.

Kanetsuna, F. & Bartoli, A. (1972). A simple chemical method to differentiate *Mycobacterium* from *Nocardia. J. gen. Microbiol.* **70,** 209.

Krassilnikov, N. A. (1938). Proactinomycetes. *Bull. Acad. Sci. URSS 1938a,* 139.

Krassilnikov, N. A. & Tausson, T. A. (1938). Variability of proactinomycetes and mycobacteria. *Mikrobiologiya* **7,** 50.

Kurup, P. V. & Schmitt, J. A. (1973). Numerical taxonomy of *Nocardia. Can. J. Microbiol.* **19,** 1035.

Kwapinski, J. B. (1963). Antigenic structure of *Actinomycetales.* VI. *J. Bact.* **86,** 179.

Kwapinski, J. B. G. (1969). Analytical serology of *Actinomycetales.* In *Analytical Serology of Microorganisms,* vol. 1. Ed. J. B. G. Kwapinski. p. 1. New York: John Wiley (Interscience).

Kwapinski, J. B. G. & Horsman, G. (1973). Cultural characterization and differentiation of nocardiae. *Can. J. Microbiol.* **19,** 895.

Lacey, J. & Goodfellow, M. (1974). A new genus of nocardioform actinomycetes? *Proc. I Intern. Conf. Biol. Nocardiae, Merida, Venezuela.* p. 22.

Lachner-Sandoval, V. (1898). Ueber Strahlenpilze. *Thesis.* Strasbourg. Abstracted in *Zentbl. Bakt. Parasitkde Abt. I.* **25,** 782.

Lanéelle, G., Asselineau, J. & Chamoiseau, G. (1971). Présence de mycosides C′ (formes simplifées de mycoside C) dans les bactéries isolées de bovins atteints du farcin. *FEBS Letters* **19,** 109.

Lanéelle, M. A., Asselineau, J. & Castelnuovo, G. (1965). Etudes sur les mycobacteries et les nocardiae. *Annls Inst. Pasteur, Paris,* **108,** 69.

Langeron, M. (1922). Les oosporoses. *Nouveau traité de médicine* **4,** 430.

Lechevalier, H. A. & Lechevalier, M. P. (1970). A critical evaluation of the genera of aerobic actinomycetes. In *The Actinomycetales.* Ed. H. Prauser. p. 393. Jena: Gustav Fischer Verlag.

Lechevalier, H. A., Lechevalier, M. P. & Becker, B. (1966). Comparison of the chemical composition of cell walls of nocardiae with that of other aerobic actinomycetes. *Int. J. Syst. Bact.* **16,** 151.

Lechevalier, H. A., Lechevalier, M. P. & Gerber, N. N. (1971). Chemical composition as a criterion in the classification of actinomycetes. *Adv. appl. Microbiol.* **14,** 47.

Lechevalier, H. A., Solotorovsky, M. & McDurmont, C. I. (1961). A new genus of the actinomycetales: *Micropolyspora* gen. nov. *J. gen. Microbiol.* **26,** 11.

Lechevalier, M. P. (1968). Identification of aerobic actinomycetes of clinical importance. *J. Lab. clin. Med.* **71,** 934.

Lechevalier, M. P. & Gerber, N. N. (1970). The identity of madurose with 3-0-methyl-D-galactose. *Carbohydr. Res.* **13,** 451.

Lechevalier, M. P., Horan, A. C. & Lechevalier, H. (1971). Lipid composition in the classification of nocardiae and mycobacteria. *J. Bact.* **105,** 313.

Lechevalier, M. P. & Lechevalier, H. (1970). Chemical composition as a criterion in the classification of aerobic actinomycetes. *Int. J. Syst. Bact.* **20,** 435.

Lechevalier, M. P. & Lechevalier, H. (1974). *Nocardia amarae* sp. nov., an actinomycete common in foaming activated sludge. *Int. J. Syst. Bact.* **24,** 278.

Lechevalier, M. P., Lechevalier, H. and Horan, A. C. (1973). Chemical characteristics and classification of nocardiae. *Can. J. Microbiol.* **19,** 965.

Lehmann, K. B. & Neumann, R. O. (1927). *Bakteriologische Diagnostik,* 7 edn. Munich. (Cited by Jensen, 1931.)

Lessel, E. F. (1960). The nomenclatural status of the generic names of the *Actinomycetales. Int. Bull. bact. Nomencl. Taxon.* **10,** 87.

Lieske, R. (1921). *Morphologie und Biologie der Strahlenpilze* (*Actinomyceten*). Leipzig: Gebruder Borntraeger.

MacKinnon, J. E. & Artagaveytia-Allende, R. C. (1956). The main species of pathogenic aerobic actinomycetes causing mycetomas. *Trans. Roy. Soc. trop. Med. Hyg.* **50,** 31.

Magnusson, M. & Mariat, F. (1968). Delineation of *Nocardia farcinica* by delayed type skin reactions in guinea pigs. *J. gen. Microbiol.* **51,** 151.

Manion, R. E., Bradley, S. G., Zinneman, H. H. & Hall, W. H. (1964). Interrelationships among mycobacteria and nocardiae. *J. Bact.* **87,** 1056.

Mariat, F. (1963). Activité uréasique des actiomycètes aérobies pathogènes. *Annls Inst. Pasteur, Paris,* **105,** 795.

Mariat, F. (1965). Etude comparative de souches de *Nocardia* isolées de mycétomes. *Annls Inst. Pasteur, Paris,* **109,** 90.

McCarter, J. & Hastings, E. G. (1935). The morphology of the mycobacteria. *J. Bact.* **29,** 503.

McClung, N. M. (1949). Morphological studies in the genus *Nocardia.* I. Developmental studies. *Lloydia* **12,** 137.

McClung, N. M. (1954). The utilization of carbon compounds by *Nocardia* species. *J. Bact.* **68,** 231.

McClung, N. M. (1974). Family VI. *Nocardiaceae* Castellani and Chalmers 1919. *Bergey's Manual of Determinative Bacteriology,* 8 edn. Eds R. E. Buchanan & N. E. Gibbons. p. 726. Baltimore: Williams and Wilkins.

Minnikin, D. E., Alshamaony, L. & Goodfellow M. (1974). Differentiation between mycobacteria, nocardiae and related taxa by analysis of whole cell methanolysates. *Proc. 1 Intern. Conf. Biol. Nocardiae, Merida, Venezuela.* p. 36.

Mordarska, H. & Mordarski, M. (1969). Comparative studies on the occurrence of lipid A, diaminopimelic acid and arabinose in *Nocardia* cells. *Arch. Immun. Ther. Exp.* **17,** 739.

Mordarska, H., Mordarski, M. & Goodfellow, M. (1972). Chemotaxonomic characters and classification of some nocardioform bacteria. *J. gen. Microbiol.* **71,** 77.

Mordarski, M. (1966). Differentiation between *Streptomyces* and *Nocardia* on the basis of arabinose content of the cell wall. *Arch. Immun. Ther. Exp.* **14,** 602.

Nikitina, E. T. & Kalakoutskii, L. V. (1971). Induction of nocardioform growth in actinomycetes on media with D-fructose. *Z. allg. Mikrobiol.* **11,** 601.

Nocard, E. (1888). Note sur la maladie de boeufs de la Guadeloupe, connue sous le nom de farcin. *Annls Inst. Pasteur, Paris* **2,** 293.

Nonomura, H. & Ohara, Y. (1971). Distribution of actinomycetes in soil. X. *J. Fement. Technol, Osaka* **49,** 895.

Ørskov, J. (1923). *Investigations into the Morphology of the Ray Fungi.* Copenhagen: Levin and Munksgaard.

Ørskov, J. (1938). Untersuchungen über Strahlenpilze, reingezüchtet aus dänischen Erdproben. *Zentbl. Bakt. Parasitkde Abt. II.* **98,** 344.

Patel, P. V. & Ratledge, C. (1973). Isolation of lipid-soluble compounds that bind ferric ions from *Nocardia* species. *Biochem. Soc. Trans.* **1,** 886.

Pier, A. C. & Fichtner, R. E. (1971). Serologic typing of *Nocardia asteroides* by immunodiffusion. *Am. Rev. resp. Dis.* **103,** 698.

Pinoy, E. (1913). Actinomycoses et mycétomes. *Bull. Inst. Pasteur* **11,** 929; 977.

Prauser, H. (1967). Contributions to the taxonomy of the *Actinomycetales. Publ. Fac. Sci. Univ. Brno* **K40,** 196.

Prauser, H. (1967–1968). State and tendencies in the taxonomy of *Actinomycetales. Spisy prir. Fak. Sci. Univ. Brno* **43,** 295.

Prauser, H. (1974). The actinomycete-genus *Nocardiodes. Proc. 1 Intern. Conf. Biol. Nocardiae, Merida, Venezuela.* p. 20.

Prauser, H. & Falta, R. (1968). Phagensensibilitat, Zellwand–Zusammensetzung und Taxonomie von Actinomyceten. *Z. allg. Mikrobiol.* **8**, 39.

Prauser, H., Lechevalier, M. P. & Lechevalier, H. (1970). Description of *Oerskovia* gen. n. to harbor Ørskov's motile *Nocardia. Appl. Microbiol.* **19,** 534.

Prévot, A.-R. (1946). Études de systématique bactérienne. VII. *Actinomycétales. Annls Inst. Pasteur, Paris* **72,** 1.

Prévot, A. R. (1961). *Traité de Systématique Bactérienne.* p. 463. Paris: Dunod.

Pridham, T. G. & Lyons, Jr., A. J. (1969). Progress in clarification of the taxonomic and nomenclatural status of some problem actinomycetes. *Devs ind. Microbiol.* **10,** 183.

Puntoni, V. & Leonardi, D. (1940). Sulla sistemica degli Attinomiceti *Asteroides* n.g. *Boll. Accad. med. Roma* **61,** 90. (Cited in Lessel, (1960) and Sforza, M. (1940). Pluralità di specie del cosidetto *Actinomyces Farcinicus. Archo ital. Sci. med. colon.* **21,** 1.

Ratledge, C. & Patel, P. V. (1974). Lipid-soluble, iron-binding compounds in *Nocardia, "M." rhodochrous* and *Mycobacterium. Proc. 1 Intern. Conf. Biol. Nocardiae, Merida, Venezuela.* p. 58.

Ratledge, C. & Snow, G. A. (1974). Isolation and structure of nocobactin NA, a lipid-soluble iron-binding compound from *Nocardia asteroides. Biochem. J.* **139,** 407.

Ridell, M. (1974*a*). Taxonomical studies on *N. farcinica. Proc. 1 Intern. Conf. Biol. Nocardiae, Merida, Venezuela.* p. 18.

Ridell, M. (1974*b*). Serological study of nocardiae and mycobacteria by using "*Mycobacterium*" *pellegrino* and *Nocardia corallina* precipitation reference systems. *Int. J. Syst. Bact.* **24,** 64.

Romano, A. H. & Sohler, A. (1956). Biochemistry of the *Actinomycetales.* II. *J. Bact.* **72,** 865.

Schaal, K. P. (1974). Use of an indirect fluorescent antibody technique in the identification of aerobic actinomycetes. *Proc. 1 Intern. Conf. Biol. Nocardiae, Merida, Venezuela.* p. 94.

Schneidau, Jr., J. D. (1956). Antigenic and cultural properties of Nocardia. Ph.D. Thesis. *Diss. Abstr.* **16,** 1319.

Schneidau, J. D. Jr. & Shaffer, M. F. (1957). Studies on *Nocardia* and other Actinomycetales. I. *Ann. Rev. Tuberc.* **76,** 770.

Schneidau, J. D. Jr., & Schaffer, M. F. (1960). Studies on *Nocardia* and other *Actinomycetales.* II. *Am. Rev. resp. Dis.* **82,** 64.

Sneath, P. H. A. (1957). The application of computers to taxonomy. *J. gen. Microbiol.* **17,** 201.

Sneath, P. H. A. (1970). Application of numerical taxonomy to *Actinomycetales*: Problems and prospects. In *The Actinomycetales.* Ed. H. Prauser. p. 371. Jena: Gustav Fischer Verlag.

Sneath, P. H. A. (1974). An evaluation of numerical taxonomic techniques in the taxonomy of *Nocardia* and allied taxa. *Proc. 1 Intern. Conf. Biol. Nocardiae, Merida, Venezuela.* p. 6.

Snow, G. A. (1970). Mycobactins: Iron-chelating growth factors from mycobacteria. *Bact. Rev.* **34,** 99.

Sohler, A., Romano, A. H. & Nickerson, W. J. (1958). Biochemistry of the *Actinomycetales.* III. *J. Bact.* **75,** 283.

Staneck, J. L. & Roberts, G. D. (1974). Simplified approach to identification of aerobic actinomycetes by thin-layer chromatography. *Appl. Microbiol.* **28,** 226.

Stanier, R. Y. (1970). Towards an evolutionary taxonomy of the bacteria. *Recent Progress in Microbiology, vol. X. X Intern. Congr. Microbiol., Mexico City.*

Strange, R. E. (1956). The structure of an amino sugar present in certain spores and bacterial cell walls. *Biochem. J.* **64,** 23P.

Šuput, J., Lechevalier, M. P. & Lechevalier, H. A. (1967). Chemical composition of variants of aerobic actinomycetes. *Appl. Microbiol.* **15,** 1356.

Tacquet, A., Plancot, M. T., Debruyne, J., Devulder, B., Joseph, M. & Losfield, J. (1971). Etudes préliminaires sur la classification numérique des mycobactéries et des nocardiae. *Annls Inst. Pasteur, Lille* **22,** 121.

Takeya, K., Nakayama, Y. & Muraoka, S. (170). Specificity in skin reaction to tuberculin protein prepared frcm rapidly growing mycobacteria and some nocardia. *Am. Rev. resp. Dis.* **102,** 982.

Tarnok, I., Röhrscheidt, E. & Bönicke, R. (1967). Basenzusammensetzung der Desoxyribonukleinsäure (DNS) von Mykobakterien und verwandten Mikroorganismen. *Rass. Patol Appar. resp.* **17,** 3.

Tewfik, E. M. & Bradley, S. G. (1967). Characterization of deoxyribonucleic acids from streptomycetes and nocardiae. *J. Bact.* **94**, 1994.

Thoen, C. O., Karlson, A. G. & Ellefson, R. D. (1971*a*). Fatty acids of *Mycobacterium kansasii. Appl. Microbiol.* **21,** 628.

Thoen, C. O., Karlson, A. G. & Ellefson, R. D. (1971*b*). Comparison by gas-liquid chromatography of the fatty acids of *Mycobacterium avium* and some other nonphotochromogenic mycobacteria. *Appl. Microbiol.* **22,** 560.

Topping, L. E. (1937). The predominant micro-organisms in soils. I. *Zentbl. Bakt. Parasitkde Abt. II.* **97,** 289.

Trevisan, V. (1889). *I generi e le specie delle Batteriacee.* p. 9. Milano.

Tsukamura, M. (1969). Numerical taxonomy of the genus *Nocardia. J. gen. Microbiol.* **56,** 265.

Tsukamura, M. (1971). Proposal of a new genus, *Gordona,* for slightly acid-fast organisms occurring in sputa of patients with pulmonary disease and in soil. *J. gen. Microbiol.* **68,** 15.

Tsukamura, M. (1974). A numerical taxonomic study of the relationship between *Mycobacterium,* rhodochrous group and *Nocardia. Proc. 1 Intern. Conf. Biol. Nocardiae, Merida, Venezuela.* p. 10.

Twort, F. W. & Ingram, G. L. Y. (1912). A method for isolating and cultivating *Mycobacterium enteritidis chronicae pseudotuberculosis bovis,* Jöhne, and some experiments on the preparation of a diagnostic vaccine for pseudo-tuberculous enteritis of bovines. *Proc. Roy. Soc. Ser. B. Biol. Sci.* **84,** 517.

Uesaka, I. (1969). Breaking up of vegetative mycelium of *Nocardia* and *Streptomyces* cultures on agar film. *Jap. J. Microbiol.* **13,** 65.

Umbreit, W. W. (1939). Studies on the proactinomyces. *J. Bact.* **38,** 73.

Vacheron, M.-J., Guinand, M., Michel, G. & Ghuysen, J.-M. (1972). Structural investigations on cell walls on *Nocardia* sp. The wall lipid and peptidoglycan moieties of *Nocardia kirovani. Eur. J. Biochem.* **29,** 156.

Vestal, J. R. & Perry, J. J. (1971). Effect of substrate on the lipids of the hydrocarbon-utilizing *Mycobacterium vaccae. Can. J. Microbiol.* **17,** 445.

Vilkas, E., Massot, J. C. & Zissmann, E. (1970). Étude des parois d'une souche de *Micromonospora.* I. Isolement d'un phosphate de glucosamine. *FEBS Letters* **7,** 77.

Waksman, S. A. (1919). Cultural studies of species of *Actinomyces. Soil Sci.* **8,** 71.

Waksman, S. A. (1957). Order V. *Actinomycetales* Buchanan, 1917. In *Bergey's Manual of Determinative Bacteriology,* 7 edn. Eds R. S. Breed, E. G. D. Murray & A. P. Hitchens. p. 694. Baltimore: Williams and Wilkins.

Waksman, S. A. & Henrici, A. T. (1943). The nomenclature and classification of the actinomycetes. *J. Bact.* **46,** 337.

Waksman, S. A. & Henrici, A. T. (1948). Family *Actinomycetaceae* Buchanan and Family *Streptomycetaceae* Waksman & Henrici. In *Bergey's Manual of Determinative Bacteriology,* 6 edn. Eds R. S. Breed, E. G. D. Murray & A. D. Hitchens. p. 892. Baltimore: Williams and Wilkins.

Wayne, L. G., Juarez, W. J. & Nichols, E. G. (1958). Aryl sulfatase activity of aerobic actinomycetales. *J. Bact.* **75,** 367.

Williams, S. T., Lacey, J., Serrano, J. A., de Serrano, A. A. & Sharples, G. P. (1974). The micromorphology and ultrastructure of nocardioform actinomycetes. *Proc. 1 Intern. Conf. Biol. Nocardiae, Merida, Venezuela.* p. 26.

Wright, J. H. (1905). The biology of microorganisms of actinomycosis. *J. med. Res.* **13,** 381.

Yamada, K. & Komagata, K. (1970). Taxonomic studies on coryneform bacteria. III. DNA base composition of coryneform bacteria. *J. gen. appl. Microbiol., Tokyo* **16,** 215.

Yamaguchi, T. (1967). Similarity in DNA of various morphologically distinct actinomycetes. *J. gen. appl. Microbiol, Tokyo* **13,** 63.

2. The "*Rhodochrous*" Complex and its Relationships with Allied Taxa

I. J. BOUSFIELD

National Collection of Industrial Bacteria, Torry Research Station, Aberdeen AB9 8DG, Scotland

and

M. GOODFELLOW

Department of Microbiology, The Medical School, The University, Newcastle-upon-Tyne NE1 7RU, England

Contents

1. Introduction

The group of bacteria known collectively as "*Mycobacterium*" *rhodochrous*, the "*rhodochrous*" group or the "*rhodochrous*" complex continues to bedevil those of us working on the classification of nocardioform and coryneform bacteria. There now seems little doubt that the *rhodochrous* group forms a recognizable taxon; where this taxon should be classified is another matter. We shall attempt to summarize here what is known about the *rhodochrous* complex and its relationship to other taxa, to explore some of the taxonomic options open to us and, hopefully, to stimulate further discussion on the classification of this difficult group.

The specific epithet "*rhodochrous*" (Zopf, 1891) was used by Gordon & Mihm (1957) for bacteria showing similarities both to the nocardiae and to the mycobacteria, but which could not be assigned confidently to either. The *rhodochrous* taxon was classified tentatively as a species in the genus *Mycobacterium* on the basis of colony morphology and appearance when stained; although some workers disagreed with this grouping (Adams & McClung, 1962; Jones & Bradley, 1964). More strains were later included in this species and its description became more comprehensive (Gordon & Mihm, 1959, 1961; Gordon, 1966). Most of these strains had been classified previously as *Mycobacterium, Nocardia, Proactinomyces* and *Corynebacterium* species, although 11 genera were represented altogether. Gordon (1966) concluded that because strains of "*Mycobacterium*" *rhodochrous* had been placed in so many different genera, the species was a generic intermediate; its provisional location in *Mycobacterium* being one of convenience rather than of taxonomic certainty. Gordon also suggested that an examination of coryneform genera such as *Arthrobacter* and *Brevibacterium* might reveal other hitherto unrecognised "*M*". *rhodochrous* strains; this subsequently proved to be the case.

2. Historical Aspects

Whereas Gordon & Mihm were perhaps the first to bring together under one taxonomic umbrella the members of the *rhodochrous* complex, the difficulties associated with distinguishing between strains within it, and between those classified in such genera as *Corynebacterium, Mycobacterium* and *Nocardia* had been known for many years. As early as 1888, Nocard said that the organism subsequently known as *Nocardia farcinica* resembled the diphtheria bacillus in laboratory culture, and Wolff & Israel (1891) made similar observations on organisms isolated from actinomycoses in cattle. The similarity between *Corynebacterium diphtheriae* and *Mycobacterium tuberculosis* was noted by Lehmann & Neumann (1896), who also suggested a possible relationship between these organisms and *Actinomyces* species. Later (1920) these authors pointed out the difficulties of distinguishing between the genera *Corynebacterium, Mycobacterium* and the actinomycetes, saying "Various only relatively acid-fast species form transitions between *Corynebacterium* and *Mycobacterium.* The border between *Mycobacterium* and *Actinomyces* is made very indefinite because of the fragility and the relative acid-fastness of many actinomycetes".

(From the English translation by R. S. Breed, 1930.) Lieske (1921) also considered the fragmenting actinomycetes to be transitional organisms. Ørskov (1923) showed that some actinomycetes (his Group IIb), after fragmentation of the primary mycelium, continued to multiply in an "angular" fashion similar to that observed in the corynebacteria and mycobacteria, and that it was impossible to distinguish them. He concluded that "... correspondence in morphology is so great that it is not possible to set up any definite boundary between the bacteria (corynebacteria and mycobacteria) examined and those ray fungi which exhibited the angular arrangement of the elements; ... "

Some years later, Jensen (1931) proposed the genus *Proactinomyces* to accommodate Ørskov's Group II actinomycetes, but most workers tended not to use this name after establishment of the priority of the names *Actinomyces* and *Nocardia* for the anaerobic and aerobic proactinomycetes respectively (Waksman & Henrici, 1943). However, some workers, notably Krasilnikov and his colleagues, have continued to use the name *Proactinomyces.* Jensen (1931, 1932, 1934) assigned several (then) recently described mycobacteria to this genus. He also commented on the acid-fastness of *Corynebacterium equi*, suggesting that it would be better named *Mycobacterium equi.* It is interesting to note that Jensen (1934) also assigned *Bacillus* (*Serratia*) *rubropertinctus* (more recently known as *Nocardia rubropertincta* and a member of the *rhodochrous* complex) to *Mycobacterium* and not to *Proactinomyces.* Thus the indeterminate nature of the generic boundaries in this region was again demonstrated.

Two groups of proactinomyces were recognised by Umbreit (1939); the α-forms which developed no aerial mycelium and the β-forms which produced an aerial mycelium. Krasilnikov (1938) similarly recognized two morphological groups. Jensen (1953) later suggested that *Proactinomyces* might be restricted to the α-forms, reserving *Nocardia* for the aerial mycelium producing β-forms.

The blurring of the boundary between *Mycobacterium* and *Proactinomyces* and the virtual interchangeability of portions of the two genera were exacerbated by the concurrent expansion of the genus *Corynebacterium*; a process deplored by Conn in 1947. All three genera were essentially "morphological" in concept and it was becoming increasingly difficult to assign new isolates unequivocally to one or other of them. This situation was clearly demonstrated by Conn & Dimmick (1947) who sent cultures of *Bacterium globiforme*, a soil organism described by Conn in 1928, to both Jensen and Krasilnikov. The former considered it to be a species of *Corynebacterium* whilst the latter was convinced that it should be classified with the mycobacteria. Conn & Dimmick said that *B. globiforme* bore a superficial resemblance to certain nocardiae, but they maintained that it was sufficiently distinct from all previously named organisms to merit a generic location of its own. Consequently they revived the name *Arthrobacter* (Fisher, 1895) and proposed *B. globiforme* as the type species of their new genus.

A classification of the actinomycetes based on the mode of cellular branching was proposed by Bisset & Moore in 1949. They described a new genus *Jensenia* (formally proposed in 1950) to accommodate the "soil diphtheroids" which they considered distinct from *Corynebacterium*, *Mycobacterium* and *Nocardia*. Jensen (1952, 1953) wondered whether *Jensenia* and *Arthrobacter* were synonymous, but it is not clear whether Bisset & Moore examined any arthrobacters. From their descriptions of colonial and cellular morphology it is possible that most of Bisset & Moore's diphtheroids were *rhodochrous* strains. Phillips (1953) suggested that *Jensenia* was synonymous with *Nocardia*; later the type species of *Jensenia*, *J. canicruria*, was considered similar to *Nocardia rubra* by Adams & McClung (1960) and to *N. erythropolis* by Adams (1970). It was reduced to synonymy with "*Mycobacterium*" *rhodochrous* by Gordon & Mihm (1961) and called *Nocardia canicruria* by Adams & McClung (1962).

Yet another genus of Gram-positive bacteria, *Brevibacterium*, was proposed by Breed in 1953. *Brevibacterium* has become a diverse collection of strains, most of which show the pleomorphism characteristic of *Corynebacterium* or *Arthrobacter* and at least one species produces a well-developed substrate mycelium (*Brevibacterium fermentans*, which is probably a species of *Oerskovia*).

Thus by the time Gordon (1966) reported her comparative study on the strains she considered to belong to the *rhodochrous* taxon, there were five genera to which rhodochrous strains might be assigned. The recently described *Gordona* (Tsukamura, 1971) has added a sixth alternative.

3. Morphological Considerations

Prior to the 1960's, the nocardioform bacteria were classified largely by the use of morphological features. This approach was illustrated and reviewed by McClung & Adams (1962) and will not be considered in detail here. Morphological differences are apparent but seem to be those of degree rather than of kind. The presence of a primary mycelium is a difficult character to handle taxonomically as was implied by Jensen (1953) when he asked at what stage did branching cells become part of a rudimentary mycelium. This point is clearly illustrated if some of Sundman's (1958) photographs of certain arthrobacters are compared with those of *Nocardia erythropolis* taken by Adams & McClung (1962). The production of nocardioform mutants from a coryneform bacterium (Jičínská, 1973) also highlights the problem of dealing with morphological characteristics. Nevertheless it was on largely morphological grounds that Gordon & Mihm (1957) excluded the partially acid-fast, "soft-growing" *rhodochrous* complex from *Nocardia,* preferring to retain the latter generic name for those strains with a well-developed aerial mycelium. Morphological considerations also led Adams & McClung (1962) to the opposite conclusion; they argued that the similar growth cycle of members of the *rhodochrous* complex and nocardiae prevented a generic separation, despite considerations of colonial morphology and acid fastness. Jones & Bradley (1964) came to the same conclusion by considering physiological features. They contructed a taxonomic tree based on overall similarity in which *rhodochrous* strains were placed on a branch originating from "typical" nocardiae. Thus opinion tended to harden in favour of the *rhodochrous* complex being absorbed into *Nocardia.* However, the application of newer taxonomic techniques has shown that the systematic position of the *rhodochrous* complex is still a matter for dispute and we shall now consider the information obtained in recent years.

4. Phenetic Studies

In a limited numerical investigation Cerbón (1967) recovered "*Mycobacterium*" *rhodochrous* in a cluster readily distinguishable from strains of *Nocardia asteroides* and *Nocardia brasiliensis.* Later, more comprehensive studies (Bradley, 1971; Goodfellow, 1971; Tacquet *et al.*, 1971; Goodfellow, Fleming & Sackin, 1972; Goodfellow *et al.*, 1974) suggested that the *rhodochrous* complex was equivalent in rank to clusters equated with the genera *Nocardia* and *Mycobacterium*, although no attempts were made to erect a new genus. However, it was found generally that the *rhodochrous* complex had a higher affinity with the nocardiae than with the mycobacteria. Tsukamura (1971) classified a few reference *rhodochrous* strains in the genus *Gordona*, which he proposed to accommodate slightly acid-fast bacteria isolated from soil and from the sputum of patients with pulmonary disease. Later (Tsukamura, 1973, 1974) additional *rhodochrous* strains were classified as *Gordona rhodochroa*, *Gordona rosea* and *Gordona rubropertincta*, although the author now questioned the suitability of *Gordona* as a generic epithet for the *rhodochrous* taxon.

In all of the numerical analyses the *rhodochrous* strains were recovered as a heterogeneous taxon. Bradley (1971) found two clusters, one containing strains of "*Mycobacterium*" *rhodochrous*, *Nocardia corallina*, *N. opaca* and *N. rubra*, the other containing strains of *N. canicruria*, *N. coeliaca*, *N. erythropolis* and *N. globerula*. He recommended that the first cluster be called either *N. corallina* or *N. rubra* and the second *N. erythropolis*. Tacquet *et al.* (1971) obtained two clusters containing strains labelled "*Mycobacterium*" *pellegrino*, "*M*" *rhodochrous*, *Nocardia kirovani*, and "*M.*" *rhodochrous*, *N. corallina* and *N. serratia* respectively. Two phena were also recovered by Jones (1975), F2 contained strains received as *Arthrobacter variabilis*, *Corynebacterium fascians*, *C. rubrum*, and "*Mycobacterium*" *rhodochrous*, while F3 included *Corynebacterium paurometabolum*, *Jensenia canicruria*, "*M.*" *rhodochrous*, *Nocardia calcarea* and *N. opaca*. Three homogeneous subclusters were found by Goodfellow (1971); 14A containing strains received as *Nocardia pellegrino*, 14C containing "*Mycobacterium*" *rhodochrous*, *N. corallina*, *N. rubra*, *N. salmonicolor*, *Proactinomyces corallinus*, *P. luteus*, *P. phenoli* and *P. ruber*, and 14D containing *Jensenia canicruria*, "*M.*" *rhodochrous*, *N. corallina*, *N. erythropolis*, *N. globerula*, *N. lutea*, *N. opaca*, *N. rubropertincta*, *P. canicruria* and *P. globerulus*. A fourth, heterogeneous subcluster included strains labelled "*M.*" *rhodochrous*, *N. aurantia*, *N. opaca*, *N. restrictus* and *Corynebacterium fascians*. Finally, in the IWGMT (International Working

Group on Mycobacterial Taxonomy) co-operative study (Goodfellow *et al.*, 1974) 80 strains provisionally classified in the *rhodochrous* complex were recovered in three homogeneous subclusters; 1A contained *Nocardia corallina, N. erythropolis, N. rubra* and *Rhodococcus rhodochrous,* IB contained *Corynebacterium rubrum, N. globerula* and numerous "*Mycobacterium*" *rhodochrous* strains, and 1C strains of *C. equi* and *N. restrictus.*

In most of these numerical studies, the *rhodochrous* taxon was defined at or around 80% similarity level and the subgroups at or above the 85% level. Because many of the test strains were incorrectly identified and as few were common to all the studies, it is not possible to equate all of the subgroups recovered by different investigators. However, it is clear that *Nocardia rubra* (Bradley, 1971) can be compared with subcluster 14C (Goodfellow, 1971), *Gordona rhodochroa* (Tsukamura, 1973) and subcluster 1A (Goodfellow *et al.*, 1974). Similarly, *Nocardia erythropolis* (Bradley, 1971) is substantially the same taxon as Goodfellow's (1971) subcluster 14D and phenon F3 (Jones, 1975).

There has been an unfortunate tendency among those interested in the coryneform bacteria largely to ignore the *rhodochrous* group or at best to regard it as a collection of "fringe" strains, worthy of only cursory consideration. Not only has this lack of attention resulted in there being no detailed investigations of the relationships between the *rhodochrous* taxon and the coryneform group, but the recent proliferation of such dubious names as *Corynebacterium hydrocarboclastus* (Iizuka & Komagata, 1964), *Arthrobacter paraffineus, Brevibacterium sterolicum* (names cited in recent patent applications) etc. might have been avoided.

Harrington (1966) carried out a numerical study, on a somewhat restricted range of strains, which led him to the conclusion that *Corynebacterium, Mycobacterium* and *Nocardia* should be merged into one genus. Nocardiae, mycobacteria and *rhodochrous* strains were recovered in one of three aggregated clusters equated with families by Davis & Newton (1969). The remaining clusters contained a number of taxa including the animal pathogenic corynebacteria and some arthrobacters. Masuo & Nakagawa (1969) did not include any *rhodochrous* strains in their study of coryneform bacteria, but they found the two *Mycobacterium* strains they used grouped with the animal pathogenic corynebacteria.

The single strain of *Jensenia canicruria* included by Stuart & Pease (1972) in a study of *Listeria, Erysipelothrix* and coryneform bacteria was ungrouped. Bousfield (1972) was able to separate *rhodochrous* strains from animal pathogenic corynebacteria and from most arthrobacters, but had difficulty in distinguishing *Jensenia canicruria*, from bacteria similar to *Brevibacterium linens*. This overlap of *J. canicruria* was noted also by Skyring & Quadling (1970). Holmberg & Hallender (1973) were able to separate *rhodochrous* strains from mycobacteria and animal pathogenic corynebacteria.

In an extensive numerical taxonomic study of coryneform bacteria Jones (1975) recovered the *rhodochrous* strains in a distinct taxon which formed part of an aggregate cluster equated with a family. The genera *Arthrobacter, Kurthia, Nocardia, Mycobacterium* and certain species of *Brevibacterium* were found in this heterogeneous cluster. All of the strains in this taxon were sharply separated from those classified in the two other aggregate clusters one of which included the animal corynebacteria, cellulomonads and propionibacteria.

Thus whilst some of the numerical studies on coryneform bacteria have been somewhat inconclusive over the relationships between *rhodochrous* strains and certain arthrobacters and *Brevibacterium linens*, Gordon's (1966) view that *Corynebacterium* would not be a suitable niche for the *rhodochrous* complex has been borne out in general. It also appears that most of *Arthrobacter* is distinct from the *rhodochrous* complex.

Although many nocardioform strains probably can be classified in one or other of the discrete groups within the *rhodochrous* taxon, the systematic position of some "soft-growing" species is evidently confused. For instance, strains labelled *Corynebacterium fascians* (a species suggested to be related to the nocardiae by Lacey (1955) and Ramamurthi (1959)) have been considered to be closely related to the animal pathogenic corynebacteria (Harrington, 1966; Yamada & Komagata, 1972), markedly different from them (da Silva & Holt, 1965), intermediate between the animal and plant pathogenic corynebacteria (Robinson, 1966), intermediate between the mycobacteria and animal corynebacteria (Masuo & Nakagawa, 1969) and to be closely related to the *rhodochrous* complex (Gordon, 1966; Goodfellow, 1971; Bousfield, 1972).

Further comparative studies using additional taxonomic criteria are obviously needed in this area.

There is some evidence that antibiotic sensitivity tests may be of value in the classification of nocardioform bacteria. Sensitivity tests have provided characters for the differentiation of mycobacteria (Silcox & David, 1971; Boisvert, 1973; Marks, 1973) and have been used in the classification of nocardiae (Bradley, 1971; Kurup & Schmitt, 1973) and coryneform bacteria (Seaman & Woodbine, 1962; Davis & Newton, 1969). Recently, Goodfellow & Orchard (1974) screeened 151 nocardioform strains for their susceptibility *in vitro* to 52 antibacterial agents. *Nocardia, Gordona* and the *rhodochrous* taxon formed a series with increasing sensitivity to the antibacterial antibiotics and several tests provided good characters for identification.

5. Serological Studies

Several serological techniques have been applied to *Nocardia, Mycobacterium, Corynebacterium* and related taxa with varying results. It has been shown that these genera have antigens in common (Cummins, 1962*a*, 1965; Kwapinski, 1956, 1964, 1969, 1970; Kwapinski & Seeliger, 1964; Schneidau & Schaffer, 1960; Castelnuovo *et al.*, 1968; Stanford & Wong, 1974; Ridell, 1974) and it has been suggested by Schleifer & Kandler (1972) that the common cell wall arabinogalactan may be responsible for many of the cross-reactions. However, several workers have shown that it is possible to differentiate between these taxa by serological methods (Castelnuovo *et al.*, 1964; Magnusson, 1962; Magnusson & Mariat, 1968; see chapter 8). Castelnuovo *et al.* (1964) demonstrated a close relationship between strains of *Nocardia corallina, N. rubra,* "*Mycobacterium*" *rhodochrous* and "*M.*" *pellegrino,* considering the taxon so formed to be more closely related to the genus *Nocardia* than to *Mycobacterium.* In a more extensive study of nocardiae by comparative immunodiffusion against antimycobacterial sera, Ridell & Norlin (1973) found that strains of "*Mycobacterium*" *pellegrino* and "*M.*" *rhodochrous* did not react to any greater extent than did any other nocardial strain, thus supporting the view that the designation *Mycobacterium* be questioned. However, they did recognize two serological groups, distinguishing *Nocardia asteroides,*

N. brasiliensis and *N. caviae* from the *rhodochrous*-like strains. Later Ridell (1974) found that several *rhodochrous*-like strains fell into two serological groups typified by reference cultures of *Nocardia corallina* and "*Mycobacterium*" *pellegrino*. In the co-operative IWGMT study (Goodfellow *et al.*, 1974) strains in the *rhodochrous* complex reacted more frequently with nocardial reference systems than with mycobacterial systems though they were serologically distinct from each. There was also little reaction between the *rhodochrous* reference system and reference strains of *Nocardia* and *Mycobacterium*.

The most extensive serological investigations have been done by Kwapinski and his colleagues who have examined wall and cytoplasmic antigens and exoantigens of many actinomycetes (see Kwapinski, 1970). Results obtained with cell and cell wall agglutinogens formed a very complex pattern as frequent cross reactions were observed between cultures of nocardiae, mycobacteria, *rhodochrous* and other actinomycete taxa (Kwapinski, 1964; 1970). Purified polysaccharide fractions from the walls of nocardiae and mycobacteria showed a narrower specificity and have been used for classification (Kwapinski & Seeliger, 1965).

The range of serological activity of the exoantigens was narrower than that of the agglutinogen preparations but even so, cross-specificity between mycobacteria, corynebacteria and the *rhodochrous* complex was observed (Kwapinski, 1966, 1970). Studies using cytoplasmic antigens (Kwapinski, 1966, 1970; Kwapinski *et al.*, 1973) have resulted in the recognition of 22 serogroups within the nocardiae, which can be recognized using 12 selected antisera. In these studies nocardiae and *rhodochrous* cultures were more closely related to the fast-growing mycobacteria and scotochromogens than to other actinomycete taxa, but the interrelationships between the nocardiae and *rhodochrous* strains were complex and will not be discussed here.

6. Chemotaxonomic Studies

Since the introduction of cell wall analysis as a taxonomic tool (Cummins & Harris, 1956) many actinomycetes and coryneform bacteria have been examined, and cell wall composition is now considered to be a major criterion in the classification of these organisms. At present cell wall analysis does not seem to be of great value in distinguishing the *rhodochrous* complex from the true corynebacteria, mycobacteria and nocardiae. A *meso*-diaminopimelic acid (DAP)-containing peptidoglycan associated with an arabinogalactan polymer is common to all four taxa (Cummins & Harris, 1956, 1958; Cummins, 1962*a,b*; Becker, Lechevalier & Lechevalier, 1965; Lechevalier, Lechevalier & Gerber, 1971). However, more detailed analysis has shown that in the case of certain mycobacteria and *Nocardia kirovani*, the muramic acid residues of the peptidoglycan are *N*-glycolated and not *N*-acetylated as in other groups (Adam *et al.*, 1969; Azuma *et al.*, 1970; Kanetsuna & San Blas, 1970; Bordet *et al.*, 1972). Bordet *et al.* (1972) suggested that the "true" nocardiae might be those containing *N*-glycolyl muramic acid but Schleifer & Kändler (1972) questioned whether this compound might be restricted to the mycobacteria, certain nocardiae and micromonosporae.

It is fairly easy to distinguish between the *rhodochrous* group and coryneforms of the *Arthrobacter* type on the basis of wall composition. Whilst *Arthrobacter* itself is a heterogeneous genus at present in terms of wall composition, there is little evidence so far to suggest that the *meso*-DAP/arabinose combination occurs. The two or three "arthrobacters" which have been shown to have this type of cell wall (Schleifer & Kändler, 1972) can be classified on other criteria either in the *rhodochrous* complex or in the genus *Corynebacterium* (Goodfellow & Bousfield, unpublished results). Arabinose has not been found in arthrobacters, although some may have *meso*-DAP. Two have LL-DAP, but the majority contain lysine as the cell wall diamino acid.

Many *Brevibacterium* species have a cell wall containing arabinose and *meso*-DAP but most of these can be reduced to synonymy with *Corynebacterium glutamicum* (Takayama, Abe & Kinoshita, 1965*a*), which is probably the only saprophytic true *Corynebacterium* known so far. A few others, such as *Brevibacterium sterolicum* perhaps should be transferred to the *rhodochrous* complex.

Although we have said that members of the *rhodochrous* taxon contain *meso*-DAP and arabinose in their walls, there appear to be some anomalies. Certain strains of *Nocardia rubra* and *N. salmonicolor* containing LL-DAP and no arabinose were mentioned by Schleifer & Kändler (1972) in their review of peptidoglycan types. We can only assume that the particular strains mentioned were not authentic.

Lipid analyses are now becoming increasingly useful in the classification and identification of actinomycetes. Much attention has been given to the analysis of the mycolic acid moiety which is associated mainly with the walls of true corynebacteria, nocardiae and mycobacteria. Mycolic acids are high molecular weight β-hydroxy acids with a long chain in the α-position (Asselineau, 1966) and they invariably occur as mixtures of homologues (Bordet & Michel, 1969; Yano *et al.*, 1972). Three kinds of mycolic acid are generally recognised; mycolic acids *sensu stricto* with a chain length of around 80 carbon atoms are associated with mycobacteria, nocardomycolic acids with a chain length of about C_{50} occur in the true nocardiae, and short-chain corynomycolic acids (C_{30}) are found in the true corynebacteria (Michel, Bordet & Lederer, 1960; Etémadi, 1967; Ioneda, Lederer & Rozanis, 1970; Maurice, Vacheron & Michel, 1971; Yano & Saito, 1972). The detection of differences in the structure of the respective methyl mycolates provides a reliable way of distinguishing between true corynebacteria, mycobacteria and nocardiae (Lanéelle, Asselineau & Castelnuovo, 1965; Lechevalier, Horan & Lechevalier, 1971; Lechevalier, Lechevalier & Horan, 1973). Mycolic acids *sensu stricto* can also be distinguished from nocardomycolic acids by a simple solubility technique (Kanetsuna & Bartoli, 1972).

Lanéelle, Asselineau & Castelnuovo, (1965) detected mycolic acids in strains of *Nocardia pellegrino* and "*Mycobacterium*" *rhodochrous* and suggested that the latter be transferred to the genus *Nocardia.* Mycolic acids of chain length C_{38}–C_{46} have been found in *Nocardia corallina* (Batt, Hodges & Robertson, 1971) and *N. rhodochrous* (Ioneda, Lederer & Rozanis, 1970) whilst mixtures ranging from C_{36} to C_{48} have been found in *rhodochrous* strains later classified in subcluster 1A of the IWGMT co-operative study (Azuma *et al.*, 1973; Goodfellow *et al.*, 1974). strains labelled *Nocardia farcinicum* (Maurice, Vacheron & Michel, 1971) and *N. erythropolis* (Yano *et al.*, 1972) had mycolic acids with chain lengths of C_{32} to C_{48}.

Using thin-layer chromatography (t.l.c.) Mordarska (1968) detected a lipid which was later named lipid LCN-A or Lipid Characteristic of *Nocardia* (Mordarska & Réthy, 1970). This lipid was found subsequently in many true nocardiae and *rhodochrous* strains, in animal pathogenic corynebacteria, but not in mycobacteria or arthrobacters (Mordarska & Mordarski, 1970; Mordarska, Mordarski & Goodfellow, 1972; Goodfellow, 1973; Goodfellow *et al.*, 1974). There are at least two analogues of LCN-A with different R_F values; LCN-A type *asteroides* is found mainly in *Nocardia asteroides*, *N. brasiliensis* and *N. caviae* and LCN-A type *calcarea* is found in *N. calcarea* and many *rhodochrous* strains (Mordarska, Mordarski & Goodfellow, 1972; Mordarska, Mordarski & Pietkiewicz, 1973; Goodfellow *et al.*, 1974). Mordarska's lipid is now known to have properties consistent with those of free mycolic acids (Goodfellow *et al.*, 1973). Minnikin, Patel & Goodfellow (1974) found that the mycolic acids of a typical *rhodochrous* strain had a significantly shorter chain length (C_{34} to C_{46}) than the major acids from *Nocardia caviae* (C_{48} to C_{56}), which probably explains the different mobilities of the LCN-A analogues in thin-layer systems. These results were confirmed in a more extensive study (Alshamaony, Goodfellow & Minnikin, 1976) in which strains of *Nocardia asteroides*, *N. brasiliensis* and *N. caviae* were found to contain nocardomycolic acids whereas strains representing the three homogeneous subgroups of the *rhodochrous* cluster (Goodfellow, 1971) contained significantly smaller mycolic acids with molecular weights ranging from C_{34} to C_{50} units. However, these workers also found that a strain received as *Nocardia opaca*, and recovered in the *rhodochrous* cluster, contained nocardomycolic acids. A similar observation was made by Maurice, Vacheron & Michel (1971) for a strain labelled *Nocardia erythropolis*. Again one of the *rhodochrous* subgroups recovered in the IWGMT co-operative study (Goodfellow *et al.*, 1974) contained LCN-A type *asteroides*, a result indicative of the presence of nocardomycolic acids. It would appear, therefore, that the *rhodochrous* taxon includes strains which have different molecular species of mycolic acids. It has been shown recently (Minnikin, Alshamaony & Goodfellow, 1975) that *rhodochrous* strains, nocardiae, mycobacteria and corynebacteria can be distinguised readily by examining the pattern of mycolic esters obtained on t.l.c. analysis of their methanolysis products.

The information gained from lipid analyses weakens the case for retaining the *rhodochrous* taxon in the genus *Mycobacterium* and further illustrates its relationships with the true nocardiae and corynebacteria. On the basis of lipid studies several workers have recommended that the *rhodochrous* taxon might be classified in the genus *Nocardia* (Lechevalier, Lechevalier & Gerber, 1971; Kanetsuna & Bartoli, 1972; Azuma *et al.*, 1973), or in the genus *Corynebacterium* (Lechevalier, Lechevalier & Horan, 1973), but acceptance of these suggestions would seem to be at variance with phenetic and serological data. (Goodfellow, 1971; Tsukamura, 1971, 1973, 1974; Goodfellow *et al.*, 1974; Jones, 1975). There is also the evidence that the mycolic acids of many *rhodochrous* strains seem to form an anomalous group with molecular weights centred around 40 carbon units (Minnikin, Patel & Goodfellow, 1974; Alshamaony, Goodfellow & Minnikin, 1975).

No systematic attempt has been made so far to determine the value of glycolipids and phospholipids as characters in the classification of the *rhodochrous* complex and related taxa, although preliminary results are encouraging (Lanéelle & Asselineau, 1970; Khuller & Brennan, 1972*a,b*; Pommier & Michel, 1972). Phosphatidylglycerol is found in small amounts, if at all, in mycobacteria and nocardiae (Yano, Furukawa & Kusunose, 1969) but it is the major phospholipid in *Corynebacterium diphtheriae* (Brennan & Lehane, 1971). Pommier & Michel (1973) suggest that whereas mycobacteria possess oligomannosides, nocardiae and the *rhodochrous* complex contain phosphatidylinositol monomannosides. However, Khuller & Brennan (1972*c*) have found dimannophosphoinositides in *Mycobacterium. Nocardia rhodochrous* and *N. asteroides* can be differentiated from each other by the type of dicorynemycolyltrehalose units ("cord factors") they contain. The structure of the molecule in the latter species appears to be identical with that found in *Corynebacterium diphtheriae* (Senn *et al.*, 1967). Other lipids, such as the menaquinones (Azerad & Cyrot-Pelletier, 1973) and carotene-like substances (Röhrscheidt & Tarnok, 1972), may also have a role in the classification of nocardioform bacteria.

Comparison of the fatty acid profiles of the *rhodochrous* complex and related taxa reveals many similarities (Bordet & Michel, 1963; Lanéelle, Asselineau & Castelnuovo, 1965; Yano, Furukawa & Kusunose, 1969), but according to Bowie *et al.* (1972) *Arthrobacter* may be distinguished from *Corynebacterium, Mycobacterium* and *Nocardia* on the basis of fatty acid composition.

7. DNA Analyses and Genetic Studies

Although the DNA guanine/cytosine content (G+C%) is an invaluable character in many areas of bacterial taxonomy, it is of limited use as a differential feature in the coryneform and nocardioform groups. G+C values for these taxa form a virtually continuous spectrum from just above 50% to over 70%. It may be possible to distinguish the true corynebacteria, since most reported values are below 60% (Bouisset, Breuillard & Michel, 1963; Takayama, Abe & Kinoshita, 1965*b*; Bowie *et al.*, 1972; Bousfield, 1972), although unusually high values have been reported for certain strains of *Corynebacterium xerosis* and *C. bovis* (Yamada & Komagata, 1970). On the other hand, there is a great deal of coincidence among G+C values for the mycobacteria, nocardiae, arthrobacters and the *rhodochrous* complex (see e.g. Tewfik & Bradley, 1967; Wayne & Gross, 1968; Hagihara & Shikada, 1969; Skyring & Quadling, 1970; Yamada & Komagata, 1970; Bowie *et al.*, 1972; Bousfield, 1972; Crombach, 1972).

Bradley (1971) pointed out that G+C values for nocardioform organisms, although covering a fairly wide range, do not form a continuum but appear to follow a bimodal distribution. The peaks of this distribution are around 61–63% and 66–68%, and members of the *rhodochrous* complex can be found at both levels. It may be noted also that G+C values for the *rhodochrous*-like *Corynebacterium equi, C. fascians* and *C. hydrocarboclastus* can be fitted to one or other of these peaks (Yamada & Komagata, 1970; Bowie *et al.*, 1972; Bousfield, 1972) thus adding weight to Bradley's (1971) assertion that there are at least two populations in the *rhodochrous* complex in respect of DNA base composition.

DNA/DNA hybridization techniques have been applied recently to various actinomycete taxa (see chapter 12). Bradley (1973) found that DNA from strains designated *Nocardia asteroides, N. brasiliensis* and *N. caviae* bound significant amounts of *N. farcinica* reference DNA, whereas a very low level of reassociation was detected with DNA from *N. erythropolis, N. rubra* and *Mycobacterium* species. Only DNA from the mycobacteria and from *Nocardia farcinica* annealed appreciably with *Mycobacterium smegmatis* reference DNA. There was good correlation between DNA reassociation results and earlier phenetic data (Bradley, 1971) in respect of the quantitative relationships found among the mycobacteria.

Using reference DNA from *Nocardia erythropolis*, Bradley, Brownell & Clark (1973) found a high degree of relative binding with DNA from strains of *N. corallina* and *N. opaca*. Binding of DNA from other nocardiae, mycobacteria and streptomycetes was found to some extent, although the duplexes formed in such cases were generally thermally labile. However, there was evidence that two strains labelled *Nocardia corallina* and *N. rubra* respectively were more closely related to *N. erythropolis* than to the other strains studied. It was concluded again that the *rhodochrous* taxon was heterogeneous and that certain strains labelled *Nocardia erythropolis*, *N. corallina* and *N. opaca* constituted a genospecies.

Genetic recombination between nocardioform bacteria of heterologous origin has been demonstrated by Adams (1964, 1968) and Brownell & Adams (1967, 1968). It was found that cultures derived from *Nocardia erythropolis* did not recombine with each other, nor did cultures derived from *N. canicruria*. However, recombinants between these taxa were fertile. Representative strains of *Jensenia canicruria* were found later to be identical with representative strains of *Nocardia erythropolis* (Adams, Adams & Brownell, 1970). It was concluded that *J. canicruria* was a mating type of *N. erythropolis* and should be considered a later synonym of the latter. Matings of *Nocardia rubra* and *N. erythropolis* and *Jensenia canicruria* were infertile and these taxa were considered to belong to different genospecies. Clark, Brownell & Bradley (1971) considered that the two *rhodochrous* subgroups equivalent to the taxa *Nocardia erythropolis* and *N. corallina* or *N. rubra* represented distinct genospecies.

The first attempt to use phages in actinomycete classification was made by Bradley & Anderson in 1958. In a later, more comprehensive study (Bradley, Anderson & Jones, 1961) 37 actinophages were tested against 25 actinomycete cultures, when it was found that several phages could lyse both nocardiae and streptomycetes. Prauser & Falta (1968) found that the phages they used were able to lyse only cells showing a specific cell wall type. They screened 20 actinomycete taxa including *Nocardia*, *Mycobacterium* and the *rhodochrous* complex.

Castelnuovo *et al.* (1964) found that some *rhodochrous* strains (*Nocardia corallina, N. rubropertincta, "Mycobacterium" pellegrino* and *"M." rhodochrous*) were susceptible to phage Pellegrino whereas strains of *N. asteroides* and *N. brasiliensis* were not. Manion *et al.* (1964) also concluded that the *rhodochrous* complex was distinct from the *Nocardia asteroides* group in respect of phage susceptibility. This view was supported by other studies in which *rhodochrous* strains were lysed by phages unable to attack nocardiae and mycobacteria (Bönicke & Juhasz, 1965; Jones & Bradley, 1964). There is evidence also from phage sensitivity studies that the *rhodochrous* taxon is heterogeneous (see chapter 10). Brownell, Adams & Bradley (1967) isolated two mating-specific phages, one (ϕC) which lysed two strains of *Nocardia canicruria* and a strain received as *N. corallina,* but which had no effect upon *N. erythropolis, N. opaca* or other *N. corallina* strains. The other phage (ϕEC) also lysed *Nocardia erythropolis* and *N. opaca*. Bradley (1971) found that the cluster he equated with *Nocardia erythropolis* contained strains which generally were susceptible to phage ϕEC.

8. Conclusions

The chaotic situation surrounding the *rhodochrous* complex *vis à vis* other nocardioform and coryneform taxa illustrates the pitfalls of relying too heavily on morphological features for identification and classification. In many cases the difficulties have been exaggerated by the use of different cultivation conditions, non-comparable media etc. for the morphological examination of different strains (see Williams *et al.*, chapter 5). However, as we have shown, there is now a wealth of other evidence available on a large number of strains and perhaps it is timely to re-evaluate the taxonomic status of the *rhodochrous* complex.

The *rhodochrous* taxon is a recognizable, if heterogeneous, taxonomic entity, but the problem of a satisfactory systematic niche remains. There appear to be at least three alternatives open to us. Firstly, the *rhodochrous* taxon could be left as it stands, all members of it being called *Mycobacterium rhodochrous.* This would seem to be the least satisfactory solution. Classification in the genus *Mycobacterium* was never intended to be more than a tentative arrangement and evidence accumulated over the past few years suggests that the *rhodochrous* complex is misplaced in this genus. In addition, many workers feel that there is more than one species within the *rhodochrous* group and that to retain a single specific epithet is unwise. In the eighth edition of Bergey's *Manual of Determinative Bacteriology* (Buchanan & Gibbons, 1974) "*M*" *rhodochrous* is excluded from the genus *Mycobacterium.* Some *rhodochrous* taxa are retained in the genus *Nocardia* with the comment that further work is needed to clarify their position.

A second alternative would be to classify the *rhodochrous* complex within an existing genus. In this event the favourite genus would appear to be *Nocardia,* primarily on morphological and chemotaxonomic grounds. Several strains have already been classified in this genus though, as we have mentioned earlier, opinion is divided as to whether or not these strains should remain in the genus *Nocardia.*

However, numerical studies have repeatedly shown that the *rhodochrous* taxon is equivalent in rank to *Nocardia,* and to *Corynebacterium* and *Mycobacterium.* In view of this, can the genus *Nocardia* and the *rhodochrous* complex justifiably be combined while these other taxa are left distinct? Would not any combination be equally valid? Alternatively, should all four be combined as has been suggested from time to time? It is our view that the inclusion of the *rhodochrous* complex within the genus *Nocardia* (or for that matter within *Corynebacterium* or *Mycobacterium*) is not justifiable in the face of recent taxonomic evidence.

The final alternative is to consider the *rhodochrous* complex as a genus in its own right, an idea cautiously intimated by Cross & Goodfellow in their 1973 classification of the actinomycetes. We now believe that serious consideration should be given to elevating the *rhodochrous* taxon to generic rank. This view was supported by an informal meeting of experts at the 1st International Conference on the Biology of the Nocardiae held in 1974. However, even if this view is accepted several problems remain. Despite recent work, it is still difficult to define such a genus unequivocally and an ambigous definition would serve no useful purpose. A revised definition of the genus *Nocardia* almost certainly would be necessary in order to exclude members of the *rhodochrous* complex. There are also nomenclatural problems. Existing names which might be considered are *Rhodococcus* (Zopf, 1891), *Proactinomyces* (Jensen, 1932),

Jensenia (Bisset & Moore, 1950), and *Gordona* (Tsukamura, 1971). Of these, *Rhodococcus* has priority. Although originally described as a genus to accommodate pink-pigmented cocci, the only authentic extant strain of the type species (*Rhodococcus rhodochrous*) is undoubtably a member of the *rhodochrous* taxon. Since *Rhodococcus* has priority over *Jensenia* and *Proactinomyces* we are unable to agree with Bradley & Bond (1974) who have resurrected the latter to accommodate part of the *rhodochrous* taxon. Finally, it is not clear whether *Gordona* can be considered synonymous with *Rhodococcus*. The numerical data indicates that some gordonae should be classified in the *rhodochrous* complex but confirmatory evidence based upon other taxonomic criteria is required to clarify the situation. However, Tsukamura (1974) classified the *rhodochrous* strains and gordonae together but questioned the epithet *Gordona* for this taxon stating that "the appropriate name for this genus appears to be *Rhodococcus*, as this is the oldest name among generic names given for the organisms belonging to this genus".

In conclusion, we would say that there is considerable evidence in favour of the *rhodochrous* complex being given generic status, and that a suitable name would be *Rhodococcus*. However, we hesitate at present to offer an unambiguous generic definition.

9. References

Adam, A., Petit, J. F., Wietzerbin-Falszpan, J., Sinay, P., Thomas, D. W. & Lederer, E. (1969). L'acide *N*-glycolylmuramique, constituent des parois de *Mycobacterium smegmatis*: identification par spectrometrie de masse. *FEBS Letters* **4,** 87.

Adams, J. N. (1964). Recombination between *Nocardia erythropolis* and *Nocardia canicruria*. *J. Bact.* **88,** 865.

Adams, J. N. (1968). Partial exclusion of the *Nocardia erythropolis* chromosome in nocardial recombinants. *J. Bact.* **96,** 1750.

Adams, J. N. & McClung, N. M. (1960). Morphological studies in the genus *Nocardia*. V. Septation in *Nocardia rubra* and *Jensenia canicruria*. *J. Bact.* **80,** 281.

Adams, J. N. & McClung, N. M. (1962). Comparison of the developmental cycles of some members of the genus *Nocardia*. *J. Bact.* **84,** 206.

Adams, M. M., Adams, J. N. & Brownell, G. H. (1970). The identification of *Jensenia canicruria* Bisset and Moore as a mating type of *Nocardia erythropolis* (Gray and Thornton) Waksman and Henrici. *Int. J. Syst. Bact.* **20,** 133.

Alshamaony, L., Goodfellow, M. & Minnikin, D. E. (1976). Free mycolic acids as criteria in the classification of *Nocardia* and the "*rhodochrous*" complex. *J. gen. Microbiol.* **92,** 188.

Asselineau, J. (1966). *The Bacterial Lipids.* Paris: Hermann.

Azerad, R. & Cyrot-Pelletier, M. O. (1973). Structure and configuration of the polyprenoid side chain of dihydromenaquinones from myco- and corynebacteria. *Biochimie* **55,** 591.

Azuma, I., Ohuchida, A., Taniyama, T., Yamamura, Y., Shoji, K., Hori, M. & Ribi, E. (1973). The mycolic acids of *Mycobacterium rhodochrous* and *Nocardia corallina. Biken J.* **17,** 1.

Azuma, I., Thomas, D. W., Adam, A., Ghuysen, J. M., Bonaly, R., Petit, J. F. & Lederer, E. (1970). Occurrence of *N*-glycolylmuramic acid in bacterial cell walls. A preliminary survey. *Biochim. biophys. Acta* **208,** 444.

Batt, R. D., Hodges, R. & Robertson, J. G. (1971). Gas chromatography and mass spectrometry of the trimethylsilyl ether methyl ester derivatives of long chain hydroxy acids from *Nocardia corallina. Biochim. biophys. Acta* **239,** 368.

Becker, B., Lechevalier, M. P. & Lechevalier, H. A. (1965). Chemical composition of cell wall preparations from strains of various form-genera of aerobic actinomycetes. *Appl. Microbiol.* **13,** 236.

Bisset, K. A. & Moore, F. W. (1949). The relationship of certain branched bacterial genera. *J. gen. Microbiol.* **3,** 387.

Bisset, K. A. & Moore, F. W. (1950). *Jensenia,* a new genus of the Actinomycetales. *J. gen. Microbiol.* **4,** 280.

Boisvert, H. (1973). Mycobactéries (*M. bovis* et "atypiques") identifiées a l'Institut Pasteur de Paris de 1960 à 1972). *Annls Soc. belge Méd. trop.* **53,** 29.

Bönicke, R. & Juhasz, S. E. (1965). Morphogenetische und biochemische Kriterien zur Unterscheidung der Gattungen *Nocardia* und *Mycobacterium. Ann. Inst. Super. Sanità* **1,** 629.

Bordet, C., Karahjoli, M., Gateau, O. & Michel, G. (1972). Cell walls of nocardiae and related actinomycetes: identification of the genus *Nocardia* by cell wall analysis. *Int. J. Syst. Bact.* **22,** 251.

Bordet, C. & Michel, G. (1963). Étude des acides-gras isolés de plusieurs espéces de *Nocardia. Biochim. biophys. Acta* **70,** 613.

Bordet, C. & Michel, G. (1969). Structure et biogenèse des lipides à haut poids moléculaire de *Nocardia asteroides. Bull. Soc. Chim. biol.* **51,** 527.

Bouisset, L., Breuillard, J. & Michel, G. (1963). Étude de l'ADN chez les *Actinomycetales. Annls Inst. Pasteur, Paris* **104,** 756.

Bousfield, I. J. (1972). A taxonomic study of some coryneform bacteria. *J. gen. Microbiol.* **71,** 441.

Bowie, I. S., Grigor, M. R., Dunckley, G. G., Loutit, M. W. & Loutit, J. S. (1972). The DNA base composition and fatty acid constitution of some Gram-positive pleomorphic soil bacteria. *Soil Biol. Biochem.* **4,** 397.

Bradley, S. G. (1971). Criteria for definition of *Mycobacterium, Nocardia* and the *rhodochrous* complex. *Adv Fronts Pl. Sci.* **28,** 349.

Bradley, S. G. (1973). Relationships among mycobacteria and nocardiae based upon deoxyribonucleic acid reassociation. *J. Bact.* **113,** 645.

Bradley, S. G. & Anderson, D. L. (1958). Taxonomic implication of actinophage host range. *Science, N.Y.* **128,** 413.

Bradley, S. G., Anderson, D. L. & Jones, L. A. (1961). Phylogeny of actinomycetes as revealed by actinophage susceptibility. *Devs ind. Microbiol.* **2,** 223.

Bradley, S. G. & Bond, J. S. (1974). Taxonomic criteria for mycobacteria and nocardiae. *Adv. appl. Microbiol.* **18,** 131.

Bradley, S. G., Brownell, G. H. & Clark, J. (1973). Genetic homologies among nocardiae and other actinomycetes. *Can. J. Microbiol.* **19,** 1007.

Breed, R. S. (1953). The *Brevibacteriaceae* fam. nov. of the order *Eubacteriales. Symposium. VI Intern. Congr. Microbiol., Roma.* p. 13.

Brennan, P. J. & Lehane, D. P. (1971). The phospholipids of corynebacteria. *Lipids* **6,** 401.

Brownell, G. H. & Adams, J. N. (1967). Linkage and segregation of unselected markers in matings of *Nocardia erythropolis* and *Nocardia canicruria. J. Bact.* **94,** 650.

Brownell, G. H. & Adams, J. N. (1968). Linkage and segregation of a mating type specific phage and resistance characters in nocardial recombinants. *Genetics, Princeton* **60,** 437.

Brownell, G. H., Adams, J. N. & Bradley, S. G. (1967). Growth and characterization of nocardiophages for *Nocardia canicruria* and *Nocardia erythropolis* mating types. *J. gen. Microbiol.* **47,** 247.

Buchanan, R. E. & Gibbons, N. E. (Eds) (1974). *Bergey's Manual of Determinative Bacteriology,* 8 edn. Baltimore: Williams and Wilkins.

Castelnuovo, G., Bellezza, G., Duncan, M. E. & Asselineau, J. (1964). Études sur les mycobactéries et les nocardiae. *Annls Inst. Pasteur, Paris* **107,** 828.

Castelnuovo, G., Bellezza, G., Guiliani, H. J. & Asselineau, J. (1968). Relations chimiques et immunologiques chez les Actinomycétales: *Annls Inst. Pasteur, Paris* **114,** 139.

Cerbón, J. (1967). Taxonomic analysis of *Nocardia. Revta lat.-am. Microbiol.* **9,** 65.

Clark, J. E., Brownell, G. H. & Bradley, S. G. (1971). Analysis of compatible nocardiae by DNA:DNA hybridizations. *Bact. Proc.* **1971,** 38.

Conn, H. J. (1928). A type of bacterium abundant in productive soils, but apparently lacking in certain soils of low productivity. *Bull. N.Y. St. agric. Exp. Stn.* No. 138.

Conn, H. J. (1947). A protest against the misuse of the generic name *Corynebacterium. J. Bact.* **54,** 10.

Conn, H. J. & Dimmick, I. (1947). Soil bacteria similar in morphology to *Mycobacterium* and *Corynebacterium. J. Bact.* **54,** 291.

Crombach, W. H. J. (1972). DNA base composition of soil arthrobacters and other coryneforms from cheese and sea fish. *Antonie van Leeuwenhoek* **38,** 105.

Cross, T. & Goodfellow, M. (1973). Taxonomy and classification of the actinomycetes. In *Actinomycetes: Characteristics and Practical Importance,* Eds F. A. Skinner & G. Sykes. p. 11. London: Academic Press.

Cummins, C. S. (1962*a*). Chemical composition and antigenic structure of cell walls of *Corynebacterium, Mycobacterium, Nocardia, Actinomyces* and *Arthrobacter. J. gen. Microbiol.* **28,** 35.

Cummins, C. S. (1962*b*). La composition chimique des parois cellulaires d'actinomycetes et son application taxonomique. *Annls Inst. Pasteur, Paris* **103,** 385.

Cummins, C. S. (1965). Chemical and antigenic studies on cell walls of mycobacteria, corynebacteria and nocardias. *Am. Rev. resp. Dis.* **92,** 63.

Cummins, C. S. & Harris, H. (1956). The chemical composition of the cell wall in some gram-positive bacteria and its possible value as a taxonomic character. *J. gen. Microbiol.* **14,** 583.

Cummins, C. S. & Harris, H. (1958). Studies on the cell wall composition and taxonomy of *Actinomycetales* and related groups. *J. gen. Microbiol.* **18,** 173.

da Silva, G. A. N. & Holt, J. G. (1965). Numerical taxonomy of certain coryneform bacteria. *J. Bact.* **90,** 921.

Davis, G. H. G. & Newton, K. G. (1969). Numerical taxonomy of some named coryneform bacteria. *J. gen. Microbiol.* **56,** 195.

Etémadi, A. H. (1967). The use of pyrolysis gas chromatography and mass spectrometry in the study of the structure of mycolic acids. *J. Gas Chromat.* **5,** 447.

Fischer, A. (1895). Untersuchungen über Bakterien. *Jb. wiss. Bot.* **27,** 1. (Cited in Conn & Dimmick, 1947.)

Goodfellow, M. (1971). Numerical taxonomy of some nocardioform bacteria. *J. gen. Microbiol.* **69,** 33.

Goodfellow, M. (1973). Characterisation of *Mycobacterium, Nocardia, Corynebacterium* and related taxa. *Annls Soc. belge Méd. trop.* **53,** 287.

Goodfellow, M., Fleming, A. & Sackin, M. J. (1972). Numerical classification of *Mycobacterium rhodochrous* and Runyon's group IV mycobacteria. *Int. J. Syst. Bact.* **22,** 81.

Goodfellow, M., Lind, A., Mordarska, H., Pattyn, S. & Tsukamura, M. (1974). A co-operative numerical analysis of cultures considered to belong to the *rhodochrous* complex. *J. gen. Microbiol.* **85,** 291.

Goodfellow, M., Minnikin, D. E., Patel, P. V. & Mordarska, H. (1973). Free nocardomycolic acids in the classification of nocardiae and strains of the *rhodochrous* complex. *J. gen. Microbiol.* **14,** 185.

Goodfellow, M. & Orchard, V. A. (1974). Antibiotic sensitivity of some nocardioform bacteria and its value as a criterion for taxonomy. *J. gen. Microbiol.* **83,** 375.

Gordon, R. E. (1966). Some strains in search of a genus—*Corynebacterium, Mycobacterium, Nocardia* or what? *J. gen. Microbiol.* **43,** 329.

Gordon, R. E. & Mihm, J. M. (1957). A comparative study of some strains received as nocardiae. *J. Bact.* **73,** 15.

Gordon, R. E. & Mihm, J. M. (1959). A comparison of four species of mycobacteria. *J. gen. Microbiol.* **21,** 736.

Gordon, R. E. & Mihm, J. M. (1961). The specific identity of *Jensenia canicruria. Can. J. Microbiol.* **7,** 108.

Hagihara, Y. & Shikada, K. (1969). The base composition of deoxyribonucleic acid isolated from mycobacteria. *Kekkaku* **44,** 321.

Harrington, B. J. (1966). A numerical taxonomical study of some corynebacteria and related organisms. *J. gen. Microbiol.* **45,** 31.

Holmberg, K. & Hallander, H. O. (1973). Numerical taxonomy and laboratory identification of *Bacterionema matruchotii, Rothia dentocariosa, Actinomyces naeslundii, Actinomyces viscosus,* and some related bacteria. *J. gen. Microbiol.* **76,** 43.

Iizuka, H. & Komagata, K. (1964). Microbiological studies on petroleum and natural gas. I. Determination of hydrocarbon-utilizing bacteria. *J. gen. appl. Microbiol., Tokyo* **10,** 207.

Ioneda, T., Lederer, E. & Rozanis, J. (1970). Sur la structure des diesters de tréhalose ("cord factors") produit par *Nocardia asteroides* et *Nocardia rhodochrous. Chem. Phys. Lipids* **4,** 375.

Jensen, H. L. (1931). Contributions to our knowledge of the *Actinomycetales.* II. The definition and subdivision of the genus *Actinomyces* with a preliminary account of Australian soil actinomycetes. *Proc. Linn. Soc. N.S.W.* **56,** 345.

Jensen, H. L. (1932). Contribution to our knowledge of the *Actinomycetales.* IV. The identity of certain species of *Mycobacterium* and *Proactinomyces. Proc. Linn. Soc. N.S.W.* **57,** 364.

Jensen, H. L. (1934). Studies on the saprophytic mycobacteria and corynebacteria. *Proc. Linn. Soc. N.S.W.* **59,** 19.

Jensen, H. L. (1952). The coryneform bacteria. *A. Rev. Microbiol.* **6,** 77.

Jensen, H. L. (1953). The genus *Nocardia* (or *Proactinomyces*) and its separation from other *Actinomycetales,* with some reflections on the phylogeny of the actinomycetes. *Symposium. VI Intern. Congr. Microbiol., Roma.* p. 69.

Jičínská, E. (1973). *Nocardia*-like mutants of a soil coryneform bacterium. *Arch. Mikrobiol.* **89,** 269.

Jones, D. (1975). A numerical taxonomic study of coryneform and related bacteria. *J. gen. Microbiol.* **87,** 52.

Jones, L. A. & Bradley, S. G. (1964). Relationships among streptomycetes, nocardiae, mycobacteria and other actinomycetes. *Mycologia* **56,** 505.

Kanetsuna, F. & Bartoli, A. (1972). A simple chemical method to differentiate *Mycobacterium* from *Nocardia. J. gen. Microbiol.* **70,** 209.

Kanetsuna, F. & San Blas, G. (1970). Chemical analysis of a mycolic acid—arabinogalactan-mucopeptide complex of mycobacteria cell walls. *Biochim. biophys. Acta* **208,** 434.

Khuller, G. K. & Brennan, P. J. (1972*a*). The polar lipids of some species of *Nocardia. J. gen. Microbiol.* **73,** 409.

Khuller, G. K. & Brennan, P. J. (1972*b*). Further studies on the lipids of corynebacteria: the mannolipids of *Corynebacterium aquaticum. Biochem. J.* **127,** 369.

Khuller, G. K. & Brennan, P. J. (1972*c*). The mannophosphoinositides of the unclassified *Mycobacterium* P_6. *Am. Rev. resp. Dis.* **106,** 892.

Krasilnikov, N. A. (1938). Ray fungi and related organisms—*Actinomycetales.* 13V. *Akad. Nauk SSSR.* Moscow.

Kurup, P. V. & Schmitt, J. A. (1973). Numerical taxonomy of *Nocardia. Can. J. Microbiol.* **19,** 1035.

Kwapinski, J. B. G. (1956). Research on the antigenic structure of the *Actinomycetales.* Antigenic relationship between strains of the genera *Mycobacterium* and *Corynebacterium. Bull. acad. pol. Sci.* **4,** 379.

Kwapinski, J. B. G. (1964). Cytoplasmic antigen relationships among the *Actinomycetales. J. Bact.* **87,** 1234.

Kwapinski, J. B. G. (1966). Antigenic structure of the *Actinomycetales.* XI. Spectra of the serological activity of the plasm antigens. *Zentbl. Bakt. ParasitKde. Abt. I.* **200,** 380.

Kwapinski, J. B. G. (1969). *Analytic Serology of Microorganisms,* vol. 1. Ed. J. B. G. Kwapinski. New York: Interscience Publishers.

Kwapinski, J. B. G. (1970). Serological Taxonomy and relationships of *Actinomycetales.* In *The Actinomycetales.* Ed. H. Prauser. Jena: Gustav Fischer.

Kwapinski, J. B. G., Kwapinski, E. H., Dowler, J. & Horsman, G. (1973). The phyloantigenic position of nocardiae revealed by examination of cytoplasmic antigens. *Can. J. Microbiol.* **19,** 955.

Kwapinski, J. B. G. & Seeliger, N. P. R. (1964). Immunological characteristics of the *Actinomycetales. Zentbl. Bakt. ParasitKde. Abt.* 1. **195,** 808.

Kwapinski, J. B. G. & Seelinger, N. P. R. (1965). Serological classification of the nocardiae with the polysaccharide fraction of their cell walls. *Mycopath. Mycol. appl.* **25,** 173.

Lacey, M. S. (1955). The cytology and relationship of *Corynebacterium fascians. Trans. Br. mycol. Soc.* **38,** 49.

Lanéelle, M. A. & Asselineau, J. (1970). Caractérization de glycolipides dans une souche de *Nocardia brasiliensis. FEBS Letters* **7,** 64.

Lanéelle, M. A., Asselineau, J. & Castelnuovo, G. (1965). Études sur les mycobactéries et les nocardiae. IV. Composition des lipides de *Mycobacterium rhodochrous, M. pellegrino* sp., et de quelques souches de nocardiae. *Annls Inst. Pasteur, Paris* **108,** 69.

Lechevalier, H. A., Lechevalier, M. P. & Gerber, N. N. (1971). Chemical composition as a criterion in the classification of actinomycetes. *Adv. appl. Microbiol.* **14,** 47.

Lechevalier, M. P., Horan, A. C. & Lechevalier, H. A. (1971). Lipid composition in the classification of nocardiae and mycobacteria. *J. Bact.* **105,** 313.

Lechevalier, M. P., Lechevalier, H. A. & Horan, A. C. (1973). Chemical characteristics and classification of nocardiae. *Can. J. Microbiol.* **19,** 965.

Lehmann, K. B. & Neumann, R. O. (1896). *Bakteriologische Diagnostik.* Munich: Lehmann Verlag.

Lehmann, K. B. & Neumann, R. O. (1920). *Bakteriologische Diagnostik,* 6 edn. Munich: Lehmann Verlag. (Cited in Jensen, 1953.)

Lehmann, K. B. & Neumann, R. O. (1930). *Bakteriologische Diagnostik,* 8 edn. Munich: Lehmann Verlag.

Lieske, R. (1921). *Morphologie und Biologie der Strahlenpilze.* Leipzig: Bebr. Borntraeger.

Magnusson, M. (1962). Specificity of sensitins. III. Further studies in guinea pigs with sensitins of various species of *Mycobacterium* and *Nocardia. Am. Rev. resp. Dis.* **86,** 395.

Magnusson, M. & Mariat, F. (1968). Delineation of *Nocardia farcinica* by delayed type skin reactions on guinea pigs. *J. gen Microbiol.* **51,** 151.

Manion, R. E., Bradley, S. G., Zinneman, H. H. & Hall, W. H. (1964). Interrelationships among mycobacteria and nocardiae. *J. Bact.* **87,** 1056.

Marks, J. (1973). Identification of mycobacteria as a clinical service. *Annls Soc. belge. Méd. trop.* **53,** 43.

Masuo, E. & Nakagawa, T. (1969). Numerical classification of bacteria. Part II. Computer analysis of coryneform bacteria (2). Comparison of group-formations obtained on two different methods of scoring data. *Agr. Biol. Chem.* **33,** 1124.

Maurice, M. T., Vacheron, M. J. & Michel, G. (1971). Isolément d'acides nocardiques de plusieurs espèces de *Nocardia. Chem. Phys. Lipids* **7,** 9.

McClung, N. M. & Adams, J. N. (1962). The morphology of actinomycetes in the genus *Nocardia. Revta lat.-am. Microbiol.* **5,** suppl. 9, 1.

Michel, G., Bordet, C. & Lederer, E. (1960). Isolément d'un nouvel acide mycologique: l'acide nocardique a partir d'une souche de *Nocardia asteroides. C.r. hedb. Séanc Acad. Sci., Paris* **250,** 3518.

Minnikin, D. E., Alshamaony, L. & Goodfellow, M. (1975). Differentiation of *Mycobacterium, Nocardia,* and related taxa by thin-layer chromatographic analysis of whole-cell methanolysates. *J. gen. Microbiol.* **88,** 200.

Minnikin, D. E., Patel, P. V. & Goodfellow, M. (1974). Mycolic acids of representative strains of *Nocardia* and the "rhodochrous" complex. *FEBS Letters* **39,** 322.

Mordarska, H. (1968). A trial of using lipids for the classification of actinomycetes. *Arch. Immun. Ther. Exp.* **16,** 45.

Mordarska, H. & Mordarski, M. (1970). Cell lipids of *Nocardia.* In *The Actinomycetales.* Ed. H. Prauser. Jena: Gustav Fischer.

Mordarska, H., Mordarski, M. & Goodfellow, M. (1972). Chemotaxonomic characters and classification of some nocardioform bacteria. *J. gen. Microbiol.* **71,** 77.

Mordarska, H., Mordarski, M. & Pietkiewicz, D. (1973). Chemical analysis of hydrolysates and cell extracts of *Nocardia pellegrino. Int. J. Syst. Bact.* **23,** 274.

Mordarska, H. & Réthy, A. (1970). Preliminary studies on the chemical character of the lipid fraction of *Nocardia. Arch. Immun. Ther. Exp.* **18,** 455.

Nocard, M. E. (1888). Note sur le maladie des boeufs de la Guadeloupe, connue sous le nom de farcin. *Annls Inst. Pasteur, Paris* **2,** 293.

Ørskov, J. (1923). *Morphology of the Ray Fungi.* Copenhagen: Levin & Munksgaard.

Phillips, N. C. (1953). Characterization of the soil globiforme bacteria. *Iowa St. Coll. J. Sci.* **27,** 240.

Pommier, M. T. & Michel, G. (1972). Isolément et caractéristique d'un nouveau glycolipide de *Nocardia caviae. C.r. hebd. Séanc. Acad. Sci., Paris* **275,** 1323.

Pommier, M. T. & Michel, G. (1973). Phospholipid and acid composition of *Nocardia* and nocardoid bacteria as criteria of classification. *Biochem. Syst.* **1,** 3.

Prauser, H. & Falta, R. (1968). Phagensensibilitat, Zellwand—Zusammensetzung und Taxonomie von Actinomyceten. *Z. allg. Mikrobiol.* **8,** 39.

Ramamurthi, C. S. (1959). Comparative studies on some Gram-positive phytopathogenic bacteria and their relationship to the corynebacteria. *Mem. Cornell Univ. agric. Exp. Stn.* No. 366.

Ridell, M. (1974). Serological study of nocardiae and mycobacteria by using "*Mycobacterium*" *pellegrino* and *Nocardia corallina* precipitation reference systems. *Int. J. Syst. Bact.* **24,** 64.

Ridell, M. & Norlin, M. (1973). Serological study of *Nocardia* by using mycobacterial precipitation reference systems. *J. Bact.* **113,** 1.

Robinson, K. (1966). An examination of *Corynebacterium* species by gel electrophoresis. *J. appl. Bact.* **29,** 179.

Röhrscheidt, E. & Tarnok, I. (1972). Untersuchungen an *Nocardia*-Pigmenten. Chromatographische Eigenschaften der Farbstoffe und ihre Bedeutung für die Differenzierung pigmentierter *Nocardia*-Stämme. *Zentbl. Bakt. ParasitKde* **221,** 221.

Schleifer, K. H. & Kändler, O. (1972). Peptidoglycan types of bacterial cell walls and their taxonomic implications. *Bact. Rev.* **36,** 407.

Schneidau, J. D. & Shaffer, M. F. (1960). Studies on *Nocardia* and other *Actinomycetales.* II. Antigenic relationships shown by slide agglutination tests. *Am. Rev. resp. Dis.* **82,** 64.

Seaman, A. & Woodbine, M. (1962). Antibiotics in bacterial classification with special reference to corynebacteria. In *Antibiotics in Agriculture.* Ed. M. Woodbine. London: Butterworths.

Senn, M., Ioneda, T., Pudles, J. & Lederer, E. (1967). Spectrometrie de masse de glycolipides. 1. Structure du "cord factor" de *Corynebacterium diphtheriae. Eur. J. Biochem.* **1,** 353.

Silcox, V. A. & David, H. L. (1971). Differential identification of *Mycobacterium kansasii* and *Mycobacterium marinum. Appl. Microbiol.* **21,** 327.

Skyring, G. W. & Quadling, C. (1970). Soil bacteria: a principal component analysis and guanine-cytosine contents of some arthrobacter-coryneform soil isolates and of some named cultures. *Can. J. Microbiol.* **16,** 95.

Stanford, J. L. & Wong, J. K. C. (1974). A study of the relationship between *Nocardia* and *Mycobacterium diernhoferi*—a typical fast growing *Mycobacterium. Br. J. exp. Path.* **55,** 291.

Stuart, M. R. & Pease, P. E. (1972). A numerical study of the relationships of *Listeria* and *Erysipelothrix. J. gen. Microbiol.* **73,** 551.

Sundman, V. (1958). Morphological comparison of some *Arthrobacter* species. *Can. J. Microbiol.* **4,** 221.

Tacquet, A., Plancot, M. T., Debruyne, J., Devulder, B., Joseph, M. & Losfeld, J. (1971). Études préliminaires sur la classification numérique des mycobactéries et des nocardias. 1) Relations taxonomique entre *Mycobacterium rhodochrous, Mycobacterium pellegrino* et les genres *Mycobacterium* et *Nocardia. Annls Inst. Pasteur, Lille* **22,** 121.

Takayama, K., Abe, S. & Kinoshita, S. (1965*a*). Taxonomical studies on glutamic acid producing bacteria. II. Classification of glutamic acid producing bacteria. *J. agric. chem. Soc., Japan* **39,** 335.

Takayama, K., Abe, S. & Kinoshita, S. (1965*b*). Taxonomical studies on glutamic acid producing bacteria. III. Base composition of DNA. *J. agric. chem. Soc., Japan* **39,** 342.

Tewfik, E. M. & Bradley, S. G. (1967). Characterization of deoxyribonucleic acid from streptomycetes and nocardiae. *J. Bact.* **94,** 1994.

Tsukamura, M. (1971). Proposal of a new genus, *Gordona,* for slightly acid-fast organisms occurring in sputa of patients with pulmonary disease and in soil. *J. gen. Microbiol.* **68,** 15.

Tsukamura, M. (1973). A taxonomic study of strains received as "*Mycobacterium*" *rhodochrous*. *Jap. J. Microbiol.* **17,** 189.

Tsukamura, M. (1974). A further numerical taxonomic study of the *rhodochrous* group. *Jap. J. Microbiol.* **18,** 37.

Umbreit, W. W. (1939). Studies on the *Proactinomyces*. *J. Bact.* **38,** 73.

Waksman, S. A. & Henrici, A. T. (1943). The nomenclature and classification of the actinomycetes. *J. Bact.* **46,** 337.

Wayne, L. G. & Gross, W. M. (1968). Base composition of deoxyribonucleic acid isolated from mycobacteria. *J. Bact.* **96,** 1915.

Wolff, M. & Israel, J. (1891). Über Reincultur des Actinomyces und seine Ubertrăgbarkeit auf Tiere. *Virchows Arch. path. Anat. Physiol.* **126,** 11. (Cited in Ørskov, 1923).

Yamada, K. & Komagata, K. (1970). Taxonomic studies on coryneform bacteria. III. DNA base composition of coryneform bacteria. *J. gen. appl. Microbiol., Tokyo* **16,** 215.

Yamada, K. & Komagata, K. (1972). Taxonomic studies on coryneform bacteria. V. Classification of coryneform bacteria. *J. gen. appl. Microbiol., Tokyo* **18,** 417.

Yano, I., Furukawa, Y. & Kusunose, M. (1969). Phospholipids of *Nocardia coeliaca*. *J. Bact.* **98,** 124.

Yano, I. & Saito, K. (1972). Gas chromatographic and mass spectrometric analysis of molecular species of corynomycolic acids from *Corynebacterium ulcerans*. *FEBS Letters* **23,** 352.

Yano, I., Saito, K., Furukawa, Y. & Kusunose, M. (1972). Structural analysis of molecular species of nocardomycolic acids from *Nocardia erythropolis* by the combined system of gas chromatography and mass spectroscopy. *FBES Letters* **21,** 215.

Zopf, W. (1891). Über Ausscheidung von Fettfarbstoffen (Lipochromen) seitens gewisser Spaltpilze. *Ber. dt. bot. Ges.* **9,** 22.

3. A Taxonomist's Obligation

RUTH E. GORDON

Waksman Institute of Microbiology, Rutgers University, The State University of New Jersey, New Brunswick, New Jersey, U.S.A.

Contents

1. Introduction

From 1921 through the midfifties, a system of classification of the genus *Salmonella* was developed on the basis of serological differences among the strains (Schütze, 1921; White, 1925, 1926; Kauffmann, 1941, 1954; Edwards & Ewing, 1955). In that classification, every serological difference among strains meant a new species name. By the 1955 date of Edwards & Ewing's publication, the number of recognized species names had reached 337, and only a large medical laboratory could afford all the antisera necessary for identifying and naming new isolations of salmonellas.

Then, in her introduction to the section on the genus *Salmonella* in the seventh edition of Bergey's *Manual of Determinative Bacteriology* (1957), Kalz wrote, "It is hardly possible to propose a classification for the salmonellas which can include all the factors established for the large number of types. However, the genus is composed of disease-producing organisms, and the first and foremost duty of any classification scheme is to make it workable under practical routine conditions".

2. Some Problems and Suggested Solutions

(a) *Classification of the actinomycetes*

Although the role of Cassandra is an unhappy thankless one, I should like to point out that the increasing practice of describing new genera among the actinomycetes on the basis of chemical differences in the cells, by morphological characteristics detectable only by an electron microscope, or by other complicated time-consuming determinations is creating a similar generally unworkable classification. There are disease-producing strains among the actinomycetes, and the classification of the actinomycetes should be workable in the routine diagnostic laboratory. Expensive equipment, long involved chemical analyses, and an electron microscope should not be necessary for the identification of strains of actinomycetes. Obviously, Kalz' observations for the classification of the salmonellas apply to any system developed for disease-producing actinomycetes.

This objection to the creation of new genera on the basis of time-consuming, expensive determinations by no means extends to any information provided by chromatography, the electron microscope and other instruments. These "newer" methods of examination add much to our knowledge of microorganisms and are of great value to the taxonomist, who can never have enough correlating data for the confirmation of his taxa. All new knowledge of microorganisms is welcomed as having potential use in taxonomy—use that must be evaluated by application to many strains.

However, when characteristics determined and established by these "newer" methods of examination are used, as they should be, in describing a taxon, they should have some correlation with characters obtained by simpler methods, methods that can be applied in the routine diagnostic laboratory. Incorporation of a chromatographically determined property in a species description does not, in my opinion, lessen the property's taxonomic value. Absence of arabinose and galactose in the cell wall of strains of

Nocardia madurae, N. pelletieri and *N. dassonvillei* is just as useful in their species description as in a generic description. When potentially pathogenic microorganisms are involved, the taxonomist should accept his responsibility to the routine diagnostic laboratory. The shepherd should not lead where his sheep cannot follow.

(b) *Pathogenicity of strains of streptomycetes*

Another worry concerns the practice of medical microbiologists of regarding strains of streptomycetes isolated from patients as harmless saprophytes. An investigator, for example, who isolated approximately equal numbers of strains of *Streptomyces griseus* and of *N. dassonvillei* from a group of patients, examined histologically only the tissues taken from the patients from whom strains of *N. dassonvillei* were isolated. This acceptance of streptomycetes as mere saprophytes seems incomprehensible. *S. somaliensis* has long been accepted as a causal agent of mycetomas, and cases of mycetomas due to strains of *S. griseus* have appeared in recent literature. Lewis, Fidler & Crumrine (1972) isolated a streptomycete from a mycetoma in a cat; Jasmin, Powell & Baucom (1972) also described a streptomycete as the causative agent of a mycetoma in a captive Bottlenose dolphin. Both these strains were subsequently identified as strains of *S. griseus* in the laboratories of the Mycology Section, Center for Disease Control, Atlanta, and in this laboratory. Bakerspigel (1973) isolated a strain that he described as an unusual strain of *Nocardia* from a pouch-like growth in a cat. This strain, received twice in this laboratory from Bakerspigel, was one of *S. griseus.* In addition, W. A. Causey (pers. comm.) described several strains of streptomycetes, including *S. griseus,* from clinically significant sources as causal agents of infections. Causey wrote, "We can no longer afford to discard blandly strains of *Streptomyces* spp. as harmless contaminants".

(c) *Proposed definition of the genus* Nocardia

What constructive suggestion can be offered for the correction of these two current problems? My suggestion has two parts: (1) a simple definition of the genus *Nocardia* and (2) the development of a standard, commercially available, micromethod for the identification of species. I would define the genus *Nocardia* as comprised of the actinomycetes that are aerobic (do not require anaerobic or microaerophilic conditions for growth), without motile forms, and that produce filamentous, branching vegetative hyphae and aerial hyphae (the aerial hyphae may or may not divide into conidia). This definition, based on *N. asteroides** as the type species (Gordon & Mihm, 1962), is workable, I believe, in the diagnostic laboratory. This simple definition would exclude *N. turbata* (*Oerskovia*), *N. salivae* (*Rothia*), all the taxa assigned to the "*rhodochrous*" group (including *Gordona*), but would encompass the genus *Streptomyces*, species included in the genus *Actinomadura* (*N. dassonvillei*, *N. madurae*, and *N. pelletieri*), other less generally known genera (Cross & Goodfellow, 1973; Lechevalier, 1974), and new taxa such as *Nocardioides* and *Saccharopolyspora*.

In the eighth edition of Bergey's *Manual of Determinative Bacteriology*, Pridham & Tresner (1974) recognized 463 species of the genus *Streptomyces*, a more unmanageable number than the 337 species of *Salmonella* in 1957. In 1965, however, Pridham, Lyons & Seckinger stated that the species of the streptomycetes could be reduced to eight. In this laboratory we have examined nearly the same number of strains of streptomycetes as of nocardiae and, in my opinion Pridham was more nearly correct in 1965. Eight or 10 or 15 more species added to the genus *Nocardia* would not be unmanageable.

With this proposed definition of the genus *Nocardia*, strains of nocardiae could be recognized in the routine diagnostic laboratory, and the mistaken belief in the nonpathogenicity of the streptomycetes would be neatly eliminated by combining the streptomycetes with the nocardiae and using the older name, *Nocardia*.

Having spent nearly a century on the taxonomy of the genus *Bacillus*, I am, of course, influenced by that experience, and there is a similarity between this proposed definition of the genus *Nocardia* and our working definition of the genus *Bacillus* (Gordon, Haynes & Pang, 1973), which is as follows: rod-shaped bacteria that aerobically form heat-resistant endospores. This definition of the genus *Bacillus* includes morphologically different strains, psychrophiles,

* The status of *N. farcinica*, that has priority over *N. asteroides*, is still unsettled. In 1974, at the First International Conference on the *Biology of the Nocardiae* in Mérida, Venezuela, a cooperative study was initiated to resolve the problem, which has been complicated by the assignment of some african strains from bovine farcy to the genus *Mycobacterium*.

mesophiles, thermophiles, strict aerobes, facultative anaerobes, motile strains, non-motile strains, Gram-positive strains, Gram-negative strains, strains containing certain lipids and strains without, and so on through a long list of characteristics. The 22 species recognized in the eighth edition of Bergey's *Manual of Determinative Bacteriology* are not, however, unmanageable, and the simple definition has not led to chaos.

(d) *Proposed standard method for species identification*

The second part of my suggestion is that a system be devised (and this will not be done over night) for the identification of strains belonging to the species of this proposed genus *Nocardia* that is similar, perhaps, to the API System for the Identification of the *Enterobacteriaceae* (Smith *et al.*, 1972), a commercially available micromethod for making 10 to 50 physiological determinations at one time. Instead of (or in conjunction with) a standard micromethod of physiological reactions, a rapid chromatographic method (Staneck & Roberts, 1974) may be useful. The possibilities are numerous. The development of such a system must be based on all useful, available information on a large number of strains of each species. Advantages of such a system, which could be easily employed in the routine diagnostic laboratory, are the rapidity with which a large number of tests, if necessary, can be made and the benefit of internationally standard media and methods for making the tests.

(e) *In the Interim*

Until a standard system for specific identification of strains of nocardiae is available, the apparent predominance of strains of *N. asteroides* among strains isolated from patients is to the advantage of the medical microbiologist. In their review of cases of nocardiosis appearing in current English literature from 1961 through 1971, Palmer, Harvey & Wheeler (1974) stated that "... more recent studies have shown that infection with *Nocardia asteroides* appears to be surprisingly frequent when appropriate diagnostic studies are performed". Ajello *et al.* (1970), personal communications from W. A. Causey and B. L. Beaman, and the experience in our laboratory with strains of nocardiae isolated from human and animal patients attest to the predominating number of strains of *N. asteroides* over the number of strains of other species of *Nocardia* isolated in medical laboratories. (For isolation of strains from clinical specimens, see Ajello *et al.* (1970), Gordon (1970), and Palmer, Harvey & Wheeler (1974).)

Strains of *N. asteroides*, believed to be the most common cause of infections by nocardiae, are easily identified. Ajello *et al.* (1970) separated strains of *N. asteroides* from strains of *N. brasiliensis*, *N. caviae*, *N. madurae*, *N. pelletieri* and *S. somaliensis* by their acid-fastness; inability to digest casein, xanthine and hypoxanthine; and their failure to form acid from arabinose and xylose. Goodfellow (1971) suggested the following diagnostic characters for recognition of strains of *N. asteroides*: lack of acid from D-xylose and D-glucose (fermentative); inability to hydrolyze casein, hypoxanthine, keratin and xanthine; utilization of sebacic acid and testosterone as sources of carbon; failure to utilize inositol and sodium tartrate as carbon sources; and resistance to lysozyme. According to Gordon *et al.* (1974), strains of *N. asteroides* are distinguished by their filamentous colonies, aerial hyphae (visible in some strains only under the microscope), and the following group of physiological reactions: failure to decompose casein, tyrosine and xanthine, and inability to form acid from inositol, mannitol and xylose. For the microbiologist who objects, with good reason, to the use of so many negative characteristics for identifying strains, resistance to lysozyme, the presence of urease, hydrolysis of esculin and growth in Sabouraud dextrose broth are positive properties of *N. asteroides*.

To anyone seeking experience in the recognition of these strains, use of properly characterized strains of *N. asteroides* as reference strains cannot be too strongly recommended. Comparison of unknown strains with known strains provides an invaluable check on the medium and method of each test applied. Ability to identify strains of *N. asteroides*, or of any species, comes only from the study of many strains. If the medical microbiologist without experience with the nocardiae isolates a strain of this taxon that is not one of *N. asteroides*, or if he has difficulty in identifying his isolation, he should seek assistance from someone with experience with these strains who can examine his isolate and provide (if possible) strains like it to be used as reference strains for further study and the acquisition of experience with the taxon.

3. Summary

This report is not intended to discourage or disparage accumulation of any kind of knowledge about the actinomycetes. It is merely to remind the taxonomist of his obligation to the worker in the diagnostic laboratory who does not have much time or expensive equipment for the identification of his isolates.

To meet a portion of this obligation and to thwart acceptance of the streptomycetes as mere saprophytes, assignment of the aerobic, nonmotile actinomycetes that form vegetative and aerial hyphae to the genus *Nocardia* is proposed, and development of a standard micromethod similar, perhaps, to the API System for Identification of the *Enterobacteriaceae* is urged for recognition of the species of this proposed genus. Until this can be done, the apparent numerical preponderance of strains of *N. asteroides* among the strains of aerobic actinomycetes isolated in medical laboratories is fortunate, because strains of this species are the most easily identified by simple microscopic and physiological examinations.

4. Acknowledgement

The work on the nocardiae and streptomycetes in this laboratory was supported in part by Public Health Service grant AI-06276 from the National Institute of Allergy and Infectious Diseases. This assistance is deeply appreciated.

5. References

Ajello, L., Georg, L. K., Kaplan, W. & Kaufman, L. (1970). Mycotic infections. In *Diagnostic Procedures for Bacterial, Mycotic and Parasitic Infections,* 5 edn. Eds H. L. Bodily, E. L. Updyke & J. O. Mason. New York: American Public Health Association.

Bakerspigel, A. (1973). An unusual strain of *Nocardia* isolated from an infected cat. *Can. J. Microbiol.* **19,** 1361.

Cross, T. & Goodfellow, M. (1973). Taxonomy and classification of the actinomycetes. In *Actinomycetes: Characteristics and Practical Importance.* Eds F. A. Skinner & G. Sykes. London: Academic Press.

Edwards, P. R. & Ewing, W. H. (1955). *Identification of Enterobacteriaceae.* Minneapolis: Burgess.

Goodfellow, M. (1971). Numerical taxonomy of some nocardioform bacteria. *J. gen. Microbiol.* **69,** 33.

Gordon, M. A. (1970). Aerobic, pathogenic actinomycetes. In *Manual of Clinical Microbiology*. Eds J. E. Blair, E. H. Lennette, & J. P. Truant. Bethesda: American Society for Microbiology.

Gordon, R. E., Barnett, D. A., Handerhan, J. E. & Pang, C. H-N. (1974). *Nocardia coeliaca, Nocardia autotrophica*, and the nocardin strain. *Int. J. Syst. Bact.* **25,** 54.

Gordon, R. E., Haynes, W. C. & Pang, C. H-N. (1973). *The Genus Bacillus*. Agriculture Handbook No. 427. U.S. Department of Agriculture.

Gordon, R. E. & Mihm, J. M. (1962). The type species of the genus *Nocardia*. *J. gen. Microbiol.* **27,** 1.

Jasmin, A. M., Powell, C. P. & Baucom, J. N. (1972). Actinomycotic mycetoma in a Bottlenose dolphin (*Tursiops truncatus*) due to *Nocardia paraguayensis*. *Vet. Med. small Anim. Clin.* **67,** 542.

Kalz, G. (1957). *Salmonelleae* Bergey, Breed and Murray, 1938. In *Bergey's Manual of Determinative Bacteriology*, 7 edn. Eds R. S. Breed, E. G. D. Murray & N. R. Smith. Baltimore: Williams and Wilkins.

Kauffmann, F. (1941). *Die Bakteriologie der Salmonella-Gruppe*. Copenhagen: Ejnar Munksgaard.

Kauffmann, F. (1954). *Enterobacteriaceae*, 2 edn. Copenhagen: Ejnar Munksgaard.

Lechevalier, H. A. (1974). Soil or oxidative actinomycetes. In *Handbook of Microbiology*, condensed edn. Eds A. I. Laskin & H. A. Lechevalier. Cleveland: CRC Press.

Lewis, G. E., Fidler, W. J. & Crumrine, M. H. (1972). Mycetoma in a cat. *J. Am. vet. med. Ass.* **161,** 500.

Palmer, D. L., Harvey, R. L. & Wheeler, J. K. (1974). Diagnostic and therapeutic considerations in *Nocardia asteroides* infection. *Medicine* **53,** 391.

Pridham, T. G., Lyons, A. J. & Seckinger, H. L. (1965). Comparison of some dried holotype and neotype specimens of streptomycetes with their living counterparts. *Int. Bull. bact. Nomencl. Taxon.* **15,** 191.

Pridham, T. G. & Tresner, H. D. (1974). *Streptomycetaceae* Waksman and Henrici, 1943. In *Bergey's Manual of Determinative Bacteriology*, 8 edn. Eds R. E. Buchanan & N. E. Gibbons. Baltimore: Williams and Wilkins.

Schütze, H. (1921). The permanence of the serological paratyphoid B types, with observations on the non-specificity of agglutination with "rough" variants. *J. Hyg., Camb.* **20,** 330.

Smith, P. B., Tomfohrde, K. M., Rhoden, D. L. & Balows, A. (1972). API system: a multitube micromethod for identification of *Enterobacteriaceae*. *Appl. Microbiol.* **24,** 449.

Staneck, J. L. & Roberts, G. D. (1974). Simplified approach to identification of aerobic actinomycetes by thin-layer chromatography. *Appl. Microbiol.* **28,** 226.

White, P. B. (1925). Serological studies with regard to the classification and behaviour of bacilli of the salmonella group. *Med. Res. Counc. Spec. Rep.* (ser.). **91,** 9.

White, P. B. (1926). Further studies of the salmonella group. A supplement to special report no. 91. *Med. Res. Counc. Spec. Rep.* (ser.). **103,** 1.

4. An Evaluation of Numerical Taxonomic Techniques in the Taxonomy of *Nocardia* and Allied Taxa

P. H. A. SNEATH

Medical Research Council Microbial Systematics Unit, University of Leicester, Leicester, England

Contents

1. Introduction

The evaluation of a technique requires material that one can evaluate and methods for doing so, with some criterion of goodness. I shall not attempt here a comprehensive review of numerical taxonomic work upon *Nocardia* and its allies, as our present knowledge has been very well summarized by Colwell (1973). Further reviews and bibliographies may be found in Sneath (1970), Bradley (1971) and Sneath & Sokal (1973); the last and Lockhart & Liston (1970) and Sneath (1972) contain accounts of numerical taxonomic methods. A general background is given by Cross & Goodfellow (1973). Nor shall I go into details of discrepancies involving named strains or species, because these can only be clarified by those working on them, especially when we know that many strains have been mislabelled. Instead I will try to evaluate numerical taxonomy as a technique when applied to this group of bacteria.

The boundaries of the group are still uncertain, so I shall discuss in the main *Nocardia* and *Mycobacterium* with some digressions upon genera that seem fairly close, such as *Streptomyces* and *Arthrobacter*.

Pioneering work in this group was carried out by Bojalil & Cerbón (1961), Cerbón & Bojalil (1961) and Bojalil, Cerbón & Trujillo (1962) on *Mycobacterium*. They reached two important conclusions: first, that numerical taxonomy supported proposals to reduce some old names (e.g. *M. lacticola, M. friedmanii*) to synonymy with well-known species; second, that there were new and undescribed species of mycobacteria which were revealed as clusters by the analysis (e.g. *M. gordonae*), a point soon confirmed by others (e.g. Wayne, 1966; Tsukamura, 1966) and many of these seem to be proving sound (e.g. *M. gordonae, M. terrae, M. gastri,* despite uncertainties over which names to apply, see Kubica & Silcox, 1973). The most suprising finding at the time, a tight cluster containing strains labelled *M. smegmatis, M. phlei* and *M. ranae* is at least partly explained by the confusion over mislabelled strains of these organisms (e.g. Gordon & Mihm, 1959).

Since these early studies on mycobacteria the concept of the Hypothetical Median Organism (Liston, Weibe & Colwell, 1963) has been followed up by several groups of workers, and this is discussed later in this paper.

Silvestri and his colleagues (Gilardi *et al.*, 1960; Hill *et al.*, 1961; Silvestri *et al.*, 1962) studied a number of genera of Actinomycetales by numerical taxonomic methods, among which were some mycobacteria and nocardiae, though the majority were streptomycetes. Despite the deficiences of the studies, which they noted, two important conclusions can be drawn. First, the genus *Streptomyces* has far more named species than are justified by modern taxonomy. Second, that numerical taxonomy can distinguish different genera, because the *Nocardia* strains nearly all fell into one major cluster (E) well-separated from most *Streptomyces* strains.

In comparing this work with that of Gordon (1967) and Gordon & Horan (1968) one finds considerable agreement. Where the same strains are listed by both groups of workers, those in Gordon's species *Streptomyces albus, S. fradiae* and *S. griseus* are found in corresponding single groups ("spheres") of Silvestri *et al.* (1962) or occasionally in adjacent ones. This shows not merely that the numerical taxonomy is good and that the species concept of Gordon is about equivalent to a major phenetic cluster: it also implies that there are only about 20 or 30 species of this kind in *Streptomyces,* so the 600 or more named species are in need of drastic synonymy.

Jones & Bradley (1964) studied more nocardiae, and Cerbón (1967) made a first numerical study of the speciation within the genus. He showed the outline of three major groups that are now generally accepted, the *Nocardia asteroides* group, the *N. brasiliensis* group, and what is usually called the "*rhodochrous*" group. But because of the small numbers of strains, and possibly the selection of tests, these studies did not give very clear-cut conclusions.

These earlier studies, and also many more recent ones, are not readily evaluated because of lack of sufficient detail on strains and methods. To perform a satisfactory evaluation one must be sure of the identity of strain numbers, and (as will be noted below) we would preferably like to test the quantitative agreement between different classes of data, rather than to rely on impressions. The studies that are suitable for this are fewer; they include those of Goodfellow (1971), Wayne *et al.* (1971), Tacquet *et al.* (1971), Kubica *et al.* (1972), Goodfellow, Fleming & Sackin (1972) and Kurup & Schmitt (1973). Those of Wayne *et al.* (1971) and Kubica *et al.* (1972) in particular, allow comparison of taxonomies based on different sets of characters in different laboratories, and with groupings obtained from serology and other data.

2. Criteria for Assessment of Numerical Taxonomies

The evaluation of numerical taxonomic techniques is achieved essentially by comparing the congruence of the phenetic groups (phenons, clusters) produced by numerical taxonomy with groups established on other grounds. Such congruence or concordance between different kinds of evidence is indeed the major criterion for assessing any sort of scientific work. One application is to test the congruence between numerical taxonomies based upon different sets of characters. Unless the selection of the characters has been strongly biased one expects that the taxonomies will agree broadly, and this is true for various groups of organisms (Sneath & Sokal, 1973, p. 97) where the discrepancy (incongruence) is usually not much more than that expected from chance accidents of sampling.

The methods used to estimate congruence are in essence to calculate the agreement between similarity values for a given pair of OTU's (Operational Taxonomic Units) based on one set of characters and similarity values for the same pair of OTU's based on a second set of characters. These similarity values may be taken from the two similarity matrices (***S***-matrices, obtained from the two character sets) or from the two phenograms. Inasmuch as the production of a phenogram (phenetic dendrogram) from a similarity matrix involves some distortion of the taxonomic relationships in summarizing these to give a phenogram, it is preferable to compare two similarity matrices. But useful information is also obtained by comparing two phenograms. The similarity value in a phenogram is taken as that implied by the level of the lowest cross-link on the path from one OTU to the other.

There are a variety of measures for congruence (Sneath & Sokal, 1973, p. 278), but the commonest is the correlation coefficient, r, and this is normally satisfactory. Thus the *matrix correlation* between ***S***-matrices is obtained by calculating the correlation coefficient between the corresponding pairs of entries. Perfect congruence gives $r = 1{\cdot}0$. Values of r over about 0·7 are usual, and if r is over about 0·8 the agreement is generally considered to be good. For random values it would be zero. However, the expected value from random division of the total characters into two sets must also be taken into account. Thus the correlation between 29 morphological characters and 76 physiological characters of *Chromobacterium* would be expected to be about 0·75, whereas the observed correlation was 0·61 (Sneath & Sokal, 1973, p. 102): the congruence is thus only a little less than the best agreement one could expect.

There is now also evidence from comparisons of numerical taxonomies with other classes of evidence, and four of these are of special interest, DNA, serology, chemotaxonomy and genetics.

These data are seldom evaluated numerically for their concordance with numerical taxonomies or with one another a deficiency that should be remedied in future work. In the succeeding sections the numerical values are given where available, but many of the assessments have to be qualitative and somewhat subjective.

There is a second criterion of this kind, that measures the agreement between the similarity values of the ***S***-matrix and those of the phenogram derived from it, the *cophenetic correlation.* This is primarily a test of whether the OTU's fall into tight homogeneous clusters. Tight clusters give r near 1·0. A diffuse swarm gives r near zero, because one cannot produce a good summary of such relationships in a phenogram. This is therefore not a test of whether a phenogram represents a good classification, because it is possible that the true relationships are indeed those of a diffuse swarm without any clusters. However, in biological taxonomy one does nearly always expect some clusters—in bacteria as in higher species—and to this extent a low cophenetic correlation is a warning that the data may be unsatisfactory. In particular the effect of test errors is to lower the cophenetic correlation even when tight clusters are present (Sneath & Johnson, 1972). Our experience is that with most groups the cophenetic correlation is 0·8 or higher (e.g. Johnson & Sneath, 1973, Figs 1–3).

3. Congruence Between Different Character Sets

Although no expressly comparative studies have yet been published, Goodfellow & Orchard (1974) showed that new tests based on antibiotic sensitivity were reasonably congruent with clusters found earlier (Goodfellow, 1971). However there is remarkable concordance between the main taxonomic structure of *Nocardia* as shown in four numerical taxonomic studies (Tsukamura, 1969; Goodfellow, 1971; Tacquet *et al.*, 1971; Kurup & Schmitt, 1973) even though mostly different strains and tests were employed. In addition Bradley (1971) found rather similar relationships, though details are few.

There is a large heterogeneous group containing *N. asteroides* and *N. farcinica. N. brasiliensis* and *N. caviae* form tight groups. There is a high level group containing close but separate clusters of *Actinomadura madurae*, and *A. pelletieri.* Holmberg & Hallander

(1973) noted that certain nocardiae from the mouth were similar to *A. pelletieri* in a numerical study. There is another heterogeneous cluster in the "*rhodochrous*" group, within which is a tight cluster, *N. pellegrino* (or at least most strains bear this name). Some other tight clusters are seen in one or more studies (e.g. *Oerskovia turbata*) and evidently there are more species represented by only one or two strains (e.g. *A. dassonvillei, N. vaccinii*). The "*asteroides*" group evidently contains several different forms. One of these, *Nocardia farcinica,* is the type species of the genus, and unfortunately it is also a source of confusion (discussed later), but the *N. farcinica* strains tend to form a subcluster in all four studies.

The "*rhodochrous*" group is currently receiving intensive study (see chapter 2). It is evident that there are several species in this area (apart from the distinctive *N. pellegrino* that was stable when new strains were added (Goodfellow *et al.*, 1972). Bradley (1971) reported two phenetic clusters, corresponding broadly to the strains bearing the epithets *rubra* or *corallina* and to those with epithets like *erythropolis* and *canicruria.* A few of his strains can be identified with those in Goodfellow's studies, and these clusters seem to correspond at least partly to phenons 14C and 14D respectively. There is some evidence of instability with different similarity coefficients (Goodfellow *et al.*, 1972), but it may well be that they are sound, and it is possible that they correspond to one or other of the groups of Tsukamura (1973). Recent collaborative work (Goodfellow *et al.*, 1974) and the paper of Tacquet *et al.* (1971) suggest that different laboratories can recover the same clusters in this area. The relation of *Gordona* (Tsukamura, 1971*c*) to these clusters is still not clear. The complex is evidently heterogeneous as shown by the low percentage of constant characters of phenon 14 in Goodfellow (1971) noted by Sneath (1974*b*) even if the extraneous coryneforms and arthrobacters are excluded.

Turning to *Mycobacterium*, the two cooperative studies (Wayne *et al.*, 1971; Kubica *et al.*, 1972) allow some interesting evaluations. Wayne and his colleagues studied 60 strains of slow growing scotochromogens. Eight laboratories were allowed a free choice of tests, and although many tests were done in all, the eight ***S***-matrices are based on substantially different data sets. They show four clear clusters, and these are clearly congruent in most of the laboratories (despite a few gaps that do not affect the general conclusions). The laboratories that show least agreement with the ***S***-matrix from the pooled data are not those that used the smallest number of characters, as one would expect from sampling theory. The worst discrepancy is the laboratory with most tests. In the absence of the matrix correlations, the congruence can be roughly assessed by the ratio of the average similarities between and within the two major clusters, 2 and 4, calculated from the values that are cited by the authors. Tight homogeneous clusters with good separation should show a low ratio (Table 1), and this does broadly agree with an assessment

Table 1

Comparison of similarity coefficients obtained on mycobacteria of two groups (Wayne et al., 1971)

Laboratory	Number of characters	Average within-group S_{SM}% for clusters 2 and 4	Average between-group S_{SM}% for clusters 2 and 4	Ratio between/within
A	9	77	60	0·78
B	98	93·5	92	0·98
C	42	86	74	0·86
D	58	81·5	75	0·92
E	17	83·5	79	0·95
F	53	85·5	78	0·91
G	24	91·5	64	0·70
H	32	92·5	67	0·72
All (pooled but excluding duplicate tests)	140	82	72	0·88

by eye of the matrices from each laboratory compared to that from pooled data. In fact one would expect that the pooled data has been somewhat degraded by the less satisfactory data. Experimental errors might be partly responsible, because they would reduce the within-group average, although they should not increase the between-group average. This does not explain the results of laboratory B, but that apart, the effect of different choice of tests could account for much of the variation between laboratories.

In the collaborative study of Kubica *et al.* (1972) similar data was provided in the form of within-group and between-group average similarities for eight laboratories. There are a number of different clusters, and in order to present a simple table for comparison with that based on the study of Wayne *et al.* (1971) the average similarities for two major blocks of similarity coefficients are presented in Table 2 (values below 65% S_{SM} were taken as 60%).

Table 2

Comparison of similarity coefficients obtained on mycobacteria by Kubica et al. (1972)

Laboratory	Number of characters	Average within-group S_{SM}% for clusters 8 plus 9	Average between-group S_{SM}% for clusters 1–5 vs clusters 8–9	Ratio between/ within
A	55	70	64	0·90
B	80	81	58	0·72
C	58	90	63	0·70
D	11	76	62	0·82
E	49	89	69	0·78
F	10	82	83	1·01
G	53	81	65	0·80
H	99	80	65	0·81

Clusters 8 and 9 are very close, whereas 1 to 5 are dissimilar to clusters 8 and 9. Again it is seen that some laboratories have a low level of contrast between the clusters, though in the case of laboratory F this may be due to the small number of characters studied. Possibly some of the differences between laboratories are due to choosing more or fewer tests that differentiate fast from slow growing mycobacteria. The study also shows how effective numerical taxonomy can be in deciding whether certain named taxa should be synonymized. Thus strains of *M. vaccae* and *M. parafortuitum* are mingled, and seem to belong to one species; their extreme closeness had been noted in earlier studies (Tsukamura, Mizuno & Tsukamura, 1968; Goodfellow, Fleming & Sackin, 1972). Agreement such as this is not invariable (e.g. compare

Nakayama, Nakayama & Takeda, 1970 with Tsukamura, Tsukamura & Mizuno, 1967, on the status of *Mycobacterium runyonii*), but commonly later workers do substantiate earlier findings (e.g. Tacquet *et al.*, 1971; Miková, Chladková & Kubín, 1973, on *M. avium* and *M. intracellulare*).

In the peripheral groups like the cornyeforms there have been only a few studies that can be directly compared. However it can again be seen that there is broad agreement on the numerical taxonomies of Davis & Newton (1969), Davis *et al.* (1969), Bousfield (1972), Holmberg & Hallander (1973) and Jones (1975). Insofar as these workers used the same strains the broad relationships were similar in all studies. Thus the animal coryneforms formed a group separate from the plant species, and *Corynebacterium fascians* came with nocardiae. Perhaps owing to the selection of tests and strains there is a tendency for the *Nocardia* and *Mycobacterium* strains to be placed closer together than one finds when studying these genera alone: this is partly because of effects of excluding invariant characters, but possibly because distinctions are less clear with distantly related groups (discussed below). The arthrobacters in all the studies lay near the ill-defined boundary of coryneforms and nocardiae.

Harrington (1966) had earlier shown the heterogeneity of *Corynebacterium*, and the relative closeness of *Mycobacterium* and *Nocardia* which is brought out by recent studies. Skyring & Quadling (1969) used a two-stage ordination on soil bacteria, including nocardioform bacteria, but possibly this introduced distortions into the taxonomic structure, because the resulting clusters are not very satisfactory as judged by other criteria.

4. Congruence Between Phenons and DNA Data

The close correspondence between the degree of DNA base pairing and phenetic similarity has been noted by many authors, and on theoretical grounds a strong case can be made for the importance of DNA base pairing as a way of estimating genomic similarity. It must be noted, however, (Sneath, 1972) that the relationship cannot be a linear one, because the theoretical curve connecting the two measures must pass through the limiting points of 0% and 100% on each axis; a probit transformation of the percent. of pairing with homologous DNA helps to linearize the relationship. DNA pairing falls to a very low level for organisms that are only moderately different phentically (Jones & Sneath, 1970); it is especially sensitive to small differences within high-level phenons (e.g. 90% within-group ***S***), and it is apt to give uncertain results at lower phenetic levels. The occurrence of quite significant experimental error in DNA techniques should always be remembered.

Information on nocardiae and their allies is still scanty. In most studies only a few reference strains are used to screen a large number of other strains, and these reference strains are commonly not those most typical of numerical taxonomic clusters. This limits the fineness of analysis and makes interpretation difficult: it is like viewing only one or two stripes across an unordered ***S***-matrix. Most of our knowledge on nocardiae comes from Bradley and his colleagues (Clark, Brownell & Bradley, 1971; Bradley, 1972, 1973; Bradley, Brownell & Clark, 1973, see chapter 12). Their results are in close accord with the conclusions from numerical and conventional taxonomy. The relationships between groups such as *Nocardia farcinica, Mycobacterium tuberculosis* and *M. smegmatis* are much as expected, with other nocardiae like *N. asteroides, brasiliensis* and *N. caviae* related closely to *N. farcinica,* for example. Similarly *M. bovis* is very close to *M. tuberculosis.* These studies also confirm the heterogeneity of the "*asteroides*" and "*rhodochrous*" groups noted earlier. The two phenetic subgroups within the rhodochrous area were found each to be homogeneous but different, and they differed also in GC ratio. The *N. rubra* group had a ratio of 66–68%, whereas the *N. erythropolis* group had a ratio of only 61–63%.

Gross & Wayne (1970) have also published on DNA pairing between two strains of *Mycobacterium* species (*M. tuberculosis* and *M. kansasii*) and various organisms, and again the results are in close accord with phenetic relationships that they present in parallel. They found that the correlation between DNA pairing and phenetic similarity was 0·85, which compares well with the figure of 0·91 found by Colwell (1970) for comparable data on *Vibrio,* and 0·83 for collected figures analysed by Sneath (1972). Elution temperatures gave similar results. This concordance is very satisfactory considering (as Gross & Wayne point out) that there are differences in genome size, and possibly both technical variation and satellite DNA may introduce discrepancies. Again, Wayne & Gross (1969) found that GC base ratios are quite constant within phenetic clusters, though they did not correlate with fast and slow growing groups. Lechevalier, Lechevalier & Gerber (1971), Bousfield (1972) and Šlosárek (1973) also confirm the consistency of GC ratios for bacteria of this area. The DNA work has tended to support the thesis that the *rhodochrous* group is closer to *Nocardia* than to *Mycobacterium.*

5. Congruence Between Phenons and Serology

The position with regard to serology is much less clear. Some serological systems pose both technical problems and problems of interpretation. Thus Prauser (1970) noted that the system of cytoplasmic antigens studied by Kwapinski (most fully described in Kwapinski, 1970) gave a network of cross-reactions that were of little taxonomic value because they related similar and dissimilar organisms in a haphazard way. It may be that the antigens in this system are largely nonspecific ones, as there are cross-reactions between prokaryotes and eukaryotes (Kwapinski *et al.*, 1973). Wayne *et al.* (1971) found little agreement between phenons and the antigenic groups defined by this system. On the other hand Stanford & Beck (1968, 1969) and Ridell & Norlin (1973) have developed immunodiffusion techniques that give specific results. The antigens detected by serological methods may not always reflect faithfully the full genotype and phenotype, so it is encouraging that the information to date is largely consistent with numerical and other studies.

Thus Ridell & Norlin found three antigens α, pα (partial alpha) and β, that distinguished several phenetic groups of nocardiae. All of 24 strains of *Nocardia asteroides, N. brasiliensis, N. caviae* and *N. farcinica* had α precipitinogen, and rarely the β antigen, whereas most of 29 strains of the *rhodochrous* group (including *N. pellegrino*) had β, or pα, but never α. It will be remembered that the organisms containing α form the core of *Nocardia sensu stricta* in numerical taxonomic studies. Pier & Fichtner (1971) confirmed serologically the relative heterogeneity of the *asteroides* group, and recent cooperative work on the *rhodochrous* group (Goodfellow *et al.*, 1974) has also indicated correspondence between serology and phenetics.

Similarly in mycobacteria the correspondence of serology and phenetics has been reasonably good (Wayne, Doubek & Diaz, 1967; Wayne *et al.*, 1971; Jenkins, Marks & Schaefer, 1971; Kubica *et al.*, 1972). Since the serological groupings are not always easily expressed as single antigens it is difficult to present such data except in the form of large tables, but if one lists the concordance of cluster membership with specificity of antigens as against nonspecificity (when antigens occur in more than one phenon) and lack of reaction (for untypable strains), it can be seen from the study of Wayne *et al.* (1971) that for the agglutination system of Schaefer (1965) about 40% of the two major clusters, 2 and 4, give specific serological reactions and only a few percent. give nonspecific ones (the rest were untypable). The great majority of these strains (90%) gave specific induced hypersensitivity reactions by the method of Magnusson (1967), but the technique of Kwapinski, Alcasid and Pulser (1970) showed only about 10% of specific reactions. Kubica *et al.* (1972) also found specific results with hypersensitivity reactions by the method of Takeya, Nakayama & Muraoka (1970) and with one of the immunodiffusion techniques, but the other gave erratic behaviour. Other papers indicating a general concordance between serology and phenetics are those of Magnusson (1973) and Nakayama, Nakayama & Takeya (1970) on hypersensitivity, and Pattyn (1970) on agglutination. Of these methods hypersensitivity seems to give the sharpest results, but Magnusson (1973) notes that although strains of a single species cross-react readily, sometimes it detects cross-reactions between only the closest of species, like *Mycobacterium avium* and *M. intracellulare.* Different serological techniques and antigens thus give serological specificities that range from being highly species-specific to those that are genus-specific or show even broader specificity. There may also be difficulty in quantifying the cross-reactions.

6. Congruence Between Phenons and Chemotaxonomy

Most characters in bacteriology are chemotaxonomic in one sense, but in this section discussion is of chemical characters (in particular characteristic compounds produced by the organisms) that can be used as markers for taxa. Reviews of this in nocardiae and mycobacteria can be found in Lechevalier, Lechevalier & Becker (1966), Lechevalier, Lechevalier & Horan (1973), Cross & Goodfellow (1973) and in chapter 7.

It must be emphasized that phenetic groups are polythetic, i.e. thay admit of a limited amount of variation on *every* character, so that one cannot require that *any* character must be invariant within a taxon. But the distribution of character frequencies in bacteria is strongly U-shaped for tight clusters and homogeneous taxa, (Sneath, 1974*b*), the great majority being almost always present or almost always absent. One may therefore expect to find some marker chemicals that are excellent for identifying taxa. This is proving particularly rewarding in the nocardioform bacteria. Nevertheless we must be prepared for exceptional strains, and possibly the absence of a marker is more common than the presence of a marker of a different taxon, a point worth some investigation.

Wall composition has proved very rewarding in taxonomy of Gram-positive bacteria. The walls of *Nocardia asteroides, N. brasiliensis* and *N. caviae* all share the pattern IV of Lechevalier, Lechevalier & Becker (1966), with arabinose, galactose and *meso*-diaminopimelic acid (*m*-DAP), and there is excellent agreement with these phenetic groups as defined by Tsukamura (1969) and Goodfellow (1971). This work has been extended by Mordarska and her colleagues, and in recent papers (Mordarska, Mordarski & Goodfellow, 1972; Mordarska, Mordarski & Pietkiewicz, 1973; Goodfellow *et al.*, 1974) the wall patterns of several new phenetic groups have been described (e.g. *Nocardia pellegrino*). It is evident that there is considerable consistency. Thus mislabelled strains of *Streptomyces* can be distinguished from *Nocardia* by the absence of arabinose and by containing L-DAP.

Strains of nocardiae also contain a characteristic lipid (Mordarska & Mordarski, 1969) the "Lipid Characteristic for Nocardia" or LCN-A. This is present in all nocardiae belonging to *N. asteroides, N. brasiliensis, N. caviae* and the *rhodochrous* group, but is absent from the *Actinomadura madurae, A. pelletieri* group with cell wall Type III and the *Oerskovia turbata* group with wall Type VI. The LCN-A from the *rhodochrous* group has a lower R_F

than that from *N. asteroides*. There is now evidence that these lipids are nocardomycolic acids, present in *Nocardia*, whereas in *Mycobacterium* their place is taken by mycolic acids (Lechevalier, Lechevalier & Horan, 1973; Goodfellow *et al.*, 1973; Minnikin, Patel & Goodfellow, 1974). Recent unpublished work by Goodfellow and his colleagues suggests that there may be a few exceptional strains whose lipid patterns are not so clear-cut, but the concordance with taxa is very good overall.

There is still some confusion over the type species of *Nocardia*, *N. farcinica*, as at least some of the strains from bovine farcy in Africa contain mycolic acid: this casts doubt on its nomenclatural validity, and hence on the validity of the genus name itself (see chapter 1). The original cultures came from bovine farcy in Guadeloupe but it is not clear whether they were of the same type as the African strains, perhaps imported with diseased cattle.

Less has been done with lipid patterns in mycobacteria. However the phenetic groups in Wayne *et al.* (1971) and Kubica *et al.* (1972) are reasonably concordant with the lipid patterns obtained by the method of Marks & Szulga (1965).

Protein electrophoresis has not been greatly used on nocardioform bacteria, but Nakayama (1967) found it useful in assessing the status of *Mycobacterium runyonii*, which had its own distinctive pattern, and Cann & Willox (1965) found that several species of mycobacteria could be distinguished by its means.

Protein sequence analysis is now proving a powerful taxonomic tool. A growing number of sequences are being obtained, mostly from Gram-negative bacteria, where there is evident correspondence between sequence matching and phenetics (Sneath, 1974*a*), so no doubt this will become a valuable technique for nocardiae. The sequences of some proteins from streptomycetes have been elucidated, but their interpretation is quite difficult, as noted by Hartley (1974).

7. Congruence Between Phenons and Genetics

The relation between taxonomy and genetics (including bacteriophage reactions) in the nocardioform bacteria has been studied for a long time, but has given rather inconclusive results. Bradley, Anderson & Jones (1961) noted that one could obtain phages that were specific to *Nocardia*, but also others that cross-reacted with *Streptomyces*. The recognition of a specific bacteriophage is clearly dependent on prior recognition of the taxon. It is thus difficult to be sure whether the specificity is simply a consequence of selecting for a phage that requires a particular surface receptor (for example), and if so whether the receptor is itself taxon specific. The general impression of the genetic studies in the area is that genetic cross-reactions of several kinds are readily achieved between numerous species and even genera of the Actinomycetales (Jones & Sneath, 1970). Their interpretation in terms of taxonomy is thus difficult. Nevertheless Kubica *et al.* (1972) noted that phage sensitivity patterns were quite congruent with the numerical taxonomic groups (see chapter 10).

It is possible that among nocardiae one can have persistent partial diploids, which would segregate variants that might differ appreciably (see Hopwood, 1973). This, and the widespread occurrence of gene exchange, would make it more difficult than we may have thought to unravel the phylogeny of these bacteria. Some of the problems in this have been discussed eleswhere (Sneath, 1974*a*), where the imperative need for numerical phyletic methods is stressed.

8. The Effects of Differences in Numerical Methods

Despite a good deal of discussion on the possibility that differences in numerical techniques may yield radically different taxonomies (Sneath & Sokal, 1973, p. 427) this has not proved a very serious matter in bacterial taxonomy. If the OTU's form reasonably compact and separated clusters then most similarity coefficients and cluster methods will reveal the structure. There are of course occasional exceptions. Probably one of the most difficult situations is when all the strains under study are single representatives of diverse clusters, so that the phenetic space is filled with a single cloud of scattered points: as there is little structure to be recovered one cannot expect very clear-cut results, and small differences in technique may yield big changes in the end-result, according to how the uniform cloud is partitioned. In practice this seldom happens. Thus Goodfellow (1971) and Goodfellow, Fleming & Sackin (1972) used the similarity coefficients $\boldsymbol{S}_J$ and $\boldsymbol{S}_{SM}$ on nocardiae, and the Single Link and Average Link (UPGMA) clustering methods, and obtained very similar results with all. This was also true for a recent cooperative study of the *rhodochrous* group (Goodfellow *et al.*, 1974). Miková, Chladková & Kubín (1973) made a similar observation. My own preference is to use S_{SM} and UPGMA, but to include also Single Link clustering as an insurance against "chaining" (recovering straggly chains of OTU's instead of compact clusters): major differences in the results of the two cluster methods require investigation.

A larger problem is that of growth rates. Goodfellow (1971) and Goodfellow, Fleming & Sackin (1972) noted that strains that resemble one another in being negative in most tests will be grouped together by S_{SM}, and that this might account for some of the differences in the results with S_J and S_{SM}. The number of negative tests at any given time of reading also depends on growth rate, and there are commonly big differences in rate of growth between members of the nocardiae and related bacteria, notably *Mycobacterium*. Differences in growth rate may therefore produce large apparent differences between strains that are otherwise very similar. The introduction of the Vigour and Pattern similarity statistics (Sneath, 1968) was an attempt to overcome this by separating the component that represents the proportion of positive tests. The vigour difference D_V is $(c-b)/n$, and the pattern difference D_P is $2\sqrt{(bc)}/n$ where b and c are the numbers of tests positive for the first OTU but not the second, and vice versa, and n is the total number of tests (all of which must be of 0, 1 type). Then pattern similarity, S_P, is $1-D_P$.

Pattern similarity has proved a useful statistic to correct for small differences in growth rate, time of reading of tests, incubation temperatures and the like, and in our laboratory we now generally use it as well as S_{SM}. The surveys we have ourselves made have seldom contained strains of very different growth rate, so the differences from S_{SM} have always been small. Nevertheless it should be of real value with the nocardioform bacteria.

Wayne (1967) first discussed the growth rate problem with mycobacteria and he suggested looking for minima on distribution curves of test reactions in order to decide on the optimum time for reading. These optima may vary, however, with different sets of organisms, so the pattern similarity may offer a more general solution, as long as the distinction between positive and negative is based on a sound reason and is not an arbitrary choice. There is still some danger that two strains with predominantly negative reactions will be grouped together, but this only occurs if the negative scores predominate very heavily in both. In such situations S_P becomes indeterminate. Goodfellow (pers. comm.) has been investigating the form S_P that is analagous to S_J, that is where the denominator is $a+b+c$ instead of n (where a is the number of positive matches), and this may offer some advantages in such situations.

There is still a weakness in numerical taxonomic work in the way discrimination is lost at lower levels of S, around 50%. There is a parallel in DNA pairing, as noted earlier. This leads to difficulty in deciding whether two groups are moderately distant (e.g. two species) or extremely distant (e.g. two families). No very useful solution is in sight, but one may wonder whether there is very much structure to be found at the higher levels of the taxonomic hierarchy; if there is none then we should not demand it.

9. The Effect of Test Errors

The influence of test errors and the lack of reproducibility of tests is only now beginning to be appreciated. Although such errors must affect any method of taxonomy their effect is particularly obvious in numerical analyses. If errors in presence-absence tests are more than about 10% they are likely to degrade the taxonomic structure severely, and this first affects tight clusters (Sneath & Johnson, 1972). Not much evidence is yet available for *Nocardia* and allied bacteria. There is no reason, however, to suppose that test errors are much more serious than in other groups. The average discrepancies between laboratories for tests on pseudomonads were about 10%, but within laboratories they were only about 2·5% (Sneath & Johnson, 1972); for propionibacteria the within-laboratory discrepancies found by Kurmann (1959) were however, higher, about 10%. Perhaps slow-growing forms may be more prone to errors of this kind. Most published work has probably not been severely affected by test errors. It is true that one report on *Streptomyces* showed only about 80% agreement on the average (Taylor, Guthrie & Shirling, 1970), but that was between different laboratories, and fortunately within one laboratory the reproducibility can generally be kept above 95% without much difficulty (Sneath, 1974*b*). If data from different laboratories are processed separately the results are generally acceptable: this is well illustrated by the note of Melville (1965) who showed that tests obtained under aerobic and anaerobic conditions gave similar phenetic relationships even though the detailed results differed a good deal under the two sets of conditions. When, however, results from different laboratories (or results from different occasions within one laboratory using different batches of media, different personnel etc.) are combined, particularly if they refer to different sets of strains, this question needs careful attention. It is now regular practice in our laboratory to include some duplicate cultures and to check the degree of discrepancy between the duplicates. If these pairs do not cluster together tightly this suggests that test errors have crept in on an appreciable scale. An examination of known duplicates of the same strain from different culture collections suggests that most published numerical taxonomic studies on nocardioform bacteria have achieved satisfactory levels of test reproducibility. An indication that errors are unacceptably large is often obtained from the phenogram, which shows no tight clusters and a ragged branching pattern. It should be noted however that the number of tests should not be reduced too far in an effort to retain only those that are most reproducible (Sneath & Johnson, 1972), because the increase in error due to sampling of characters can easily outweigh those due to experimental factors.

10. Distinctness of Clusters

Effective methods for confirming the distinctness of two clusters are only slowly being developed, as this is quite a difficult statistical problem. There have been a number of attempts at this. Tsukamura (1967*b*) first suggested a *t* test for the significance of the difference between intergroup and intragroup similarities, but this is analagous to testing the difference between two centroids, and this can be significant for the centroids of two overlapping populations. He has recently (Tsukamura, 1971*a*) replaced it by constructing histograms of the similarities of strains of one taxon from the centre of another taxon. As a result he has combined some groups that he had earlier considered distinct, and this method may help to dispel some of the confusion over the distinctness of species in the *Mycobacterium terrae* area for example (Tsukamura, 1967*a*; Kubica *et al.*, 1970; Tsukamura, 1971*b*; Kubica & Silcox, 1973; Miková, Chladková & Kubin, 1973), and to prevent the proliferation of new species based on insecure foundations.

This principle of empirical plots of distributions was initiated by Hutchinson, Johnston & White (1965). Most bacteriologists have used the Hypothetical Median Organism (HMO) of Liston, Weibe & Colwell (1963) as the central point of a taxon, but Tsukamura & Mizuno (1968) employ a modified version. There may well be some advantages in using the centroid of a cluster instead (Sneath, 1972).

As a result of the work of Tsukamura & Mizuno (1968), Kubica & Silcox (1973) and Kubica, Silcox & Hall (1973) there is now a growing body of knowledge about the relationships between the HMO's of mycobacteria. This is proving very valuable, and Kubica and his colleagues are preparing histograms of the distributions, with an aim to deciding whether some named species are distinct. However, HMO's only define the centres of taxa, and the similarities between them are complementary to distances (Tsukamura, 1970). For the limits of taxa one needs also their radii derived from the histograms, and a model based on this is described in Sneath & Sokal (1973, p. 395) and Sneath (1974*b*). Kurup & Schmitt (1973) have used HMO's in *Nocardia*. A large number of papers in this area are preoccupied with the question of whether certain named species are sufficiently distinct to be worth recognition and naming, and there can be little doubt that numerical methods can reveal minor discontinuities within apparently homogeneous clusters (e.g. Bogdănescu & Racotta, 1967). This does not necessarily make it worthwhile to recognize these variants in a formal taxonomy.

11. Discussion

The introduction of numerical taxonomic methods has led to rapid changes in the taxonomy of groups to which it has been applied. Some of this has resulted in extensive synonymization of existing named species, but it is clear from the work on *Mycobacterium* that there are also new species to discover and describe.

Goodfellow (1971) has noted the value of numerical methods in assessing the validity of new genera such as *Actinomadura* and *Oerskovia.* It would be interesting to use the relatively standard set of tests suggested by Colwell & Weibe (1970) for heterotrophic bacteria. The value of quantitation produced by the numerical approach can also be seen from two examples. If one considers the i.r. spectra given by Arai (1970) and tries to relate them to one another and to Arai's descriptions one has a difficult task (indeed Prauser, 1970, noted that in his hands the spectra were not meaningful taxonomically). Yet if the resemblances could be quantified one might obtain a powerful new method, and related work on protein electrophoresis (Kersters & De Ley, pers. comm., Feltham & Sneath, 1974) suggests there may be promise in this. From another angle numerical methods offer sensitive ways of identification of unknown strains (Bascomb *et al.*, 1973). Bogdănescu & Vlădescu (1973) have noted this with mycobacteria, and also the potential for detecting mixed cultures, which may occasionally cause difficulties in this group of organisms.

We have seen how numerical taxonomies in the nocardioform bacteria have generally been in close accord with data from other sources, and the method can be expected to give reliable classifications if the experimenter is willing to provide carefully collected and complete test data. Although other techniques obviously have their place, phenetic data has big advantages in giving a well-balanced representation of the phenotype and genotype that allows the taxonomic structure to be elucidated with a minimum of ambiguity and effort. If serology and DNA pairing are to provide data of this quality and quantity they must give the relationships between all pairs of OTU's and also express these relationships as resemblance coefficients that have suitable properties of additivity that will allow them to be related to the phenetic hyperspaces.

There can be little doubt that numerical taxonomy will revolutionize the taxonomy of *Streptomyces* when it is seriously applied to this genus, as it is doing for *Nocardia* and has done for *Mycobacterium*, and that it will provide requisite information for taxonomy and identification in quantity not approached by ordinary techniques for serological or DNA studies. There is need for more care with performing tests, and this emphasis on quality becomes particularly important as collaborative studies are extended; in this field the workers on *Streptomyces* have taken a leading part (a good summary is given by Gottlieb, 1973), and we may hope to benefit in our turn from their experiences in this area of taxonomic endeavour.

12. Summary

The evaluation of numerical taxonomic techniques is achieved essentially by comparing the agreement (or congruence) of the phenetic groups produced by numerical taxonomy with groups established on other grounds, or calculated numerically from other sets of characters. Methods of testing such congruence have been discussed.

Comparison of numerical taxonomies from different sets of taxa has not yet been extensively studied in *Nocardia* and allied taxa, although the available evidence points to reasonably good agreement. Instead there is more evidence from comparing numerical taxonomies with four major classes of alternative ways for grouping bacteria together—DNA, serology, chemotaxonomy and gene exchange.

The congruence between phenetic similarity and DNA base pairing appears to be good, and within the expected limits of the experimental error or statistical sampling error. The data on GC ratios also generally shows agreement with phenetic groups.

Serological data gives poorer agreement, perhaps because serology does not always reflect accurately the full phenetic relationships; it is sometimes the expression of relatively few antigens, and the specificity of the cross-reactions varies considerably with the antigens and techniques.

The possession of characteristic chemical substances by many taxa is reflected in the good congruence between chemotaxonomy and numerical taxonomy. It must be remembered that phenetic groups are polythetic, that is they admit a limited amount of variation on every character, and therefore no character can be taken as completely invariant. But this limited variation permits the recognition of marker chemicals, particularly in cell-wall constituents and lipids, that agree well with phenetic groupings.

Gene exchange has proved less conclusive, partly because criteria for using genetic tests of relationship are heavily dependent on prior recognition of taxa, and hence can seldom be used as critical evidence for the goodness of these taxa. Nevertheless the agreement with phenetic groupings is reasonably good.

A pervasive problem with *Nocardia* and allied taxa is the effect of different rates of growth and metabolism upon the similarity coefficients used in numerical taxonomy. The introduction of vigour and pattern statistics has been of considerable use in avoiding such problems. There still remains some weakness in distinguishing groupings around the 50% similarity level, i.e. there is not very clear discrimination between moderately different organisms (e.g. two genera of a family) and extremely different ones (e.g. bacteria from different orders).

The influence of errors due to inadequate techniques and lack of reproducibility of tests is only now beginning to be appreciated. Errors in present-absent tests are likely to give misleading numerical taxonomies if they rise much above 10%, but fortunately they can be kept to lower values by careful attention to technique. Errors can also lead to serious failures in identification.

Effective methods for deciding whether two phenetic clusters are distinct are only now being developed, and this is an area where promising new advances can be made.

13. References

Arai, T. (1970). Infra red spectrophotometry as an aid to the identification of actinomycetes. In *The Actinomycetales.* Ed. H. Prauser. p. 273. Jena: Gustav Fischer.

Bascomb, S., Lapage, S. P., Curtis, M. A. & Willcox, W. R. (1973). Identification of bacteria by computer: identification of reference strains. *J. gen. Microbiol.* **77,** 291.

Bogdănescu, V. & Racotta, R. (1967). Identification of mycobacteria by overall similarity analysis. *J. gen. Microbiol.* **48,** 111.

Bogdănescu, V. & Vlădeşcu, D. (1973). Experiences concerning the differentiation of atypical mycobacteria isolated from patients in Roumania. In *Atypical Mycobacteria.* Ed. J. G. Weiszfeiler. p. 39. Budapest: Académiai Kiadó.

Bojalil, L. F. & Cerbón, J. (1961). Taxonomic analysis of nonpigmented, rapidly growing mycobacteria. *J. Bact.* **81,** 338.

Bojalil, L. F., Cerbón, J. & Trujillo, A. (1962). Adansonian classification of mycobacteria. *J. gen. Microbiol.* **28,** 333.

Bousfield, I. J. (1972). A taxonomic study of some coryneform bacteria. *J. gen. Microbiol.* **71,** 441.

Bradley, S. G. (1971). Criteria for definition of *Mycobacterium*, *Nocardia* and the rhodochrous group. *Adv. Fronts. Pl. Sci.* **28,** 349.

Bradley, S. G. (1972). Reassociation of deoxyribonucleic acid from selected mycobacteria with that from *Mycobacterium bovis* and *Mycobacterium farcinica. Am. Rev. resp. Dis.* **106,** 122.

Bradley, S. G. (1973). Relationships among mycobacteria and nocardiae based upon deoxyribonucleic acid reassociation. *J. Bact.* **113,** 645.

Bradley, S. G., Anderson, D. L. & Jones, L. A. (1961). Phylogeny of actinomycetes as revealed by susceptibility to actinophage. *Devs ind. Microbiol.* **2,** 223.

Bradley, S. G., Brownell, G. H. & Clark, J. (1973). Genetic homologies among nocardiae and other actinomycetes. *Can. J. Microbiol.* **19,** 1007.

Cann, D. C. & Willox, M. E. (1965). Analysis of multimolecular enzymes as an aid to the identification of certain rapidly growing mycobacteria, using starch gel electrophoresis. *J. appl. Bact.* **28,** 165.

Cerbón, J. (1967). Taxonomic analysis of *Nocardia. Revta lat.-am Microbiol.* **9,** 65.

Cerbón, J. & Bojalil, L. F. (1961). Physiological relationships of rapidly growing mycobacteria. Adansonian classification. *J. gen. Microbiol.* **25,** 7.

Clark, J. E., Brownell, G. H. & Bradley, S. G. (1971). Analysis of compatible nocardiae by DNA : DNA hybridization. *Bact. Proc.* **1971,** 38.

Colwell, R. R. (1970). Polyphasic taxonomy of the genus *Vibrio*: numerical taxonomy of *Vibrio cholerae*, *Vibrio parahaemolyticus*, and related *Vibrio* species. *J. Bact.* **104,** 410.

Colwell, R. R. (1973). Genetic and phenetic classification of bacteria. *Adv. appl. Microbiol.* **16,** 137.

Colwell, R. R. & Weibe, W. J. (1970). "Core" characteristics for use in classifying aerobic, heterotrophic bacteria by numerical taxonomy. *Bull. Georgia Acad. Sci.* **28,** 165.

Cross, T. & Goodfellow, M. (1973). Taxonomy and classification of the actinomycetes. In *Actinomycetales: Characteristics and Practical Importance.* Eds G. Sykes & F. A. Skinner. p. 11. London: Academic Press.

Davis, G. H. G., Fomin, L., Wilson, E. & Newton, K. G. (1969. Numerical taxonomy of *Listeria*, streptococci and possibly related bacteria. *J. gen. Microbiol.* **57,** 333.

Davis, G. H. G. & Newton, K. G. (1969). Numerical taxonomy of some named coryneform bacteria. *J. gen. Microbiol.* **56,** 195.

Feltham, R. K. A. & Sneath, P. H. A. (1974). A comparison of numerical and electrophoretic studies of the Enterobacteriaceae. *Proc. Soc. gen. Microbiol.* **1,** 46.

Gilardi, E., Hill, L. R., Turri, M. & Silvestri, L. (1960). Quantitative methods in the systematics of Actinomycetales. I. *G. Microbiol.* **8,** 203.

Goodfellow, M. (1971). Numerical taxonomy of some nocardioform bacteria. *J. gen. Microbiol.* **69,** 33.

Goodfellow, M., Fleming, A. & Sackin, M. J. (1972). Numerical classification of "*Mycobacterium*" *rhodochrous* and Runyon's group IV mycobacteria. *Int. J. Syst. Bact.* **22,** 81.

Goodfellow, M., Lind, A., Mordarska, H., Pattyn, S. & Tsukamura, M. (1974). A cooperative numerical analysis of cultures considered to belong to the "*rhodochrous*" complex. *J. gen. Microbiol.* **85,** 291.

Goodfellow, M., Minnikin, D. E., Patel, P. V. & Mordarska, H. (1973). Free nocardomycolic acids in the classification of nocardias and strains of the "*rhodochrous*" complex. *J. gen. Microbiol.* **74,** 185.

Goodfellow, M. & Orchard, V. (1974). Antibiotic sensitivity as a criterion in the taxonomy of some nocardioform bacteria. *J. gen. Microbiol.* **83,** 375.

Gordon, R. E. (1967). The taxonomy of soil bacteria. In *The Ecology of Soil Bacteria: An International Symposium.* Eds. T. R. G. Gray & D. Parkinson. p. 293. Liverpool: Liverpool University Press.

Gordon, R. E. & Horan, A. C. (1968). A piecemeal description of *Streptomyces griseus* (Krainsky) Waksman and Henrici. *J. gen. Microbiol.* **50,** 223.

Gordon, R. E. & Mihm, J. M. (1959). A comparison of four species of mycobacteria. *J. gen. Microbiol.* **21,** 736.

Gottlieb, D. (1973). General considerations and implications of the Actinomycetales. In *Actinomycetales: Characteristics and Practical Importance.* Eds G. Sykes & F. A. Skinner. p. 1. London: Academic Press.

Gross, W. M. & Wayne, L. G. (1970). Nucleic acid homology in the genus *Mycobacterium. J. Bact.* **104,** 630.

Harrington, B. J. (1966). A numerical taxonomical study of some corynebacteria and related organisms. *J. gen. Microbiol.* **45,** 31.

Hartley, B. S. (1974). Enzyme families. *Symp. Soc. gen. Microbiol.* **24,** 151.

Hill, L. R., Turri, M., Gilardi, E. & Silvestri, L. (1961). Quantitative methods in the systematics of Actinomycetales. II. *G. Microbiol.* **9,** 56.

Holmberg, K. & Hallander, H. O. (1973). Numerical taxonomy and laboratory identification of *Bacterionema matruchotii, Rothia dentocariosa, Actinomyces naeslundii, Actinomyces viscosus* and some related bacteria. *J. gen. Microbiol.* **76,** 43.

Hopwood, D. A. (1973). Genetics of the Actinomycetales. In *Actinomycetales: Characteristics and Practical Importance.* Eds G. Sykes & F. A. Skinner. p. 131. London: Academic Press.

Hutchinson, M., Johnson, K. I. & White, D. (1965). The taxonomy of certain thiobacilli. *J. gen. Microbiol.* **41,** 357.

Jenkins, P. A., Marks, J. & Schaefer, W. B. (1971). Lipid chromatography and seroagglutination in the classification of rapidly growing mycobacteria. *Am. Rev. resp. Dis.* **103,** 179.

Johnson, R. & Sneath, P. H. A. (1973). Taxonomy of *Bordetella* and related organisms of the families Achromobacteraceae, Brucellaceae, and Neisseriaceae. *Int. J. Syst. Bact.* **23,** 381.

Jones, D. (1975). A numerical taxonomic study of coryneform and related bacteria. *J. gen. Microbiol.* **87,** 52.

Jones, D. & Sneath, P. H. A. (1970). Genetic transfer and bacterial taxonomy. *Bact. Rev.* **34,** 40.

Jones, L. A. & Bradley, S. G. (1964). Relationships among streptomycetes, nocardiae, mycobacteria and other actinomycetes. *Mycologia* **56,** 505.

Kubica, G. P., Baess, I., Gordon, R. E., Jenkins, P. A., Kwapinski, J. B. G., McDurmont, C., Pattyn, S. R., Saito, H., Silcox, V., Stanford, J. L., Takeya, K. & Tsukamura, K. (1972). A cooperative numerical analysis of rapidly-growing mycobacteria. *J. gen. Microbiol.* **73,** 55.

Kubica, G. P. & Silcox, V. A. (1973). Numerical taxonomic analysis of some slowly growing mycobacteria using hypothetical median strain patterns. *J. gen. Microbiol.* **74,** 149.

Kubica, G. P., Silcox, V. A. & Hall, E. (1973). Numerical taxonomy of selected slowly growing mycobacteria. *J. gen. Microbiol.* **74,** 159.

Kubica, G. P., Silcox, V. A., Kilburn, J. O., Smithwick, R. W., Beam, R. E., Jones, W. D., Jr. & Stottmeir, K. D. (1970). Differential identification of mycobacteria. VI. *Mycobacterium triviale* Kubica sp. nov. *Int. J. Syst. Bact.* **20,** 161.

Kurmann, J. (1959). Ein Beitrag zur Systematik der Propionsäurebakterien. I. Prüfung und Konstanz der zur Klassifizierung heraugezogenen Artmerkmale an 22 Stammen. *Landw. Jbr.* **73,** 35.

Kurup, P. V. & Schmitt, J. A. (1973). Numerical taxonomy of *Nocardia. Can. J. Microbiol.* **19,** 1035.

Kwapinski, J. B. G. (1970). Serological taxonomy and relationships of Actinomycetales. In *The Actinomycetales.* Ed. H. Prauser. p. 345. Jena: Gustav Fischer.

Kwapinski, J. B. G., Alcasid, A. & Palser, H. (1970). Serologic relationships of endoplasm antigens of saprophytic mycobacteria. *Can. J. Microbiol.* **16,** 871.

Kwapinski, J. B. G., Kwapinski, E. H., Dowler, J. & Horsman, G. (1973). The phyloantigenic position of nocardiae revealed by examination of cytoplasmic antigens. *Can. J. Microbiol.* **19,** 955.

Lechevalier, H., Lechevalier, M. P. & Becker, B. (1966). Comparison of the chemical composition of cell-walls of nocardiae with that of other aerobic actinomycetes. *Int. J. Syst. Bact.* **16,** 151.

Lechevalier, H. A., Lechevalier, M. P., & Gerber, N. N. (1971). Chemical composition as a criterion in the classification of actinomycetes. *Adv. appl. Microbiol.* **14,** 47.

Lechevalier, M. P., Lechevalier, H. & Horan, A. C. (1973). Chemical characteristics and classification of nocardiae. *Can. J. Microbiol.* **19,** 965.

Liston, J., Weibe, W. & Colwell, R. R. (1963). Quantitative approach to the study of bacterial species. *J. Bact.* **85,** 1061.

Lockhart, W. R. & Liston, J. (Eds). (1970). *Methods for Numerical Taxonomy.* Bethesda, Maryland: American Society for Microbiology.

Magnusson, M. (1967). Identification of species of Mycobacterium on the basis of the specificity of the delayed type reaction in guinea pigs. *Z. Tuberk.* **127,** 55.

Magnusson, M. (1973). Taxonomic usefulness of skin tests on guinea pigs with sensitins of mycobacteria. In *Atypical Mycobacteria.* Ed. J. G. Weiszfeilder. p. 23. Budapest: Akadémiai Kiadó.

Marks, J. & Szulga, T. (1965). Thin-layer chromatography of mycobacterial lipids as an aid to classification; technical procedures: *Mycobacterium fortuitum. Tubercle, Lond.* **46,** 400.

Melville, T. H. (1965). A study of the overall similarity of certain actinomycetes mainly of oral origin. *J. gen. Microbiol.* **40,** 309.

Miková, Z., Chladková, D. & Kubín, M. (1973). Numerical taxonomy of Runyon Group III mycobacteria. In *Atypical Mycobacteria.* Ed. J. G. Weiszfeilder. p. 57. Budapest: Akadémiai Kiadó.

Minnikin, D. E., Patel, P. V. & Goodfellow, M. (1974). Mycolic acids of representative strains of *Nocardia* and the "rhodochrous" complex. *FEBS Letters* **39,** 322.
Mordarska, H. & Mordarski, M. (1969). Comparative studies on the occurrence of lipid A, diaminopimelic acid and arabinose in *Nocardia* cells. *Arch. Immun. Ther. Exp.* **17,** 739.
Mordarska, H., Mordarski, M. & Goodfellow, M. (1972). Chemotaxonomic characters and classification of some nocardioform bacteria. *J. gen. Microbiol.* **71,** 77.
Mordarska, H., Mordarski, M. & Pietkiewicz, D. (1973). Chemical analysis of hydrolysates and cell extracts of *Nocardia pellegrino. Int. J. Syst. Bact.* **23,** 274.
Nakayama, Y. (1967). The electrophoretical analysis of esterase and catalase and its use in taxonomical studies of mycobacteria. *Jap. J. Microbiol.* **11,** 95.
Nakayama, Y., Nakayama, H. & Takeya, K. (1970). Studies of the relationship between *Mycobacterium fortuitum* and *Mycobacterium runyonii. Am. Rev. resp. Dis.* **101,** 558.
Pattyn, S. R. (1970). Agglutination with rapidly growing (Runyon's Group IV) mycobacteria. *Zentbl. Bakt. Parasitkde. Abt.* 1 **215,** 99.
Pier, A. C. & Fichtner, R. E. (1971). Serological typing of *Nocardia asteroides* by immunodiffusion. *Am. Rev. resp. Dis.* **103,** 698.
Prauser, H. (1970). Characters and genera arrangement in the Actinomycetales. In *The Actinomycetales.* Ed. H. Prauser. p. 407. Jena: Gustav Fischer.
Ridell, M. & Norlin, M. (1973). Serological study of *Nocardia* by using mycobacterial precipitation reference systems. *J. Bact.* **113,** 1.
Schaefer, W. B. (1965). Serologic identification and classification of the atypical mycobacteria by their agglutination. *Am. Rev. resp. Dis.* **92** suppl., 85.
Silvestri, L., Turri, M., Hill, L. R. & Gilardi, E. (1962). *Symp. Soc. gen. Microbiol.* **12,** 333.
Skyring, G. W. & Quadling, C. (1969). Soil bacteria: principal component analysis of descriptions of named cultures. *Can. J. Microbiol.* **15,** 141.
Šlosárek, M. (1973). DNA base composition in *Mycobacterium avium/intracellulare.* In *Atypical Mycobacteria.* Ed. J. G. Weiszfeiler. p. 27. Budapest: Akadémiai Kiadó.
Sneath, P. H. A. (1968). Vigour and pattern in taxonomy. *J. gen. Microbiol.* **54,** 1.
Sneath, P. H. A. (1970). Application of numerical taxonomy to Actinomycetales: problems and prospects. In *The Actinomycetales.* Ed. H. Prauser. p. 371. Jena: Gustav Fischer.
Sneath, P. H. A. (1972). Computer taxonomy. In *Methods in Microbiology,* vol. 7A. Eds J. R. Norris & D. W. Ribbons. p. 29. London: Academic Press.
Sneath, P. H. A. (1974*a*). Phylogeny of micro-organisms. *Symp. Soc. gen. Microbiol.* **24,** 1.
Sneath, P. H. A. (1974*b*). Test reproducibility in relation to identification. *Int. J. Syst. Bact.* **24,** 508.
Sneath, P. H. A. & Johnson, R. (1972). The influence on numerical taxonomic similarities of errors in microbiological tests. *J. gen. Microbiol.* **72,** 377.
Sneath, P. H. A. & Sokal, R. R. (1973). *Numerical Taxonomy: the Principles and Practice of Numerical Classification.* San Francisco: W. H. Freeman.

Stanford, J. L. & Beck, A. (1968). An antigenic analysis of the mycobacteria, *Mycobacterium fortuitum, Myco. kansasii, Myco. phlei, Myco. smegmatis* and *Myco. tuberculosis*. *J. Path. Bact.* **95,** 131.

Stanford, J. L. & Beck, A. (1969). Bacteriological and serological studies of fast growing mycobacteria identified as *Mycobacterium friedmanii*. *J. gen. Microbiol.* **58,** 99.

Tacquet, A., Plancot, M. T., Debruyne, J., Devulder, B., Joseph, M. & Losfeld, J. (1971). Études préliminaires sur la classification numérique des mycobactéries et des nocardias. 1. Relations taxonomiques entre *Mycobacterium rhodochrous, Mycobacterium pellegrino* et les genres *Mycobacterium* et *Nocardia*. *Annls Inst. Pasteur, Lille* **22,** 121.

Takeya, K., Nakayama, Y. & Muraoka, S. (1970). Specificity in skin reaction to tuberculin protein prepared from rapidly growing mycobacteria and some nocardia. *Am. Rev. resp. Dis.* **102,** 982.

Taylor, G. R., Guthrie, R. K. & Shirling, E. B. (1970). Serological characteristics of *Streptomyces* species using cell wall immunizing antigens. *Can. J. Microbiol.* **16,** 107.

Tsukamura, M. (1966). Adansonian classification of mycobacteria. *J. gen. Microbiol.* **45,** 253.

Tsukamura, M. (1967*a*). Two types of slowing growing nonphotochromogenic mycobacteria obtained from soil by the mouse passage method: *Mycobacterium terrae* and *Mycobacterium novum*. *Jap. J. Microbiol.* **11,** 163.

Tsukamura, M. (1967*b*). A statistical approach to the definition of bacterial species. *Jap. J. Microbiol.* **11,** 213.

Tsukamura, M. (1969). Numerical taxonomy of the genus *Nocardia*. *J. gen. Microbiol.* **56,** 265.

Tsukamura, M. (1970). Relationship between *Mycobacterium* and *Nocardia*. *Jap. J. Microbiol.* **14,** 187.

Tsukamura, M. (1971*a*). Some considerations on classification of mycobacteria. Definition of bacterial species by introduction of the concept of "Hypothetical Median or Mean Organism". *Jap. J. Tuberc.* **17,** 18.

Tsukamura, M. (1971*b*). Relationship between *Mycobacterium nonchromogenicum, Mycobacterium terrae, Mycobacterium novum* and subgroup "V" (*Mycobacterium triviale*). *Jap. J. Microbiol.* **15,** 229.

Tsukamura, M. (1971*c*). Proposal of a new genus, *Gordona*, for slightly acid-fast organisms occurring in sputa of patients with pulmonary disease and in soil. *J. gen. Microbiol.* **66,** 15.

Tsukamura, M. (1973). A taxonomic study of strains received as "*Mycobacterium*" *rhodochrous*. Description of *Gordona rhodochroa* (Zopf; Overbeck; Gordon et Mihm) Tsukamura *comb. nov.* *Jap. J. Microbiol.* **17,** 189.

Tsukamura, M. & Mizuno, S. (1968). "Hypothetical Mean Organisms" of mycobacteria. A study of classification of mycobacteria. *Jap. J. Microbiol.* **12,** 371.

Tsukamura, M., Mizuno, S. & Tsukamura, S. (1968). Classification of rapidly growing mycobacteria. *Jap. J. Microbiol.* **12,** 151.

Tsukamura, M., Tsukamura, S. & Mizuno, S. (1967). Numerical taxonomy of *Mycobacterium fortuitum*. *Jap. J. Microbiol.* **11,** 243.
Wayne, L. G. (1966). Classification and identification of mycobacteria. III. Species within group III. *Am. Rev. resp. Dis.* **93,** 919.
Wayne, L. G. (1967). Selection of characters for an Adansonian analysis of mycobacterial taxonomy. *J. Bact.* **93,** 1382.
Wayne, L. G., Dietz, T. M., Gernez-Rieux, C., Jenkins, P. A., Käppler, W., Kubica, G. P., Kwapinski, J. B. G., Meissner, G., Pattyn, S. R., Runyon, E. H., Schröder, K. H., Silcox, V. A., Tacquet, A., Tsukamura, M. & Wolinsky, E. (1971). A cooperative numerical analysis of scotochromogenic slowly growing mycobacteria. *J. gen. Microbiol.* **66,** 255.
Wayne, L. G., Doubek, J. R. & Diaz, G. A. (1967). Classification and identification of mycobacteria. IV. Some important scotochromogens. *Am. Rev. resp. Dis.* **96,** 88.
Wayne, L. G. & Gross, W. M. (1969). Base composition of deoxyribonucleic acid isolated from mycobacteria. *J. Bact.* **96,** 1915.

5. The Micromorphology and Fine Structure of Nocardioform Organisms

S. T. WILLIAMS and G. P. SHARPLES

Department of Botany, University of Liverpool, Liverpool, England

J. A. SERRANO and A. A. SERRANO

Centro de Microscopia Electronica, Universidad de Los Andes, Merida, Venezuela

and

J. LACEY

Department of Plant Pathology, Rothamsted Experimental Station, Harpenden, England

Contents

1. Introduction

The taxonomy of *Nocardia* and related taxa has been greatly changed recently as a result of studies relying more on chemical and physiological characters than on morphology (Lechevalier & Lechevalier, 1970*a*; Cross & Goodfellow, 1973). However, individual strains within these newly defined taxa often show morphological differences which have been induced, in part, by the growth medium and cultivation conditions. Many such variants have been named as different species but to minimize confusion we have used names currently accepted by most taxonomists. In this review we have attempted to survey the morphology and fine structure of nocardioform organisms and the methods used in their study. In addition to *Nocardia*, the genera *Actinomadura, Micropolyspora, Saccharopolyspora, Mycobacterium, Oerskovia, Pseudonocardia, Dermatophilus, Geodermatophilus* and the "*rhodochrous*" taxon will be considered. These may be regarded as "nocardioform" (Prauser, 1967), having regular fragmentation of their primary mycelium.

2. Culture Methods and Media

The consistency and composition of the growth medium can have a profound effect on the growth and stability of hyphae of nocardioform organisms. The growth requirements of different species and the preferences of different investigators have resulted in use of many media such as glycerol-nutrient agar (McClung, 1949, 1950, 1954, 1955), semi-synthetic medium (McClung, 1954), nutrient agar plus 1% fructose (Webb & Clark, 1957), oatmeal agar (Waksman, 1961), potato carrot agar (Cross, Lechevalier & Lechevalier, 1963), half-strength nutrient and V-8 vegetable juice agars (Cross, Maciver & Lacey, 1968), soil extract agar (Gordon & Smith, 1955; Gordon & Mihm, 1957) and liquid or semi-solid media (Bojalil & Cérbon, 1959; Kwapinski & Horsman, 1973). A good general purpose medium (liquid or solid) is the yeast extract-malt extract medium of Pridham *et al.* (1957).

The choice of culture method must depend on the aims of the investigation. The morphology of undisturbed growth on agar media may be studied by light microscopy, a ×20 or ×40 long-working distance objective (Vickers Instruments, York, U.K.) being helpful for examining aerial growth. Blocks from such cultures can be used for scanning electron microscopy and for preparing ultra-thin sections. Broth cultures have been used for studies of fragmentation and colony morphology (Adams, 1966; Kwapinski & Horsman, 1973). They are also a convenient source of vegetative hyphae which can be pelleted by centrifugation before thin-sectioning (Kawata & Inoue, 1965; Farshtchi & McClung, 1967).

Cultures grown on slides or coverslips are often convenient for staining and light microscopy (Williams & Cross, 1971). They can also be examined by scanning electron microscopy (Williams & Davies, 1967; Williams, 1970). Nocardioform organisms have often been studied growing in drops of liquid medium or on agar blocks on slides, incubated in a moist chamber (McClung, 1949; Gordon & Smith, 1955; Brown & Clark, 1966).

Carbon replicas can be prepared from slide cultures (Williams *et al.*, 1972), while cultures grown on sterile cellophane film laid on agar are convenient for sectioning or direct examination with the electron microscope. Growth can then be scraped off or small pieces can be cut from the film and treated intact, frequently allowing lengths of hyphae to be cut longitudinally.

3. Light Microscopy

The micromorphology of a culture should first be examined unstained on plate, slide or coverslip cultures. It is then possible to determine the extent of the disruption caused to fixed, stained smears. Autolysed hyphae can sometimes disrupt during mounting in a way that can be confused with the characteristic fragmentation of nocardioform organisms.

Staining techniques form two broad groups. One group is used to increase the contrast between the cells and their background, and can also provide useful diagnostic information. Such stains include crystal violet, methylene blue, carbol fuschin, the acid-fast and Gram stains. Nocardioform cells are generally Gram-positive and may be acid-fast (e.g. *Mycobacterium*), partially acid-fast (e.g. *Nocardia*) or non-acid fast (e.g. *Actinomadura*). The degree of acid-fastness has been reported to be greater after growth on organic than on inorganic media (McClung & Uesaka, 1961), and in newly isolated strains (Gordon & Mihm, 1957). A second group of stains consists of those used to stain parts of the cell selectively such as the walls, lipid inclusions, metachromatic granules and nuclear material. Walls and septa have been stained with Webb's (1954) tannic acid-crystal violet (Webb, Clark & Chance, 1954; Adams & McClung, 1960, 1962*a*; Uesaka, 1969). Other stains for walls include Robinow's (1945), Giemsa stain-tannic acid-crystal violet, and alcoholic Victoria blue 4R (Adams & McClung, 1962*a*). We have found that only Webb's procedure consistently demonstrated walls and septa in *rhodochrous* strains. Lipids have been demonstrated with Burdon's (1946) Sudan Black B technique or modifications of it (McClung, 1950; Clark & Aldridge, 1960; Adams & McClung, 1962*a*), and metachromatic granules with basic aniline dyes (McClung, 1950) or Loeffler's methylene blue (Adams & McClung, 1962*a*). Claims to have stained nuclear material have not always been substantiated. Adams & McClung (1962*a*) and Adams (1963) were most successful using an acid hydrolysis-Giemsa stain technique. Adams & McClung (1962*a*) also evaluated many staining techniques on a *rhodochrous* strain (*Nocardia rubra*), showing the value of successive staining procedures.

Clinical material may be prepared by standard cytological methods (Emmons, Binford & Utz, 1970). The Gomori-methenamine-silver technique is useful if the staining time is extended (Emmons, Binford & Utz, 1970).

4. Transmission Electron Microscopy

(a) *Whole cell silhouettes*

Washed cell suspensions are pipetted onto coated grids or, for aerial growth, the grids are touched on to a colony surface. Specimens may then be examined without further treatment as silhouettes that show the size, shape and ornamentation of cells or spores, and indicate areas within them which are unusually electron-light or dense (Webley, 1954; Adams & McClung, 1962*a*). If whole cells are shadowed with uranium oxide, fibrillar surface layers may be seen.

(b) *Flagella*

These may be seen in untreated whole cell mounts, but are clearer if cells are fixed to reduce distortion and then metal shadowed after mounting (Luedemann, 1968; Sukapure *et al.*, 1970). Detail can be further increased by stains such as 3% (w/v) phosphotungstic acid at pH 7·0 (Gordon, 1964).

(c) *Negative staining*

The structure of cell surfaces may be studied by negative staining and sometimes mesosomes can be located in intact cells. Stains are applied to whole or disrupted cells before or after mounting on coated grids. Stains used on nocardioform organisms include 2% (w/v) sodium silicotungstate at pH 7·5 (Serrano, Serrano & Tablante, 1971; Serrano *et al.*, 1972), ammonium molybdate (Beaman *et al.*, 1971), 0·5% uranyl acetate at pH 4·2 (Imaeda, Kanetsuna & Galindo, 1968), and 1% sodium phosphotungstate at pH 7·0 (Bradshaw, pers. comm.). Stains can be mixed with an equal volume of cell suspension, a drop of the mixture placed on a grid, and excess liquid drained off with filter paper. Alternatively, a drop of stain is placed onto grids on which the cells are present. Addition of 0·005% (v/v) glycerol to the stain helps spreading when cells are hydrophobic. These techniques have been used on nocardioform organisms by Imaeda *et al.* (1968, 1969), Beaman *et al.*, (1971), Draper (1971). Serrano *et al.* (1972) and Draper & Rees (1973).

(d) *Replicas*

Replicas of electron-dense surfaces prepared with relatively electron-transparent materials such as carbon can provide detailed information on surface structure. Specimens are coated with carbon or carbon–platinum, and the cell material is digested away with chromic acid or sodium hydroxide. The replicas are then washed, mounted and shadowed with metal. Replicas may also be prepared from freeze-fractured or freeze-etched cells of fixed or unfixed cells. Internal cell structure is better seen if cells are suspended in 20% (v/v) glycerol before freezing. Cells are first frozen in Freon 12 or 22 and then in liquid nitrogen. Etching or fracturing is done at −100°C to show the internal surfaces of the cell. Replicas of nocardioform organisms have been examined by Beaman & Shankel (1969), Draper & Rees (1973) Serrano & Serrano (1974) and Williams, Sharples & Bradshaw (1974).

(e) *Ultra-thin sections*

Internal structures and developmental stages are best revealed by study of ultra-thin sections with the transmission electron microscope. Growth from solid or liquid media is harvested, fixed, dehydrated and embedded for sectioning. Several alternative procedures exist and it is difficult to select any one suitable for all organisms and structures.

Nocardioform bacteria are often fixed with 1% (w/v) osmium tetroxide, following the procedure of Kellenberger, Ryter & Séchaud (1958) (Kawata & Inoue, 1965; Dorokhova *et al.*, 1969). Other fixatives include glutaraldehyde, formaldehyde and potassium permanganate either used alone or successively (Serrano *et al.*, 1972). Contrast is enhanced by staining with lead citrate and by uranyl acetate which also stains chromatin. Studies on the fine structure of nocardioform organisms include those of Kawata & Inoue (1965), Macotela-Ruiz & Gonzalez-Angulo (1966), Farshtchi & McClung (1967), Imaeda, Kanetsuna & Galindo (1968), and Serrano *et al.* (1972).

(f) *Cytochemical techniques*

Serrano *et al.* (1974) have used two cytochemical techniques in the electron microscopy of *Nocardia asteroides*. The first method demonstrates polysaccharides in ultra-thin sections and is based on the procedure of Seligman *et al.* (1965). Sections fixed in glutaraldehyde are treated for 15 min with 0·5% (w/v) periodic acid which oxidises polysaccharides exposing aldehyde groups. The sections are then treated for 1–2 h with thiocarbohydrazide (TCH), 2-pyrrolecarboxaldehydethiocarboxydrazone (PTCH) or thiosemicarbazide (TSC), each at 1% (w/v) in 25% (w/v) acetic acid. The thio-radicals are then made visible by treatment with either osmium tetroxide vapour (1 h at 60°C) or 1% (w/v) silver proteinate in the dark. Distribution of electron-dense particles in the sections indicates the probable locations of polysaccharides.

The second method (OTO method) increases the electron-density of lipid-containing regions of the cell (Seligman, Wasserkrug & Hanker, 1966). Material is fixed with osmium tetroxide and sections are treated with a hot aqueous 1% (w/v) solution of TCH for 1 h at 50°C. One end of the TCH molecule attaches to the osmium in the section. Further treatment with osmium tetroxide vapour (1 h at 60°C) results in more osmium binding to the TCH and hence increases the electron-density of these sites.

5. Scanning Electron Microscopy

The scanning electron microscope allows examination of surfaces at magnifications from 20 to 50 000 times, with a greater depth of focus than the light microscope. It is most useful for examining the morphology of colonies, sporing structures and spore formation. Few examples of nocardioform actinomycetes have been examined previously (Kormendy & Wayman, 1972).

Methods for preparing microbes were reviewed by Williams, Veldkamp & Robinson (1973*b*). Actinomycetes grown in broth culture can be prepared by washing, fixing in 10% (w/v) neutralised formalin, re-washing and air-drying on specimen stubs or coverslips. Growth on agar media, or on slide and coverslip cultures, is best prepared by quenching at low temperature (e.g. in iso-pentane at −150°C), followed by freeze-drying. This procedure minimizes disruption. All specimens must be coated with a thin film of metal (e.g. gold–palladium) before examination.

6. Micromorphology

The micromorphology of nocardioform actinomycetes varies greatly, with respect to substrate growth, aerial growth and spore production (Table 1). This section considers variation within and between genera, and the influence of external factors on growth form.

(a) *Morphology of substrate hyphae*

Fragmentation of hyphae is characteristic of nocardioform actinomycetes, but there are large differences in the extent, timing and type of fragmentation within the group. Fragmentation can

Table 1
Micromorphology of some nocardioform genera

Genus	Substrate hyphae fragmenting before autolysis	Aerial hyphae formed	Spores produced On substrate hyphae	On aerial hyphae
Pseudonocardia	±	+	−	+
Actinomadura	±	±	−	±
Micropolyspora	±	+	+	+
Saccharopolyspora	+	+	−	+
Nocardia	+	±	−	±
Oerskovia	+*	−	−	−
"*Rhodochrous*" taxon	+	−	−	−
Mycobacterium	+ (often no hyphal phase)	−	−	−
Dermatophilus	+*	−	(+)	−
Geodermatophilus	+*	−	(+)	−

+ Character always present.
± Character sometimes present.
− Character never present.
* Motile cells produced.
( ) May be hyphal fragments.

also occur in genera not normally regarded as "nocardioform". For instance, Nikitina & Kalakoutskii (1971) found that 6% of the *Streptomyces* species examined had fragmented substrate hyphae when D-fructose was included in the medium, fragmentation being most marked in *Streptomyces roseoflavus* var. *roseofungini* (Nikitina, Kasakova & Kalakoutskii, 1971). The appearance of the substrate growth of the nocardioform genera included in recent taxonomic studies is given in Table 2. The substrate hyphae of *Pseudonocardia* have a zig-zag appearance due to the angular displacement of cells by acropetal budding of the hyphae (Henssen & Schäfer, 1971). This process probably differs from fragmentation by having constrictions rather than cross-wall formation.

Fragmentation occurs in *Micropolyspora* species, but is usually less pronounced and less frequent than in most other nocardioform genera; the substrate hyphae also bear spore chains. The type species, *Micropolyspora brevicatena*, does not fragment *in situ*, but breaks easily during preparation of smears (Lechevalier, Solotorovsky & McDurmont, 1961). *Micropolyspora faeni* fragments

Table 2

Characteristics of substrate hyphae in some nocardioform genera (from Goodfellow, 1971; Goodfellow, Fleming & Sackin, 1972; Tsukamura, 1969; 1970; 1973)

	Actinomadura	*Nocardia*	*Oerskovia*	"*Rhodochrous*" taxon	*Mycobacterium*
Hyphae formed	+	+	+	±	±
Hyphae permanent	±	–	–	–	–
Rods and cocci formed	±	+	+*	+	+

* Motile elements.

occasionally in slide culture (Cross, Maciver & Lacey, 1968). Guzeva, Agre & Sokolov (1972), who reported that *M. rectivirgula* and other species fragmented less frequently than *Nocardia brasiliensis*, suggested that the breaking of hyphae in *Micropolyspora* was partly due to localised area of autolysis. It is important to distinguish between this latter process and the fragmentation at cross-walls in healthy active hyphae which occurs in *Nocardia* (Uesaka, 1969). Autolysis can occur in *Streptomyces*, and indeed in any actinomycete; it is indicated by clear areas in stained preparations.

Although many hyphae of *Saccharopolyspora hirsuta* are stable, some fragment like those of *Nocardia* species, producing long chains of cells in angular apposition (Lacey & Goodfellow, 1975).

Species of *Actinomadura* differ greatly in the stability of their substrate hyphae; those of *A. dassonvillei* readily fragment (Lechevalier & Lechevalier, 1970*b*; Guzeva, Agre & Sokolov, 1972). By contrast, hyphae of *A. pelletieri* are usually stable, but 25% of the *A. madurae* strains examined by Goodfellow (1971) fragmented to rods and cocci. Nonomura & Ohara (1971) did not describe substrate hyphal behaviour in their new species of *Actinomadura*, but we have not detected fragmentation in *A. pusilla*, *A. roseoviolacea*, *A. spadix* or *A. verrucospora*.

Nocardia asteroides has been shown to be heterogeneous (Tsukamura, 1969; Goodfellow, 1971; Kurup & Schmitt, 1973) and it is not surprising that morphological differences between strains exist. Isolates have been described which differ in the degree of fragmentation of substrate hyphae. The resulting elements may be in the form of cocci, rods of various lengths, short branching filaments or combinations of these forms (Gordon & Mihm, 1957, 1962*a*; Goodfellow, 1971). McClung (1949, 1954) showed that rods and cocci germinated slowly to produce branching hyphae, but fragmentation was usually scarce during the first 90 h of growth in culture and sometimes never occurred. Adams & McClung (1962*b*) observed that septation in hyphae began after 48 h but was not complete until 72 h. We have classified the mycelial morphology of mature *N. asteroides* cultures grown on half-strength nutrient agar as follows: (i) long, branched, stable hyphae (Fig. 1), (ii) zig-zag, fragmented hyphae or more complex funiculose structures formed by repeated fragmentation (Fig. 2), (iii) fragments often spherical in short branches from a main hypha, resembling sporulation in *Micropolyspora* (Fig. 3), (iv) very short, repeated branching (Fig. 4).

N. brasiliensis forms a relatively homogeneous taxon (Tsukamura, 1969; Goodfellow, 1971; Kurup & Schmitt, 1973) but descriptions of the morphology of different strains vary. Only 23% fragmented into rods, cocci and short filaments (Gordon & Mihm, 1959, 1962*a*), while all of Goodfellow's (1971) 10 strains formed rods and cocci. In *N. caviae* only 4 out of 15 strains examined by Gordon & Mihm (1962*b*) fragmented, but all of Goodfellow's (1971) 10 strains did so. Such variations are partly caused by differences in media, age of cultures and the observation technique. When intact colonies are examined, less fragmentation is observed; when stained smears are used, fragmentation is more readily detected and also probably induced. The taxonomic value of this and other morphological features of nocardioform actinomycetes would be increased if culture conditions and observation techniques were standardised.

LM—light micrograph; SEM—scanning electron micrograph; TEM—transmission electron micrograph (ultra-thin section); TEM-S—transmission electron micrograph (silhouette); FE—freeze etched specimen; NS—negatively stained specimen; CR—carbon replica.

Strain designations: A & N, Dr M. Goodfellow, Microbiology Dept., Newcastle University, Newcastle upon Tyne, England; CUB, Dr. T. Cross, Biological Sciences Dept., Bradford University, Bradford, England; S, Dr. J. A. Serrano, Centro de Microscopia Electronica, Universidad de Los Andes, Merida, Venezuela; ATCC, American Type Culture Collection, Rockville, Maryland, U.S.A.; IP, Institut Pasteur, Paris, France; NCIB, National Collection of Industrial Bacteria, Torry Research Station, Aberdeen, U.K.; NCTC, National Collection of Type Cultures, Colindale, London, U.K.

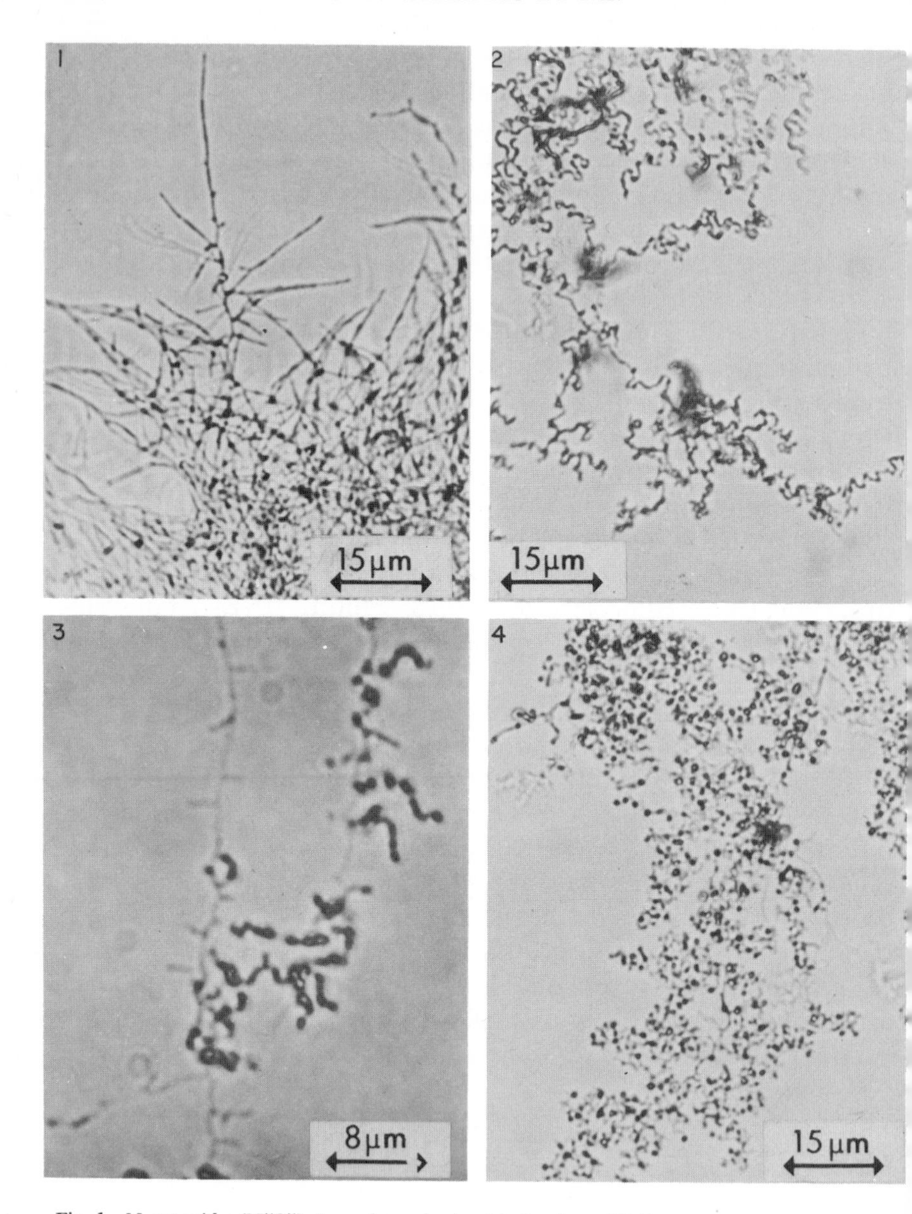

Fig. 1. *N. asteroides* (N595). Long branched stable hyphae. (LM.)

Fig. 2. *N. asteroides* (N518). Zig-zag, fragmenting hyphae. (LM.)

Fig. 3. *N. asteroides* (N558). Spherical fragments on short branches. (LM.)

Fig. 4. *N. asteroides* (N483). Hyphae with short, repeated branching. (LM.)

The taxonomic status of most other *Nocardia* species is doubtful and there is little information on their morphology. Strains originally classified as *Nocardia rubra, N. corallina* and other names are now placed in the *rhodochrous* taxon. The primary mycelium of most *rhodochrous* strains quickly fragments into rods and cocci. We grew 97 isolates on yeast-glucose agar and after 18 h found that 40% consisted of mixtures of branched filaments, rods and cocci, 37% of rods only and 23% of rods and cocci. After 5 days, a third of the cultures contained only cocci and the remainder consisted of a mixture of rods and cocci. The rods and cocci of this group germinate quicker than the comparable elements produced by *N. asteroides*; the cells elongate then branch from lateral protuberances. McClung (1949, 1954) reported that developing hyphae may bend before fragmenting, but breaks also occur at the junctions of branches (McClung, 1949); he grouped young colonies into those which fragment before or after branching. Scanning electron microscopy shows slimy extra-cellular material on the surface of *rhodochrous* colonies, varying from a few strands to large sheets (Fig. 5). The growth cycle of organisms included in the *rhodochrous* complex has been controversial, partly because misinterpretation of results from certain selective staining procedures has led to confusion. Initially, rates of hyphal growth and nuclear division exceed septum formation, resulting in coenocytic filaments, which then fragment into units containing one or more nuclear elements (Adams, 1963). There are three phases in the growth cycle: (i) active growth, with increasing cell mass but constant viable count, (ii) fragmentation with constant mass but increasing viable count, (iii) decreasing viability and mass (Adams, 1966). Cells in phase (i) show a large increase in the ratio of surface area to volume, increases in endogenous respiration and the ability to oxidise fructose (Brown & Reda, 1967). During this phase, glucose become less able to stimulate oxygen uptake, perhaps because of changes in permeability, which may then effect morphogenesis.

Mycobacterium strains usually exist as rods but occasionally branched filaments or even a mycelium are formed. Goodfellow, Fleming & Sackin (1972) detected temporary mycelium in only 14% of strains of mycobacteria in Group IV (Runyon, 1955), compared with 100% of strains in the *rhodochrous* taxon. Tsukamura

(1973) concluded that lack of even a temporary mycelium distinguished *Mycobacterium* from *rhodochrous* strains.

The genus *Oerskovia*, containing the species originally named *Nocardia turbata* (Prauser, Lechevalier & Lechevalier, 1970), forms abundant branched hyphae which eventually fragment into motile rods. Other genera which fragment into motile units are *Dermatophilus* and *Geodermatophilus*. In these, the filaments are divided by longitudinal and transverse septa to produce motile cocci in *Dermatophilus* and both motile rods and non-motile cocci in *Geodermatophilus* (Ishiguro & Wolfe, 1970).

Table 3

Some factors influencing the behaviour of substrate hyphae of nocardioform organisms

Factors	Period of stability Increased	Period of stability Decreased	
Polysaccharides as C source	+		*
Nitrate as N source	+		*
Hydrocarbons as C source	+		Webley (1954)
Mn deficiency	+		Webley (1960)
K, Na salts, CO_2	+		Roberts (1963, 1964)
Mono- and divalent ions	+		Ishiguro & Wolfe (1974)
High humidity		+	*
Low temperature		+	*
Low surface tension		+	*
NH_4 as N source		+	*
Glucose as C source		+	Webb & Clark (1957)
D-fructose as C source		+	Nikitina & Kalakoutskii (1971); Webb & Clark (1957)
Agents in medium		+	Brown & Clark (1966)
Large inoculum		+	Brown & Clark (1966)
Increased aeration		+	Webb & Clark (1957)

* Present authors.

The above summary emphasizes how hyphal morphology can vary within a species. Such variations can be attributed to genetic differences between strains in heterogeneous taxa, to various environmental effects (Table 3), or even to the size of the initial inoculum (Brown & Clark, 1966).

(b) *Aerial hyphae and spores*

Table 1 shows that the nocardioform genera which regularly produce aerial hyphae are *Pseudonocardia, Micropolyspora, Actinomadura, Saccharopolyspora* and *Nocardia.* Aerial growth may be abundant enough to detect with the naked eye or so sparse that it is difficult to see even with the scanning electron microscope. The production of aerial mycelium is influenced by the composition of the growth medium, incubation temperature and several other factors, which partly explains the different observations of succeeding investigators (Table 4).

Aerial hyphae of *Pseudonocardia* produce chains of long, cylindrical spores. According to Henssen & Schäfer (1971), blastospores are produced acropetally by constriction of the hyphae. Fragmentation spores may also be produced. Lechevalier & Lechevalier (1970*b*) observed more or less simultaneous division of the hyphae into long elements which divided further in older cultures.

In the genus *Micropolyspora,* aerial hyphae are usually formed, and bear short chains of spores. *M. brevicatena* forms sparse aerial growth but long hyphae may grow into the air and arch back into the medium like stolons. Spores occur in short chains (2–10 spores) on short branches of aerial and substrate hyphae (Lechevalier, Solotorovsky & McDurmont, 1961). Aerial growth of *M. faeni* is abundant on some media, with short hyphae bearing lateral chains of up to five spores and terminal ones with up to 10 spores. The spores form basipetally, and stained preparations show unstained areas between them (Cross, Maciver & Lacey, 1968). Aerial growth in *M. rectivirgula* is variable, with chains of up to 10 spores formed basipetally on short lateral branches (Krassilnikov & Agre, 1964; Dorokhova *et al.*, 1970). Several other *Micropolyspora* species have been described by Russian workers. Our examination of two of these, *M. caesia* and *M. viride,* has shown that they mostly form single spores and are indistinguishable from *Saccharomonospora viridis.* In *Saccharopolyspora hirsuta,* aerial growth usually arises as dense tufts near the colony centre which develop into long chains of bead-like spores, often separated by short lengths of apparently empty hyphae. The chains may be straight but at the edges of the tufts they are often in the form of loops or spirals. Electron micrographs show that the spores are covered by a hairy sheath (Lacey & Goodfellow, 1975).

Table 4

Frequency of formation of aerial hyphae by nocardioform taxa

	Nocardia asteroides	*N. caviae*	*N. brasiliensis*	*Actino-madura madurae*	*A. pelletieri*	*Oerskovia turbata*	*rhodochrous* taxon	*Saccharopolyspora hirsuta*
	(% of cultures forming aerial hyphae)							
A	24	9	17	0	0	—	0	—
B	98	90	100	23	72	0	39	—
C	100	100	100	45	7	—	—	—
D	96	100	100	83	50	—	25	—
E	80	100	50	—	—	—	—	100
F	100	100	100	80	50	—	8	—

A, Tsukamura (1969)—by eye.
B, Goodfellow (1971)—by light microscopy.
C, Gordon & Mihm (1962*a*, 1962*b*) and Gordon (1966)—by light microscopy.
D, Berd (1973*a*)—by light microscopy.
E, Present authors—by light microscopy.
F, Present authors—by scanning electron microscopy.
—, Not examined.

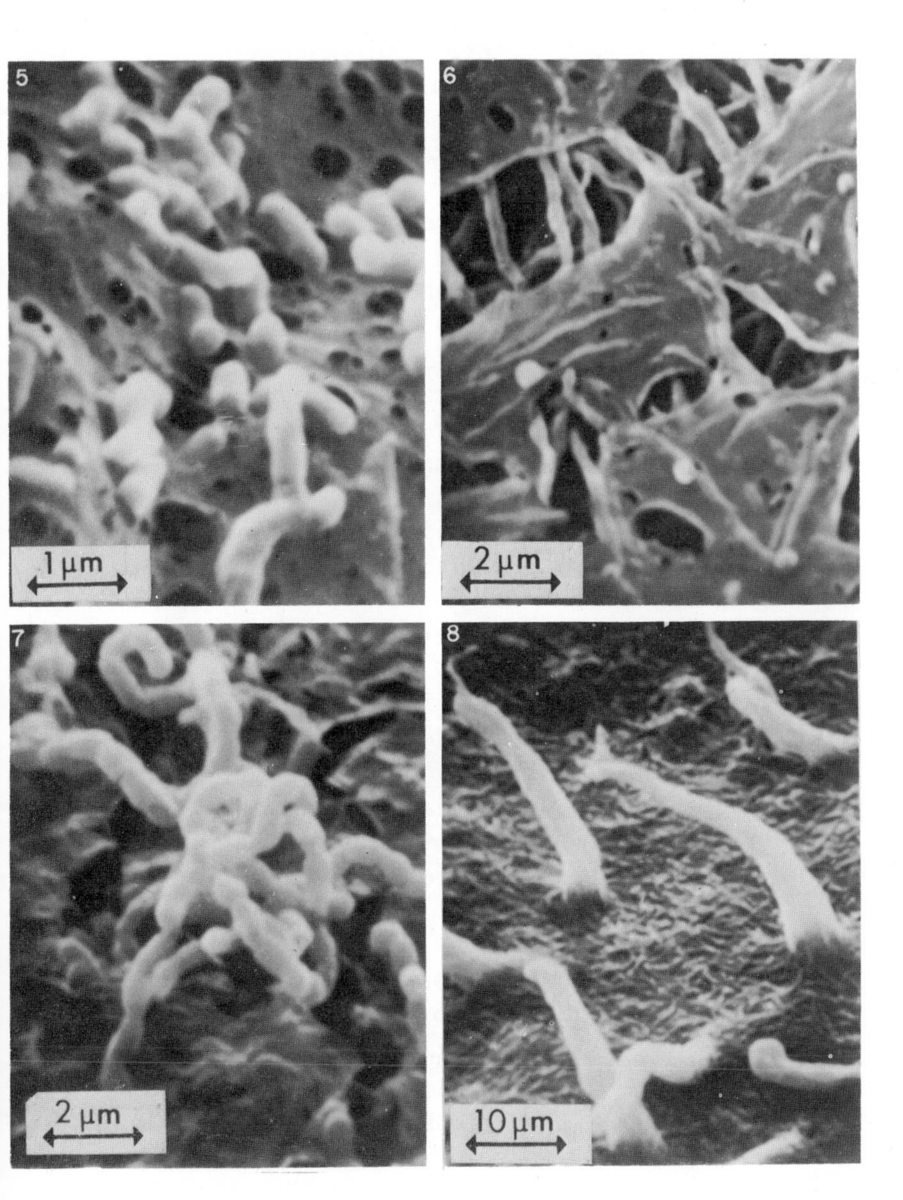

Fig. 5. *rhodochrous* strain (NCIB8863). Rods coated with slime. (SEM.)

Fig. 6. *A. madurae* (IP364). Colony surface covered by slime, with isolated spore-like bodies. (SEM.)

Fig. 7. *A. madurae* (N263). Spores chains produced on hyphae growing up through slime. (SEM.)

Fig. 8. *A. pelletieri* (NCTC3026). Coremia on colony surface. (SEM.)

Species of *Actinomadura* differ greatly in the amount and form of aerial growth. *Actinomadura madurae* usually has little or no aerial growth (Table 4) (Gordon, 1966). Short spore chains may be formed on aerial hyphae but Gordon (1966) found spores on only 6 of 44 isolates examined, and Emmons, Binford & Utz (1970) and Berd (1973*a*) also reported infrequent sporulation. However, a strain of *A. madurae* with short aerial hyphae bearing looped chains of 4–8 spores was described by Dementjeva (1970) and isolates from barley grain (Lacey, 1971), placed in *A. madurae* by Goodfellow (1971), also often formed long, straight to looped spore chains. We have used scanning electron microscopy to examine the surfaces of colonies of *A. madurae*, grown on oatmeal, yeast-extract and potato-carrot media to encourage aerial growth. Most strains produced a few aerial hyphae (Table 4), often coated by slime, sometimes with single spore-like structures on the surface (Fig. 6). In one strain, looped spore chains were formed on hyphae growing up through the slime (Fig. 7).

A. pelletieri rarely produces aerial hyphae (Table 4), only one of the 15 isolates examined by Gordon (1966) produced short, sparse, irregularly branched hyphae. Sporulation is infrequent (Lechevalier & Lechevalier, 1970*b*) or absent (Berd, 1973*a*). Using the methods described for *A. madurae*, the scanning electron microscope shows that in some isolates, hyphae did not penetrate through the extensive slime covering the colony surface but coremia were sometimes produced (Fig. 8). One isolate, grown on oatmeal agar, produced aerial hyphae which showed some sporulation.

Unlike the two previous species, *A. dassonvillei* usually forms an abundant aerial mycelium (Gordon & Horan, 1968; Lechevalier & Lechevalier, 1970*b*) macroscopically resembling *Streptomyces griseus* (Gordon & Horan, 1968). Aerial hyphae frequently form chains of spores of variable length and during spore formation hyphae have a zig-zag appearance, due to developing spores subtending angles of varying degree to their neighbouring spores (Williams, Sharples & Bradshaw, 1974).

Other *Actinomadura* species, such as *A. pusilla* and *A. roseoviolacea* (Nonomura & Ohara, 1971) produce an aerial mycelium bearing spore chains that form tight spirals (Fig. 9). Most of the spores are smooth but those of *A. verrucospora* and the supporting hyphae bear small warts (Fig. 10).

Aerial growth is usual in *Nocardia* (Table 4) but varies in amount and form from isolate to isolate and with culture conditions. Gordon & Mihm (1962*b*) reported that all their 219 cultures of *N. asteroides, N. brasiliensis* and *N. caviae* developed aerial hyphae although on 26 it grew sparsely. We found aerial hyphae in 43 of 54 *N. asteroides* isolates, and Goodfellow (1971) on 90% of his isolates; the incidence of abundant aerial growth varied greatly from 0 to 90% in the five sub-groups distinguished by Goodfellow. The form of aerial growth varies from rudimentary, to long with coalesced hyphae (Gordon & Mihm, 1957, 1959, 1962*a*). There was more aerial mycelium produced on a glucose-salts medium than on a glycerol-salts medium (McClung & Uesaka, 1961). The frequency with which aerial hyphae divided into chains of spores has varied. Gordon & Mihm (1958) detected spores in about 25% of 79 cultures, and we found them in 37% of 54 cultures; Bradley (1959) and Berd (1973*b*) recorded spores less frequently. The form of aerial growth of cultures we examined by light and scanning electron microscopy varied from short outgrowths from the colony surface (Fig. 11), through longer fragmented hyphae, to long, stable hyphae forming ropes with short side branches (Fig. 12). Sometimes hyphae formed spirals (Fig. 13), or formed spores in short branches (Fig. 14). Fragmentation (or sporing) of aerial mycelium was usually greater in cultures grown on nutrient agar than on yeast extract-malt extract agar.

Aerial growth is also common in *N. brasiliensis* (Table 4). Hyphae vary from rudimentary to long and branched, with round or irregular masses of coalesced hyphae scattered amongst the aerial mycelium of some strains (Gordon & Mihm, 1959). We also found aggregates of partially fragmented hyphae, but in other cultures long, stable hyphae were organised into thick ropes (Fig. 15). Transmission electron micrographs showed that cells were smooth and cylindrical, sometimes remaining joined together (Fig. 16).

Aerial hyphae in *N. caviae* showed similar morphological variability (Table 4) (Gordon & Mihm, 1962*b*). Sporulation was found in one of Gordon & Mihm's (1962*b*) 14 isolates and in one of Berd's (1973*a*) eight isolates. In our five cultures, aerial hyphae varied from very sparse with much slime development, to abundant with branched hyphae. Sporulation was detected to two strains (Fig. 17). Members of the *rhodochrous* group normally produce little or no aerial growth (Table 4), but once primary hyphae have fragmented

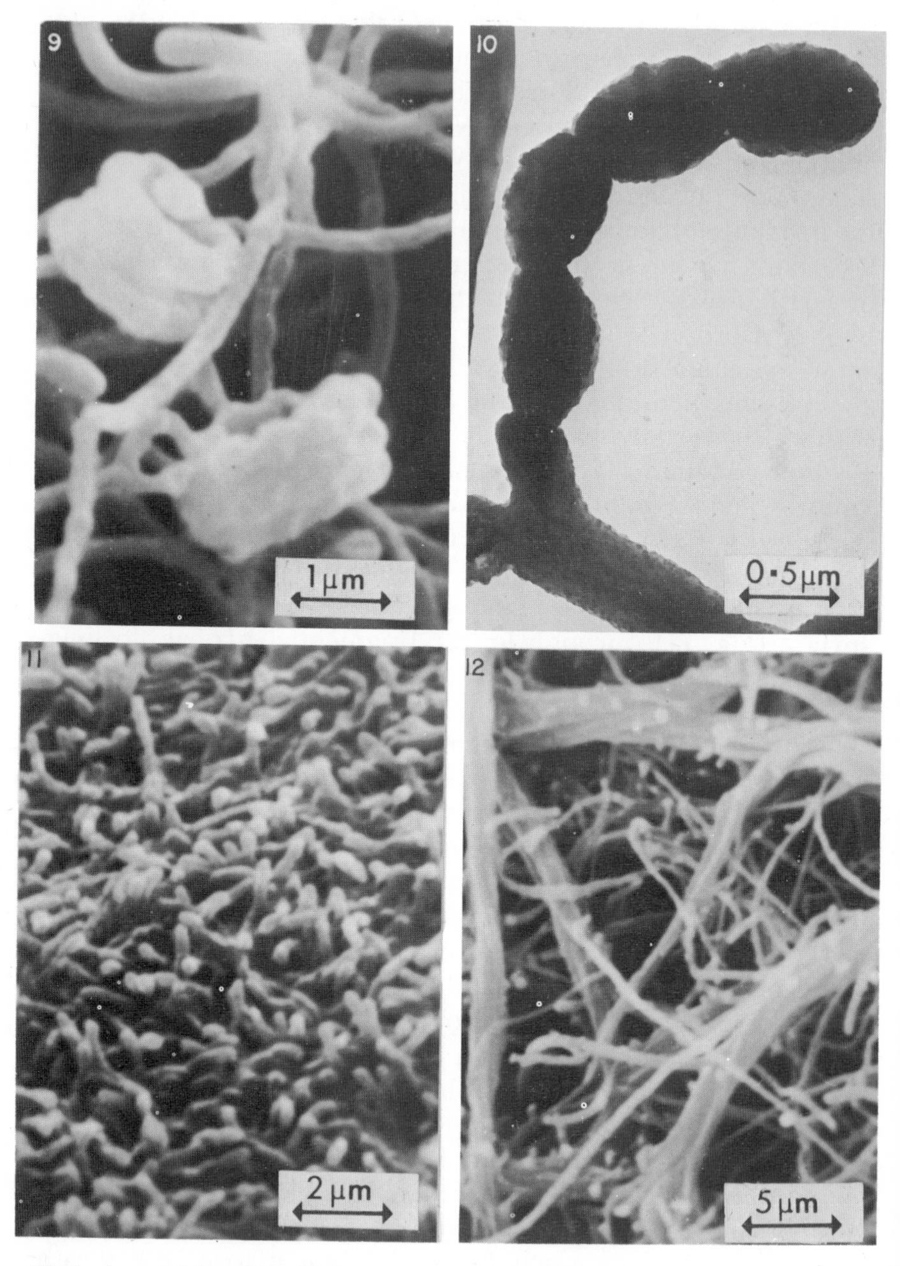

Fig. 9. *A. roseoviolacea* (A2). Tightly spiralled spore chains. (SEM.)

Fig. 10. *A. verrucospora* (A1259). Warts on spores and aerial hyphae. (TEM-S.)

Fig. 11. *N. asteroides* (N512). Very restricted development of aerial hyphae. (SEM.)

Fig. 12. *N. asteroides* (N317). Long aerial hyphae forming ropes. (SEM.)

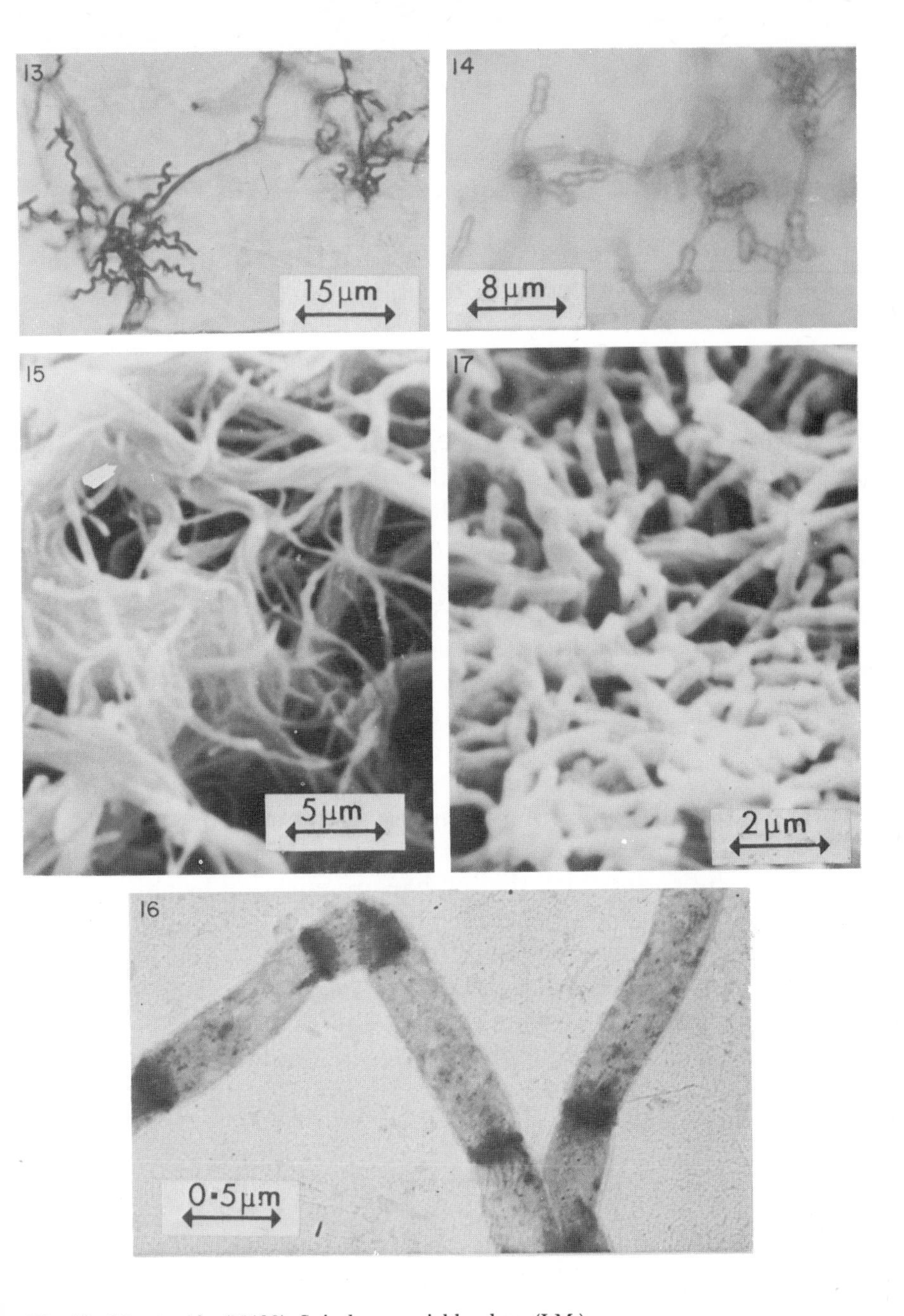

Fig. 13. *N. asteroides* (N489). Spirals on aerial hyphae. (LM.)

Fig. 14. *N. asteroides* (N525). Spore-like bodies on branches of aerial hyphae. (LM.)

Fig. 15. *N. brasiliensis* (*N*438). Long aerial hyphae forming ropes. (SEM.)

Fig. 16. *N. brasiliensis* (N37). Elongated elements in aerial hyphae. (TEM-S.)

Fig. 17. *N. caviae* (N371). Abundant aerial hyphae with fragmentation. (SEM.)

it is difficult to distinguish between aerial and substrate growth. Goodfellow, Fleming & Sackin (1972) reported that 42% of their isolates produced sparse and 3% moderate aerial growth, while none was detected in isolates of *Mycobacterium* Group IV. Possible sparse aerial hyphae occurred on one of our 12 isolates (Fig. 18).

(c) *Morphology in clinical material*

Nocardioform organisms infecting man or animals can be distinguished to some extent by the presence, colour and morphology of granules in the pus. Granules from *Actinomadura madurae* infections are usually white or yellow, soft or pasty, spherical with a lobed surface and 1–5 mm in diameter. They are surrounded by a 50 μm fringe of hyphae; this fringe is lacking in *A. pelletieri* granules which are smaller (300–500 μm diameter), bright red and more resistant to crushing (Emmons, Binford & Utz, 1970).

Nocardia asteroides occasionally forms granules (Blair, Lennette & Truant, 1970). Branched hyphae are usually sparse in infected tissue, although loose clusters are occasionally found. The granules of *N. brasiliensis* are irregularly spherical, often lobed, white to yellow and less than 1 mm diameter (Emmons, Binford & Utz, 1970). Their centre is a mass of tangled hyphae surrounded by lobes or "clubs", in which the hyphae are enclosed by amorphous material thought to be calcium phosphate. The "clubs" may be surrounded by polymorphonuclear leucocytes from the host (Macotela-Ruiz & Gonzalez-Angulo, 1966). *N. caviae* granules from artificially infected mice were variable in shape, 100–500 μm in diameter and lacked "clubs" (Kurup, Randhawa & Mishra, 1970).

7. Cytology and Fine Structure

(a) *Cell structure*

Cells of all nocardioform organisms are prokaryotic, lacking nuclear membranes, mitochondria and a reticulum with polyribosomes. The only membrane is the plasmamembrane and its ramifications (e.g. mesosomes). This is a typical "unit membrane" about 10 nm across and is usually closely associated with the cell wall. Nuclear material, seen as fine fibrils, is usually located centrally and numerous ribosomes (12 nm diameter) are present in actively growing cells.

(b) *Cytoplasmic inclusions*

The most characteristic and frequent inclusions are lipid globules and polyphosphate granules, which have been detected by both light and electron microscopy. Lipid globules are frequent in cells of *Mycobacterium* and the *rhodochrous* taxon. The globules stain with Sudan black B (McClung, 1950; Clark & Aldridge, 1960; Adams & McClung, 1962*a*; Schaefer & Lewis, 1965) and in thin sections appear as uniform, rather electron-transparent areas with or without an enclosing membrane (Schaefer & Lewis, 1965; Beaman & Shankel, 1969). They are present initially in cells of *rhodochrous* strains but disappear on germination, to re-appear later in longer filaments (Clark & Aldridge, 1960), sometimes close to cell walls (Adams & McClung, 1962*a*). Lipid globules have also been observed in *Nocardia asteroides* by Farshtchi & McClung (1967) and by Serrano *et al.* (1974) using the technique of Seligman, Wasserkrug & Marker (1966) (Fig. 19), and in *Saccharopolyspora* (Lacey & Jones, pers. comm.). They were also present in *N. brasiliensis* growing in infected tissue (Macotela-Ruiz & Gonzalez-Angulo, 1966).

There was much controversy over the identity of granules staining with basic aniline dyes which are common in *Mycobacterium*, the *rhodochrous* taxon and several other nocardioform organisms (Table 5). They were often thought to be nuclei (Drechsler, 1919; Morris, 1951; Webb, Clark & Chance, 1954; McClung, 1955; Webb & Clark, 1957; Clark & Frady, 1957), but with the demonstration of the nuclear material by electron microscopy (Hagedorn, 1959*a*,*b*), were identified as metachromatic granules (Adams & McClung, 1962*a*) containing polyphosphates and nucleic acid (Arai, Kuroda & Koyama, 1961). They are extremely electron-dense, often in polar regions of the cell (Fig. 20), and tend to volatilize under the beam of the electron microscope, leaving characteristic holes (Drews, 1960; Harold, 1966). They also account for the dark spots visible in silhouettes (Fig. 21) (Webley, 1954; Rees, Valentine & Wong, 1960; Adams & McClung, 1962*a*). Drews (1960) suggested that they are deposited onto a protein matrix in *Mycobacterium* and they have also been found associated with lipid globules of strains

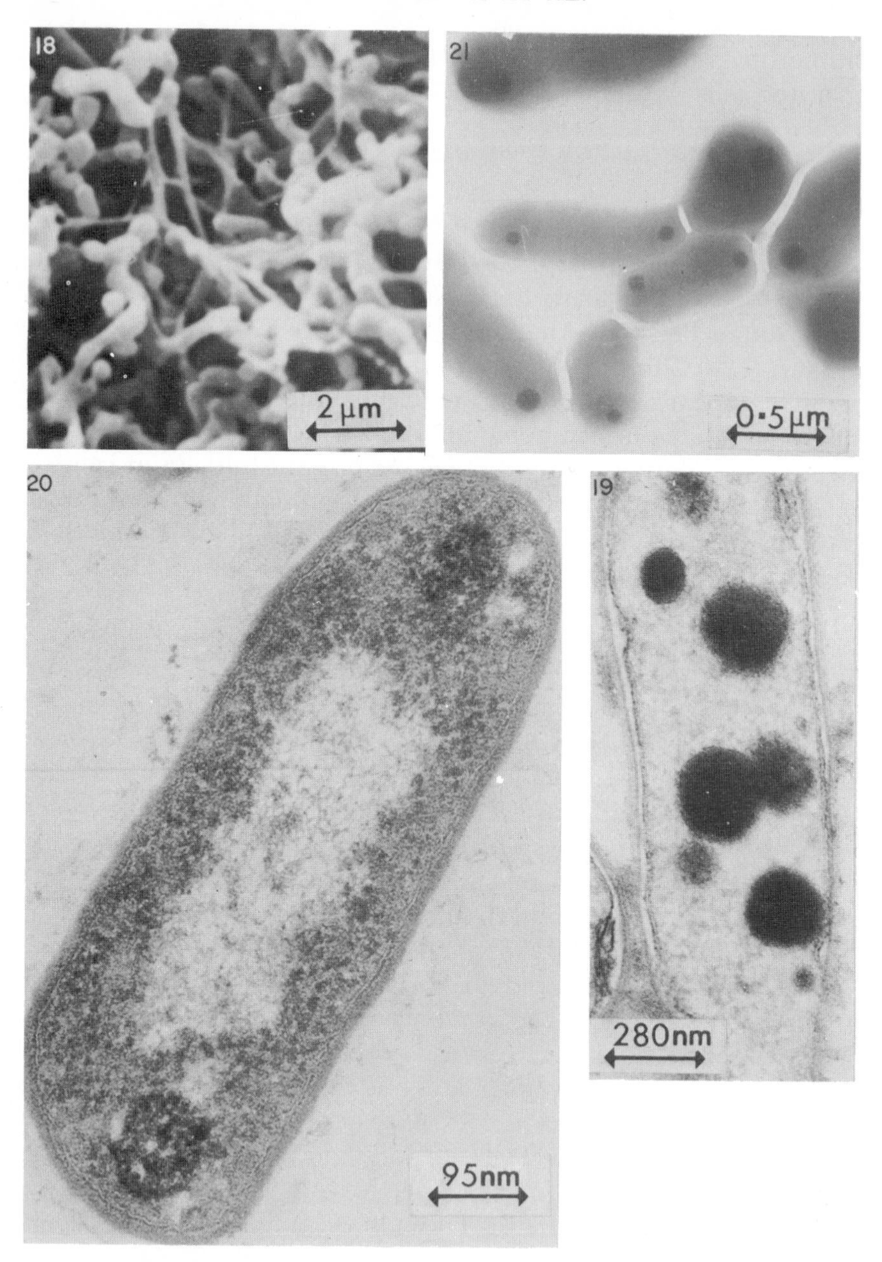

Fig. 18. *rhodochrous* strain (CUB670). Possible development of sparse aerial growth. (SEM.)

Fig. 19. *N. asteroides* (S-244). Lipid bodies after treatment with OsO_4–T.C.H. (TEM.)

Fig. 20. *rhodochrous* strain (*N. canicruria*, N53). Polyphosphate granules. (TEM.)

Fig. 21. *N. asteroides* (N487). Polyphosphate granules visible as electron dense spots. (TEM-S.)

grown on fatty acids (Schaefer & Lewis, 1965). The granules are most frequent in *rhodochrous* cells during the late log and stationary growth phases. Similar trends have been noted in mycobacteria (Mudd, Yoshida & Koike, 1958; Winder & Denneny, 1957). When growth is resumed, they may be used as a phosphorus source for synthesis of nucleic acids and lipids (Winder & Denneny, 1957). Increases in polyphosphates have been induced by zinc deficiency (Winder & O'Hara 1962). In freeze-etched cells of *rhodochrous* strains and *Nocardia asteroides*, cross-fractured round bodies are commonly observed (Fig. 22). These lack the stranded or particulate surface of β-hydroxybutyrate granules (Dunlop & Robards, 1973) but resemble the inclusions in *Synechococcus lividus* that Holt & Edwards (1972) suggested could be polyphosphate or lipid granules.

Table 5
Certain and probable occurrence of polyphosphate granules in nocardioform organisms

Organism	Authors
Nocardia asteroides	Kawata & Inoue (1965)*
N. caviae	*
rhodochrous strain (*N. rubra*)	Adams & McClung (1962*a*)
rhodochrous strain (*Corynebacterium rubrum*)	Serrano *et al.* (1972)
rhodochrous strain (*N. opaca*)	Webley (1954)
rhodochrous strain (*N. rubra*)	*
Mycobacterium lepraemurium	Rees, Valentine & Wong (1960)
M. kansasii	Schaefer & Lewis (1965)
M. phlei	Drews (1960)
Actinomadura dassonvillei	Williams, Sharples & Bradshaw (1974)
Saccharopolyspora hirsuta	*

* Present authors.

We have observed elongated electron-light regions bounded by a single-track membrane in *Nocardia caviae*, which resemble inclusions in *Microellobosporia* (Williams, Sharples & Bradshaw, 1973), but whose nature and function are unknown. Serrano *et al.* (1974) used the method of Seligman *et al.* (1965) for demonstrating polysaccharides in thin sections of *Nocardia asteroides*. As with *Escherichia coli* (Thiéry, 1967), positive reactions were given by the

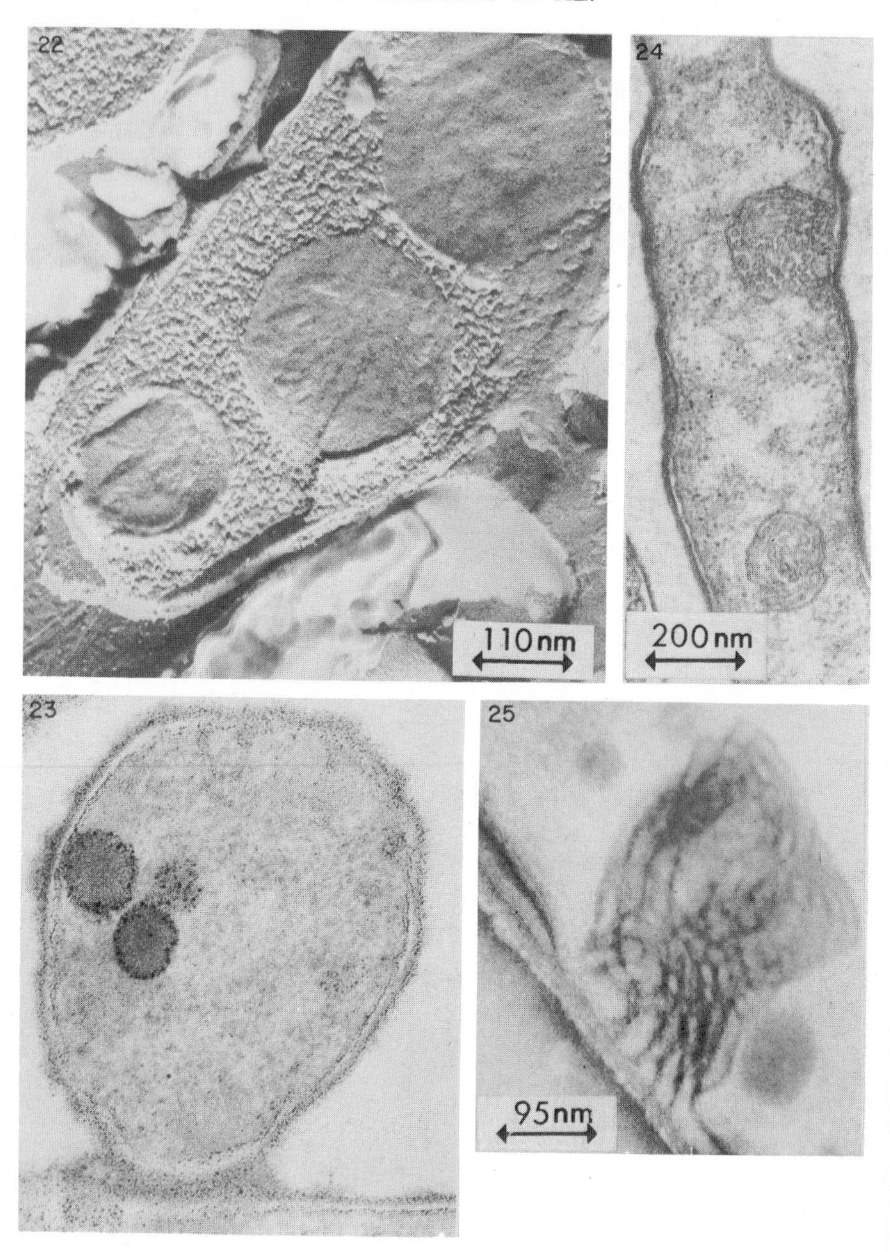

Fig. 22. *rhodochrous* strain (*N. erythropolis* ATCC4277). Spherical bodies, probably lipid or polyphosphate granules. (FE.)

Fig. 23. *N. asteroides* (S-244). Distribution of polysaccharides, indicated by occurrence of electron-dense grains after treatment of Seligman *et al.* (1965). (TEM.)

Fig. 24. *rhodochrous* strain (*N. canicruria*, N457). Tubular-vesicular mesosomes. (TEM.)

Fig. 25. *N. asteroides* (N13). Mesosomes connected with plasmamembrane (NS.)

cell wall, plasmamembrane, some cytoplasmic granules and by membranes surrounding vesicles (Fig. 23).

(c) *Mesosomes*

These structures occur in many actinomycetes (Williams, Sharples & Bradshaw, 1973) in lamellar and tubular-vesicular forms. The latter are common in nocardioform cells (Fig. 24), where they appear to develop by successive invaginations of the plasmamembrane to which they may remain connected (Fig. 25) (Imaeda & Ogura, 1963). They are often associated with developing cross-walls (Fig. 26) and sometimes appear to be in contact with the nuclear material. Occasionally the formation of membrane-bound vesicles is associated with a mesosome. It has been suggested that mesosomes may be centres for electron transport systems but sites of potassium tellurite reduction in *Nocardia asteroides* were not associated with mesosomes (Kawata & Inoue, 1965).

(d) *Cell walls and external layers*

The wall of nocardioform organisms is typical of Gram-positive bacteria. Sections show one or two electron-dense layers may be present and the mucopeptide yields alanine, glutamic acid, glucosamine and muramic acid on hydrolysis. In *Mycobacterium*, the wall has layers which are partly fibrillar and contain glycolipids (Takeya, Hisatune & Inoue, 1963; Imaeda, Kanetsuna & Galindo, 1968; Imaeda *et al.*, 1969; Draper, 1971; Draper & Rees, 1973). Imaeda *et al.* (1968, 1969) distinguished three wall layers in *M. tuberculosis* and *M. smegmatis* (Table 6). The outer lipopolysaccharide layer was prominent in young cells but tended to become detached during the stationary phase. Fibrillar elements, possibly lipopolysaccharides, occurred in each layer. Cells of *M. lepraemurium* growing in host tissue were surrounded by a single electron-transparent layer (Draper, 1971; Draper & Rees, 1973). This layer was composed largely of parallel fibrils of peptidoglycolipid wrapped longitudinally around the cells, forming a capsule that protected them from host lysosomes.

Fibrillar material is usually visible on the surface of negatively stained or freeze-etched cells of *rhodochrous* strains. It can be partially removed by alkaline ethanol (Beaman & Shankel, 1969) and appears to consist of lipid associated with peptide or protein

Table 6

The structure and composition of the Mycobacterium *cell wall (from Imaeda, Kanetsuna & Galindo, 1968)*

	Chemical treatments	Chemical composition	Structure
Outer layer	Extracted with T.C.A. Hydrophilic after phenol	Arabinose, galactose, glucose, lipids (containing 20% mycolic acid)	Fibrils and homogeneous spheres
Mid layer	Phenol-philic	Amino acids, glucose, mannose, galactose, arabinose, lipids (containing 30% mycolic acid)	Fibrillar network on membranous layer
Inner layer	Phenol and water insoluble	A.D.P., alanine. glutamic acid, glucosamine, muramic acid, mycolic acid	Thick with ramified fibrils

T.C.A. = trichlor acetic acid.
A.D.P. = adenosine diphosphate.

(Beaman *et al.*, 1971). Lipid groups were detected on the cell surface by Douglas, Ruddick & Williams (1970). In *Corynebacterium rubrum* (a *rhodochrous* strain), the outer fibrillar layer could be partially removed by sonication, 5% (w/v) trichloroacetic acid, pancreatin or neutral solvents, the latter giving rise to crystal-like structures (Serrano *et al.*, 1972). Under the outer layer was another containing fibrils of 8–10 nm diameter and beneath this an inner layer with fibrils of 4 nm diameter. The chemical analysis of phenol-treated walls suggested the presence of an arabinogalactan-peptidoglycan complex. No mycolic acids were detected in this fraction.

Extra-cellular material often coats the wall of *Nocardia* species, holding together dividing elements (Farshtchi & McClung, 1967) (Fig. 27) and disappearing following cell separation (Kawata & Inoue, 1965). The fibrillar nature of the surface of both substrate and aerial hyphae is clearly shown by carbon replication or negative staining (Figs 28 and 29), but, apart from some evidence that it contains nocardomycolic acids (Serrano, pers. comm.), little is known of its chemical composition.

The spores on the aerial growth of *Actinomadura dassonvillei* and perhaps other *Actinomadura* species form within an extra-cellular

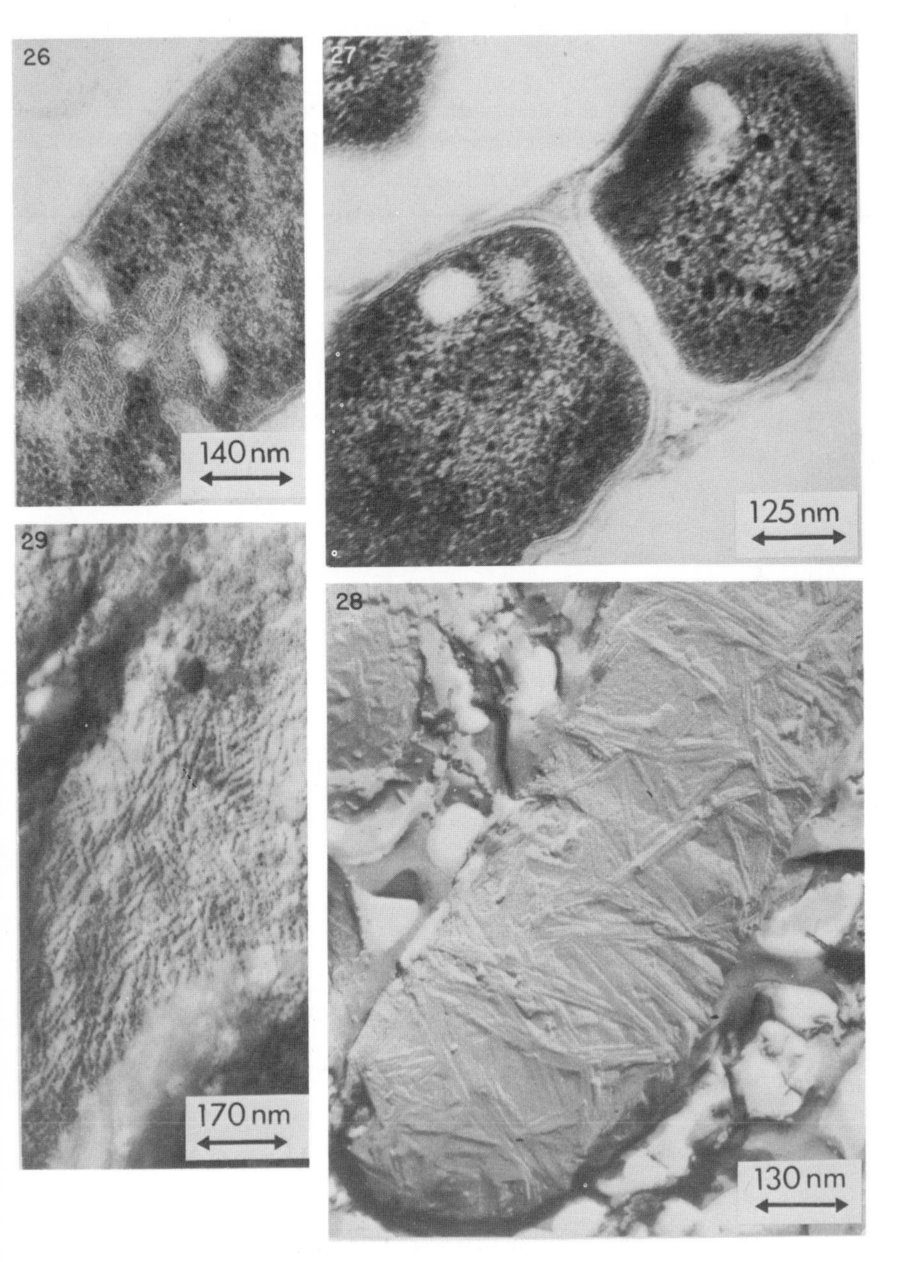

Fig. 26. *rhodochrous* strain (*N. canicruria*, N457). Mesosomes associated with developing Type 2 septum. (TEM.)

Fig. 27. *N. asteroides* (N483). Divided cell showing continuity of outer wall and extra-cellular material. (TEM.)

Fig. 28. *N. asteroides* (S-243). Fibrillar surface layer (FE.)

Fig. 29. *N. asteroides* (N317). Fibrillar surface of aerial hyphae. (CR.)

sheath (Lechevalier & Lechevalier, 1970*b*) with a fibrillar structure similar to that of *Streptomyces griseus* (Williams *et al.*, 1972, 1974). Spore chains on aerial hyphae of *Micropolyspora faeni* have a prominent sheath but this is less evident on chains formed in the substrate growth (Williams, Sharples & Bradshaw, 1973). *M. rectivirgula* also has a prominent, multilayered sheath over its developing spore chains (Dorokhova *et al.*, 1969).

The rod stage of *Geodermatophilus* has a two-layered wall, while cocci have an additional fibrous layer holding cells together (Ishiguro & Wolfe, 1970). This possibly resembles the capsule-like material observed in *Dermatophilus* (Richard, Richie & Pier, 1967).

(e) *Flagella*

Table 7 summarizes observations of flagella that occur on motile stages in nocardioform organisms.

Table 7
Flagella of nocardioform organisms

Genus	Flagella	Authors
Oerskovia	Monotrichous on short cells Peritrichous on long cells	Sukapure *et al.* (1970)
Oerskovia	Sub-polar tufts, 1 to 3	Higgins, Lechevalier & Lechevalier (1967)
Dermatophilus	1 to several	Gordon (1964)
Dermatophilus	Few to 50	Richard, Ritchie & Pier (1967)
Dermatophilus	Tuft, 5 to 7	Higgins, Lechevalier & Lechevalier (1967)
Geodermatophilus	Many	Luedemann (1968)
Geodermatophilus	4, polar	Ishiguro & Wolfe (1970)

(f) *Fragmentation and spore formation*

We have adopted the widely accepted distinction between fragmentation of substrate hyphae and spore formation by aerial hyphae. There is a trend of decreasing fragmentation of substrate hyphae and increasing spore production on aerial hyphae through the series *Mycobacterium*, the *rhodochrous* taxon, *Nocardia*,

Saccharopolyspora, Actinomadura, Micropolyspora and *Pseudonocardia* (Table 1). However, it is necessary to consider the distinction between these two processes in greater detail.

Spore formation by actinomycetes may be categorised as: (i) fragmentation of a sheathless hypha. (ii) fragmentation of a sheathed hypha. (iii) endogenous spore formation (Williams, Sharples & Bradshaw, 1973). Most nocardioform organisms have extra-cellular material sheathing their cell walls, but its composition may not be the same in all genera. Therefore, both fragmentation and spore formation by nocardioform organisms involve fragmentation of a sheathed hypha. The cross-walls which delimit both fragments and spores are often of Type II (Williams, Sharples & Bradshaw, 1973). They form as a double ingrowth from the inner part of the hyphal wall (Fig. 26), with a common growing zone, but otherwise the two components remain separate throughout cross-wall development (Fig. 30). This type of cross-wall occurs in many nocardioform strains, including *rhodochrous* strains (Beaman & Shankel, 1969; Serrano *et al.*, 1972) and *Mycobacterium* (Imaeda & Ogura, 1963). In *Mycobacterium*, the *rhodochrous* taxon and *Nocardia*, the delimited cells often remain attached at one end by the outer layer of the parent hyphal wall (Fig. 31) to give V-forms or angular growth. Sometimes there is a stage similar to the constriction division of many Gram-negative bacteria, caused by the angle of the double ingrowth being wider than usual. While the Type II cross-wall results in the separation of cells, the Type I septum develops from a single ingrowth and does not necessarily lead to division (Williams, Sharples & Bradshaw, 1973). Type I septa occur in the many actinomycetes which have stable substrate hyphae. These septa may isolate autolysing sections of the hypha and must be distinguished from true fragmentation septa. As the initial stages in the formation of fragments and spores are so similar, they must be distinguished by the timing and position of cross-wall formation, together with the changes in shape and thickness of walls that occur after completion of the septa. In the cells of *Mycobacterium* and the hyphae of *rhodochrous* strains and *Nocardia*, the positioning and sequence of cross-wall formation are irregular (Fig. 32) so size of units is variable. Usually there is no increase in wall thickness of the fragmented units. The aerial hyphae of *Nocardia* also develop many cross-walls simultaneously

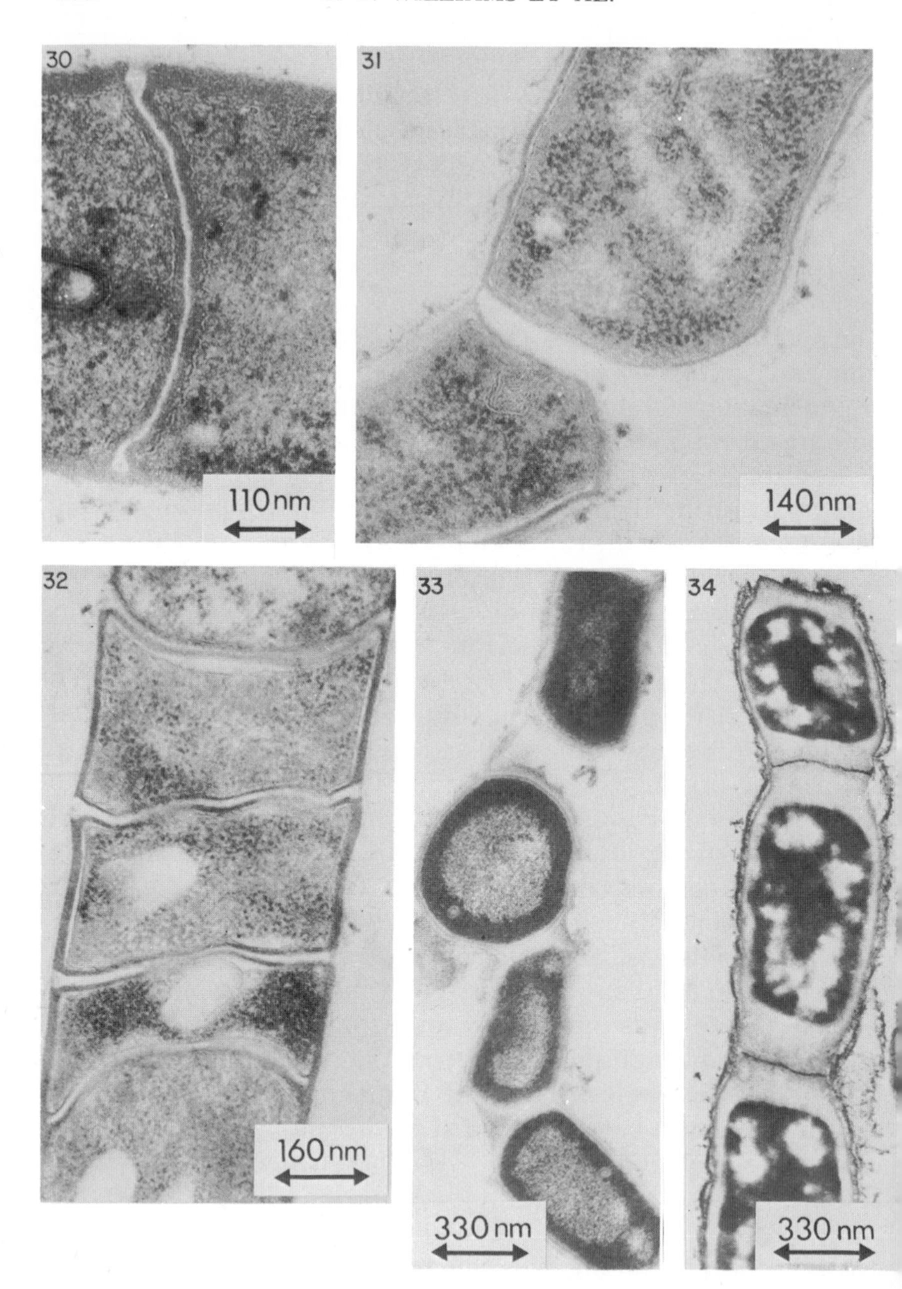

Fig. 30. *rhodochrous* strain (*N. erythropolis*, ATCC4277). Type 2 septum. (TEM.)

Fig. 31. *rhodochrous* strain (*N. erythropolis*, ATCC4277). V-form fragments held together by outer layer of wall and extracellular material. (TEM.)

to produce various sized elements within a sheath; these may round off but show little increase in wall thickness (Fig. 33).

Spores of *Actinomadura* species examined so far have usually been rounded (Fig. 7). In *A. dassonvillei*, the aerial hyphae divide into spores of variable length, and with considerably increased wall thickness, particularly in the region of the cross-walls (Fig. 34). Similar changes occur in *Saccharopolyspora hirsuta* (Lacey & Goodfellow, 1975), *Micropolyspora rectivirgula* (Dorokhova *et al.*, 1969; 1970) and in *M. faeni* where short spore chains are delimited in an orderly basipetal sequence in defined regions of the hyphae. Increases in thickness of septa and lateral walls also occur in the nocardioform fructose variant of *Streptomyces roseoflavus* var. *roseofungini* (Cherny *et al.*, 1972*a*,*b*).

Fragments of substrate hyphae may thus be distinguished from spores by developments subsequent to cross-wall formation. However, although it is possible to distinguish between the rods or cocci of *Mycobacterium* and the spores of *Micropolyspora*, the fragments and spore-like bodies of *Nocardia* species are less easily distinguished.

8. Conclusions

Morphology has been important in differentiating genera of actinomycetes, but has led to much confusion in the taxonomy of nocardioform organisms. Recent studies, emphasizing chemical and physiological characters, have brought some order. We have not attempted to reinstate morphology as a major taxonomic criterion, only to catalogue and discuss the morphology of these organisms, in the framework of these recent taxonomic groupings. While the morphological variability of nocardioform actinomycetes confounds the taxonomist, it offers interesting opportunities for studying morphogenesis. If the factors governing morphogenesis in actinomycetes are to be better understood, it may

---

Fig. 32. *N. asteroides* (N457). Delimitation of various sized fragments in substrate hypha. (TEM.)

Fig. 33. *N. caviae* (N371). Formation of various shaped elements in aerial hypha. (TEM.)

Fig. 34. *A. dassonvillei* (NCTC10488). Thick walled spores within prominent sheath. (TEM.)

well be by studying organisms such as rhodochrous strains or *Nocardia* grown in liquid media under the controlled conditions of the chemostat.

9. Acknowledgements

We acknowledge a research grant from the Science Research Council and from Consejo Nacional de Investigaciones Cientificas y Tecnologigas (CONICIT), Venezuela. The help of Mr P. Jones, Mrs S. Roberts, Miss D. H. Williams, Mr B. Webster, Mrs B. Webster, Mrs P. Evans and Mr Sulbaran is gratefully acknowledged.

10. References

Adams, J. N. (1963). Nuclear morphogenesis during the developmental cycles of some members of the genus *Nocardia*. *J. gen. Microbiol.* **33,** 429.

Adams, J. N. (1966). Studies on the fragmentation of *Nocardia erythropolis* in liquid media. *Can. J. Microbiol.* **12,** 433.

Adams, J. N. & McClung, N. M. (1960). Morphological studies in the genus *Nocardia*. V. Septation in *Nocardia rubra* and *Jensenia canicruria*. *J. Bact.* **80,** 281.

Adams, J. N. & McClung, N. M. (1962*a*). On the nature of cytoplasmic inclusions of *Nocardia rubra*. *J. gen. Microbiol.* **28,** 231.

Adams, J. N. & McClung, N. M. (1962*b*). Comparison of the developmental cycles of some members of the genus *Nocardia*. *J. Bact.* **84,** 206.

Arai, T., Kuroda, S. & Kayama, Y. (1961). Isolation and characterisation of electron scattering granules in *Nocardia*. *Annual Report Institute Food Microbiology, Chiba University* **14,** 62.

Beaman, B. L., Kim, K. S., Salton, M. R. J. & Barksdale, L. (1971). Amino acids of the cell wall of *Nocardia rubra*. *J. Bact.* **108,** 941.

Beaman, B. L. & Shankel, D. M. (1969). Ultrastructure of *Nocardia* cell growth and development on defined and complex agar media. *J. Bact.* **99,** 876.

Berd, D. (1973*a*). Laboratory identification of clinically important aerobic actinomycetes. *Appl. Microbiol.* **25,** 665.

Berd, D. (1973*b*). *Nocardia asteroides*. A taxonomic study with clinical correlations. *Am. Rev. resp. Dis.* **108,** 909.

Blair, J. E., Lennette, E. H. & Truant, J. P. (1970). *Manual of Clinical Microbiology*. Bethesda: American Society of Microbiology.

Bojalil, L. F. & Cérbon, J. (1959). Scheme for the differentiation of *Nocardia asteroides* and *Nocardia brasiliensis*. *J. Bact.* **78,** 852.

Bradley, S. G. (1959). Sporulation by some strains of nocardiae and streptomycetes. *Appl. Microbiol.* **7,** 89.

Brown, O. R. & Clark, J. B. (1966). Fragmentation in *Nocardia corallina*. *J. gen. Microbiol.* **45,** 525.

Brown, O. R. & Reda, S. (1967). Enzyme and permeability changes during morphogenesis of *Nocardia corallina*. *J. gen. Microbiol.* **47,** 199.

Burdon, K. L. (1946). Fatty material in bacteria and fungi revealed by staining dried, fixed slide preparations. *J. Bact.* **52,** 665.

Cherny, N. E., Tikhonenko, A. S., Nikitina, E. T. & Kalakoutskii, L. V. (1972*a*). Ultrastructure of *Streptomyces roseoflavus* var. *roseofungini* and its stable nocardioform "fructose" variant. *Cytobios* **5,** 7.

Cherny, N. E., Tikhonenko, A. S., Nikitina, E. T. & Kalakoutskii, L. V. (1972*b*). Intra-cytoplasmic membrane system in "fructose" variant of *Streptomyces roseoflavus* var. *roseofungini* grown in meat peptone agar. *Cytobios* **5,** 101.

Clark, J. B. & Aldridge, C. (1960). Fat bodies in *Nocardia corallina*. *J. Bact.* **79,** 756.

Clark, J. B. & Frady, J. (1957). Secondary life cycle of *Nocardia corallina*. *J. Bact.* **74,** 698.

Cross, T. & Goodfellow, M. (1973). Taxonomy and classification of the actinomycetes. In *The Actinomycetales, Characteristics & Practical Importance*. Eds F. Skinner & G. Sykes. p. 11. London and New York: Academic Press.

Cross, T., Lechevalier, M. P. & Lechevalier, H. (1963). A new genus of the Actinomycetales: *Microellobosporia* gen. nov. *J. gen. Microbiol.* **31,** 421.

Cross, T., Maciver, A. M. & Lacey, J. (1968). The thermophilic actinomycetes in mouldy hay: *Micropolyspora faeni* sp. nov. *J. gen. Microbiol.* **50,** 351.

Dementjeva, G. R. (1970). Studies on a case of actinomycetoma pedis in Queensland. *Sabouraudia,* **8,** 81.

Dorokhova, L. A., Agre, N. S., Kalakoutskii, L. V. & Krassilnikov, N. A. (1969). Fine structure of sporulating hyphae and spores in a thermophilic actinomycete, *Micropolyspora rectivirgula*. *J. Microscopie* **8,** 845.

Dorokhova, L. A., Agre, N. S., Kalakoutskii, L. V. & Krassilnikov, N. A. (1970). A study of the morphology of two cultures belonging to the genus *Micropolyspora*. *Mikrobiologiya* **39,** 79. (English translation.)

Douglas, H. W., Ruddick, S. M. & Williams, S. T. (1970). A study of the electrokinetic properties of some actinomycete spores. *J. gen. Microbiol.* **63,** 289.

Draper, P. (1971). The walls of *Mycobacterium lepraemurium*: chemistry and ultrastructure. *J. gen. Microbiol.* **69,** 313.

Draper, P. & Rees, R. J. W. (1973). The nature of the electron-transparent zone that surrounds *Mycobacterium lepraemurium* inside host cells. *J. gen. Microbiol.* **77,** 79.

Drechsler, C. (1919). Morphology of genus *Actinomyces*. Bot. Gaz. **67,** 65–83, 147.

Drews, G. (1960). Untersuchungen zum Polyphosphatstoffwechsel und der Bildung metachromatischer Granula bei *Mycobacterium phlei*. *Arch. Mikrobiol.* **36,** 387.

Dunlop, W. F. & Robards, A. W. (1973). Ultrastructural study of poly-B-hydroxybutyrate granules from *Bacillus cereus*. *J. Bact.* **114,** 1271.

Emmons, C. W., Binford, C. H. & Utz, J. P. (1970). *Medical Mycology*, 2 edn. Philadelphia: Lea & Febriger.

Farshtchi, D. & McClung, N. M. (1967). Fine structure of *Nocardia asteroides* grown in a chemically defined medium. *J. Bact.* **94,** 255.

Goodfellow, M. (1971). Numerical taxonomy of some nocardioform bacteria. *J. gen. Microbiol.* **69,** 33.

Goodfellow, M., Fleming, A. & Sackin, M. J. (1972). Numerical classification of "*Mycobacterium*" *rhodochrous* and Runyon's group IV mycobacteria. *Int. J. Syst. Bact.* **22,** 81.

Gordon, M. A. (1964). The genus *Dermatophilus. J. Bact.* **88,** 509.

Gordon, R. E. (1966). Some criteria for the recognition of *Nocardia madurae* (Vincent) Blanchard. *J. gen. Microbiol.* **45,** 355.

Gordon, R. E. & Horan, A. C. (1968). *Nocardia dassonvillei,* a macroscopic replica of *Streptomyces griseus. J. gen. Microbiol.* **50,** 223.

Gordon, R. E. & Mihm, J. M. (1957). A comparative study of some strains received as nocardiae. *J. Bact.* **73,** 15.

Gordon, R. E. & Mihm, J. M. (1958). Sporulation by two strains of *Nocardia asteroides. J. Bact.* **75,** 239.

Gordon, R. E. & Mihm, J. M. (1959). A comparison of *Nocardia asteroides* and *Nocardia brasiliensis. J. gen. Microbiol.* **20,** 129.

Gordon, R. E. & Mihm, J. M. (1962*a*). The type species of the genus *Nocardia. J. gen. Microbiol.* **27,** 1.

Gordon, R. E. & Mihm, J. M. (1962*b*). Identification of *Nocardia caviae* (Erikson) nov. comb. *Ann. N.Y. Acad. Sci.* **98,** 628.

Gordon, R. E. & Smith, M. M. (1955). Proposed group of characters for the separation of *Streptomyces* and *Nocardia. J. Bact.* **69,** 147.

Guzeva, L. N., Agre, N. S. & Sokolov, A. A. (1972). Taxonomy of actinomycetes forming catenate spores. *Mikrobiologiya* **41,** 957. (English translation).

Hagedorn, H. (1959*a*). Licht-und elektronenmikroskopische Untersuchungen en *Nocardia corallina* (Bergey *et al.* 1923). *Zentbl. Bakt. ParasitKde Abt. II* **112,** 214.

Hagedorn, H. (1959*b*). Electronenmikroskopische Untersuchungen uber den Teilungs verlauf bei *Nocardia corallina* (Bergey *et al.,* 1923) *Zentbl. Bakt. ParasitKde Abt. II* **112,** 359.

Harold, F. M. (1966). Inorganic polyphosphates in biology: structure, metabolism and function. *Bact. Rev.* **30,** 772.

Henssen, A. & Schäfer, D. (1971). Emended description of the genus *Pseudonocardia* Henssen and description of a new species *Pseudonocardia spinosa* Schäfer. *Int. J. Syst. Bact.* **21,** 29.

Higgins, M. L., Lechevalier, M. P. & Lechevalier, H. A. (1967). Flagellated actinomycetes. *J. Bact.* **93,** 1446.

Holt, S. C. & Edwards, M. R. (1972). Fine structure of the thermophilic blue-green alga *Synechococcus lividus* Copeland. A study of frozen-fractured-etched cells. *Can. J. Microbiol.* **18,** 175.

Imaeda, T., Kanetsuna, F. & Galindo, B. (1968). Ultrastructure of cell wall of genus *Mycobacterium. J. Ultrastruct. Res.* **25,** 46.

Imaeda, T., Kanetsuna, F., Galindo, G., Rieber, M. & Cesari, I. (1969). Ultrastructural characteristics of mycobacterial growth. *J. med. Microbiol.* **2,** 181.

Imaeda, T. & Ogura, M. (1963). Formation of intracytoplasmic membrane system of mycobacteria related to cell division. *J. Bact.* **85,** 150.

Ishiguro, E. E. & Wolfe, R. S. (1970). Control of morphogenesis in *Geodermatophilus*: ultrastructural studies. *J. Bact.* **104,** 566.

Ishiguro, E. E. & Wolfe, R. S. (1974). Induction of morphogenesis in *Geodermatophilus* by inorganic cations and by organic nitrogenous cations. *J. Bact.* **117,** 189.

Kawata, T. & Inoue, T. (1965). Ultrastructure of *Nocardia asteroides* as revealed by electron microscopy. *Jap. J. Microbiol.* **9,** 101.

Kellenberger, E., Ryter, A. & Séchaud, J. (1958). Electron microscope study of DNA-containing plasms. II. Vegetative and mature phage DNA as compared with normal bacterial nucleoids in different physiological states: *J. biophys. biochem. Cytol.* **4,** 671.

Kormendy, A. C. & Wayman, M. (1972). Scanning electron microscopy of microorganisms. *Micron* **3,** 33.

Krassilnikov, N. A. & Agre, N. S. (1964). On two new species of *Thermopolyspora. Hindustan Antibiot. Bull.* **6,** 97.

Kurup, P. V., Randhawa, H. S. & Mishra, S. K. (1970). Use of paraffin bait technique in the isolation of *Nocardia asteroides* from sputum. *Mycopath. Mycol. appl.* **40,** 363.

Kurup, P. V. & Schmitt, J. A. (1973). Numerical taxonomy of *Nocardia. Can. J. Microbiol.* **19,** 1035.

Kwapinski, J. B. G. & Horsman, G. (1973). Cultural characterization and differentiation of nocardiae. *Can. J. Microbiol.* **19,** 895.

Lacey, J. (1971). The microbiology of moist barley storage in unsealed silos. *Ann. appl. Biol.* **69,** 187.

Lacey, J. & Goodfellow, M. (1975). A novel actinomycete from sugar cane bagasse: *Saccharopolyspora hirsuta* gen. et sp. nov. *J. gen. Microbiol.* **88,** 75.

Lechevalier, M. P. & Lechevalier, H. (1970*a*). Chemical composition as a criterion in the classification of aerobic actinomycetes. *Int. J. Syst. Bact.* **20,** 435.

Lechevalier, H. A. & Lechevalier, M. P. (1970*b*). A critical evaluation of the genera of aerobic actinomycetes. In *The Actinomycetales.* Ed. H. Prauser. Jena: Gustav Fischer.

Lechevalier, H. A., Solotorovsky, M. & McDurmont, C. I. (1961). A new genus of the Actinomycetales, *Micropolyspora g.n. J. gen. Microbiol.* **26,** 11.

Luedemann, G. M. (1968). *Geodermatophilus,* a new genus of the *Dermatophilaceae* (Actinomycetales). *J. Bact.* **96,** 1848.

Macotela-Ruiz, E. & Gonzalez-Angulo, A. (1966). Electron microscopic studies of granules of *Nocardia brasiliensis* in man. *Sabouraudia* **5,** 92.

McClung, N. M. (1949). Morphological studies in the genus *Nocardia.* I. Developmental studies. *Lloydia* **12,** 137.

McClung, N. M. (1950). Morphological studies in the genus *Nocardia.* II. Cytological studies. *J. Bact.* **59,** 589.

McClung, N. M. (1954). The utilization of carbon compounds by *Nocardia* species. *J. Bact.* **68,** 231.

McClung, N. M. (1955). Morphological studies in the genus *Nocardia.* IV. Bright phase contrast observations of living cells. *Trans. Kans. Acad. Sci.* **58,** 50.

McClung, N. M. & Uesaka, J. (1961). Morphological studies in the genus *Nocardia.* VI. Aerial hyphal production and acid-fastness of *N. asteroides* isolates. *Revta lat.-am. Microbiol.* **4,** 97.

Morris, E. O. (1951). Observations of the life cycle of the *Nocardia. J. Hyg., Camb.* **49,** 175.

Mudd, S., Yoshida, A. & Koike, M. (1958). Polyphosphate as accumulator of phosphorus and energy. *J. Bact.* **75,** 224.

Nikitina, E. T. & Kalakoutskii, L. V. (1971). Induction of nocardioform growth in actinomycetes on media with D-fructose. *Z. allg. Mikrobiol.* **11,** 601.

Nikitina, E. T., Kasakova, G. G. & Kalakoutskii, L. V. (1971). Induction of unusual development in *Actinomyces roseoflavus* var. *roseofungini* on the media with fructose. *Dokl. Akad. Nauk SSSR* **196,** 448.

Nonomura, H. & Ohara, Y. (1971). Distribution of actinomycetes in soil. XI. Some new species of the genus *Actinomadura* Lechevalier *et al. J. Fermen. Technol., Osaka* **49,** 904.

Prauser, H. (1967). Contributions to the taxonomy of the Actinomycetales. *Publ. Fac. Sci. Univ. Brno* **40,** 196.

Prauser, H., Lechevalier, M. P. & Lechevalier, H. (1970). Description of *Oerskovia* gen. nov. to harbor Oerskov's motile *Nocardia. Appl. Microbiol.* **19,** 534.

Pridham, T. G., Anderson, P., Foley, C., Lindenfelser, L. A., Hesseltine, C. W. & Benedict, R. G. (1957). A selection of media for maintenance and taxonomic study of *Streptomyces. Antibiotics* A. **1956/57,** 947.

Rees, R. J. W., Valentine, R. C. & Wong, P. C. (1960). Application of qualitative electron microscopy to the study of *Mycobacterium lepraemurium* and *Mycobacterium leprae. J. gen. Microbiol.* **22,** 443.

Richard, J. L., Ritchie, A. E. & Pier, A. C. (1967). Electron microscopic anatomy of motile-phase and germinating cells of *Dermatophilus congolensis. J. gen. Microbiol.* **49,** 23.

Roberts, D. S. (1963). The influence of carbon dioxide on the growth and sporulation of *Dermatophilus dermatonomus. Aust. J. agric. Res.* **14,** 412.

Roberts, D. S. (1964). Synergistic effects of salts and carbon dioxide on *Dermatophilus dermatonomus. J. gen. Microbiol.* **37,** 403.

Robinow, C. F. (1945). Nuclear apparatus and cell structure of rod-shaped bacteria. In addendum to *The Bacterial Cell.* R. J. Dubos. Cambridge, Massachusetts: Harvard University Press.

Runyon, E. H. (1955). Veterans Administration-National Tuberculosis Association Co-operative Study of mycobacteria. *Am. Rev. resp Dis.* **72,** 866.

Schaeffer, W. B. & Lewis, C. W. (1965). Effect of oleic acid on growth and cell structure of mycobacteria. *J. Bact.,* **90,** 1438.

Seligman, A. M., Hanker, J. S., Wasserkrug, H., Dmochowski, H. & Katzoff, L. (1965). Histochemical demonstration of some oxidised macromolecules with thiocarbohydrazide (T.C.H.) or thiosemicarbazide (T.S.C.) and osmium tetroxide. *J. Histochem. Cytochem.* **13,** 629.

Seligman, A. M., Wasserkrug, H. L. & Hanker, J. S. (1966). A new staining method (OTO) for enhancing contrast of lipid-containing membranes and droplets in osmium tetroxide-fixed tissue with osmophilic thiocarbohydrazide (TCH). *J. Cell Biol.* **30,** 424.

Serrano, J. A. & Serrano, A. A. (1974). Nocardial ultrastructure as observed by freeze-etching. *Proc. 1 Intern. Conf. Biol. of Nocardiae, Merida, Venezuela.* p. 32.

Serrano, J. A., Serrano, A. A., Plapinger, R. E. & Seligman, A. M. (1974). Ultrastructural cytochemistry of *Nocardia asteroides* S-244. *Proc. 2 Lat. Am. Electron Microscopy Congr., Sao Paulo, Brazil.* p. 106.

Serrano, J. A., Serrano, A. A. & Tablante, R. V. (1971). Ultrastructure of *Nocardia asteroides. 29 Ann. Proc. Electron Microscopy Soc. Am.* Ed. C. J. Arceneaux, Boston, Mass.

Serrano, J. A., Tablante, R. V., Serrano, A. A., San Blas, G. & Imaeda, T. (1972). Physiological, chemical and ultrastructural characteristics of *Corynebacterium rubrum. J. gen. Microbiol.* **70,** 339.

Sukapure, R. S., Lechevalier, M. P., Reber, H., Higgins, H. L., Lechevalier, H. A. & Prauser, H. (1970). Motile nocardioid Actinomycetales. *Appl. Microbiol.* **19,** 527.

Takeya, K., Hisatune, K. & Inoue, Y. (1963). Mycobacterial cell walls. II. Chemical composition of the "basal layer". *J. Bact.* **85,** 24.

Thiéry, J. P. (1967). Mise en évidence des polysaccharides sur coupes fines en microscopie électronique. *J. Microscopie* **6,** 987.

Tsukamura, M. (1969). Numerical taxonomy of the genus *Nocardia. J. gen. Microbiol.* **56,** 265.

Tsukamura, M. (1973). A taxonomic study of strains received as "*Mycobacterium*" *rhodochrous. Jap. J. Microbiol.* **17,** 189.

Uesaka, I. (1969). Breaking up of vegetative mycelium of *Nocardia* and *Streptomyces* cultures on agar film. *Jap. J. Microbiol.* **13,** 65.

Waksman, S. A. (1961). *The Actinomycetes,* vol. II. *Classification, identification and description of genera and species.* Baltimore: Williams and Wilkins.

Webb, R. B. (1954). A useful bacterial cell wall stain. *J. Bact.* **67,** 252.

Webb, R. B. & Clark, J. B. (1957). A cytogenetic study of *Nocardia corallina. J. Bact.* **74,** 31.

Webb, R. B., Clark, J. B. & Chance, H. L. (1954). A cytological study of *Nocardia corallina* and other actinomycetes. *J. Bact.* **67,** 498.

Webley, D. M. (1954). The morphology of *Nocardia opaca* Waksman & Henrici when grown on hydrocarbons, vegetable oils, fatty acids and related sustances. *J. gen. Microbiol.* **11,** 420.

Webley, D. M. (1960). The effect of deficiency of iron, zinc and manganese on the growth and morphology of *Nocardia opaca. J. gen. Microbiol.* **23,** 87.

Welshimer, H. J. & Robinow, C. F. (1949). The lysis of *Bacillus megaterium* by lysozyme. *J. Bact.* **57,** 489.

Williams, S. T. (1970). Further investigations of actinomycetes by scanning electron microscopy. *J. gen. Microbiol.* **62,** 67.

Williams, S. T., Bradshaw, R. M., Costerton, J. W. & Forge A. (1972). Fine structure of the spore sheath of several *Streptomyces* species. *J. gen. Microbiol.* **72,** 249.

Williams, S. T. & Cross, T. (1971). Isolation, purification, cultivation and preservation of actinomycetes. In *Methods in Microbiology,* vol. 4. Ed. C. Booth, p. 295. London and New York: Academic Press.

Williams, S. T. & Davies, F. L. (1967). Use of a scanning electron microscope for the examination of actinomycetes. *J. gen. Microbiol.* **48,** 171.

Williams, S. T., Veldkamp, C. J. & Robinson, C. S. (1973). Preparation of microbes for scanning electron microscopy. In *Scanning Electron Microscopy.* Eds O. Johari & I. Corvin. p. 735. Chicago: I.I.T.R.I.

Williams, S. T., Sharples, G. P. & Bradshaw, R. M. (1973). The fine structure of the Actinomycetales. In *The Actinomycetales, Characteristics & Practical Importance.* Eds F. Skinner & G. Sykes. p. 113. London and New York: Academic Press.

Williams, S. T., Sharples, G. P. & Bradshaw, R. M. (1974). Spore formation in *Actinomadura dassonvillei* (Brocq-Rousseu) Lechevalier and Lechevalier. *J. gen. Microbiol.* **84,** 415.

Winder, F. G. & Denneny, J. M. (1957). The metabolism of inorganic polyphosphate in mycobacteria. *J. gen. Microbiol.* **17,** 573.

Winder, F. G. & O'Hara, C. (1962). Effect of iron deficiency and of zinc deficiency on composition of *Mycobacterium smegmatis. Biochem. J.* **82,** 98.

6. Cell Walls of Nocardiae

G. MICHEL and CLAUDETTE BORDET

Laboratoire de Biochimie Microbienne, Université de Lyon I, 43 Boulevard du 11 Novembre 1918, 69621 Villeurbanne, France

Contents

1. Introduction

The first investigations on the composition of bacterial cell walls were made by Salton (1953). Soon after, a large variety of bacteria were examined (Cummins & Harris, 1956; Cummins, Glendenning & Harris, 1957). In the field of Actinomycetales, the relations between cell wall composition and taxonomy were studied (Cummins & Harris, 1958).

More recently, purified walls from various strains of Actinomycetales were prepared for a more precise study of their macromolecular components. In this review, we will summarize the work carried out to determine the structure of the walls of nocardiae. Some related microorganisms will be occasionally introduced when their walls show some similarities with those of nocardiae. The value of these studies will be outlined by a statement of the biological properties of these walls and of their use in the classification of nocardiae and related actinomycetes.

2. Composition of the Walls

The walls of nocardiae have a complex structure which consists of a peptidoglycan, lipid constituents and other polysaccharide or polypeptide compounds.

(a) *Peptidoglycan*

Peptidoglycan (mucopeptide or murein) is a constituent common to the walls of all bacteria. It has been studied in a great number of bacteria and several books and reviews summarize the various structures of peptidoglycans (Rogers & Perkins, 1968; Salton, 1964; Ghuysen, Strominger & Tipper, 1968; Ghuysen, 1968; Schleifer & Kandler, 1972).

The glycan part is a linear strand which consists of disaccharide units β-D-*N*-acetylglucosamine 1 → 4 *N*-acetylmuramic acid linked together by 1 → 4 bonds. Muramic acid is bound by its carboxyl to a peptide, generally a tetrapeptide chain with the sequence L-Ala-D Glu-X-D Ala. X is a diaminated amino acid, L lysine or diaminopimelic acid.

Two peptide subunits can be linked together either directly by a D Ala-X (ε-NH_2) linkage when X is diaminopimelic acid or by the means of an amino acid or an oligopeptide when X is L lysine (Fig. 1). The whole complex has a reticulated structure.

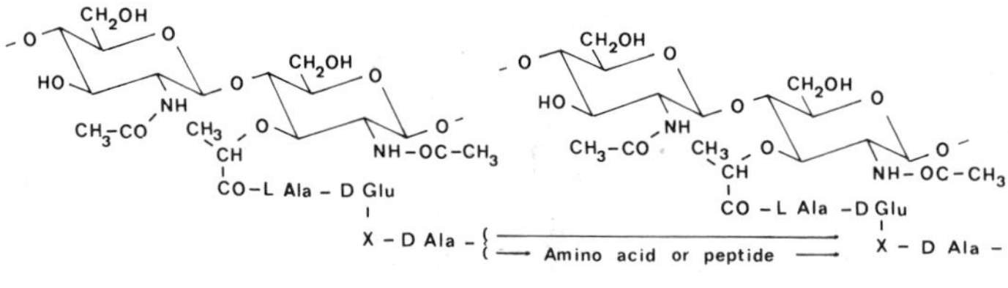

Fig. 1. Structure of a peptidoglycan unit from a bacterial cell wall.

Initial studies concerning the composition of peptidoglycan from Actinomycetales were only related to the qualitative analysis of amino acids (Becker, Lechevalier & Lechevalier, 1965; Cummins, 1962*a,b*; Szaniszlo & Gooder, 1967; Yamaguchi, 1965). Determination of the structure of peptidoglycans was first carried out upon streptomycetes (Nakamura, Tamura & Arima, 1967; Arima, Nakamura & Tamura, 1968; Leyh-Bouille *et al.*, 1970).

They found that the amino acid X was LL-diaminopimelic acid and that peptide subunits were linked together by means of glycine.

Peptidoglycan from mycobacteria was studied by Lederer and his coworkers (see the review by Lederer, 1971). The glycan moiety isolated from *Mycobacterium smegmatis* contains *N*-glycolyl-muramic acid instead of *N*-acetylmuramic acid (Petit *et al.*, 1969; Adam *et al.*, 1969). The peptide subunits in *Mycobacterium smegmatis* and *Mycobacterium phlei* are tetrapeptides with *meso*-2,6-diaminopimelic acid (Dpm) (Wietzerbin-Falszpan *et al.*, 1970).

It was known that the walls of nocardiae and mycobacteria have the same qualitative composition in amino acids (Romano & Sohler, 1956; Hoare & Work, 1957; Cummins & Harris, 1958). Therefore, an analogous structure for the peptidoglycans from mycobacteria and nocardiae could be assumed.

In our laboratory, we studied the peptidoglycan of *Nocardia kirovani*. After being isolated and purified, this peptidoglycan was analysed and found to contain the amino acids *meso*-Dpm, Glu, Ala and hexosamines in the molar ratios 1:1,2:1,65:1,8 (Guinand, Vacheron & Michel, 1970). A structural study of the peptidoglycan was carried out by hydrolysis with specific enzymes: lysozyme, *N*-acetylmuramyl-L-alanine amidase and Myxobacter AL_1 protease (Ghuysen, 1968; Ghuysen *et al.*, 1969; Ensign & Wolfe, 1965, 1966). The degraded products were separated by filtration on two linked Sephadex G 50 and Sephadex G 25 columns. The glycan part gave a disaccharide and the peptide part gave tripeptides and tetrapeptides which are monomer units and hepta- and octapeptides which are dimer units.

The glycan moiety is constituted by disaccharide units whose structure was determined by mass spectrometry of the permethylated derivative (Fig. 2). The molecular peak at $m/e = 638$ and fragments at $m/e = 231$, 362, 535 are in agreement with the structure shown on Fig. 2. In this derivative, muramic acid is *N*-substituted by a glycolyl group. The spectrum is identical to that of the permethylated disaccharide from *Mycobacterium smegmatis* (Adam *et al.*, 1969). On the other hand, the chromatographic behaviour of the disaccharides from *Nocardia kirovani* and *Mycobacterium smegmatis* are analogous and they are different from the disaccharide of *Micrococcus luteus* whose muramic acid is *N*-acetylated (Sharon *et al.*, 1966; Leyh-Bouille *et al.*, 1966).

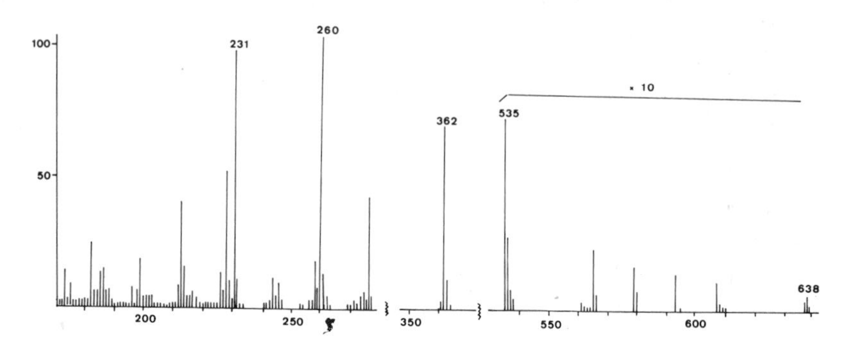

Fig. 2. Mass spectrum of permethylated disaccharide from the cell wall of *Nocardia.*

The peptide moiety gave rise to several fractions: monomers, dimers, oligomers and polymers which were separated by gel filtration on Sephadex.

Monomers were purified by chromatography on Beckman PA 35 resin and gave essentially one compound which had the same behaviour as a diamidated tetrapeptide by t.l.c. and by paper electrophoresis (Vacheron *et al.*, 1972). The peptide sequence determined by chemical methods was confirmed by mass spectrometry of the permethylated and perdeuteriomethylated derivatives (Fig. 3). Fragmentations shown in Fig. 3 demonstrate the peptide sequence. The peak at $m/e = 484$ after elimination of the lateral chain of Dpm shows that C-terminal alanine and glutamic acid are bound to the same asymetric carbon of diaminopimelic acid. Such a structure has also been demonstrated in *Corynebacterium diphtheriae* (Kato, Strominger & Kotani, 1968), *Escherichia coli* and *Bacillus megaterium* (Van Heijenoort *et al.*, 1969; Dézelée & Bricas, 1970). The mass spectrum also shows the presence of two amide groups located on the carboxyls of glutamic acid and diaminopimelic acid not engaged in peptide bonds.

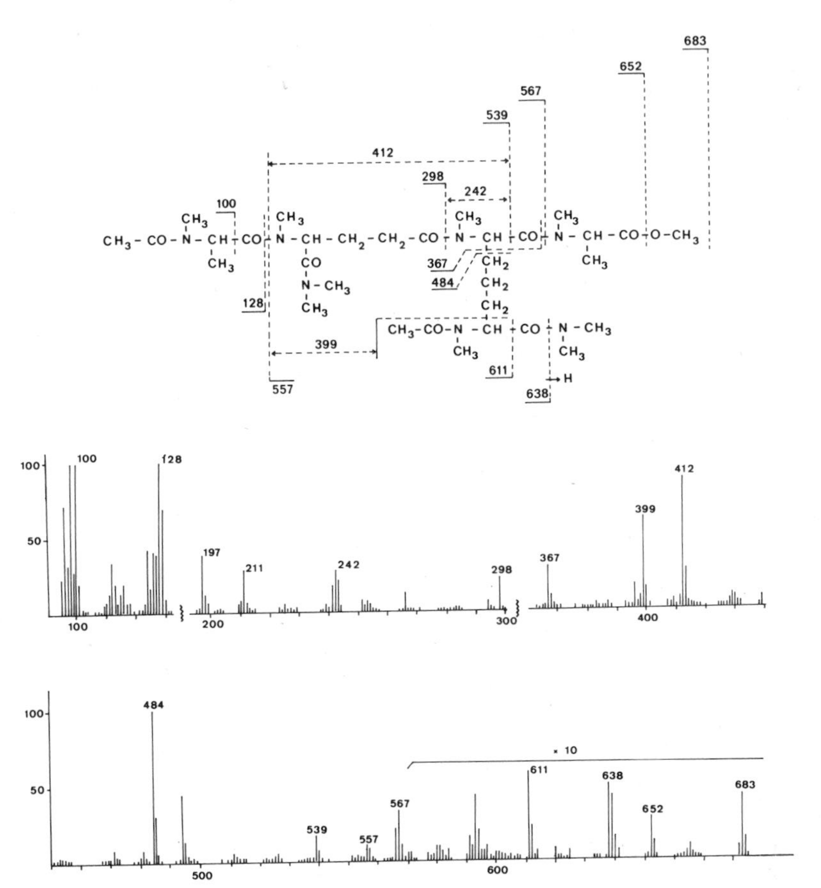

Fig. 3. Mass spectrum of the *N*-acetylated and permethylated tetrapeptide from the cell wall of *Nocardia*.

Thus, the structure shown in formula 1 was proposed for the dimer isolated from the peptidoglycan of *Nocardia kirovani* cell walls (Vacheron *et al.*, 1972).

-O CH2OH O CH2OH O -O CH2OH O CH2OH O
HO NH O O HO NH O O
CH3-CO CH3 O NH-CO-CH2OH CH3-CO CH3 O NH-CO-CH2OH
CH L Ala iso D Gln CH L Ala iso D Gln
CO-NH-CH-CO-NH-CH-CO-NH2 CO-NH-CH-CO-NH-CH-CO-NH2
CH3 CH2 CH3 CH2
CH2 meso Dpm (L) D Ala CH2 meso Dpm (L) D Ala
CO-NH-CH-CO-NH-CH-CO CO-NH-CH-CO-NH-CH-COOH
(CH2)3 CH3 (CH2)3 CH3
H2N-CH-CO-NH2 HN-CH-CO-NH2
(D) (D) (1)

More recently, we studied peptidoglycans from other species of *Nocardia*: *Nocardia asteroides* R 399 and *Nocardia caviae* IM 1381 (Bordet *et al.*, 1972). We found the same structure with the presence of *N*-glycolylmuramic acid and of diamidated tetrapeptide units.

(b) *Lipid constituents*

The cell wall of *Nocardia kirovani* contains 18% of free lipids. Upon saponification, a new bound lipid fraction is obtained, corresponding to 23% of the walls (Guinand, Vacheron & Michel, 1970).

(i) *Bound lipids*

These were mainly nocardic acids (Vacheron *et al.*, 1972). These acids are α-branched, β-hydroxylated long chain fatty acids of the mycolic type. Nocardic acids of *Nocardia asteroides* were first described in 1960 (Michel, Bordet & Lederer, 1960) and their structures determined by mass spectrometry (Bordet *et al.*, 1965). These acids exist as a mixture of mono, di and tri-unsaturated compounds whose structure is shown in formula 2 (Bordet & Michel, 1969).

$$R\text{-}CH{=}CH\text{-}(CH_2)_n\text{-}\underset{\substack{|\\(CH_2)_m\\|\\CH_3}}{CH}OH\text{-}COOH \qquad n = 13, 15 \quad m = 11, 13$$

(a) Monoethylenic

$C_{46}H_{90}O_3$ $C_{48}H_{94}O_3$ $C_{50}H_{98}O_3$ $C_{52}H_{102}O_3$

$R = CH_3\text{-}(CH_2)_n$ $\qquad n = 13$ to 16

(b) Diethylenic $\qquad$ (2)

$C_{50}H_{96}O_3$ $C_{52}H_{100}O_3$ $C_{54}H_{104}O_3$

$R = CH_3\text{-}(CH_2)_{n_1}\text{-}CH{=}CH\text{-}(CH_2)_{n_2}$ $\qquad n_1 = 6$ to 9, $n_2 = 4$ to 10

(c) Triethylenic

$C_{52}H_{98}O_3$ $C_{54}H_{102}O_3$ $C_{56}H_{106}O_3$

$R = CH_3\text{-}(CH_2)_{n_1}\text{-}CH{=}CH\text{-}(CH_2)_{n_2}\text{-}CH{=}CH\text{-}(CH_2)_{n_3}$ $\qquad n_1 = 6$ to 9, $n_2 + n_3 = 6$ to 15

More recently, nocardic acids ranging from C_{44} to C_{60} were isolated from *Nocardia brasiliensis* (Lanéelle & Asselineau, 1970), *Nocardia opaca* (Etémadi, Markovits & Pinte, 1966), *Nocardia asteroides* (Ioneda, Lederer & Rozanis, 1970) and other species of

Nocardia (Maurice, Vacheron & Michel, 1971). Curiously, in a strain of *Nocardia asteroides*, nocardic acids from C_{50} to C_{58} were found only in the walls whereas corynomycolic acids were present in the free lipids (Ioneda, Lederer & Rozanis, 1970). Corynomycolic acids have the same type of structure but a smaller size with 28 to 36 carbon atoms.

An important part of the nocardic acids of the walls is bound to a polysaccharide. A strain of *Nocardia brasiliensis* contains arabinose nocardates and formula 3 was proposed for these compounds (Lanéelle & Asselineau, 1970).

$$CH_3-(CH_2)_a-CH=CH-(CH_2)_b-CH=CH-(CH_2)_c- \quad a+b+c = 32, 34$$

$$CH_3-(CH_2)_a-CH=CH-(CH_2)_b-CH=CH-(CH_2)_c-CH=CH-(CH_2)_d- \quad a+b+c+d = 32, 34, 36$$

$$-CHOH-CH(C_{14}H_{29} \text{ or } C_{16}H_{33})-C(=O)-O-CH_2-\text{(arabinose ring)}$$

(3)

Arabinose mycolates have been isolated from the walls of various strains of mycobacteria (Azuma & Yamamura, 1962, 1963; Acharya, Senn & Lederer, 1967; Bruneteau & Michel, 1968; Kanetsuna, Imaeda & Cunto, 1969; Markovits & Vilkas, 1969). In such compounds, nocardic acids are replaced by mycolic acids of higher molecular weight with 80 to 90 carbon atoms.

(ii) *Free lipids*

Several types of free lipids were found in the walls of *Nocardia kirovani*. Ketones represent about 10% of the total free lipids of the walls. Their structure was determined by mass spectrometry (Vacheron *et al.*, 1972). They are tri- and tetra-unsaturated straight chains of 57 to 63 carbon atoms (formula 4).

$$CH_3-(CH_2)_a-CH=CH-(CH_2)_b-CH=CH-(CH_2)_c-CH=CH-(CH_2)_d-CO-(CH_2)_n-CH$$

$$a+b+c+d = 34, 36, 38 \qquad n = 14, 16$$

$$CH_3-(CH_2)_a-CH=CH-(CH_2)_b-CH=CH-(CH_2)_c-CH=CH-(CH_2)_d-CH=CH-(CH_2)_e-CO-(CH_2)_n-CH_3 \qquad (4)$$

$$a+b+c+d+e = 34, 36, 38 \qquad n = 14, 16$$

Such structures are similar to those of the nocardones and nocardols previously isolated from *Nocardia asteroides* (Bordet & Michel, 1966*a,b*).

Glycerides were also found in the walls of *Nocardia kirovani*. They are C_{14}, C_{16} and C_{18} saturated and unsaturated fatty acids and high molecular weight poly-unsaturated acids from C_{35} to C_{45}. Their structures were studied by mass spectrometry, the most abundant acids have odd carbon atom numbers (Vacheron & Michel, 1971).

It is well known that many strains of Nocardia are pigmented. The walls of *Nocardia kirovani* are salmon pink. The main pigment was isolated and its structure determined (Vacheron, Arpin & Michel, 1970). It is a carotenoid, a phleixanthophylle palmitate (formula 5(**a**)) mixed with a small amount of 4-keto-phleixanthophylle palmitate (formula 5(**b**)).

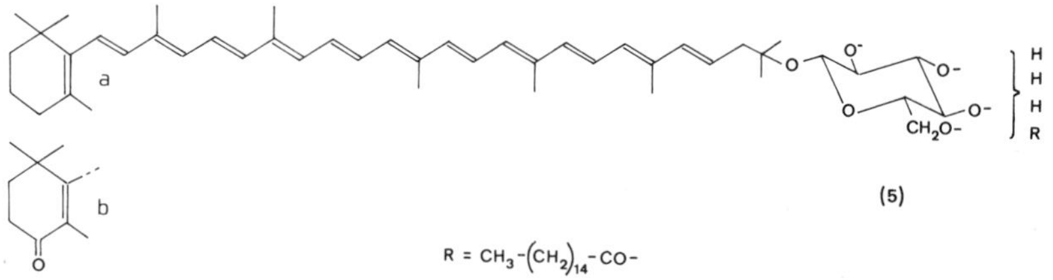

(5)

$R = CH_3-(CH_2)_{14}-CO-$

(c) *Other constituents*

(i) *Polysaccharide*

The presence of sugars in the walls of nocardiae was reported as early as the first studies on cell wall composition (Cummins & Harris, 1958; Romano & Sohler, 1956), and a polysaccharide which contained arabinose and galactose was isolated from *Nocardia asteroides* (Bishop & Blank, 1958). The presence of arabinose and galactose seems to be a common feature for all nocardiae. In addition, some strains of *Nocardia* (*Actinomadura*) contain madurose (Lechevalier, 1968) which is 3-O-methyl-D-galactose (Lechevalier & Gerber, 1970). Glucose has also been isolated from various strains of *Nocardia* (Cummins & Harris, 1958; Lechevalier, 1968; Murray & Proctor, 1965). In a strain of *Nocardia asteroides*, we found glucose, galactose and arabinose in the molar ratios (1 : 3 : 12). (A. Voiland, pers. comm). This polysaccharide is most likely to be related to the arabinogalactan isolated from mycobacteria (Misaki & Yukawa, 1966; Misaki *et al.*, 1970; Azuma, Yamamura & Misaki, 1969; Misaki, Seto & Azuma, 1974; Amar–Nacasch &

Vilkas, 1969, 1970). This polysaccharide consists of 1 → 5 linked D-arabinofuranose and D-galactofuranose units (Vilkas *et al.*, 1971). It has a branched structure and is esterified by mycolic acids (Kanetsuna, Imaeda & Cunto, 1969). The glycolipid is linked to the peptidoglycan either by a glycosidic linkage with glucosamine or, more probably, by a phosphodiester linkage between the glycolipid and muramic acid-6-phosphate contained in the glycan backbone of the peptidoglycan. (Lederer, 1971; Cunto, Kanetsuna & Imaeda, 1969; Kanetsuna & San Blas, 1970; Kanetsuna, 1972; Sohler, Romano & Nickerson, 1958).

The polysaccharide of nocardiae might possess a structure related to that of mycobacteria. However, some differences must exist since we have found glucose as a constituent sugar in addition to galactose and arabinose. At the present time, its structure is being studied.

(ii) *Polypeptide fraction*

The wall of nocardiae contains amino acids which do not belong to the peptidoglycan. This fact was first ascertained by Sohler, Romano & Nickerson (1958) in the walls of *Nocardia rubra* strain 3639 where they found quite important amounts of arginine, aspartic acid, glycine, lysine, serine, threonine, valine, phenylalanine, leucine and isoleucine. In various strains of *Nocardia*, the presence of many amino acids was noticed (Snyder *et al.*, 1967). More recently, the walls of *Nocardia rubra* strain 721 A were studied after treatment with trypsin and pepsin with or without further extraction with alkaline ethanol (Beaman *et al.*, 1971). In the walls exposed only to trypsin and pepsin, significant amounts of glycine, threonine, serine, valine, phenylalanine, isoleucine and leucine were present. After extraction with alkaline ethanol, these amino acids were lost. The examination of walls by electron microscopy also showed that alkaline extraction removed a part of the outer envelope. From these results Beaman *et al.* (1971) concluded the existence of a peptide or a protein associated with a lipid of the outer envelope of *Nocardia rubra.*

When working upon the cell walls of *Nocardia kirovani*, we noticed the presence of such amino acids (Vacheron *et al.*, 1972). A systematic study of various strains of Nocardia: *N. asteroides*, *N. brasiliensis*, *N. farcinica*, *N. erythropolis* and *N. caviae* showed that aspartic acid, threonine, serine, glycine, valine, leucine and phenylalanine were present beside alanine, glutamic acid and diaminopimelic acid (Karahjoli, 1972). The walls of *N. caviae* were studied by Sandoval (1974). After purification by several extractions with sodium dodecyl sulphate, these walls still contained non-peptidoglycanic amino acids. A further treatment with lysozyme gave an insoluble fraction enriched in polypeptides with the following amino acids: aspartic acid, glutamic acid, threonine, serine, glycine, alanine, valine, leucine, phenylalanine, lysine and arginine (Sandoval, 1974). The appearance of such a polypeptide or proteic fraction was recently pointed out in the walls of various bacteria. Thus, *Mycobacterium tuberculosis* var. *bovis* strain BCG contains a polymer of L-glutamic acid (Wietzerbin-Falszpan *et al.*, 1973) and one of the outer envelopes of *Escherichia coli* consists of a lipoprotein linked to the peptidoglycan (Braun & Rehn, 1969; Braun & Sieglin, 1970; Braun & Wolff, 1970; Bosch & Braun, 1973). The structure of this lipoprotein has been determined (Braun & Bosch, 1972).

3. Biological Activity

The biological effects of mycobacterial walls are well known. They stimulate the production of antibodies in animal organisms and increase the resistance to infections (Lederer, 1971). An immunization against tuberculosis with oil-treated walls from mycobacteria has been described (Anacker *et al.*, 1969*a*; Ribi *et al.*, 1971). These walls were inactivated by solvent extraction or alkali treatment (Anacker *et al.*, 1969*b*).

A stimulation of non-specific resistance to staphylococcal infections by wall preparations of *Mycobacterium phlei* and BCG has been observed (Fox *et al.*, 1966; Misaki *et al.*, 1966) and walls of mycobacteria gave analogous results against infections with *Klebsiella pneumoniae* (Lederer, 1971). Furthermore, the walls of *Mycobacterium kansasii* showed an antitumour activity against Ehrlich carcinoma and syngeneic lymphoid leukaemia in mice (Chedid *et al.*, 1973).

Another biological effect common to mycobacteria and nocardiae has been known for a long time: it is their adjuvant action. Freund's adjuvant consists of cells of mycobacteria or nocardiae in a water in oil emulsion. The antigen is contained in the water phase (Freund, 1956). The studies of Lederer upon mycobacteria and nocardiae showed precisely which fraction was responsible for the adjuvant activity. First this activity was located in the walls (Azuma *et al.*, 1971; Adam *et al.*, 1972) and then in a water-soluble fraction obtained after lysozyme treatment of the delipidated cells from *M. smegmatis* and *N. opaca* (Adam *et al.*, 1972; Chedid *et al.*, 1972; Adam *et al.*, 1973). More recently, in various bacteria, it has been shown that peptidoglycan or even peptidoglycanic fractions are responsible for adjuvant action (Nauciel *et al.*, 1973; Adam *et al.*, 1974*a,b*). The smallest active structure found so far is the *N*-acetyl (or *N*-glycolyl) muramyl-L-alanyl-D-isoglutamine (formula 6) Ellouz *et al.*, 1974).

CH_2OH, HO, O, CH_3, O, OH, CH, NH-CO-R

$CO-NH-CH(CH_3)-CO-NH-CH(CH_2-CH_2-COOH)-CO-NH_2$ (6)

R = CH_3 or CH_2OH

A water-soluble extract from the walls of *Nocardia opaca* was found to possess a strong mitogenic activity (Bona *et al.*, 1974; Bona, Damais & Chedid, 1974). This extract induced a stimulation of rabbit and mouse spleen lymphocytes and increased thymidine incorporation by these lymphocytes. This mitogenic activity seems to be specific for *Nocardia* because an analogous water-soluble extract from *M. smegmatis* and the lipopolysaccharide of *Salmonella enteritidis* did not possess any activity. The structure responsible for this mitogenic activity is not yet known.

4. Walls and Taxonomy

The use of wall composition for the classification of actinomycetes is not a new idea. Since the first works of Romano & Sohler (1956) and Cummins & Harris (1958), amino acid analysis of the walls of various actinomycetes has been proposed as a criterion for identification.

The investigations of Becker, Lechevalier & Lechevalier (1965) and Yamaguchi (1965) resulted in the recognition of four types of walls, I to IV. The walls of nocardiae belong to Type IV which is characterized by the presence of *meso*-2,6-diaminopimelic acid, arabinose and galactose whereas the walls of streptomycetes belong to Type I with LL-diaminopimelic acid instead of the *meso* isomer (Lechevalier, 1968; Lechevalier & Lechevalier, 1970; Lechevalier, Lechevalier & Gerber, 1971). The differentiation between *meso* and LL isomers of diaminopimelic acid is very easy by paper chromatography (Becker *et al.*, 1964). Differentiation between oerskoviae and nocardiae is also possible using the composition of walls. The motile nocardiae described by Ørskov (1938) were first called *Nocardia turbata* (Erikson, 1954). A qualitative analysis of the wall components of oerskoviae showed that these strains did not contain the characteristic amino acids of *Nocardia* walls. They contained aspartic acid beside alanine, glutamic acid and lysine (Sukapure *et al.*, 1970). Thus, Prauser, Lechevalier & Lechevalier (1970) proposed to place these bacteria in a new genus called *Oerskovia.*

At the present time however, these criteria do not seem quite sufficient for the classification of all nocardioform bacteria. Thus, some strains of *Streptomyces mediterranei* were transferred to the genus *Nocardia* because their cell walls contained *meso*-diaminopimelic acid (Thiemann, Zucco & Pelizza, 1969) and so they seemed to belong to wall Type IV. More recently, a more precise study of *S. mediterranei* cell walls showed some structural differences with the walls of true nocardiae (Bordet *et al.*, 1972). In fact, the peptide moiety of the peptidoglycan is monoamidated on diaminopimelic acid whereas in nocardiae it is diamidated on diaminopimelic acid and on glutamic acid. Furthermore, in *S. mediterranei*, muramic acid is *N*-acetylated whereas in true nocardiae it is *N*-glycolylated.

Another chemical criterion could be that of lipid composition (see chapter 7). Lechevalier, Horan & Lechevalier (1971) noticed that nocardic acids were present in all the strains of nocardiae that were studied whereas mycobacteria contained mycolic acids. So the presence of nocardic acids seems to be specific of the genus *Nocardia.* Actually, the strains of *S. mediterranei* mentioned above did not contain any nocardic acids or other acids of mycolic type (Pommier & Michel, 1973).

Thus, it seems that wall composition can be a valuable tool for the identification of nocardiae. However, some other structural criteria must be added to the classical criteria: presence of *meso*-diaminopimelic acid, arabinose and galactose. These new criteria are the presence of *N*-glycolylated muramic acid and the presence of two amide groups in the peptide moiety of peptidoglycan on the carboxyls of both glutamic and diaminopimelic acids.

All of these criteria are common to nocardiae and mycobacteria and do not permit the distinction between these two genera. The only difference is the nature of the α-branched β-hydroxylated acids bound to the polysaccharide which are nocardic acids in nocardiae and mycolic acids in mycobacteria. Another difference could be the nature of the polypeptide fraction of the cell walls.

It can be said in conclusion that although the structure of the peptidoglycan is now well-known, the same cannot be said for the other compounds of the walls: polysaccharide, polypeptide etc. which are still under investigation. The interest of these structural studies is increased given the important biological effects of the walls of *Nocardia*.

5. References

Acharya, N. P. V., Senn, M. & Lederer, E. (1967). Sur la présence et la structure de mycolates d'arabinose dans les lipides liés de deux souches de mycobactéries. *C.r. hebd. Séanc. Acad. Sci., Paris* **264,** 2173.

Adam, A., Amar, C., Ciorbaru, R., Lederer, E., Petit, J. F. & Vilkas, E. (1974*a*). Activité adjuvante des peptidoglycanes de mycobactéries. *C.r. hebd. Séanc. Acad. Sci., Paris* **278,** 799.

Adam, A., Ciorbaru, R., Ellouz, F., Petit, J. F. & Lederer, E. (1974*b*). Adjuvant activity of monomeric bacterial cell wall peptidoglycans. *Biochem. biophys. Res. Commun.* **56,** 561.

Adam, A., Ciorbaru, R., Petit, J. F. & Lederer, E. (1972). Isolation and properties of a macromolecular, water-soluble, immuno-adjuvant fraction from the cell wall of *Mycobacterium smegmatis*. *Proc. natn. Acad. Sci. U.S.A.* **69,** 851.

Adam, A., Ciorbaru, R., Petit, J. F., Lederer, E., Chedid, L., Lamensans, A., Parant, F., Parant, M., Rosselet, J. P. & Berger, F. M. (1973). Preparation and biological properties of water-soluble adjuvant fractions from delipidated cells of *Mycobacterium smegmatis* and *Nocardia opaca*. *Infect. Immun.* **7,** 855.

Adam, A., Petit, J. F., Wietzerbin-Falszpan, J., Sinay, P., Thomas, D. W. & Lederer, E. (1969). L'acide *N*-glycolylmuramique, constituant des parois de *Mycobacterium smegmatis*: identification par spectrométrie de masse. *Fed. Eur. Biochem. Soc. Lett.* **4,** 87.

Amar-Nacasch, C. & Vilkas, E. (1969). Etude des parois d'une souche humaine virulente de *Mycobacterium tuberculosis.* (1) Préparation et analyse chimique. *Bull. Soc. Chim. biol.* **51,** 613.

Amar-Nacasch, C. & Vilkas, E. (1970). Etude des parois de *Mycobacterium tuberculosis.* (II) Mise en évidence d'un mycolate d'arabinobiose et d'un glucane dans les parois de *M. tuberculosis* H 37 Ra. *Bull. Soc. Chim. biol.* **52,** 145.

Anacker, R. L., Barclay, W. R., Brehmer, W., Goode, G., List, R. H., Ribi, E. & Tarmina, D. F. (1969*a*). Effectiveness of cell walls of *Mycobacterium bovis* strain BCG administered by various routes and in different adjuvants in protecting mice against airborne pulmonary infection with *Mycobacterium tuberculosis* strain H 37 Rv. *Am. Rev. resp. Dis.* **99,** 242.

Anacker, R. L., Bickel, W. D., Brehmer, W., Niwa, M., Ribi, E. & Tarmina, D. F. (1969*b*). Immunization of mice by combinations of inactive fractions of *Mycobacterium bovis* strain BCG. *Proc. Soc. exp. Biol. Med.* **130,** 723.

Arima, K., Nakamura, T. & Tamura, G. (1968). Chemical structure of the mucopeptide of *Streptomyces roseochromogenes* cell wall. *Agr. Biol. Chem.* **32,** 530.

Azuma, I., Kishimoto, S., Yamamura, Y. & Petit, J. F. (1971). Adjuvanticity of mycobacterial cell walls. *Jap. J. Microbiol.* **15,** 193.

Azuma, I. & Yamamura, Y. (1962). Studies on the firmly bound lipids of human tubercle bacillus. I. Isolation of arabinose mycolate. *J. Biochem., Tokyo* **52,** 200.

Azuma, I. & Yamamura, Y. (1963). Studies on the firmly bound lipids of human tubercle bacilli. II. Isolation of arabinose mycolate and identification of its chemical structure. *J. Biochem., Tokyo* **53,** 275.

Azuma, I., Yamamura, Y. & Misaki, A. (1969). Isolation and characterization of arabinose mycolate from firmly bound lipids of mycobacteria. *J. Bact.* **98,** 331.

Beaman, B. L., Kim, K. S., Salton, M. R. J. & Barksdale, L. (1971). Amino acids of the cell wall of *Nocardia rubra. J. Bact.* **108,** 941.

Becker, B., Lechevalier, M. P., Gordon, R. E. & Lechevalier, H. A. (1964). Rapid differentiation between *Nocardia* and *Streptomyces* by paper chromatography of whole cell hydrolysates. *Appl. Microbiol.* **12,** 421.

Becker, B., Lechevalier, M. P. & Lechevalier, H. A. (1965). Chemical composition of cell wall preparations from strains of various form genera of aerobic actinomycetes. *Appl. Microbiol.* **13,** 236.

Bishop, C. T. & Blank, F. (1958). The chemical composition of the actinomycetales: isolation of a polysaccharide containing D-arabinose and D-galactose from *Nocardia asteroides. Can. J. Microbiol.* **4,** 35.

Bona, C., Damais, C. & Chedid, L. (1974). Blastic transformation of mouse spleen lymphocytes by a water-soluble mitogen extracted from Nocardia. *Proc. natn. Acad. Sci. U.S.A.* **71,** 1602.

Bona, C., Damais, C., Dimitriu, A., Chedid, L., Ciorbaru, R., Adam, A., Petit, J. F., Lederer, E. & Rosselet, J. P. (1974). Mitogenic effect of a water-soluble extract of *Nocardia opaca*: a comparative study with some bacterial adjuvants on spleen and peripheral lymphocytes of four mammalian species. *J. Immun.* **112,** 2028.

Bordet, C., Etémadi, A. H., Michel, G. & Lederer, E. (1965). Structure des acides nocardiques de *Nocardia asteroides. Bull. Soc. chim. Fr.* p. 234.

Bordet, C., Karahjoli, M., Gateau, O. & Michel, G. (1972). Cell walls of nocardiae and related actinomycetes. Identification of the genus *Nocardia* by cell wall analysis. *Int. J. Syst. Bact.* **22,** 251.

Bordet, C. & Michel, G. (1966*a*). Les nocardols, alcools à haut poids moléculaire de *Nocardia asteroides. C.r. hebd. Séanc. Acad. Sci., Paris* **262,** 1294.

Bordet, C. & Michel, G. (1966*b*). Isolement et structure des nocardones, cétones à haut poids moléculaire de *Nocardia asteroides. C.r. hebd. Séanc. Acad. Sci., Paris* **262,** 1810.

Bordet, C. & Michel, G. (1969). Structure et biogenèse des lipides à haut poids moléculaire de *Nocardia asteroides. Bull. Soc. Chim. biol.* **51,** 527.

Bosch, V. & Braun, V. (1973). Distribution of murein-lipoprotein between the cytoplasmic and outer membrane of *Escherichia coli. Fed. Eur. Biochem. Soc. Lett.* **34,** 307.

Braun, V. & Bosch, V. (1972). Sequence of the murein-lipoprotein and the attachment site of the lipid. *Eur. J. Biochem.* **28,** 51.

Braun, V. & Rehn, J. (1969). Chemical characterization, spatial distribution and function of a lipoprotein (murein-lipoprotein) of the *E. coli* cell wall. The specific effect of trypsin on the membrane structure. *Eur. J. Biochem.* **10,** 426.

Braun, V. & Sieglin, U. (1970). The covalent murein-lipoprotein structure of the *Escherichia coli* cell wall. The attachment site of the lipoprotein on the murein. *Eur. J. Biochem.* **13,** 336.

Braun, V. & Wolff, H. (1970). The murein-lipoprotein linkage in the cell wall of *Escherichia coli. Eur. J. Biochem.* **14,** 387.

Bruneteau, M. & Michel, G. (1968). Structure d'un dimycolate d'arabinose isolé de *Mycobacterium marianum. Chem. Phys. Lipids* **2,** 229.

Chedid, L., Lamensans, A., Parant, F., Parant, M., Adam, A., Petit, J. F. & Lederer, E. (1973). Protective effect of delipidated mycobacterial cells and purified cell walls against Ehrlich carcinoma and a syngeneic lymphoid leukemia in mice. *Cancer. Res.* **33,** 2187.

Chedid, L., Parant, M., Parant, F., Gustafson, R. H. & Berger, F. M. (1972). Biological study of a non-toxic, water-soluble immuno-adjuvant from mycobacterial cell walls. *Proc. natn. Acad. Sci. U.S.A.* **69,** 855.

Cummins, C. W. (1962*a*). La composition chimique des parois cellulaires d'actinomycètes et son application taxonomique. *Annls Inst. Pasteur, Paris* **103,** 385.

Cummins, C. S. (1962*b*). Chemical composition and antigenic structure of the cell walls of *Corynebacterium, Mycobacterium, Nocardia, Actinomyces* and *Arthrobacter. J. gen. Microbiol.* **28,** 35.

Cummins, C. S., Glendenning, O. & Harris, H. (1957). Composition of the cell wall of *Lactobacillus bifidus. Nature, Lond.* **180,** 337.

Cummins, C. S. & Harris, H. (1956). The chemical composition of the cell wall in some gram-positive bacteria and its possible value as a taxonomic character. *J. gen. Microbiol.* **14,** 583.

Cummins, C. S. & Harris, H. (1958). Studies on the cell wall composition and taxonomy of actinomycetales and related groups. *J. gen. Microbiol.* **18,** 173.

Cunto, G., Kanetsuna, F. & Imaeda, T. (1969). Chemical analysis of the mucopeptide of *Mycobacterium smegmatis. Biochim. biophys. Acta* **192,** 358.

Dézelée, P. & Bricas, E. (1970). Structure of the peptidoglycan in *Escherichia coli* B and *Bacillus megaterium* KM. Stereospecific synthesis of two *meso*-diaminopimelic acid peptides with the tetrapeptide subunit of bacterial cell wall peptidoglycan. *Biochemistry, N.Y.* **9,** 823.

Ellouz, F., Adam, A, Ciorbaru, R. & Lederer, E. (1974). Minimal structural requirements for adjuvant activity of bacterial peptidoglycan derivatives. *Biochem. biophys. Res. Commun.* **59,** 1317.

Ensign, J. C. & Wolfe, R. S. (1965). Lysis of bacterial cell walls by an enzyme isolated from a myxobacter. *J. Bact.* **90,** 395.

Ensign, J. C. & Wolfe, R. S. (1966). Characterization of a small proteolytic enzyme which lyses bacterial cell walls. *J. Bact.* **91,** 524.

Erikson, D. (1954). Factors promoting cell division in a "soft" mycelial type of Nocardia: *Nocardia turbata. J. gen. Microbiol.* **11,** 198.

Etémadi, A. H., Markovits, J. & Pinte, F. (1966). Sur les différents types d'acides nocardiques isolés de *Nocardia opaca. C.r. hebd. Séanc. Acad. Sci., Paris* **263,** 835.

Fox, A. E., Evans, G. L., Turner, F. J., Schwartz, B. S. & Blaustein, A. (1966). Stimulation of non-specific resistance to infection by a crude cell wall preparation from *Mycobacterium phlei. J. Bact.* **92,** 1.

Freund, J. (1956). The mode of action of immunologic adjuvants. *Adv. Tuberc. Res.* **7,** 130.

Ghuysen, J. M. (1968). Use of bacteriolytic enzymes in determination of wall structure and their role in cell metabolism. *Bact. Rev.* **32,** 425.

Ghuysen, J. M., Dierickx, L., Coyette, J., Leyh-Bouille, M., Guinand, M. & Campbell, J. N. (1969). An improved technique for the preparation of Streptomyces peptidases and *N*-acetyl-muramyl-L-alanine amidase active on bacterial wall peptidoglycans. *Biochemistry, N.Y.* **8,** 213.

Ghuysen, J. M., Strominger, J. L. & Tipper, D. J. (1968). Bacterial cell walls. In *Comprehensive Biochemistry,* vol. 26A. Eds M. Florkin & E. H. Stotz, p. 53. New York, American Elsevier Publishing Co.

Guinand, M., Vacheron, M. J. & Michel, G. (1970). Structure des parois cellulaires des Nocardia. I. Isolement et composition des parois de *Nocardia kirovani. Fed. Eur. Biochem. Soc. Lett.* **6,** 37.

Hoare, D. S. & Work, E. (1957). The stereoisomers of α-ε-diaminopimelic acid. 2. Their distribution in the bacterial order Actinomycetales and in certain Eubacteriales. *Biochem. J.* **65,** 441.

Ioneda, T., Lederer, E. & Rozanis, J. (1970). Sur la structure des diesters de tréhalose ("cord factors") produits par *Nocardia asteroides* et *Nocardia rhodochrous. Chem. Phys. Lipids* **4,** 375.

Kanetsuna, F. (1972). Biochemical study of mycobacterial cell wall. *Acta cient. venez.* **23,** 62.

Kanetsuna, K., Imaeda, T. & Cunto, G. (1969). On the linkage between mycolic acid and arabinogalactan in phenol-treated mycobacterial cell walls. *Biochim. biophys. Acta* **173,** 341.

Kanetsuna, F. & San Blas, G. (1970). Chemical analysis of a mycolic acid-arabinogalactan-mucopeptide complex of mycobacterial cell wall. *Biochim. biophys. Acta* **208,** 434.

Karahjoli, M. (1972). Etude de la paroi des Nocardia: peptidoglycane et fraction polypeptidique de *Nocardia caviae.* Doctorat de l'Université, Lyon, France.

Kato, K., Strominger, J. L. & Kotani, S. (1968). Structure of the cell wall of *Corynebacterium diphtheriae.* I. Mechanism of hydrolysis by the L-3 enzyme and the structure of the peptide. *Biochemistry, N.Y.* **7,** 2762.

Lanéelle, M. A. & Asselineau, J. (1970). Caractérisation de glycolipides dans une souche de *Nocardia brasiliensis. Fed. Eur. Biochem. Soc. Lett.* **7,** 64.

Lechevalier, M. P. (1968). Identification of aerobic actinomycetes of clinical importance. *J. Lab. clin. Med.* **71,** 934.

Lechevalier, M. P. & Gerber, N. N. (1970). The identity of madurose with 3-O-methyl-D-galactose. *Carbohyd. Res.* **13,** 451.

Lechevalier, M. P., Horan, A. C. & Lechevalier, H. A. (1971). Lipid composition in the classification of nocardiae and mycobacteria. *J. Bact.* **105,** 313.

Lechevalier, M. P. & Lechevalier, H. A. (1970). In *The Actinomycetales.* Ed. H. Prauser. p. 311. Jena: Gustav Fischer.

Lechevalier, H. A., Lechevalier, M. P. & Gerber, N. N. (1971). Chemical composition as a criterion in the classification of actinomycetes. In *Advances in Applied Microbiology.* **14,** 47. Ed. D. Perlman. New York: Academic Press Inc.

Lederer, E. (1971). The mycobacterial cell wall. *Pure appl. Chem.* **25,** 135.

Leyh-Bouille, M., Bonaly, R., Ghuysen, J. M., Tinelli, R. & Tipper, D. J. (1970). L,L-diaminopimelic acid containing peptidoglycans in walls of *Streptomyces spec.* and *Clostridium perfringens* (type A). *Biochemistry, N.Y.* **9,** 2944.

Leyh-Bouille, M., Ghuysen, J. M., Tipper, D. J. & Strominger, J. L. (1966). Structure of the cell wall of *Micrococcus lysodeikticus.* I. Study of the structure of the glycan. *Biochemistry, N.Y.* **5,** 3079.

Markovits, J. & Vilkas, E. (1969). Etude des cires D d'une souche humaine virulente de *Mycobacterium tuberculosis. Biochim. biophys. Acta* **192,** 49.

Maurice, M. T., Vacheron, M. J. & Michel, G. (1971). Isolement d'acides nocardiques de plusieurs espèces de Nocardia. *Chem. Phys. Lipids* **7,** 9.

Michel, G., Bordet, C. & Lederer, E. (1960). Isolement d'un nouvel acide mycolique: l'acide nocardique, à partir d'une souche de *Nocardia asteroides. C.r. hebd. Séanc. Acad. Sci., Paris* **250,** 3518.

Misaki, A., Ikawa, N., Kato, T. & Kotani, S. (1970). Cell wall arabinogalactan of *Mycobacterium phlei. Biochim. biophys. Acta* **215,** 405.

Misaki, A., Seto, N. & Azuma, I. (1974). Structure and immunological properties of D-arabino-D-galactans isolated from cell walls of *Mycobacterium* species. *J. Biochem., Tokyo* **76,** 15.

Misaki, A. & Yukawa, S. (1966). Studies on cell walls of mycobacteria. II. Constitution of polysaccharides from BCG cell walls. *J. Biochem., Tokyo* **59,** 511.

Misaki, A., Yukawa, S., Tschuiya, K. & Yamasaki, T. (1966). Studies on the cell walls of mycobacteria. Chemical and biological properties of the cell walls and mucopeptide of BCG. *J. Biochem., Tokyo* **59,** 388.

Murray, I. G. & Proctor, A. G. I. (1965). Paper chromatography as an aid to the identification of *Nocardia* species. *J. gen. Microbiol.* **41,** 163.

Nakamura, T., Tamura, G. & Arima, K. (1967). Structure of the cell walls of *Streptomyces*. *J. Ferment. Technol., Osaka* **45,** 869.

Nauciel, C., Fleck, J., Mock, M. & Martin, J. P. (1973). Activité adjuvante de fractions monomériques de peptidoglycanes bactériens dans l'hypersensibilité de type retardé. *C.r. hebd. Séanc. Acad. Sci., Paris* **277,** 2841.

Ørskov, J. (1938). Untersuchungen über Strahlenpilze, reingezüchtet aus dänischen Erdproben. *Zenbl. Bakt. ParasitKde Abt.* I **98,** 344.

Petit, J. F., Adam, A., Wietzerbin-Falszpan, J., Lederer, E. & Ghuysen, J. M. (1969). Chemical structure of the cell wall of *Mycobacterium smegmatis*. I. Isolation and partial characterization of the peptidoglycan. *Biochem. biophys. Res. Commun.* **35,** 478.

Pommier, M. T. & Michel, G. (1973). Phospholipid and acid composition of *Nocardia* and nocardoid bacteria as criteria of classification. *Biochem. Syst.* **1,** 3.

Prauser, H., Lechevalier, M. P. & Lechevalier, H. A. (1970). Description of *Oerskovia* gen. n. to harbor Ørskov's motile *Nocardia*. *Appl. Microbiol.* **19,** 534.

Ribi, E., Anacker, R. L., Barclay, W. R., Brehmer, W., Harris, S. C., Leif, W. R. & Simmons, J. (1971). Efficacy of mycobacterial cell walls as a vaccine against airborne tuberculosis in the rhesus monkey. *J. infect. Dis.* **123,** 527.

Rogers, H. J. & Perkins, H. R. (1968). *Cell Walls and Membranes.* Spon's Biochemical Monographs. London: E. and F. N. Spon Ltd.

Romano, A. H. & Sohler, A. (1956). Biochemistry of the actinomycetales. II. A comparison of the cell wall composition of species of the genera *Streptomyces* and *Nocardia*. *J. Bact.* **72,** 865.

Salton, M. R. J. (1953). Studies on the bacterial cell wall. IV. The composition of the cell walls of some gram-positive and gram-negative bacteria. *Biochim. biophys. Acta* **10,** 512.

Salton, M. R. J. (1964). *The Bacterial Cell Wall.* New York: American Elsevier Publishing Co.

Sandoval, H. (1974). Etude d'une fraction polypeptidique isolée des parois de *Nocardia caviae*. Doctorat de l'Université, Lyon, France.

Schleifer, K. H. & Kandler, O. (1972). Peptidoglycan types of bacterial cell walls and their taxonomic implications. *Bact. Rev.* **36,** 407.

Sharon, N., Osawa, T., Flowers, H. M. & Jeanloz, R. W. (1966). Isolation and study of the chemical structure of a disaccharide from *Micrococcus lysodeikticus* cell walls. *J. biol. Chem.* **241,** 223.

Snyder, M. L., Slawson, M. S., Bullock, W. & Parker, R. B. (1967). Studies on the oral filamentous bacteria. I. Cell wall composition of *Actinomyces, Nocardia, Bacterionema* and *Leptotrichia*. *J. infect. Dis.* **117,** 332.

Sohler, A., Romano, A. H. & Nickerson, W. J. (1958). Biochemistry of the *Actinomycetales*. III. Cell wall composition and the action of lysozyme upon cells and cell walls of the actinomycetales. *J. Bact.* **75,** 283.

Sukapure, R. S., Lechevalier, M. P., Reber, H., Higgins, M. L., Lechevalier, H. A. & Prauser, H. (1970). Motile nocardoid Actinomycetales. *Appl. Microbiol.* **19,** 527.

Szaniszlo, P. J. & Gooder, H. (1967). Cell wall composition in relation to the taxonomy of some *Actinoplanaceae*. *J. Bact.* **94,** 2037.

Thiemann, J. E., Zucco, G. & Pelizza, G. (1969). A proposal for the transfer of *Streptomyces mediterranei* Margalith and Beretta 1960 to the genus *Nocardia* as *Nocardia mediterranea* (Margalith and Beretta) comb. nov. *Arch. Mikrobiol.* **67,** 147.

Vacheron, M. J., Arpin, N. & Michel, G. (1970). Isolement d'esters de phleixanthophylle de *Nocardia kirovani. C.r. hebd. Séanc. Acad. Sci., Paris* **271,** 881.

Vacheron, M. J., Guinand, M., Michel, G. & Ghuysen, J. M. (1972). Structural investigations on cell walls of *Nocardia* sp. The wall lipid and peptidoglycan moieties of *Nocardia kirovani. Eur. J. Biochem.* **29,** 156.

Vacheron, M. J. & Michel, G. (1971). Présence d'acides de poids moléculaire élevé dans les triglycérides isolés des parois cellulaires de *Nocardia kirovani. C.r. hebd. Séanc. Acad. Sci., Paris* **273,** 1778.

Van Heijenoort, J., Elbaz, L., Dézelée, P., Petit, J. F., Bricas, E. & Ghuysen, J. M. (1969). Structure of the *meso*-diaminopimelic acid containing peptidoglycans in *Escherichia coli* B and *Bacillus megaterium* KM. *Biochemistry, N.Y.* **8,** 207.

Vilkas, E., Markovits, J., Amar-Nacasch, C. & Lederer, E. (1971). Sur la présence d'unités de D-galactofurannose dans l'arabinogalactane des parois et des cires D de souches humaines de *Mycobacterium tuberculosis. C.r. hebd. Séanc. Acad. Sci., Paris* **273,** 845.

Wietzerbin-Falszpan, J., Das, B. C., Azuma, I., Adam, A., Petit, J. F. & Lederer, E. (1970). Isolation and mass spectrometric identification of the peptide subunits of mycobacterial cell walls. *Biochem. biophys. Res. Commun.* **40,** 57.

Wietzerbin-Falszpan, J., Das, B. C., Gros, C., Petit, J. F. & Lederer, E. (1973). The amino acids of the cell wall of *Mycobacterium tuberculosis* var. bovis, strain BCG. Presence of a poly (L-glutamic acid). *Eur. J. Biochem.* **32,** 525.

Yamaguchi, T. (1965). Comparison of the cell wall composition of morphologically distinct actinomycetes. *J. Bact.* **89,** 444.

7. Lipid Composition in the Classification and Identification of Nocardiae and Related Taxa

D. E. MINNIKIN AND M. GOODFELLOW

Departments of Organic Chemistry and Microbiology, University of Newcastle-upon-Tyne, Newcastle-upon-Tyne NE1 7RU, England

Contents

1. Introduction

Systematists have always experienced difficulties in separating the genera *Nocardia* and *Mycobacterium* both from one another and from related actinomycete and coryneform taxa using conventional morphological, staining and biochemical properties (see chapter 2). Despite the recent improvements that have been made in the classification of these taxa (Cross & Goodfellow, 1973; Jones, 1975), there is still a need for good characters for both classification and identification. Chemotaxonomic characters are being increasingly used by taxonomists and have been partly responsible for improvements in the classification of actinomycetes and coryneform bacteria (Lechevalier, Lechevalier & Gerber, 1971; see chapter 1). Chemical tests used in chemotaxonomy should be selected, as other tests, that is for their ease of performance, exactness of results and stability to changes in growth conditions; they should not require the use of chemicals or equipment beyond the means of diagnostic laboratories.

Chemotaxonomic methods involving analyses of deoxyribonucleic acid base ratios (Hill, 1966) and whole organism and wall sugar and amino-acid composition (Cummins, 1962; Cummins & Harris, 1956; Boone & Pine, 1968; Schleifer & Kandler, 1972) are now well established. The use of lipids as chemical markers has not received detailed consideration although the results, to date, are encouraging (Asselineau, 1966; Goldfine, 1972).

Analyses of the wall amino-acid and sugar content of actinomycetes and some coryneform bacteria have shown that these organisms can be placed into several large groups on the basis of the limited distribution of certain major components of the peptidoglycan layer (Bordet *et al.*, 1972; Lechevalier & Lechevalier, 1970; see chapter 1). On the basis of such data many strains formerly assigned to the genus *Nocardia* have been transferred to the new genera *Actinomadura, Oerskovia* and *Rothia* (Cross & Goodfellow, 1973; see chapter 1). One of the major problems remaining in the taxonomy of the nocardioform bacteria is the separation of the genera *Nocardia, Mycobacterium* and *Corynebacterium* from one another, and their relationship with strains classified in the genus *Gordona* (Tsukamura, 1971), and the "*rhodochrous*" complex. The strains in these taxa all have walls of Type IV (see chapters 1 and 2). There is some evidence from studies of deoxyribonucleic acid base ratios which seems to provide strong evidence for maintaining the separation of *Corynebacterium* from *Nocardia* and *Mycobacterium* (Bouisset, Breuilland & Michel, 1963; Bousfield, 1972) though the results of Yamada & Komagata (1970) cast some doubt on this view.

The lipids of actinomycetes have great inherent potential as chemotaxonomic characters since they contain components which have rather unusual and unique structures. In particular some of the long-chain components of the lipids of actinomycetes are remarkable since they contain exceptionally high numbers of carbon atoms in the apolar alkyl chains. The polar lipids of some actinomycetes also include lipids, such as the phosphatidylinositol mannosides which are not found in any other genera. The mycobacteria, in particular, contain characteristic lipids such as the mycosides which have complicated structures containing unusual sugars and amino acids. This complexity of the lipids of actinomycetes is well-illustrated by reference to the monograph by Asselineau (1966).

There are many ways in which a study of lipids may be approached with a view to their use as chemotaxonomic characters

Table 1
Phospholipid structures

Structure	
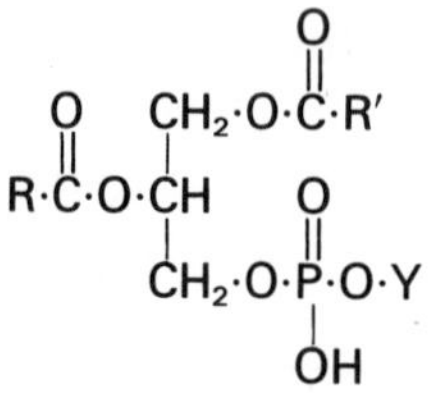 $R\cdot CO\cdot O\cdot CH(CH_2\cdot O\cdot CO\cdot R')\cdot CH_2\cdot O\cdot P(=O)(OH)\cdot O\cdot Y$	R,R′ = alkyl groups
Y =	
$-CH_2\cdot CH\cdot (NH_2)\cdot COOH$	Phosphatidylserine
$-CH_2\cdot CH_2\cdot NH_2$	Phosphatidylethanolamine
$-CH_2\cdot CH_2\cdot N^+\cdot (CH_3)_3$	Phosphatidylcholine
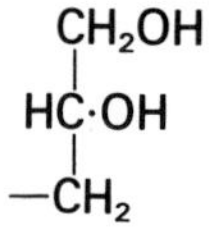 $-CH_2\cdot HC(OH)\cdot CH_2OH$	Phosphatidylglycerol
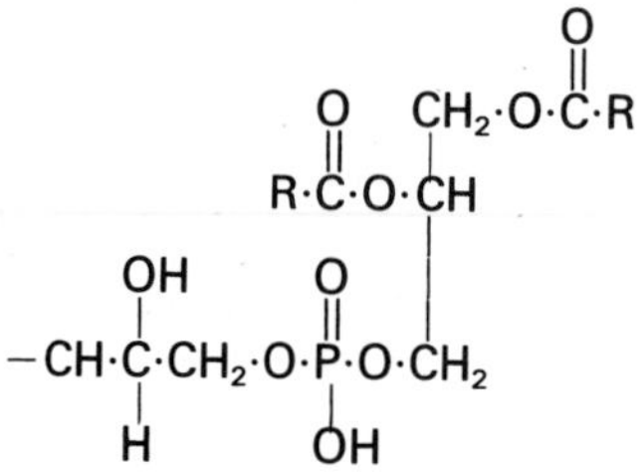 $-CH\cdot C(OH)(H)\cdot CH_2\cdot O\cdot P(=O)(OH)\cdot O\cdot CH_2\cdot CH(O\cdot CO\cdot R)\cdot CH_2\cdot O\cdot CO\cdot R$	Diphosphatidylglycerol (Cardiolipin)
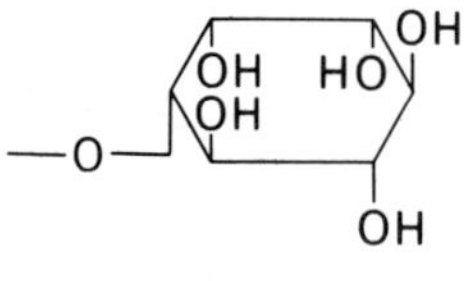 $-O-$ inositol ring bearing five OH groups	Phosphatidylinositol
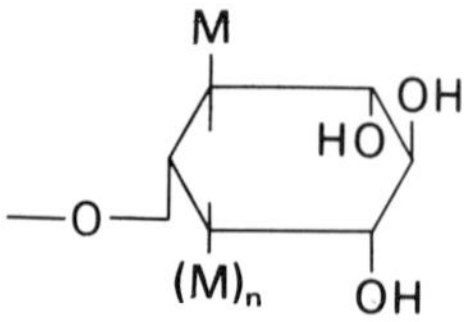 $-O-$ inositol ring bearing M, $(M)_n$, HO, OH, OH	Phosphatidylinositol mannosides (M = mannopyranosyl)

and in order to understand these it is necessary to consider the different categories into which they may be placed. Lipids by definition, are soluble in organic solvents and are either terpenoid (for example steroids, carotenoid pigments, respiratory quinones) or they contain long aliphatic chains of carbon atoms. Exhaustive extraction of cells with suitable organic solvents yields "free" lipids which are not covalently bound in the cell envelope structure. Alternatively, "bound" lipids are chemically linked to polymers in the envelope structure and for their release require the use of chemical agents capable of cleaving such linkages.

Free lipids may be broadly divided into polar and non-polar subgroups which are best considered separately. Polar lipids are amphipathetic, which means that each molecule contains a balance of polar and apolar portions. The most common polar lipids are phospholipids; the main structural types occurring in nocardioform bacteria are shown in Table 1. The other polar lipid types encountered in actinomycetes are glycolipids and ornithine or lysine amides. These lipids are not as well-characterised as the phospholipids and a detailed discussion of their structure would be superfluous at present. These polar lipids are essentially similar in character to those found in other bacteria but mycobacteria, nocardiae and related organisms also contain some rather complex lipids which are more "semi-polar" in nature. The best examples of the latter are the so-called "cord-factors" which are dimycolyl trehalose esters (Asselineau, 1966). Non-polar lipids are those in which the majority of the molecule is of an apolar nature. Acylglycerols, especially triacyglycerols, are good examples of non-polar lipids; free fatty acids, respiratory quinones and carotenoid pigments are also included in this category. The lipophilic components released from bound lipids are usually long-chain fatty acids. The different classes of lipids and individual lipid components all have a possible role as chemotaxonomic characters.

2. The General Nature of the Lipids of Nocardiae and Related Bacteria

In the wake of the intensive investigations of the lipids of mycobacteria, initiated by R. J. Anderson in the United States and continued by N. Polgar in England, and J. Asselineau & E. Lederer in France (see Asselineau, 1966), interest turned to the lipids found in strains of the genus *Nocardia*. These investigations were stimulated by the isolation of a novel peptidolipid from a strain of *Nocardia asteroides* (ATCC9969) by Guinand, Michel & Lederer (1958). Subsequent studies mainly under the direction of G. Michel have established the general nature of the lipids of *Nocardia*. The peptidolipid mentioned above was investigated in detail (Guinand, Michel & Lederer, 1964; Barber *et al.*, 1965*b*; Guinand & Michel, 1966) and the structure (I) proposed.

(I)

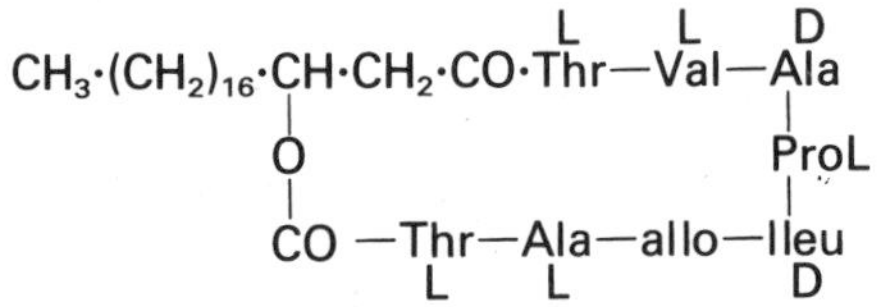

```
                             L     L    D
CH3·(CH2)16·CH·CH2·CO·Thr—Val—Ala
            |                   |
            O                   ProL
            |                   |
            CO —Thr—Ala—allo—Ileu
                  L     L       D
```

The isolation of a characteristic long-chain fatty acid from *N. asteroides* ATCC9969 and its characterisation as a 3-hydroxy acid having a long chain in the 2-position was particularly significant (Michel, Bordet & Lederer, 1960). Fatty acids of this general nature had previously been isolated from the lipids of a variety of mycobacteria and corynebacteria and had been named mycolic acids (see Asselineau, 1966). In addition to the 3-hydroxy acid unit, the mycolic acids of mycobacteria contained components having other oxygen functions, for example, keto, carboxy or methoxyl functions (Asselineau, 1966). The mycolic acids from nocardiae were found to be relatively simple in nature, being composed of homologous series with varying number of double bonds. Extensive studies involving mass spectrometry led to the structures (II(a), (b), (c)) for the main components of the mycolic acids (nocar-

(II)

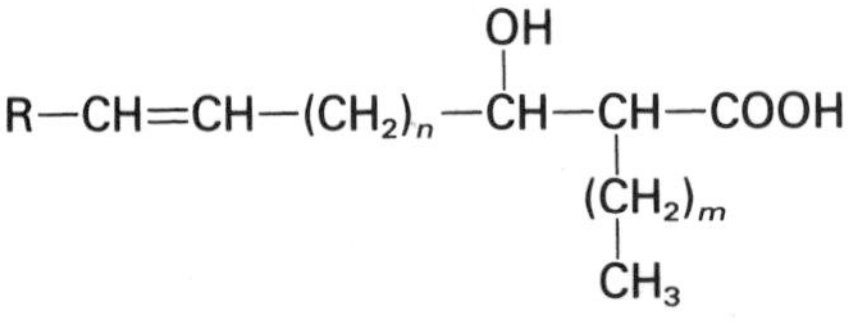

```
                        OH
                        |
R—CH=CH—(CH2)n—CH—CH—COOH          n = 13, 15
                            |
                          (CH2)m                m = 11, 13
                            |
                           CH3
```

II(a) $R = CH_3—(CH_2)_{n_1}$ $n_1 = 13$ to 16

II(b) $R = CH_3—(CH_2)_{n_1}—CH{=}CH—(CH_2)_{n_2}$ $n_1 = 6$ to 9, $n_2 = 4$ to 10

II(c) $R = CH_3—(CH_2)_{n_1}—CH{=}CH—(CH_2)_{n_1}—CH{=}CH—(CH_2)_{n_3}$
$n_1 = 6$ to 9
$n_2 + n_3 = 6$ to 15

dic acids) of *N. asteroides* ATCC9969 Bordet *et al.*, 1965; Bordet & Michel, 1969). In addition to the mycolic acids, related series of

$$\text{(III)}\quad R{-}CH{=}CH{-}(CH_2)_n\overset{\overset{\displaystyle OH}{|}}{C}H(CH_2)_m{-}CH_3 \qquad n = 11, 13, 15, \quad m = 12, 14$$

$$\text{III(a)}\quad R = CH_3{-}(CH_2)_{n_1} \qquad n_1 = 13 \text{ to } 25$$

$$\text{III(b)}\quad R = CH_3{-}(CH_2)_{n_1}{-}CH{=}CH{-}(CH_2)_{n_2} \qquad n_1 = 7 \text{ to } 16,\ n_2 = 7 \text{ to } 14$$

$$\text{III(c)}\quad R^1 = CH_3{-}(CH_2)_{n_1}{-}CH{=}CH{-}(CH_2)_{n_2}{-}CH{=}CH{-}(CH_2)_{n_3} \qquad n_1 = 9 \text{ to } 16,\ n_2 \text{ or } n_3 = 6$$

long-chain alcohols (nocardols) (III(a), (b), (c)) and ketones (nocardones) (IV(a), (b), (c)) have been characterized as components of *N. asteroides* ATCC9969 (Michel & Lederer, 1962; Bordet & Michel,

$$\text{(IV)}\quad R{-}CH{=}CH{-}(CH_2)_n{-}CO{-}(CH_2)_m{-}CH_3 \qquad n = 11, 13, 15, m = 12, 14$$

$$\text{IV(a)}\quad CH_3{-}(CH_2)_{n_1} \qquad n_1 = 13 \text{ to } 23$$

$$\text{IV(b)}\quad CH_3{-}(CH_2)_{n_1}{-}CH{=}CH{-}(CH_2)_{n_2} \qquad n_1 = 9 \text{ to } 14,\ n_2 = 7 \text{ to } 14$$

$$\text{IV(c)}\quad CH_3{-}(CH_2)_{n_1}{-}CH{=}CH{-}(CH_2)_{n_2}{-}CH{=}CH{-}(CH_2)_{n_3} \qquad n = 9 \text{ to } 14,\ n_2 = n_3 = 7 \text{ to } 11$$

1966a,b; 1969). Nocardones were also isolated from the lipids of *Nocardia brasiliensis* 705 (Lanéelle, Asselineau & Castelnuovo, 1965*a*; Lanéelle, 1966). A saturated long-chain alcohol, 16-hentriacontanol (V) was characterized as a component of the lipids of *N. brasiliensis* ATCC733 (Bordet & Michel, 1964) and *N. brasiliensis* 337 (Lanéelle, Asselineau & Castelnuovo, 1965) and the corresponding ketone (palmitone, VI) was isolated from *N. asteroides* ATCC9969 (Bordet & Michel, 1966*b*, 1969). Mycolic

$$\text{(V)}\quad CH_3{-}(CH_2)_{14}{-}\overset{\overset{\displaystyle OH}{|}}{C}H{-}(CH_2)_{14}{-}CH_3$$

$$\text{(VI)}\quad CH_3{-}(CH_2)_{14}{-}CO{-}(CH_2)_{14}{-}CH_3$$

acids of the same general type as those from *N. asteroides* ATCC9969 have been characterized from the lipids of a variety of nocardiae (Lanéelle, Asselineau & Castelnuovo, 1965; Etémadi, Markovits & Pinte, 1966; Maurice, Vacheron & Michel, 1971); these and other studies which are concerned with the use of these lipids as chemotaxonomic characters will be considered in detail later.

The great majority of bacteria contain fatty acids between 12 and 20 carbon atoms in their long chains. These fatty acids are usually derived from the apolar portion of the polar lipids and those from actinomycetes fall into the general structural types shown below:

$$CH_3{-}(CH_2)_n{-}COOH \qquad \text{e.g. palmitic } (n = 14)$$

Unsaturated

$$CH_3{-}(CH_2)_n{-}\underset{cis}{CH{=}CH}{-}(CH_2)_m{-}COOH \qquad \text{e.g. oleic } (n = m = 7)$$

Methyl branched

$$CH_3{-}(CH_2)_n{-}\overset{\displaystyle CH_3\atop |}{CH}{-}(CH_2)_m{-}COOH \qquad \begin{array}{l}\text{e.g. tuberculostearic } n = 7, m = 8\\ \text{iso-acids } n = 0\\ \text{anteiso-acids } n = 1\end{array}$$

A remarkable series of homologous fatty acids were recovered from the triacylglycerols of the walls of *Nocardia kirovani* (Vacheron & Michel, 1971). These acids contained 3 or 4 double bonds and ranged in size from 35 to 45 carbon atoms; the main components, C_{37}, C_{39}, C_{41} and C_{43}, contained odd numbers of carbons. Simple hydroxy acids have been detected as major components in *Nocardia leishmanii* (Yano, Furukawa & Kusunose, 1970*b*). These were found to be 2-hydroxy 14- and 16-carbon straight-chain acids and 2-hydroxy iso- and anteisopentadecanoic acids. A 3-hydroxy acid was present in *N. asteroides* ATCC9969 as a component of the peptidolipid (I) (Guinand & Michel, 1966).

The polar lipids of nocardioform bacteria have been the subject of several studies (Lanéelle, Asselineau & Castelnuovo, 1965; Yano, Furukawa & Kusunose, 1968, 1969*a*, 1970*b*, 1971*b*; Khuller & Brennan, 1972*a*; Pommier & Michel, 1972, 1973). Diphosphatidylglycerol, phosphatidylethanolamine, phosphatidylinositol and phosphatidylinositol mannosides (see Table 1) were found in most of the nocardioform bacteria studied; these lipids are also common in mycobacteria (Asselineau, 1966). Phosphatidylglycerol occurs in an organism named *N. leishmanii* (Yano, Furukawa & Kusunose, 1970*b*) and phosphatidylcholine has been detected in strains of *Nocardia coeliaca* (Yano, Furukawa & Kusunose, 1969*a*; Khuller & Brennan, 1972*a*). These latter workers also presented evidence for the presence in a strain of *Nocardia polychromogenes* of a phosphatidylglucosyl diacylglycerol similar in properties to that isolated from *Pseudomonas diminuta* (Wilkinson & Bell, 1971). The distribution of these phospholipids and other polar lipids will be considered in detail later.

The distribution of polar glycolipids in nocardioform bacteria has not been extensively investigated but Khuller & Brennan (1972*a*) detected the presence of a diacylglucose in a strain labelled *N. coeliaca* ATCC13181 and a triacylglucose in *Nocardia polychromogenes* Jensen. The latter also contained a glycolipid whose alkaline deacylation product co-chromatographed with a diglucosylglycerol presumably derived from the diglucosyl diacylglycerol of staphylococci or bacilli (Shaw, 1970). The presence of a very unusual acidic glycolipid in *Nocardia caviae* IM1381 was reported by Pommier & Michel (1972); details of the exact structure are as yet unpublished but the constituents were unidentified acidic polyol, glucose and fatty acids.

The best-studied glycolipids encountered in the actinomycetes are long-chain esters of trehalose. The first example of these lipids, isolated from *Mycobacterium tuberculosis* by Bloch (1950), was the so-called "cord-factor" which was shown to be a 6,6′-dimycoloyl ester of trehalose (VII) (Noll *et al.*, 1956). Similar lipids were

R = mycolate

(VII)

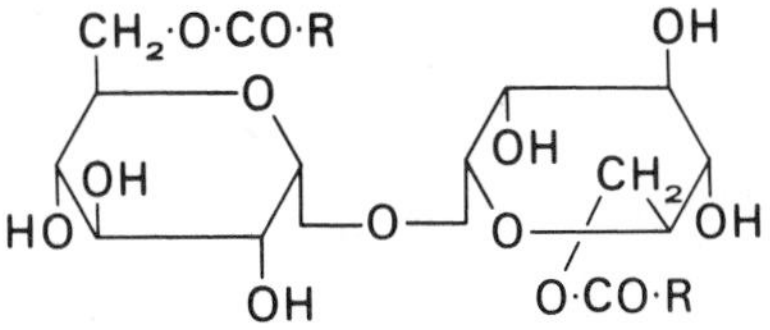

$CH_2 \cdot O \cdot CO \cdot R$ OH O OH OH CH_2 HO O O OH OH $O \cdot CO \cdot R$

isolated from a strain of *N. asteroides* and another labelled *Nocardia rhodochrous* (Ioneda, Lederer & Rozanis, 1970). Both lipids were diesters of trehalose but while that from the *rhodochrous* strain contained mycolic acids (C_{38} to C_{46}) similar to those expected in whole organisms (see later), that from *N. asteroides* has mycolic acids of much lower molecular weight (C_{32} to C_{36}) than those isolated from the walls (C_{50}–C_{58}). Diesters of trehalose, which co-chromatographed with cord factor from *M. tuberculosis*, were found to be abundant in the lipids of several glucose-grown *Nocardia* and *rhodochrous* strains (Yano, Furukawa & Kusunose,

1971*b*). Organisms using glycerol as carbon source contained only small amounts of this lipid but under both conditions of growth a lipid suspected to be a monoester of trehalose was present.

3. The Distribution of Lipid Types in Nocardiae and Related Bacteria

It is convenient to consider separately, in three different sections, the distribution of lipids in bacteria. The first section concerns the long-chain lipophilic constituents alone and the other two involve considerations of the free and bound lipids separately.

(a) *Long-chain constituents*

The great majority of lipid types encountered in bacteria contain hydrophobic portions which are composed of long chains of a hydrocarbon nature. Long-chain fatty acids having usually from 12 to 20 carbons are almost universally encountered in bacteria. Certain actinomycetes, however, contain mycolic acids, other characteristic fatty acids and long-chain alcohols or ketones. It is possible to investigate the total mixture of long-chain constituents derived from whole organisms or from individual or classes of lipid components.

(i) *Simple fatty acids*

Analyses of the simple long-chain fatty acids have become almost routine following the introduction of gas-liquid chromatography (g.l.c.), the acids being usually analysed after conversion to their methyl esters. The distribution of fatty acids in nocardiae and related bacteria is summarised in Tables 2, 3 and 4. Two broad

Table 2
Simple fatty acids of nocardiae

Taxon	Straight chain	Unsaturated	Tuberculo-stearic	Iso- and anteiso	
Nocardia asteroides ATCC9969, 443-2 and 535	++	++	++	+	Bordet & Michel, 1963
N. brasiliensis ATCC733 and 426	++	+	++	+	Bordet & Michel, 1963
N. rugosa	++	+		+	Bordet & Michel, 1963
N. lurida	++	++		++	Bordet & Michel, 1963

Table 2 continued

Taxon	Straight chain	Unsaturated	Tuberculo-stearic	Iso- and anteiso	
N. brasiliensis 705	++	++	+		Lanéelle, Asselineau & Castelnuovo, 1965
"*Mycobacterium*" *rhodochrous* 8139 and 8154	++	++	+		Lanéelle, Asselineau & Castelnuovo, 1965
"Pellegrino strain"	++	++	+		
N. brasiliensis 705	++	++	+		Ballio & Barcellona, 1968
N. pelletieri NCTC9999	++	++	+	+	Ballio & Barcellona, 1968
N. pellegrino ATCC15989	++	++	++		Ballio & Barcellona, 1968
N. rubropertincta ClCM5077	++	++	+		Ballio & Barcellona, 1968
Nocardia sp. ClCM7060	++	++	++		Ballio & Barcellona, 1968
N. polychromogenes	++	++	+		Yano, Furukawa & Kusunose, 1968
N. coeliaca	++		Traces	+++	Yano, Furukawa & Kusunose, 1969
*N. leishmanii**	++		Traces	+++	Yano, Furukawa & Kusunose, 1969*b*, 1970*b*
N. kirovani	++	++†			Guinand, Vacheron & Michel, 1970
N. farcinica ATCC13781	++‡	++	+		Chamoiseau & Asselineau, 1970
"*N. farcinica*"	++‡	+	++		Asselineau, Lanéelle & Chamoiseau, 1969
N. asteroides 124, 131, 161, 176, 180 and 185	++	++	++	Traces	Farshtchi & McClung, 1970
N. farcinica NCTC4524, *N. opaca*, *N. corallina*	++	++			Bowie *et al.*, 1972
N. asteroides ATCC9969 and R399	++	++	++	Traces	Pommier & Michel, 1973
N. caviae IM1381	++	++	++	Traces	Pommier & Michel, 1973
N. farcinica IM1377	++	++	+	Traces	Pommier & Michel, 1973
N. erythropolis A_4	++	++	+	Traces	Pommier & Michel, 1973
"*M.*" *rhodochrous* NCTC7510	++	++	?§	?§	Whiteside, DeSiervo & Salton, 1971

* This organism also contained substantial proportions of 2-hydroxyiso- and anteiso-acids (C_{14}, C_{15} and C_{16}).
† Triacylglycerols from this organism contained tri- and tetraunsaturated acids having 35 to 45 carbon atoms (Vacheron & Michel, 1971).
‡ Contain fatty acids up to 24 carbons.
§ The exact identity of the branched-chain esters present was not established.

Table 3

Simple fatty acids of actinomycetes excluding nocardiae

Taxon	Straight-chain	Unsaturated	Tuberculo-stearic	Iso- and anteiso	Hydroxy	
Mycobacterium spp.	++*	++	++	+†		Asselineau, 1966; Okuyama, Kankura & Nojima, 1967; Brennan & Ballou, 1967; Brennan, Rooney & Winder, 1970; Thoen, Karlson & Ellefson, 1971*a*,*b*, 1972
Micromonospora (5 strains)‡	+	Traces	Traces	++		Ballio, Barcellona & Boniforti, 1965; Ballio & Barcellona, 1968
Streptomyces (13 strains)‡	+	Traces	Traces	++		
Microbispora (5 strains)	+	Traces	+	++		
Thermomonospora (2 strains)	+	Traces		++		
Thermoactinomyces (1 strain)	+	Traces	Traces	++		
Thermopolyspora (2 strains)	+	Traces	Traces	++		
Actinoplanaceae (9 strains)‡	++	?	Traces	++		
Streptomyces erythrus	+	+	Traces	++		Hofheinz & Grisebach, 1965
S. halstedii	+	+	Traces	++		
Waksmania rosea					Traces	Yano, Furukawa & Kusunose, 1969*b*
Streptosporangium roseum		Not investigated			+	
Arthrobacter simplex					++	
Streptomyces sioyaensis		Not investigated			+	Kawanami & Otsuka, 1969
S. toyocaensis		Not investigated			+	Kawanami, 1971
S. aureofaciens	+	+		++		Běhal, Prochazkova & Vanek, 1968
Streptomyces (4 strains)				++	++	Lanéelle, Asselineau & Castelnuova, 1968
"*Nocardia*" *gardneri*				++	++	
Micromonospora	+		+	++		Tabaud, Tisnovska & Vilkas, 1971
Streptomyces mediterranei	+			++	++	Pommier & Michel, 1973
Oerskovia turbata	++	+		++		

* C_{12} to C_{26}.

† Positively characterized from *Mycobacterium phlei* (Campbell & Naworal, 1969).

Table 4
Simple fatty acids of coryneform bacteria

Taxon	Straight-chain	Unsaturated	Iso- and anteiso	
Corynebacterium diphtheriae, C. ovis	++	++		Asselineau, 1961
C. parvum			++	Etémadi, 1963
C. diphtheriae	++	++	+	Brennan & Lehane, 1971
C. equi, C. ovis	+		++	
C. acnes			++	Voss, 1970
Anaerobic corynebacteria (6 species, 9 strains) *Propionibacterium* (7 species, 40 strains)	+	Traces	++	Moss *et al.*, 1967, 1969; Moss & Cherry, 1968
Propionibacterium arabinosum *P. freudenreichii, P. shermanii*	+		++	Shaw & Dinglinger, 1969
P. freudenreichii	+	Traces	++	Lanéelle & Asselineau, 1968
C. fascians *Arthrobacter tumescens* *Brevibacterium ammoniagenes*	++	++		Bowie *et al.*, 1972
C. michiganense *C. sepedonicum* *C. insidiosum* *B. linens, B. sulphureum* *Arthrobacter* spp. (15 strains)	+		++	
A. globiformis	+		++	Walker & Fagerson, 1965
Arthrobacter spp. (3 strains)	+		++	Shaw & Stead, 1971
A. crystallopoietes	+		++	Kostiw, Boylen & Dyson, 1972
A. simplex	+	+	++	Fujii & Fukui, 1969
*C. simplex**	+	+	+	Yanagawa *et al.*, 1972
C. coelicolor	+		++	Whiteside, De Siervo & Salton, 1971
C. poinsettiae	+	+	++	Tadayon & Carroll, 1971
C. pseudodiphthericitum, C. equi	+	+	+	
Erysipelothrix rhusiopathiae	++	++		
Listeria monocytogenes	+	+	++	Raines *et al.*, 1968; Tadayon & Carroll, 1971
Microbacterium thermosphactum	+		++	Shaw & Stead, 1970
M. ammoniaphilum	++	++		Shibukawa, Kurima & Ohuchi, 1970
Brevibacterium lactofermentum	++	++		Takinami *et al.*, 1968
B. thiogenitalis	++	++		Okazaki, Kaszaki & Fukuda, 1968
B. flavum	++	++		Otsuka & Shiio, 1968

* This strain also contained tuberculostearic acid.

patterns of fatty acids are observed, those containing high proportions of straight-chain and unsaturated acids and those being composed mainly of iso- and anteiso-fatty acids. Tuberculostearic acid is found in high proportions in strains of mycobacteria and nocardiae and low proportions of a cyclopropane ring-containing fatty acid have been detected in certain actinomycetes (Table 3).

(ii) *Mycolic acids*

Analyses of mycolic acid composition have made important contributions to the taxonomy of nocardioform and related bacteria and have been particularly useful in distinguishing between nocardiae and mycobacteria. Several procedures, involving quite different methodology, have been proposed and these will be described separately. Many types of mycolic acids have relatively high molecular weights and their methyl esters cannot readily be analysed by g.l.c. However, homogeneous samples of the esters of mycolic acids may be obtained by thin-layer chromatography (t.l.c.) and information concerning their structures may be found by application of a selection of modern chemical techniques. In particular gas chromatography and mass spectrometric methods have been found to be very useful because of the particular properties of mycolic esters.

Analysis of mycolic esters by pyrolytic procedures. A very characteristic property of mycolic acids and esters is their facile cleavage on pyrolysis into fatty acid, or its ester, and a long-chain aldehyde (Stodola, Lesuk & Anderson, 1938; Asselineau & Lederer, 1950) according to the following scheme (Etémadi, 1964):

$$R{-}CH(OH){-}CH(R'){-}C({=}O){-}OX \xrightarrow{\text{pyrolysis}} R{-}CH{=}O + R'{\cdot}CH{=}C(OH){-}OX \rightarrow R'{\cdot}CH_2{-}C({=}O){-}OX$$

mycolic acid or ester → meroaldehyde + fatty acid or ester

$X = H$ or CH_3

The long-chain aldehydes produced by pyrolysis of mycolates are conveniently referred to as meroaldehydes (Morgan & Polgar,

1957; Etémadi, Okuda & Lederer, 1964). Pyrolysis of mycolates can be carried out by heating *in vacuo* at approximately 300°C and the products isolated and studied. It was discovered, however, that methyl esters of mycolic acids could be pyrolysed in the inlet of a gas chromatograph and the composition of the fatty acid ester released studied in the same experiment (Etémadi, 1964, 1967*a*). Mycolic esters of relatively low molecular weight (30–40 carbons) produce meroaldehydes which also may be analysed by gas chromatography (Etémadi, 1964, 1967*a*).

Studies on the nature of the long-chain acids or esters released by pyrolysis of mycolic acids or esters has been found to be of great value in clarifying the relationship between nocardiae and mycobacteria. It was noted in initial investigations (Michel, Bordet & Lederer, 1960) that the mycolic acids from *N. asteroides* ATCC9969 released, on pyrolysis, fatty acids having 14 and 16 carbon atoms. A strain of "*Mycobacterium*" *rhodochrous* 8139 produced C_{16} and C_{18} fatty acids on pyrolysis of its mycolic acids and *N. brasiliensis* 705 gave C_{14}, C_{16} and C_{18} acids from its mycolic acid (Lanéelle, Asselineau & Castelnuovo, 1965). It was proposed that the *rhodochrous* strain should be classified in the genus *Nocardia* and not in the genus *Mycobacterium.*

The generality of the pyrolysis gas chromatographic procedure was developed by Etémadi (1964, 1967*a*). Analysis of a sample of mycolic acid methyl ester from *Corynebacterium diphtheriae* gave peaks in the gas chromatogram corresponding to methyl palmitate (C_{16}) and palmitic aldehyde (C_{16}); a sample of the mycolic esters from *N. asteroides* ATCC9969 gave peaks attributable to C_{12}, C_{14} C_{16} esters and meroaldehydes (32–38 carbons). Pyrolysis of mycolic ester from a strain of *Mycobacterium smegmatis*, however, gave a peak corresponding to a C_{24} ester. It was, therefore, possible to distinguish a strain of *Nocardia* from one of *Mycobacterium* by this pyrolysis gas chromatography technique.

An important application of pyrolysis gas chromatography concerned the identification of bacteria responsible for cases of bovine farcy in zebu cattle (Chamoiseau, 1969). This actinomycete was found to produce mycolic acids whose esters on pyrolysis gas chromatography gave a single peak corresponding to a C_{24} ester, thus strongly suggesting that it was a strain of *Mycobacterium* (Asselineau, Lanéelle & Chamoiseau, 1969). This discovery was of particular interest since the causative agent of bovine farcy was previously considered to be *Nocardia farcinica*, the type species of the genus *Nocardia* (Nocard, 1888; Trevisian, 1889). Analysis of the mycolates of a strain of so-called *N. farcinica* (ATCC13781) by pyrolysis gas chromatography showed that a 24 carbon ester was the major component released (Chamoiseau & Asselineau, 1970); this strain was also, as a consequence, considered to be a representative of *Mycobacterium*. Recently, Chamoiseau (1973) has proposed the name *Mycobacterium farcinogenes* for the strains isolated from bovine farcy.

Extensive studies involving pyrolysis gas chromatography were performed on mycolic esters from a wide selection of strains bearing the labels *Mycobacterium*, *Nocardia* and *Corynebacterium* (Lechevalier, Horan & Lechevalier, 1971; Lechevalier, Lechevalier & Horan, 1973). The strains studied could be divided into two major groups depending on the chain lengths of esters released on pyrolysis of the methyl esters of their mycolic esters. All strains received as mycobacteria, with the exception of *Mycobacterium brevicale*, *M. rhodochrous*, *M. thamnopheos* and so-called Tarshis variants of *M. tuberculosis* yielded C_{22} to C_{26} fatty acid esters. *Rhodochrous* strains, those listed above, and *Nocardia sensu stricto* contained mycolic acids whose esters on pyrolysis gave C_{12} to C_{18} esters. A strain of *Corynebacterium pseudotuberculosis* contained mycolates which on pyrolysis released C_{16} esters but a strain of *Corynebacterium pyogenes* contained no mycolic acids (Lechevalier, Lechevalier & Horan, 1973). Mycolic esters isolated from authentic strains of corynebacteria could be distinguished from those of nocardiae and *rhodochrous* strains by the presence in the pyrolytic chromatograms of peaks corresponding to meroaldehydes as noted previously by Etémadi (1967*a*). Eleven strains received as *Nocardia farcinica* were found to contain mycolic acids whose esters on pyrolysis gave C_{22} and C_{24} esters thus suggesting that these strains should be assigned to the genus *Mycobacterium*.

Pyrolysis gas chromatography of the methyl esters of the mycolates of a *rhodochrous* strain (*Nocardia corallina*) released C_{12}, C_{14} and C_{16} esters and five peaks with longer retention times presumably corresponding to aldehydes were detected (Batt, Hodges & Robertson, 1971). Similar studies on a strain of *Nocardia erythropolis* gave a series of C_{10}, C_{12}, C_{14} and C_{16} esters and long-chain aldehydes ranging from C_{20} to C_{28} (Yano *et al.*, 1972). Esters of comparable chain lengths were released on pyrolysis gas chromatography of the mycolates of *Nocardia caviae* NCTC1934 (C_{14} and C_{16}) and a *rhodochrous* strain (*Nocardia calcarea* NC1B8863) (C_{12}, C_{14} and C_{16}) (Goodfellow *et al.*, 1973). Pyrolysis gas chromatography of mycolates from a strain of "*Mycobacterium*" *rhodochrous* ATCC271 gave C_{12}, C_{14} and C_{16} esters (Azuma *et al.*, 1974).

Strains of the recently described species *Nocardia amarae*, isolated from sewage-treatment plants, contained mycolic acids whose esters on pyrolysis gas chromatography released monounsaturated esters having 16 and 18 carbon atoms (Lechevalier & Lechevalier, 1974). These workers also stated that similar mycolic esters occurred in *Nocardia vaccinii* 1MRU3500 and *Nocardia carnea* 3419 though no detailed evidence was presented. Mycolic acids having an unsaturated chain in the 2-position have also been detected by other methods in *Brevibacterium thiogenitalis* (Okazaki *et al.*, 1969), *Corynebacterium hofmanii* (Welby-Gieusse, Lanéelle & Asselineau, 1970), *Nocardia erythropolis* (Yano *et al.*, 1972) and from the free lipids of a strain of *N. asteroides* (Ioneda, Lederer & Rozanis, 1970). Lechevalier & Lechevalier (1974) were unable to confirm the latter result.

The occurrence of free mycolic acids (lipid LCN-A) in ethanol-diethyl ether extracts of nocardiae and related bacteria. Analysis of ethanol-diethyl ether (1 : 1, v/v) extracts of certain *Nocardia* and *Streptomyces* strains by t.l.c. revealed the presence in the former organisms of characteristic lipids (Mordarska, 1968). One of the lipid spots was found to be present in all well-characterized strains of *Nocardia* (Mordarska & Mordarski, 1969). Preliminary studies on the nature of the lipid characteristic of nocardiae (LCN-A) showed that it was distinct from simple fatty acids and glycerides (Mordarska & Réthy, 1970). Lipid LCN-A in addition to being present in all strains of the genus *Nocardia sensu stricto* was also observed in extracts of representatives of the *rhodochrous* complex though in certain cases the characteristic spot had a slightly lower R_F value than that of the reference lipid from *N. asteroides* (Mordarska, Mordarski & Goodfellow, 1972; Goodfellow, 1973; Goodfellow *et al.*, 1974). Strains classified in the genera *Actinomadura*,

Mycobacterium, Oerskovia, Rothia, Streptomyces and the coryneform taxon *Arthrobacter* did not contain lipid LCN-A. A relatively low R_F for the LCN-A in extracts of strains labelled *Nocardia pellegrino* was also noted (Mordarska, Mordarski & Pietkiewicz, 1973), suggesting that these organisms have similarities with strains of the *rhodochrous* complex. Lipid LCN-A, in these studies was positively identified by its characteristic immobility on developing chromatograms with methanol after reversible detection of separated components with iodine vapour. Lipid LCN-A has also been found in a few strains of animal corynebacteria but not in most of the plant pathogenic strains examined (Mordarska & Mordarski, 1970; Goodfellow, 1973). An example of the thin-layer chromatographic detection of lipid LCN-A is shown in Fig. 1 which is taken from the study of Mordarska, Mordarski & Goodfellow (1972).

The essential nature of lipid LCN-A was established by Goodfellow *et al.* (1973) who showed that the characteristic lipids were in fact free mycolic acids. The presence of free mycolic acids in ethanol-diethyl ether extracts of nocardioform bacteria and in some corynebacteria was explained by their solubility in this solvent; the mycolic acids of mycobacteria being generally of higher molecular weight are relatively insoluble in ethanol-diethyl ether mixtures (Kanetsuna & Bartoli, 1972). It has been previously noted by Lanéelle, Asselineau & Castelnuovo (1965) that ethanol-diethyl ether extracts of strains of *N. asteroides, N. brasiliensis* and *rhodochrous* strains yielded mycolic acids on saponification; a subsequent extraction with chloroform is normally necessary to remove mycolic acid-containing lipids from mycobacteria (Asselineau, 1966). Extracts from strains of the genera *Actinobifida* and *Micropolyspora* did not yield chromatograms containing spots attributable to lipid LCN-A (Mordarska, Guzeva & Agre, 1973).

A probable explanation for the difference in mobility on thin-layer chromatograms of free mycolic acids (LCN-A's) was obtained by analysis of the molecular weights of the components of the methyl esters of the free mycolic acids of *N. caviae* NCTC1934 and *N. calcarea* NC1B8863 by mass spectroscopy (Minnikin, Patel & Goodfellow, 1974). It was found that the free mycolic acids of the *N. caviae* strain contained from 48 to 56 carbon atoms but those from the culture of a typical rhodochrous strain, *N. calcarea*, had only 34 to 46 carbons. A lower proportion of the apolar part of the

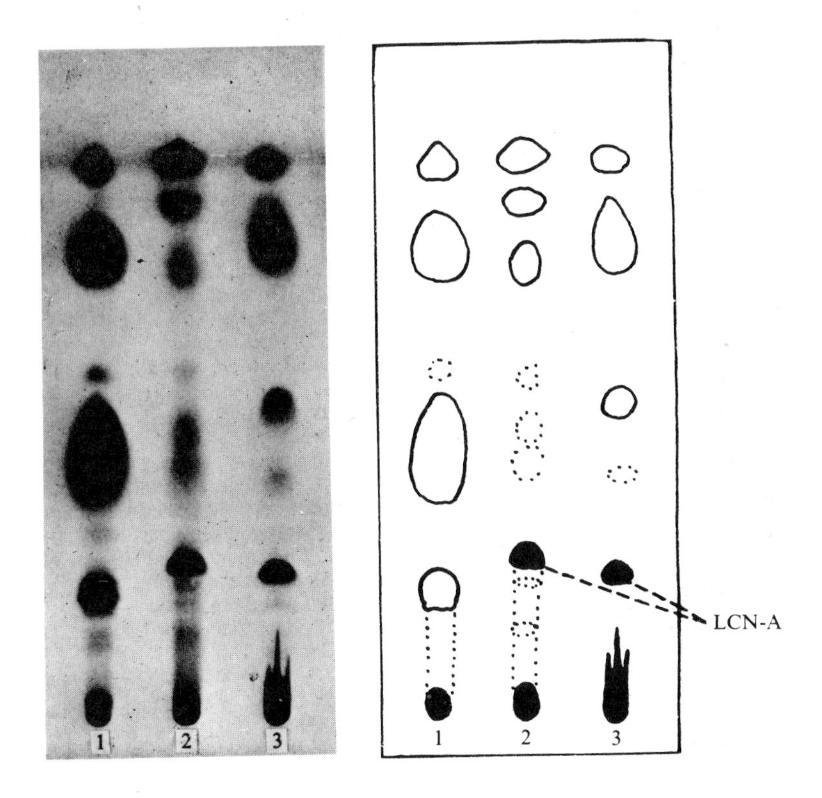

Fig. 1. Thin-layer chromatogram of ethanol-diethyl ether (1 : 1) extracts of 1 *Streptomyces griseus* 22; 2 *Nocardia asteroides* USA; 3 *N. calcarea* IMET 7018. Solvent system: petroleum ether—diethyl ether—glacial acetic acid (85 : 15 : 1). (Reproduced, with permission, from the *Journal of General Microbiology* (Mordarska, Mordarski & Goodfellow, 1972).)

molecule of a mycolic acid would be expected to increase the affinity of the acid for the silica gel layer and result in an R_F value, low relative to that of an acid with a higher proportion of the apolar chains. This molecular weight difference was therefore consistent with the LCN-A's of certain *rhodochrous* strains having lower R_F's than LCN-A's from strains of *Nocardia sensu stricto* (Mordarska, Mordarski & Goodfellow, 1972; Mordarska, Mordarski & Pietkiewicz, 1973; Goodfellow *et al.*, 1973). This general relationship between molecular size of mycolic acids and their behaviour on thin-layer chromatograms is supported by mass spectral analyses of a representative selection of *Nocardia* and *rhodochrous* strains (Alshamaony, Goodfellow & Minnikin, 1976; Alshamaony *et al.*,

1976). These results will be discussed in detail in a subsequent section on the mass spectrometry of mycolates.

Differentiation between Mycobacterium *and* Nocardia *by mycolic acid solubility.* The characteristic insolubilities of the mycolic acids of mycobacteria in ethanol-diethyl ether mixtures was used as the basis of a convenient procedure for distinguishing mycobacteria from nocardiae (Kanetsuna & Bartoli; 1972). Bacteria were saponified and the extracted lipids dissolved in a small quantity of diethyl ether and the solution treated with an equal volume of ethanol. Strains of mycobacteria gave large amounts of white precipitates (>5 mg/g wet bacteria) having a melting point between 45° and 70°C whereas nocardiae and *rhodochrous* strains gave little or no precipitate (<2 mg/g wet bacteria) which when present did not melt below 150°C or melted near 80°C. Since strains classified in the *rhodochrous* complex gave no precipitate they were classified in the genus *Nocardia.* This procedure distinguishes clearly between strains of mycobacteria and all other bacteria but it does not allow distinctions to be made between other actinomycete taxa.

Mycolic acid composition by t.l.c. analysis of chemical degradation products of whole organisms. Whole organisms, or fragments thereof, may be degraded chemically to yield mixtures of long-chain components which include mycolic acids or their derivatives or degradation products. Mycolic acids were originally discovered in the mycobacteria as components of an unsaponifiable fraction (Stodola, Lesuk & Anderson, 1938) and alkaline hydrolyses have continued to be a popular means of obtaining preparations of these acids (Asselineau, 1966).

It had been thought for some time (Asselineau, 1966) that mycobacteria produced various types of mycolic acids which might be distinguished by the presence or absence of oxygen functions in addition to the 3-hydroxyacid unit. The exact complexity of the mycolic acid composition of mycobacteria, however, was only revealed by the advent of t.l.c. The mycolic acids of *Mycobacterium paratuberculosis* were shown by Lanéelle (1963) to have three main components, a simple mycolate, another having a keto function and a dicarboxylic acid. Practically all subsequent studies on the mycolic acid composition of mycobacteria and related organisms have involved the use of t.l.c. of the methyl esters of mycolic acids.

The mycolic acids of nocardioform bacteria and corynebacteria have all been found to contain only a single component as judged by t.l.c. of their methyl esters (Lanéelle, Asselineau & Castelnuovo, 1965; Maurice, Vacheron & Michel, 1971; Minnikin, Patel & Goodfellow, 1974). It was apparent, therefore, that simple t.l.c. analysis of the methyl esters of the total mycolic acids of whole organisms might result in patterns having taxonomic value. Exploitation of this possibility has resulted in the development of a convenient taxonomic technique involving t.l.c. of acid methanolysates of a wide range of actinomycetes (Minnikin, Alshamaony & Goodfellow, 1975). Dry whole organisms were treated with a mixture of methanol, toluene and sulphuric acid, hexane added in order to extract long-chain components and a portion of this extract examined by t.l.c. A typical chromatogram resulting from such a study is shown in Fig. 2. It can be seen that the extracts of corynebacteria, nocardiae and *rhodochrous* strains gave a relatively simple pattern of spots but those from the representatives of *Mycobacterium* are more complex. The components with R_F values of 0·1–0·5 usually correspond to methyl esters of mycolic acids whereas those with higher R_F values (0·8–1·0) are attributable to the methyl esters of non-hydroxylated fatty acids and other long-chain components as yet uncharacterized. Slight differences in chromatographic mobility of the mycolates from corynebacteria, *rhodochrous* strains, nocardiae and gordonae are due to differences in molecular weights of these esters (Alshamaony, Goodfellow & Minnikin, 1976; Alshamaony *et al.*, 1976), as explained for the differences in t.l.c. mobility of free mycolic acids. The strain of gordonae studied also contained a characteristic component, not a mycolic ester, of relatively low R_F (*c.* 0·2). The immediate value of this methanolysis technique is that it may be possible to distinguish strains of nocardioform bacteria, with a simple mycolate pattern, from mycobacteria having a complex pattern. All of these organisms can, of course, be clearly separated from bacteria containing no mycolic acids; genera falling into this group were found to include *Actinomadura, Rothia, Saccharopolyspora* and *Streptomyces.*

An extension of the simple t.l.c. technique, involving the use of adsorbent impregnated with silver nitrate, allows separation of

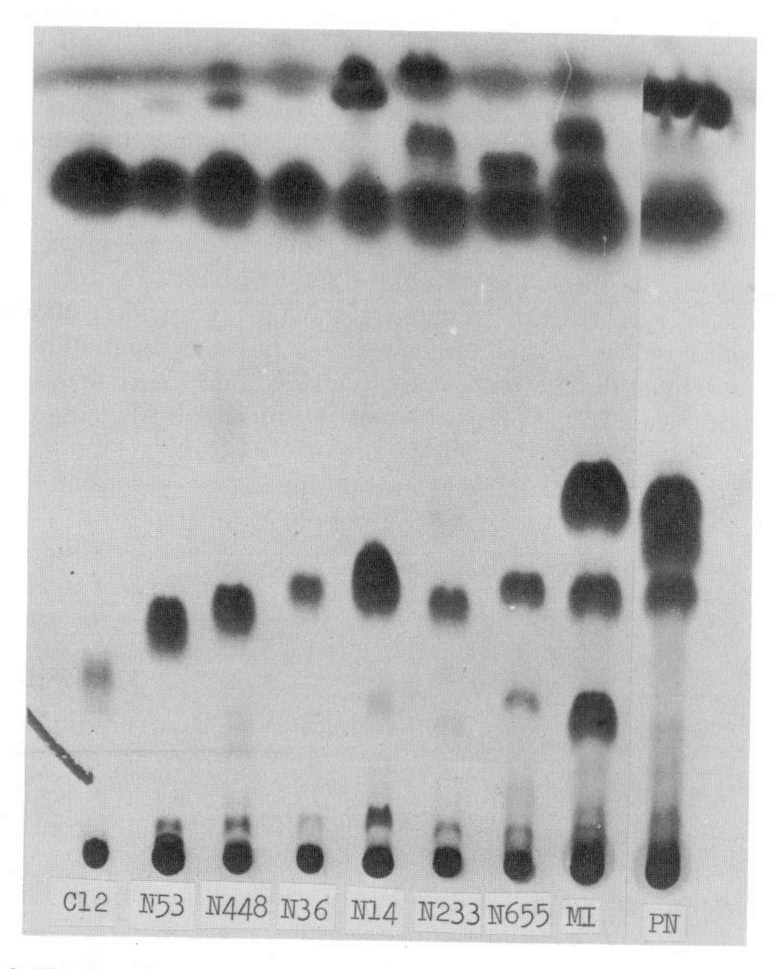

Fig. 2. Thin-layer chromatography of whole-organism methanolysates of selected bacteria. Abbreviations: C12, *Corynebacterium bovis*; N53 and N448, *rhodochrous* strains; N36, N14 and N233, *Nocardia caviae. N. brasiliensis* and *N. asteroides*, respectively; N655, *Gordona rubra*; MI, *Mycobacterium intracellulare*; PN, *Mycobacterium tuberculosis.* Solvent system: petroleum ether—diethyl ether (85 : 15). (Reproduced, with permission, from the *Journal of General Microbiology* (Minnikin, Alshamaony & Goodfellow, 1975).)

long-chain esters into zones according to the number of double bonds they contain (Morris, 1964). Application of this technique to the mycolic esters of a strain labelled *Nocardia opaca* revealed the presence of three components containing from 0–2 double bonds (Etémadi, Markovits & Pinte, 1966). The methyl esters of the mycolic acids from *Nocardia asteroides* ATCC9969 gave three bands on argentation chromatography corresponding to components containing 1–3 double bonds (Bordet & Michel, 1969). It was found (Minnikin, Patel & Goodfellow, 1974) that the mycolic esters from *Nocardia caviae* NCTC1934 and *Nocardia calcarea* NC1B8863 (*rhodochrous*) each separated into three bands on argentation chromatography but the mycolates from the former strain had from 1–3 double bonds and the latter 0–2 double bonds.

Analysis of mycolic acid methyl esters by mass and other spectroscopic techniques. Mass spectrometry is a particularly important technique in the investigation of the structure of mycolic esters since these compounds are generally too large for analysis by gas chromatography. Methyl mycolates from nocardioform bacteria produce a characteristic fragmentation pattern on mass spectrometry (Etémadi, 1967*b*; Bordet & Michel, 1969; Maurice, Vacheron & Michel, 1971; Minnikin, Patel & Goodfellow, 1974) which is summarized in the following scheme:

$$\underset{\text{mycolate}}{R_2\!-\!\underset{\,}{\overset{OH}{\overset{|}{CH}}}\!-\!\underset{R_1}{\underset{|}{CH}}\!-\!COOCH_3} \;(\text{cleavage } a^1\!\leftrightarrow\! a \text{ between C-3 and C-2};\ b \text{ between } R_2 \text{ and C-3}) \xrightarrow{-H_2O} \underset{\text{anhydromycolate}}{R_2\!-\!CH\!=\!\underset{R_1}{\underset{|}{C}}\!-\!COOCH_3}$$

Cleavage at *a* and *a*[1] involves net transfer of the hydroxyl hydrogen to carbon 2 to produce a meroaldehyde and a straight-chain ester. The fragments due to cleavage *b* complement those at *a* and allow the mass of the side chain in the 2-position (R_1) to be calculated. The highest peaks in the mass spectra of methyl mycolates correspond to anhydromycolates formed by elimination of the elements of water from the parent molecule. A good example is the partial mass spectrum of the methyl esters of the free mycolic acids of the type strain of *Nocardia asteroides* (ATCC19247) taken from the study of Alshamaony, Goodfellow & Minnikin (1976) and shown in Fig. 3. The series of peaks centred around *m/e* 796 correspond to anhydromycolates, those around *m/e* 544 to meroaldehydes, those at *m/e* 270 and 298 to fragmentation at *a* and those at *m/e* 271 and 299 to cleavage *b* (see above scheme). Mass

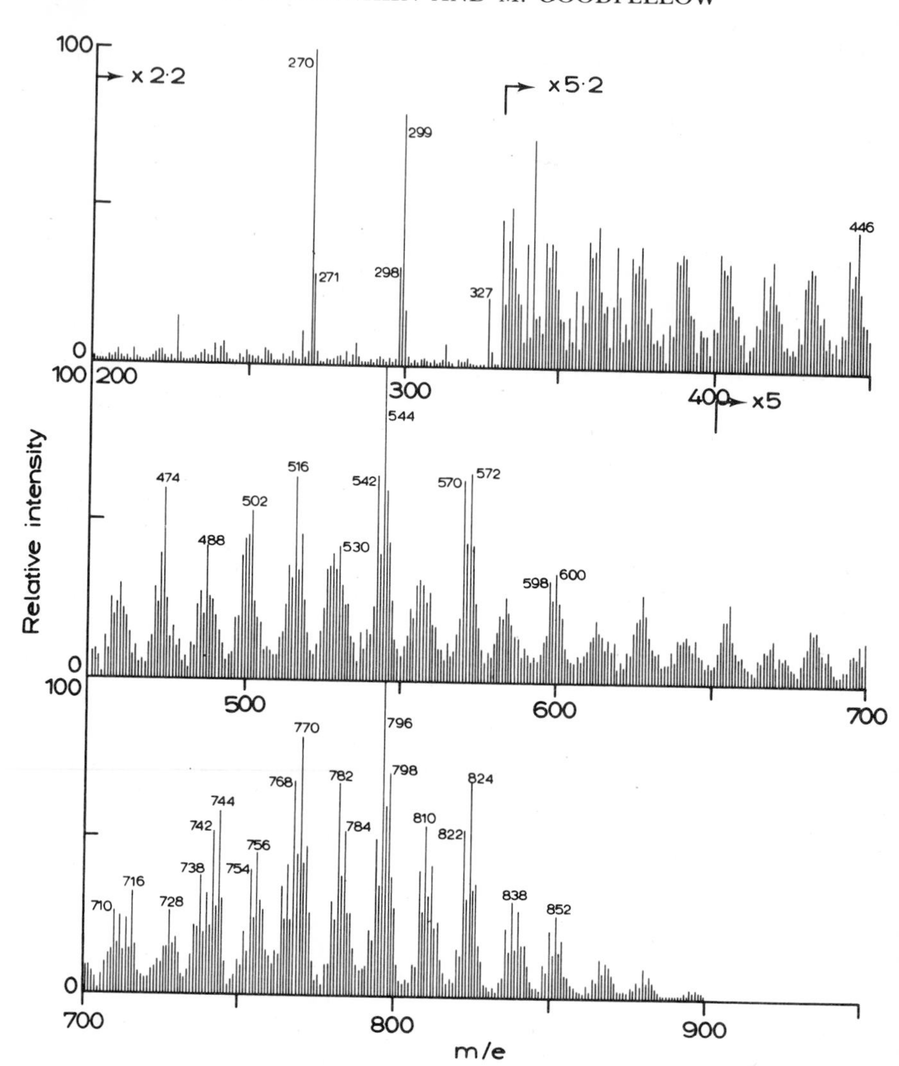

Fig. 3. Partial mass spectrum of the methyl esters of the free mycolic acids from the type strain of *Nocardia asteroides* (ATCC 19247). The base peak in this spectrum was at *m/e* 55. (Reproduced, with permission, from the *Journal of General Microbiology* (Alshamaony, Goodfellow & Minnikin, 1976).)

spectrometry of methyl esters of mycolic acids, therefore, provides direct information on the overall elemental composition and also the composition of the main side chains. The information obtained regarding the composition of the side chains complements that provided by pyrolysis gas chromatography; this latter technique is, however, more sensitive to structural features such as unsaturation (Lechevalier & Lechevalier, 1974) or chain branching. The composition of the side-chains of the mycolates of nocardiae and related bacteria derived from mass spectrometric experiments are in close agreement with those obtained by pyrolysis techniques.

Analysis of the mycolic acids using a variety of spectroscopic techniques have shown that components having oxygen functions in addition to the 3-hydroxyacid system are usually present (Etémadi, 1967*b*; Minnikin & Polgar, 1967; Kusamran, Polgar & Minnikin, 1972). The mycolic acids from mycobacteria which do not contain such oxygen functions are usually diunsaturated and may contain various combinations of double bonds, cyclopropane rings and methyl branches (Etémadi, 1967*b*; Kusamran, Polgar & Minnikin, 1972). The majority of the mycolic acids have around 80 or more carbon atoms but there are two groups of acids having lower molecular weights. Mono- and diunsaturated acids (C_{62} to C_{68}) were isolated from *Mycobacterium smegmatis* (Krembel & Etémadi, 1966; Kunesch, Ferluga & Etémadi, 1966) and a similar monounsaturated acid was obtained from a strain labelled *M. butyricum* (Kunesch, Ferluga & Etémadi, 1966). Dicarboxylic acids, isolated from *M. phlei* (Markovits, Pinte & Etémadi, 1966; Kusamran, Polgar & Minnikin, 1972) contain from 59 to 61 carbon atoms. These mycolates show in their mass spectra, however, peaks characteristic of the longer side chains of the mycolates of all mycobacteria.

The mycolic acids of genera other than mycobacteria have so far been found to be composed of components having differing numbers of double bonds and varying widely in molecular weight. The molecular size and degree of unsaturation of mycolic acids from strains of nocardiae, corynebacteria and the *rhodochrous* complex is summarized in Table 5. It is difficult to separate the strains listed in Table 5 into distinct groups according to their mycolic acid composition but it is apparent that many organisms produce mycolates having around 40 carbon atoms. This group of mycolic acids, intermediate in size between those found in strains of *Nocardia* and *Corynebacterium*, were noted as being anomalous by Ioneda, Lederer & Rozanis (1970) and are associated with organisms of the *rhodochrous* complex (see chapter 2). The results of systematic

Table 5

Overall size and structureal features of mycolic acids from nocardiae and related bacteria

Taxon	Total number of of carbons	Number of double bonds	Acid released on pyrolysis (No. of carbons)	
Nocardia asteroides ATCC9969	46–58	1, 2, 3	14, 16	Bordet & Michel, 1969
N. asteroides 131	48–60	1, 2, 3	12, 14	Azuma *et al.*, 1973
N. brasiliensis	52–60	2, 3	16, 18	Lanéelle & Asselineau, 1970
N. caviae 1M1381	50–56	2, 3	14, 16, 18	
N. sumatra 1M1375	52–58	2, 3	16, 18	
N. erythropolis	48–52	2, 3	14, 16	Maurice, Vacheron & Michel, 1971
N. kirovani 1M1374	58–66	3, 4	16	
N. farcinica 1M1377	34–48	0, 1, 2	12, 14, 16	
N. asteroides	50–58	0, 1, 2, 3	16, 18	Ioneda, Lederer & Rozanis, 1970
	28–38*	1, 2	14, 16, 18†	
N. rhodochrous	40–46	0, 1	14, 16	
N. opaca	44–50	0, 1, 2		Etémadi, Markovits & Pinte, 1966
N. corallina	38–48	0, 1, 2	12, 14, 16	Batt, Hodges & Robertson, 1971
N. erythropolis	32–46	0, 1	10, 12, 14, 16†	Yano *et al.*, 1972
N. caviae NCTC1934	48–56	1, 2, 3	12, 14, 16	Minnikin, Patel & Goodfellow, 1974
N. calcarea NC1B8863	34–46	0, 1, 2	12, 14, 16	
M. rhodochrous ATCC271, 999 and 13808; *N. corallina* ATCC4273	36–48	0, 1	12, 14, 16	Azuma *et al.*, 1974
Mycobacterium paraffinicum	32–44		12, 14, 16	Krasilnikov, Koronelli & Rozynov, 1972
M. lacticolum var. *aliphaticum*	30–34	0, 2	12–16	Krasilnikov *et al.*, 1973
Corynebacterium diphtheriae	32	0, 1	16	Lederer *et al.*, 1952; Pudles & Lederer, 1954
Corynebacterium 506	34	0, 1	16	Etémadi, Gasche & Sifferlen, 1965
C. ovis	32‡	0	16	Diara & Pudles, 1959
C. hofmanii	32–36	0, 1, 2	16, 18†	Welby-Gieusse, Lanéelle & Asselineau, 1970
C. ulcerans	20–32	0, 1, 2	14, 16	Yano & Saito, 1972
Corynebacterium 6.083–9	29–37?	0, 1	14, 16, 18†	Results cited by Etémadi, 1967 *b,c*
C. rubrum	28–46?	0, 1	14–19	
Brevibacterium thiogenitalis	36	2	18†	Okazaki *et al.*, 1969
Arthrobacter paraffineus	28–38?	?	9–14	Suzuki *et al.*, 1969

* From cord factor lipid (acylated trehalose).

† Including components having a single double bond in the branch in the 2-position.

‡ An acid having the same composition was isolated from acyl glucoses of *C. diphtheriae* and *M. smegmatis* (Brennan, Lehane & Thomas, 1970).

analyses (Alshamaony, Goodfellow & Minnikin, 1976; Alshamaony *et al.*, 1976) of the molecular weights of the free mycolic acids of *Nocardia*, representatives of the clusters of the *rhodochrous* complex recovered by Goodfellow (1971), strains of the genus *Gordona* (Tsukamura, 1971) and a strain labelled *Nocardia kirovani*, studied previously (Maurice, Vacheron & Michel, 1971), are shown in Table 6. This study shows that the mycolates of most well-authenticated *rhodochrous* strains are distinct in structure from those of *Nocardia*; one *rhodochrous* strain (cluster 14E, Goodfellow, 1971) had, however, mycolates similar in size to those of *Nocardia*. The mycolates from representatives of *Gordona* and a strain of *N. kirovani* were distinctly larger than those of *Nocardia sensu stricto*.

Table 6

Overall size and structural features of mycolic acids of Nocardia, *the* rhodochrous *complex and* Gordona *(Ashamaony, Goodfellow & Minnikin, 1976; Alshamaony* et al., *1976)*

Taxon	Cluster (Goodfellow, 1971)	Total number of carbons	Number of* double bonds	Acid released on pyrolysis (No. of carbons)
Nocardia asteroides	1(4 strains)	46–58	0, 1, 2, 3	12, 14, 16, 18
N. caviae	2(4 strains)	48–58	0, 1, 2, 3	12, 14, 16, 18
N. brasiliensis	5(3 strains)	46–58	0, 1, 2, 3	14, 16
Rhodochrous strains	14A (2 strains)	40–50	0, 1, 2	12, 14, 16
	14B (1 strain)	38–48	0, 1, 2	12, 14, 16
	14C (5 strains)	36–50	0, 1, 2	12, 14, 16
	14D (3 strains)	34–48	0, 1, 2	12, 14, 16
	14E (1 strain)	46–57	1, 2, 3	14, 16
Gordona bronchialis	(2 strains)	54–66	1, 2, 3, 4	16, 18
G. rubra	(3 strains)	52–62	1, 2, 3, 4	16, 18
G. terrae	(3 strains)	52–64	1, 2, 3, 4	16, 18
N. kirovani	(1 strain)	59–65	2, 3, 4	16, 18

* Numbers of double bonds estimated from analysis of mass spectral data.

Analysis of mycolic acid derivatives by gas chromatography-mass spectrometry. The pyrolysis of mycolic acids or esters to yield meroaldehydes and straight-chain acids or esters, respectively, depends on the presence of a free hydroxyl group in the 3-position. Conversion of the hydroxyl function to its trimethylsilyl ether derivative prevents this pyrolysis and allows analysis of intact molecules of relatively small size by gas chromatography. If a mass spectrometer is linked to the outlet of the gas chromatograph it is possible to obtain structural information concerning the separated components.

The mycolic acids of two strains of nocardioform bacteria have been investigated by combined gas chromatography-mass spectrometry of the trimethylsilyl ether derivatives of the methyl esters. Mycolic ester derivatives from a strain of *Nocardia corallina* (*rhodochrous* strain) gave five major peaks on gas chromatography (Batt, Hodges & Robertson, 1971) but those from an organism labelled *Nocardia erythropolis* were successively separated into as many as 15 components (Yano *et al.*, 1972).

Trimethylsilyl ethers of mycolic esters show a characteristic fragmentation pattern on mass spectrometry. Peaks due to loss of a methyl group (15 mass units) are usually more intense that the molecular ion and lead to the molecular weight of the mycolate. Fragmentation of the molecule is dominated by cleavages adjacent to the trimethylsilyl ether function (see formula below) which enable the composition of the main chain and the side chain in the 2-position to be determined.

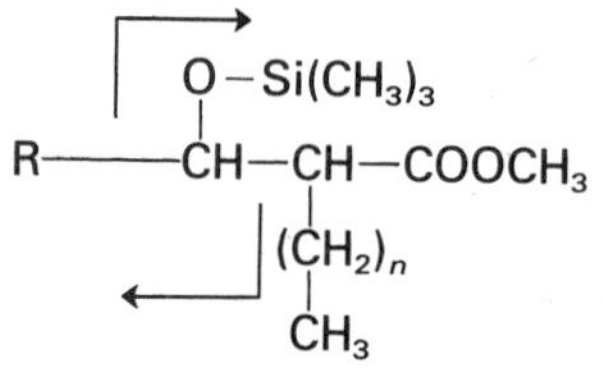

$$\mathrm{R{-}CH(O{-}Si(CH_3)_3){-}CH((CH_2)_n CH_3){-}COOCH_3}$$

The mycolic acids of *N. corallina* (Batt, Hodges & Robertson, 1971) were found to contain even numbers of carbon atoms (38 to 48) and from 0–3 double bonds; the side chains contained 10, 12 or 14 carbon atoms. The components of the mycolic acids from *N. erythropolis* were mainly saturated and include substantial proportions of acids containing odd numbers of carbon atoms; the overall range was from 32 to 46 carbons (Yano *et al.*, 1972). The side chains of these latter mycolic acids had from 8 to 14 carbons; it was particularly interesting that certain of these side chains were unsaturated.

The trimethylsilyl ether derivatives of the mycolic esters from *Corynebacterium hofmanii* were separated by gas chromatography into three components having 32, 34 and 36 carbon atoms (Welby-Gieusse, Lanéelle & Asselineau, 1970). Similar derivatives of the mycolates from *C. ulcerans* were separated into 12 components according to their degree of unsaturation (0, 1 and 2 double bonds) as well as the number of carbon atoms (C_{20}–C_{32}) present and mass spectra recorded for each peak (Yano & Saito, 1972).

The occurrence of mycolic acids in actinomycetes and related bacteria. It is apparent from the preceding sections that mycolic acids are found regularly as components of representatives of *Mycobacterium, Nocardia sensu stricto, rhodochrous* strains, *Gordona* and certain corynebacteria. The possible presence of mycolic acids in other actinomycete and coryneform genera has been investigated in a wide range of studies and the results are summarized in Table 7. Analyses of whole organism hydrolysates give a clear indication whether or not mycolic acids are present. A positive result from the LCN-A procedure, which involves detection of free mycolic acids in ethanol-diethyl ether extracts of bacteria (Mordarska, Mordarski & Goodfellow, 1972; Goodfellow *et al.*, 1973) confirms the presence of mycolic acids but a negative finding might be due to the insolubility of the mycolic acids in ethanol-diethyl mixtures as is the case for those of mycobacteria.

(iii) *Other long-chain components*

No systematic investigations of the distribution of non-carboxylic long-chain compounds in nocardiae and related bacteria have been made but, as noted in a previous section, some very characteristic long-chain alcohols (III) and ketones (IV) have been described. Table 8 shows the distribution of comparable long-chain alcohols and ketones in nocardiae, corynebacteria and mycobacteria. Characteristic long-chain diols, the phthiocerols, have been isolated only from *Mycobacterium tuberculosis* and 2-eicosanol was a component of the lipids of *M. avium* and *M. phlei* (Asselineau, 1966).

(b) *Free lipids*

Lipids, freely extractable with organic solvents, may be broadly divided into polar and non-polar classes; in addition certain lipid types, for example the cord factors, do not readily fall into these categories and must be considered separately.

Table 7

The occurrence of mycolic acids in actinomycetes and related bacteria

Taxon	Wall sugar and amino acid type*	Mycolic acid in whole organism hydrolysate		Free mycolic acid in ethanol-diethyl ether extract	
Mycobacterium	IV	+	(Asselineau, 1966; Etémadi, 1967*b*)	–	(Mordarska, Mordarski & Goodfellow, 1972)
Nocardia sensu stricto	IV	+	(Maurice, Vacheron & Michel, 1971; Lechevalier, Horan & Lechevalier, 1971; Lechevalier, Lechevalier & Horan, 1973; Lechevalier & Lechevalier, 1974)	+	(Mordarska, Mordarski & Goodfellow, 1972; Alshamaony, Goodfellow & & Minnikin, 1976)
Rhodochrous	IV	+	(Lanéelle, Asselineau & Castelnuovo, 1965; Minnikin, Patel & Goodfellow, 1974; Minnikin, Alshamaony & Goodfellow, 1975; Lechevalier, Horan & Lechevalier, 1971; Lechevalier, Lechevalier & Horan, 1973)	+	(Mordarska, Mordarski & Goodfellow, 1972; Mordarska, Mordarski & Pietkiewicz, 1973; Alshamaony, Goodfellow & Minnikin, 1976)
Gordona	IV	+	(Minnikin, Alshamaony & Goodfellow, 1975)	+	(Alshamaony *et al.*, 1976)
Corynebacterium sensuo stricto	IV	+	(Asselineau, 1966; Etémadi, 1967*b*)	+	(Mordarska & Mordarski, 1970; Goodfellow, 1973)
N. petroleophila	IV	ND†		+	(Mordarska, Mordarski & Goodfellow, 1972)
N. vaccinii	IV	ND		+	
N. aerocologenes, *N. coeliaca*, *N. orientalis*, *N. rugosa*, *N. tenuis*	IV	ND		–	(Mordarska, Mordarski & Goodfellow, 1972)
Nocardia spp. (4 strains) *Corynebacterium pyogenes*	IV	–	(Lechevalier, Horan & Lechevalier, 1971; Lechevalier, Lechevalier & Horan, 1973)	ND	
Micropolyspora	IV	ND			

			Goodfellow, 1975)		
Streptomyces (*Nocardia*) *mediterranei*	IV	–	(Pommier & Michel, 1973)	ND	
N. formica‡, *N. italica*, *N. lurida*§, *N. polychromogenes*§, *N. saturnea*	Uncertain	ND		–	(Mordarska, Mordarski & Goodfellow, 1972)
N. gardneri‖	?	–	(Lanéelle, Asselineau & Castelnuovo, 1968)	+	(Mordarska, Mordarski & Goodfellow, 1972)**
Streptomyces sensu stricto	I	–	(Lanéelle, Asselineau & Castelnuovo, 1968)	–	(Mordarska, Mordarski & Goodfellow, 1972)
Other coryneform taxa	Various		Not systematically investigated	–	(Goodfellow, 1973)
Corynebacterium parvum	Uncertain	–	(Etémadi, 1963)	ND	
Rothia	VI	–	(Minnikin, Alshamaony & Goodfellow, 1975)	–	(Goodfellow, 1973)
Oerskovia	VI	–	(Pommier & Michel, 1973)	–	(Mordarska, Mordarski & Goodfellow, 1972)
Actinomadura	III	–	(Minnikin, Alshamaony & Goodfellow, 1975)	–	(Mordarska, Mordarski & Pietkiewicz, 1973)
Thermoactinomyces (*Actinobifida*††), *Thermomonospora*	III	ND		–	(Mordarska & Mordarski, 1970; Mordarska, Guzeva & Agre, 1973)
Micromonospora	II	–	(Tabaud, Tisnovka & Vilkas, 1971)	ND	
Brevibacterium thiogenitalis	?	+	(Okazaki *et al.*, 1969)	ND	
Arthrobacter paraffineus	?	+	(Suzuki *et al.*, 1969)	ND	
Mycobacterium spp.	?	+	(Krasilnikov *et al.*, Krassilnikov, Koronelli & Rozynov, 1972, 1973)	ND	

* Lechevalier & Lechevalier, 1970.
† Not determined.
‡ Possibly wall Type I but reported to contain galactose (Pridham & Lyons, 1969).
§ Certain strains have wall Type IV (Pridham & Lyons, 1969).
‖ Wall Type I according to Pridham & Lyons (1969).
¶ NCTC6531.
** IFO3385.
†† Cross & Goodfellow, 1973.

Table 8

Distribution of related long-chain alcohols and ketones in nocardiae, corynebacteria and mycobacteria

Taxon	Ketones		Alcohols	
Corynebacterium diphtheriae	Palmitone (C_{31}) Palmitenone (C_{31})	(Pudles & Lederer, 1954)		
C. ovis	Palmitone (C_{31})	(Lacave, Asselineau & Toubiana, 1967)		
Nocardia asteroides ATCC9969	Palmitone (C_{31}) Nocardones (C_{47}–C_{59}, 1, 2 and 3 double bonds)	(Bordet & Michel, 1966*b*, 1969)	Nocardols (C_{47}–C_{59}, 1,2 and 3 double bonds)	(Bordet & Michel, 1966*a*, 1969)
Nocardia brasiliensis ATCC733			Hentriacontan-16-ol (C_{31})*	(Bordet & Michel, 1964)
N. brasiliensis 377			Hentriacontan-16-ol (C_{31})*	(Lanéelle, Asselineau & Castelnuovo, 1965)
N. brasiliensis 705	Nocardones (C_{51}–C_{57}, 2 and 3 double bonds)	(Lanéelle, Asselineau & Castelnuovo, 1965; Lanéelle, 1966)		
Mycobacterium kansasii	Kansamycolones (C_{77}, C_{78}, C_{81}; 2 cyclopropane rings)	(Miquel, Das & Etémadi, 1966)		

* Hentriacontan-16-ol was not detected in the lipids of *N. brasiliensis* 705, "*Mycobacterium*" *rhodochrous* 8154 and "*Mycobacterium*" *pellegrino* (Lanéelle, Asselineau & Castelnuovo, 1965).

(i) *Non-polar lipids*

The most commonly encountered non-polar lipid types are triacylglycerols, free fatty acids, isoprenoid quinones and pigments; the occurrence of free mycolic acids has been discussed in a previous section. Sterols are considered to be generally absent from the lipids of bacteria but positive evidence that *Streptomyces olivaceus* could synthesize cholesterol was produced by Schubert, Rose & Hörhold (1967). On chromatographic evidence alone, sterols were reported to be present in the lipids of *Corynebacterium simplex* (Yanagawa *et al.*, 1972).

The overall distribution of acylglycerols and free fatty acids in actinomycetes and coryneform bacteria is summarised in Table 9. Triacylglycerols, common constituents of eukaryotic organisms, are also found in many actinomycetes but are rare in coryneform bacteria and other prokaryotic taxa (Asselineau, 1966). Free fatty acids and mono- and diacylglycerols are possibly derived from breakdown of complex lipids but are frequently found as components of the non-polar lipids of actinomycetes and coryneform bacteria.

Isoprenoid quinones occurring in bacteria are of two distinct types, menaquinones (VIII) related to vitamin K and ubiquinones (co-enzyme Q) (IX). Menaquinones are more commonly found in

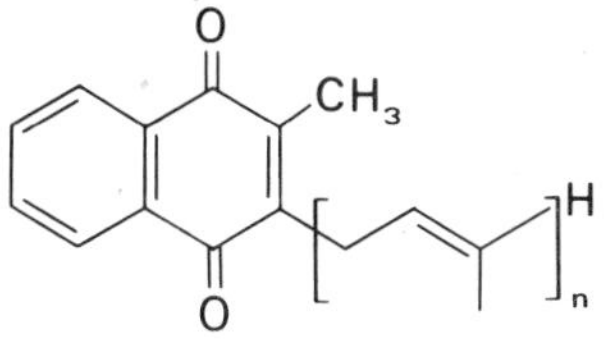

(VIII)

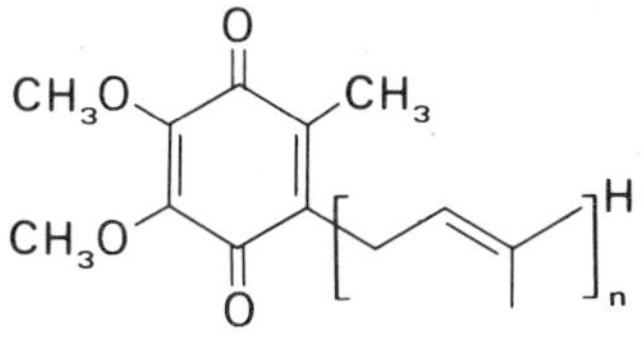

(IX)

Gram-positive bacteria and Gram-negative bacteria often contain ubiquinones alone although certain organisms, for example *Escherichia coli*, contain both types (Pennock, 1966). The distribution of isoprenoid quinones in actinomycetes and coryneform bacteria is detailed in Table 10. It is notable that partially hydrogenated menaquinones having eight (MK_8) or nine (MK_9) isoprene units are the prevalent types; in a menaquinone from a species of *Streptomyces* up to eight additional hydrogens were found (Dunphy, Phillips & Brodie, 1971). The isoprenoid quinone composition of *Nocardia* and strains of the *rhodochrous* complex has not

Table 9
Distribution of non-polar lipid types in actinomycetes and coryneform bacteria

Taxon	Acylglycerols	Free fatty acids	
Mycobacterium spp.	TAG, DAG, MAG	+	Asselineau, 1966
M. smegmatis	TAG		Walker, Barakat & Hung, 1970
M. bovis (BCG)	+	+	Brennan, Rooney & Winder, 1970
Nocardia spp. *Oerskovia, Streptomyces mediterranei*	TAG	+	Pommier & Michel, 1973
N. kirovani	TAG		Vacheron & Michel, 1971
N. farcinica	TAG		Asselineau, Lanéelle & Chamoiseau, 1969
N. brasiliensis, "*M*" *rhodochrous*	TAG	+	Lanéelle, Asselineau & Castelnuovo, 1965
Micromonospora	DAG	+	Tabaud, Tisnovska & Vilkas, 1971
Streptomyces	TAG	+	Lanéelle, Asselineau & Castelnuovo, 1968
S. sioyaensis	TAG		Kimura, Kawanami & Otsuka, 1967*b*
Chainia olivacea	TAG		Lechevalier, Lechevalier & Heintz, 1973
Corynebacterium diphtheriae	+	+	Asselineau, 1961; 1966
C. simplex	TAG	+	Yanagawa *et al.*, 1972
C. ovis	TAG	+	Lacave, Asselineau & Toubiana, 1967
Arthrobacter crystallopoietes	MAG, DAG	+	Kostiw, Boylen & Tyson, 1972
Listeria monocytogenes	DAG		Carroll, Cutts & Murray, 1968
Propionibacterium freudenreichii	+	+	Lanéelle & Asselineau, 1968
Microbacterium ammoniaphilum	TAG	+	Okabe, Shibukawa & Ohsawa, 1967

Abbreviations: TAG, triacylglycerol; DAG, diacylglycerol; MAG, monoacylglycerol.

been investigated. The remarkable report of high proportions of ubiquinone in *Arthrobacter crystallopoietes* (Kostiw, Boylen & Tyson, 1972), if substantiated, suggests that quinone analysis may be valuable for the classification and identification of coryneform bacteria.

Since bacteria often have characteristic colours, studies of the chemical nature of bacterial pigments is likely to be of value in chemotaxonomy. Considerable progress has been made in this

Table 10
Isoprenoid quinones of actinomycetes and coryneform bacteria

Taxon	Isoprenoid quinone	
Mycobacterium phlei	$MK_9(H_2)$	Gale *et al.*, 1963
M. phlei, M. fortuitum, M. avium, M. butyricum, M. tuberculosis	$MK_9(H_2)$	Beau, Azerad & Lederer, 1966
Corynebacterium diphtheriae, *C. rubrum*	$MK_8(H_2)$	
C. diphtheriae	$MK_8(H_2)$	Scholes & King, 1965
M. phlei	$MK_8(H_2)$, $\underline{MK_9(H_2)}$, $MK_{10}(H_2)$	Campbell & Bentley, 1968
M. phlei	$MK_8(H_2)$, $\underline{MK_9(H_2)}$	Dunphy, Phillips & Brodie, 1971
M. tuberculosis	MK_7, MK_8, MK_9, $MK_8(H_2)$, $\underline{MK_9(H_2)}$	
C. createnovorans	$MK_8(H_2)$, $\underline{MK_9(H_2)}$, $MK_{10}(H_2)$	
Streptomyces spp.	MK_9, $MK_9(H_2)$, $MK_9(H_4)$, $\underline{MK_9(H_6)}$, $MK_9(H_8)$	
Propionibacterium arabinosum, P. shermanii	$MK_9(H_4)$	Sone, 1974; Schwartz, 1973
Brevibacterium thiogenitalis	$MK_8(H_2)$, $\underline{MK_9(H_2)}$	Kanzaki *et al.*, 1974
B. vitarumen	$\underline{MK_8(H_2)}$, $MK_9(H_2)$	
Arthrobacter crystallopoietes	Q_n	Kostiw, Boylen & Tyson, 1972

Abbreviations: $MK_n(H_m)$, menaquinone (VIII) having n isoprene units with the addition of m hydrogen atoms; Q_n, ubiquinone (IX) having n isoprene units. The major component of any series is underlined.

area although it should be stressed that pigmentation is not a very stable character (Lechevalier, Lechevalier & Gerber, 1971). Pigments may fall into both non-polar (e.g. certain carotenoids) and polar (e.g. carotenoid glycosides) lipid categories but it is convenient to discuss all types in this section. The general nature of the pigments of actinomycetes and coryneform bacteria is demonstrated in Table 11. Mycobacteria, nocardiae and corynebacteria can to some extent be distinguished by the nature of the carotenoid pigments they contain. Prodiginine pigments having characteristic structures have been isolated from a streptomycete and strains of *Actinomadura madurae* and *A. pelletieri*. *Actinomadura dassonvillei*, however, some other actinomycetes and a strain of *Brevibacterium* are characterised by the presence of phenazine pigments. Analyses

Table 11
Pigments of actinomycetes and corynebacteria

Taxon	Pigment type	
Mycobacterium spp.	Aryl carotenoids	Jensen & Weedon, 1964
M. phlei	Carotenoid glucoside	Hertzberg & Jensen, 1967
Nocardia kirovani	Acyl carotenoid glucoside	Vacheron, Arpin & Michel, 1970
Corynebacterium erythrogenes	Carotenoid glycoside	Weeks & Andrews, 1970
C. poinsettiae	Carotenoids	Norgård, Aasen & Jensen, 1970
Mycobacterium spp.	Carotenoids (Lycopene, α- and β-carotenes)	Tárnok & Tárnok, 1970; 1971
Nocardia spp.	Carotenoids (distinct from those of *Mycobacterium*)	Röhrscheidt & Tárnok, 1972
N. corallina	Red nitrogenous pigment	Stees, Brown & Grigsby, 1969
Actinomadura madurae	Nonylprodiginine and cyclononylprodiginine	For a review see Lechevalier, Lechevalier & Gerber, 1971 and Gerber 1971, 1973 for further studies.
A. pelletieri	Undecylprodiginine and methylcyclodecylprodiginine.	
Streptomyces longisporus ruber	Undecylprodiginine and metacycloprodigiosin	
Microbispora spp.	Phenazines	For a review see Lechevalier, Lechevalier & Gerber, 1971
S. thioluteus		
Brevibacterium iodinium		
A. dassonvillei		
Streptosporangium amethylstogenes		

of the chemical nature of pigments are, therefore, of value in the classification of actinomycetes and related bacteria and the development of standard methods is to be encouraged.

(ii) *Polar lipids*

The first systematic study of the polar lipids of strains of *Nocardia* and the *rhodochrous* complex was carried out by Lanéelle, Asselineau & Castelnuovo (1965). It was found that their phospholipid composition was similar to that observed previously in strains of mycobacteria (Asselineau, 1966; Pangbourn, 1968). Subsequent studies have supported these results though certain strains labelled *Nocardia* have been found to have rather distinct phospholipid compositions. The results of the work on nocardiae is summarized in Table 12. The main points to note are the presence of diphosphatidylglycerol, phosphatidylethanolamine, phosphatidylinositol and phosphatidylinositol mannosides in most of

Table 12
Phospholipid composition of Nocardia

Taxon	DPG	PG	PA	PE	PC	PI	PIM	
Nocardia brasiliensis								
337 and 705	+		+	+			+	Lanéelle, Asselineau Castelnuovo, 1965
"*Nocardia*" *rhodochrous* 8154	+		+	+			+	
Nocardia polychromogenes	+			+		+	+*	Yano, Furukawa & Kusunose, 1968
N. polychromogenes								
NRRLB-151	+			+			+*	Katakoa & Nojima, 1967
Nocardia coeliaca	+			+	+¶	+	+*	Yano, Fukukawa & Kusunose, 1969*a*
Nocardia leishmanii	+	+		+‡		+	+*	Yano, Fukukawa & Kusunose, 1970*b*
N. polychromogenes								
Jensen§ and NRRLB-1513	+			+		+	Trace†	Khuller & Brennan, 1972*a*
N. coeliaca ATCC13181	+			+	+	+	+†	
Nocardia asteroides R399								
N. asteroides ATCC9969								
Nocardia caviae 1M1381	+			+		+	+*	Pommier & Michel, 1973
Nocardia erythropolis A_4								
Nocardia farcinica 1M1377								
"*M.*" *rhodochrous* NCTC7510	+	+					+?‖	Whiteside, DeSiervo & Salton, 1971

Abbreviations: DPG, diphosphatidylglycerol; PG, phosphatidylglycerol; PA, phosphatidic acid; PE, phosphatidylethanolamine; PC, phosphatidylcholine; PI, phosphatidylinositol; PIM, phosphatidylinositol mannosides (*, mono- and †dimannosides).
‡ These phosphatidylethanolamines include a second component containing hydroxyacids.
§ This strain contained a phosphoglycolipid, possibly a phosphatidyl monoglycosyl diacylglycerol.
‖ Uncharacterized phospholipids containing vicinal hydroxyl groups.
¶ Phosphatidylcholine was absent from the lipids of 10 other strains of nocardioform bacteria.

the well-authenticated strains of *Nocardia* and the *rhodochrous* complex.

The exact structures of the phosphatidylinositol mannosides of nocardioform bacteria are not conclusively established. The mannoside lipids from *N. polychromogenes* (Yano, Furukawa & Kusunose, 1968; 1969*a*), *N. coeliaca* (Yano, Furukawa & Kusunose, 1969*a*) and *N. leishmanii* (Yano, Furukawa & Kusunose, 1970*b*) were considered to be phosphatidylinositol monomannosides. In contrast, phosphatidylinositol dimannosides were positively identified as components of the lipids of *N. polychromogenes* (Jensen and NRRLB-1513) and *N. coeliaca* ATCC13181 by Khuller & Brennan (1972*a*). these strains are probably not good representatives of the genus *Nocardia*; a systematic study of some well-established taxa, *N. asteroides* and *N. caviae*, and strains labelled *N. erythropolis* and *N. farcinica* revealed the presence of a lipid identified as a phosphatidylinositol monomannoside (Pommier & Michel, 1973).

The two strains labelled *Nocardia coeliaca* and *Nocardia leishmanii* have very distinct polar lipid compositions, the former containing high proportions of phosphatidylcholine (Yano, Furukawa & Kusunose, 1969*a*; Khuller & Brennan, 1972*a*) and the latter phosphatidylglycerol and a second phosphatidylethanolamine incorporating hydroxy fatty acids (Yano, Furukawa & Kusunose, 1970*b*).

Phospholipid compositions of other actinomycete genera are summarized, for comparison, in Table 13. Strains of *Streptomyces mediterranei* and *Oerskovia turbata* had phospholipid patterns which supported their separation from the genus *Nocardia* (Pommier & Michel, 1973), the former lacking phosphatidylinositol and phosphatidylethanolamine being absent in the latter. Phosphatidylinositol is apparently absent in all strains of *Streptomyces* studied so far although these bacteria contain phosphatidylinositol mannosides (Katakoa & Nojima, 1967; Kimura, Kawanami & Otsuka, 1967*a,b*; Lanéelle, Asselineau & Castelnuovo, 1968; Kawanami, 1971). *Microbispora chromogenes* M-22 similarly lacked phosphatidylinositol (Katakoa & Nojima, 1967) but this lipid was present in a strain of *Micromonospora* (Tabaud, Tisnovska & Vilkas, 1971).

The phospholipid composition of coryneform bacteria is summarized in Table 14. Phosphatidylinositol dimannosides are found

Table 13
Phospholipid composition of actinomycete genera other than Nocardia

Taxon	DPG	PE	PI	PIM	
Mycobacterium spp.	+	+	+	+*	For reviews see Asselineau, 1966; Pangbourn, 1968
Oerskovia turbata 891 and 7130 *Streptomyces mediterranei* ME/R4688 and ME83/973	+ +	 +‡	+ 	+† +†	Pommier & Michel, 1973
Micromonospora SpF3	+	+	+	+†	Tabaud, Tisnovska & Vilkas, 1971
Microbispora chromogenes M-22 *Streptomyces griseus* 127-2	+	+		+†	Katakoa & Nojima, 1967
S. aureofaciens 5017 and ATCC10762 *S. gelaticus* 5010 *S. griseus* 5048 *S. (Nocardia) gardneri* NCTC6531	+?§	+‡		+	Lanéelle, Asselineau & Castelnuovo, 1968
S. sioyaensis H-690-3	+	+‡		+†	Kimura, Kawanami & Otsuka, 1967*a*,*b*
S. toyocaensis S-637	+	+‡		+	Kawanami, 1971
Actinomyces viscosus	+	+	+		Yribarren, Vilkas & Rozanis, 1974

Abbreviations are identical to those shown in Table 12.
* Phosphatidylinositol di-to pentamannosides.
† Phosphatidylinositol monomannosides.
‡ These phosphatidylethanolamines include a second component containing hydroxy acids.
% Not characterized as diphosphatidylglycerol but had chromatographic behaviour consistent with such an assignment.

in most animal corynebacteria (Brennan & Lehane, 1971) but monomannosides are present in the lipids of *Corynebacterium aquaticum* (Khuller & Brennan, 1972*b*) and *Propionibacterium shermanii* (Brennan & Ballou, 1968; Prottey & Ballou, 1968). Phosphatidylethanolamine occurs only in some strains of *Microbacterium* (Shibukawa, Kurima & Ohuchi, 1970; Shaw & Stead, 1970) and certain corynebacteria (Khuller & Brennan, 1972*b*; Kikuchi & Nakao, 1973).

Ninhydrin-positive phosphate-free polar lipids have been isolated from extracts of certain streptomycetes (Kimura, Kawanami & Otsuka, 1967*b*; Kawanami, 1971), a strain labelled *Actinomyces* 660-15 (Batrakov, Pilipenko & Bergelson, 1971), and *Mycobacterium bovis* BCG (Promé, Lacave & Lanéelle, 1969). These lipids

Table 14
Phospholipid composition of coryneform bacteria

Taxon	DPG	PG	PA	PE	PI	PIM	
Corynebacterium alkanolyticum	+	+	+	+	+		Kikuchi & Nakao, 1973
C. aquaticum	+	+		+		+*	Khuller & Brennan, 1972*b*
C. diphtheriae	+	+			+	+†	Brennan & Lehane, 1971
C. xerosis	+				+	+†	Brennan & Lehane, 1971
C. equi, C. ovis	?				+	+†	Brennan & Lehane, 1971
C. ovis						+?‡	Lacave, Asselineau & Toubiana, 1967
C. diphtheriae						+	Gomes, Ioneda & Pudles, 1966
Propionibacterium shermanii		+§			+	+*	Brennan & Ballou, 1968; Prottey & Ballou, 1968
P freudenreichii						+	Lanéelle & Asselineau, 1968
Microbacterium ammoniaphilum	+		+	+‖	+		Shibukawa, Kurima & Ohuchi, 1970
M. thermosphactum	+	+		+			Shaw & Stead, 1970
M. lacticum	+	+					Shaw, 1968
Arthrobacter crystallopoietes	+	+¶	+		+	+?**	Kostiw, Boylen & Tyson, 1972
A. simplex	+	+††					Yano, Furukawa & Kusunose, 1970*a*, 1971*a*
A. crystallopoietes, A. pascens, A. globiformis	+	+			+		Shaw & Stead, 1971
Listeria monocytogenes	+	+‡‡					Kosaric & Carroll, 1971

Abbreviations are identical to those shown in Table 12.
* Phosphatidylinositol monomannoside.
† Phosphatidylinositol dimannoside.
‡ Contain inositol and arabinose.
§ Phosphatidylglycerophosphate also possibly present.
‖ Phosphatidylcholine also present.
¶ Two spots on thin-layer chromatography, different salt forms suggested.
** Unidentified but contained inositol, mannose and glycerol.
†† Two spots on thin-layer chromatography, one containing 2-hydroxy acids.
‡‡ Bisphosphatidylglycerol phosphate also tentatively identified.

were shown in the above studies to be amides of ornithine or lysine. An extensive study of the distribution of ninhydrin-positive polar lipids in streptomycetes and a few other actinomycetes was made by Kimura, Kawanami & Otsuka (1967*a*); this and other studies are summarized in Table 15. Ornithine or lysine amides have not been detected in the lipids of nocardioform bacteria.

Table 15

Distribution of ninhydrin-positive polar lipids among strains in Actinomycetales (Kimura, Kawanami & Otsuka, 1967a; Kawanami, 1971)

Taxon	PE-1	PE-2	LA
Streptomyces griseus, *S. virginiae*, *S. albiverticuli*, *S. showdoensis* C-224 and C209A, *S. aburaviensis*, *S. minoensis* 1-523, *S. erythrus* C-233-4 and C-233-6, *S.* M-88, *S. bostroemi*, *S. orchidaceus*, *S. echinatus*, *S. netropsis*, *Streptomyces* N-946, *Streptomyces* 2-242, *S. rubireticuli*, *S. reticuli*, *S. luteoverticillatus*, *S. thiolutens*, *S. roseoverticillatus*; *Nocardia asteroides*, *N. erythropolis*; *Actinoplanes phillipinensis*; *Micromonospora* T-12	+	−	−
Streptomyces G-193, N-329 and N58, *S. olivoverticillatus*	+*	+*	−
S. sioyaensis H-690-3	+*	+	+†
S. toyocaenis S-637	+	+	+†
Actinomyces 660-15‡	?	?	+
Mycobacterium bovis BCG§	?	?	+

Abbreviations: PE-1, phosphatidylethanolamine containing non-hydroxylated fatty acids; PE-2, phosphatidylethanolamine possibly containing hydroxy fatty acids; LA, fatty acid amides of ornithine and/or lysine.

* See also Tables 12 and 13 for further examples of actinomycetes having two such phosphatidylethanolamines.

† For details of the structures of these lipids refer to Kawanami (1971) and references cited therein.

‡ Batrakov, Pilipenko & Bergelson, 1971.

§ Promé, Lacave & Lanéelle, 1969; see also Lanéelle, Lanéelle & Asselineau, 1963.

Glycolipids, as noted previously, may be divided into two general classes according to their degree of polarity. Polar glycolipids are similar in their general properties to phospholipids but other glycolipids such as the "cord factors" (VII) are more semi-polar in character since their acyl substituents are much larger than those of polar lipids in general. The distribution of polar glycolipids in *Nocardia* and related bacteria is shown in Table 16. Glycosyl diacylglycerols are found in the lipids of many coryneform bacteria (for a review see Shaw, 1970) but have not been commonly found in

Table 16

Polar glycolipids of Nocardia *and related bacteria*

Taxon	Glycosyl diacylglycerols	Acylated sugars	Acidic glycolipids	
Nocardia polychromogenes	+*	+†		Khuller & Brennan, 1972*a*
N. coeliaca‡‡		+†		
N. caviae			+	Pommier & Michel, 1972
Mycobacterium fortuitum		+‡		Vilkas & Rojas, 1964; Vilkas, Adam & Senn, 1968
Actinomyces viscosus	+§			Yribarren, Vilkas & Rozanis, 1974
Micromonospora		+‡		Tabaud, Tisnovska & Vilkas, 1971
Streptomyces LA7017			+	Bergelson, Batrakov & Pilipenko, 1970
Corynebacterium aquaticum	+‖			Khuller & Brennan, 1972*b*
C. xerosis		+†		Brennan & Lehane, 1969
C. coelicolor	?§§			Whiteside, DeSiervo & Salton, 1971
Propionibacterium shermanii		+‡+¶		Prottey & Ballou, 1968
P. shermanii, P. freudenreichii, P. arabinosum		+¶		Shaw & Dinglinger, 1969
Arthrobacter globiformis	+§+‖+**			Walker & Bastl, 1967
A. globiformis, A. pascens, A. crystallopoietes	+§+‖+**			Shaw & Stead, 1971
A. crystallopoietes	+§+**			Kostiw, Boylen & Tyson, 1972
Listeria monocytogenes	+††			Carroll, Cutts & Murray, 1968
Microbacterium lacticum	+‖			Shaw, 1968
M. thermosphactum	Trace‖	+†		Shaw & Stead, 1970

* Diglucosyl diacylglycerol.
† Acylated glucose.
‡ Acylated trehalose.
§ Monogalactosyl diacylglycerol.
‖ Dimannosyl diacylglycerol.
¶ Diacylated inositol mannoside.
** Digalactosyl diacylglycerol.
†† Possibly galactosylglucosyl diacylglycerol.
‡‡ An unknown glycolipid was detected in *N. coeliaca* by Yano, Furukawa & Kusunose (1969*a*).

actinomycetes. Polar trehalose esters have been isolated from a variety of sources and diacylated inositol mannosides are characteristic of propionibacteria. Acidic glycolipids are uncommon but no systematic investigations of the possible presence of these lipids in actinomycetes have been made.

(iii) *Other characteristic lipids*

The semi-polar mycolic esters of trehalose (VII), the so-called "cord factors", are found only in strains of certain actinomycetes and coryneform bacteria. Systematic studies of the distribution and nature of cord factors have not been made but the results obtained to date show that analyses of these lipids may be useful for the classification of the bacteria containing them. The mycolic acid substituents of the cord factor lipids from mycobacteria all have high molecular weight (C_{70}–C_{90}) and are similar to those isolated from whole bacteria (Lederer, 1967). The cord factors from *Mycobacterium tuberculosis* contained methoxymycolic acids only but those from the BCG strain had a mixture of keto- and dicyclopropyl mycolic acids, and simple unsaturated mycolic acids were obtained from the cord factors of *M. butyricum* (Adam *et al.*, 1967).

The cord factor from *Corynebacterium diphtheriae* contained mycolic acids of small size (C_{30}–C_{34}) similar to the mycolates from whole organisms (Senn *et al.*, 1967), and a strain of "*Nocardia*" *rhodochrous* had similar mycolic acids (C_{38}–C_{46}) both in whole organisms and its cord factor lipid (Ioneda, Lederer & Rozanis, 1970). In contrast the mycolic acids from the cord factor of *Nocardia asteroides* had lower molecular weights (C_{32}–C_{36}) than those from the walls (C_{50}–C_{58}) (Ioneda, Lederer & Rozanis, 1970). Trehalose dimycolates and monomycolates were isolated from several strains of nocardiae (Yano, Furukawa & Kusunose, 1971*b*) but the structures of the mycolic acid residues were not determined. A diester of trehalose, having mycolic acid residues (C_{28}–C_{38}) in undetermined positions, was observed in the emulsion layer of *Arthrobacter paraffineus* grown on hydrocarbons (Suzuki *et al.*, 1969) and the presence of similar lipids in extracts of a range of other coryneform bacteria and a single species of *Nocardia* noted.

A sugar-containing lipid isolated from the culture broth of *Brevibacterium thiogenitalis* was found to be composed of a mycolic acid (C_{36}) esterified to the 6-position of glucose (Okazaki *et al.*, 1969). Acylated glucoses also having mycolic acids esterified to the

6-position accumulated when *Corynebacterium diphtheriae*, *Mycobacterium smegmatis* and *M. bovis* (BCG) were grown in the presence of glucose (Brennan, Lehane & Thomas, 1970*a*). The mycolic acid substituent of the glycolipids from *C. diphtheriae* and *M. smegmatis* was the same and contained 32 carbon atoms; the structure of the acid from the lipid of the BCG strain was not determined. It is interesting to note that the mycolic acid from the glycolipid of *M. smegmatis* was much smaller than those (C_{70}–C_{80}) from whole organisms (Etémadi, 1967*b*).

A very characteristic peptidolipid (I), having a cyclic structure, was isolated from a single strain of *Nocardia asteroides* (ATCC9969) (Guinand, Michel & Lederer, 1958, 1964; Barber *et al.*, 1965*b*; Guinand & Michel, 1966). Linear structures have been determined for the peptidolipids from *Mycobacterium johnei* (Lanéelle *et al.*, 1965) and *M. fortuitum* (Barber *et al.*, 1965*a*). A complex peptidolipid of unknown structure was isolated from the lipids of *Corynebacterium ovis* (Lacave, Asselineau & Toubiana, 1967).

Certain strains of *Mycobacterium* produce complex peptidoglycolipids (mycosides C) (Voiland, Bruneteau & Michel, 1971 and references cited therein; Goren, 1972) but no other genera have been found to produce such lipids. It is interesting that a similar mycoside has been isolated from bacteria responsible for cases of bovine farcy thus supporting the assignment of these organisms to the genus *Mycobacterium* (Lanéelle, Asselineau & Chamoiseau, 1971; Chamoiseau, 1973).

(c) *Bound lipids*

The walls of certain actinomycetes and corynebacteria contain chemically-bound mycolic acids. Mild acid hydrolysis of defatted walls of mycobacteria releases mycolic acids esterified to arabinose (Azuma & Yamamura, 1962, 1963; Acharya, Senn & Lederer, 1967). The same glycolipids are released by similar hydrolysis (Azuma, Kimura & Yamamura, 1965) of a complex peptidoglycolipid known as "wax D" which is a possible breakdown product of the wall structure (Kanetsuna & San Blas, 1970; Markovits, Vilkas & Lederer, 1971). An arabinose mycolate was also isolated both from the walls and a wax D fraction of *Nocardia brasiliensis* by Lanéelle & Asselineau (1970). Systematic comparisons of the free and bound fatty acid compositions of the bacteria under consideration have not been carried out.

4. Conclusions

The lipids of nocardiae and related bacteria by their very complexity are potentially valuable as chemotaxonomic characters and the studies summarized above demonstrate that lipid analyses can be useful in the classification and identification of nocardioform bacteria. To date attention has been concentrated on certain very characteristic lipids particularly the mycolic acids. These acids, which are 3-hydroxy acids with a long chain in the 2-position, have only been found in mycobacteria, nocardiae, gordonae, certain corynebacteria and in *rhodochrous* strains. All of these strains are classified in taxa with a wall Type IV (Lechevalier & Lechevalier, 1970) (see Table 7). Strains labelled *Brevibacterium thiogenitalis* (Okazaki *et al.*, 1969), *Arthrobacter paraffineus* (Suzuki *et al.*, 1969), *Mycobacterium paraffinicum* (Krasilnikov, Koronelli & Rozynov, 1972) and *M. lacticolum* var. *aliphaticum* (Krasilnikov *et al.*, 1973) (Tables 5 and 7) probably also contain *meso*-diaminopimelic acid, arabinose and galactose. The classification of the bacteria which are known to have a wall Type IV but apparently contain no mycolic acids (Table 7) should be the subject of further detailed studies. The strains labelled *Nocardia formica, N. italica, N. lurida, N. polychromogenes* and *N. saturnae,* which do not readily fit into one of the established wall types, and apparently lack mycolic acids (Mordarska, Mordarski & Goodfellow, 1972), are also an anomalous group. Representatives of "*Nocardia*" *gardneri* should also be collected and studied since one strain (NCTC6531) was found to lack mycolic acids (Lanéelle, Asselineau & Castelnuovo, 1968) while another (IFO3385) apparently produced free mycolic acids (LCN-A) (Mordarska, Mordarski & Goodfellow, 1972).

Strains of mycobacteria can be clearly distinguished from allied taxa by the nature of the mycolic acids they contain. These mycolic acids usually consist of mixtures containing components having additional oxygen functions and therefore give complex patterns on t.l.c. of their methyl esters whereas single spots are obtained for the mycolates of other taxa (Minnikin, Alshamaony & Goodfellow, 1975). The size of the fatty acid ester (C_{22}–C_{26}) released on pyrolysis of the methyl esters of the mycolic acids of mycobacteria clearly distinguishes these bacteria from all other mycolic acid-containing strains whose mycolates release C_{12}–C_{18} esters (Etémadi, 1967*a,b*; Lechevalier, Horan & Lechevalier, 1971; Lechevalier, Lechevalier & Horan, 1973). Mycobacterial mycolic acids are precipitated from ethereal solution by addition of ethanol but those of other strains remain in solution (Kanetsuna & Bartoli, 1972). Free mycolic acids of nocardioform and some corynebacteria are extracted by ethanol-diethyl ether mixtures and may be detected by t.l.c. (Mordarska, Mordarski & Goodfellow, 1972). The difference in solubility between the mycolates of mycobacteria and those from other taxa is a reflection of their molecular weights; mass spectral analyses have confirmed that the mycolic acids of mycobacteria (C_{60}–C_{90}) are in general larger than those of other taxa (C_{30}–C_{70}) (Etémadi, 1967*b*).

The mycolic acids of *Nocardia sensu stricto* (0–3 double bonds, C_{46}–C_{58}) are distinguishable from those of *Gordona* (1–4 double bonds, C_{52}–C_{66}) and many *rhodochrous* strains (0–2 double bonds, C_{34}–C_{50}) (Table 6) (Alshamaony, Goodfellow & Minnikin, 1976; Alshamaony *et al.*, 1976). One *rhodochrous* strain, labelled *Nocardia opaca* (Goodfellow, 1971; cluster 14E), had mycolates of a similar size (C_{46}–C_{57}) to those of *Nocardia* (Alshamaony, Goodfellow & Minnikin, 1976), and the mycolates of a strain labelled *Nocardia kirovani* were similar in size to those of *Gordona* (Maurice, Vacheron & Michel, 1971; Alshamaony *et al.*, 1976). The mycolic acids of coryneform bacteria have been shown to have molecular weights centred around 32 carbon atoms (Table 5).

The mycolates from strains of a recently proposed species, *Nocardia amarae*, were found by pyrolysis gas chromatography to have unsaturated side chains (Lechevalier & Lechevalier, 1974). Unsaturation has also been detected in the side chains of the mycolic acids from several nocardioform and corynebacteria (Table 5); this feature should therefore be explored further in order to determine if it has any chemotaxonomic value.

Analyses of simple long-chain fatty acids have shown (Tables 2, 3 and 4) that while complex mixtures are encountered two main patterns can be recognized. Mixtures in which unsaturated and related acids (for example, tuberculostearic acid) predominate are found in all mycolic acid-containing genera and in certain representatives of coryneform taxa (Table 4) though the latter have not been investigated systematically for the presence of mycolic acids. It is very interesting that certain strains bearing the label *Nocardia*, but whose relation to strains of *Nocardia sensu stricto* is not clear (see Table 7) have fatty acid patterns of the other main type, containing high proportions of branched-chain *iso*- and *anteiso*-acids and low amounts of unsaturated acids. *Nocardia coeliaca*, for example, an organism having cell wall Type IV but apparently lacking mycolic acids (Table 7) (Mordarska, Mordarski & Goodfellow, 1972), was rich in branched-chain acids (Table 2) (Yano, Furukawa & Kusunose, 1969*a*). The fatty acids of strains of *Streptomyces mediterranei*, another organism having wall Type IV and no mycolic acids (Table 7), also are mainly of the branched chain type (Table 3) (Pommier & Michel, 1973). The fatty acids of certain strains of *Nocardia lurida* and *Nocardia rugosa* apparently contain substantial proportions of both unsaturated and branched-chain acids (Bordet & Michel, 1963) which is interesting given the apparent absence of mycolic acids (Table 7) (Mordarska, Mordarski & Goodfellow, 1972) from other representatives of these species. A strain labelled *Nocardia polychromogenes*, however, has the fatty acid pattern typical of *Nocardia sensu stricto* (Yano, Furukawa & Kusunose, 1968) but another representative contained no free mycolic acids and did not have a wall Type IV (Mordarska, Mordarski & Goodfellow, 1972). The fatty acids of *Nocardia leishmanii* contained no unsaturated components (Yano, Furukawa & Kusunose, 1969*b*, 1970) but included branched-chain hydroxy acids the overall pattern being similar to that given by certain streptomycetes (Lanéelle, Asselineau & Castelnuovo, 1968). The fatty acids of most other genera of actinomycetes (see Table 3) are predominantly of the branched-chain variety.

Polar lipids are likely to be less reliable as chemotaxonomic characters since it has been demonstrated that the relative proportions of these lipids in certain bacteria may change dramatically with alterations in growth environment (Minnikin, Abdolrahimzadeh & Baddiley, 1974; Minnikin & Abdolrahimzadeh, 1974). A particular class of polar lipids, the phosphatidylinositol mannosides, are found only in certain actinomycetes and therefore have chemotaxonomic value. These lipids have been found in representatives of *Mycobacterium*, *Nocardia*, *Corynebacterium* and some related taxa (see Tables 12, 13 and 14) but further systematic studies are needed since, for example, in some nocardiae it is uncertain whether monomannosides (Pommier & Michel, 1973) or dimannosides (Khuller & Brennan, 1972*a*) are the characteristic types. The presence of phosphatidylcholine, a lipid rare in bacteria, in strains of *Nocardia coeliaca* (Yano, Furukawa & Kusunose, 1969*a*; Khuller & Brennan, 1972*a*) supports the case for removing this species from *Nocardia sensu stricto.* Ninhydrin-positive polar lipids, because of their ease of detection on chromatograms, are potentially very valuable as chemotaxonomic characters (Table 15) though the effect of changes in growth environment on their presence or absence has still to be taken into account (Minnikin & Abdolrahimzadeh, 1974). Systematic studies on glycolipid composition (see Table 16) must also be made under standard conditions since these lipids are to some extent interchangeable with phospholipids (Minnikin, Abdolrahimzadeh & Baddiley, 1974).

A range of other lipids have still to be studied systematically to determine their potential as chemotaxonomic characters for the classification of nocardiae and related bacteria. These include the mycolic acid esters of trehalose (the so-called cord factor lipids), triacylglycerols and other non polar lipids (Table 9), isoprenoid quinones (Table 10) and pigments (Table 11). Profitable studies might also be carried out by comparing the long-chain components of the bound lipids with those of the previously extracted free lipids.

The exploitation of lipids as chemotaxonomic characters is still in its infancy; the necessary techniques, mainly chromatographic and spectroscopic in nature, are available and they only require systematic application. The results achieved so far, however, are substantial and very encouraging. Analyses of wall sugar and amino acid composition (Lechevalier & Lechevalier, 1970) have provided characters which allowed actinomycetes to be classified into distinct groups (Table 7) (see chapter 1). Lipid studies have been concentrated on bacteria having wall Type IV and suggest that this group can be split into three coherent sub-groups. The first of these contains the mycobacteria which possess large mycolic acids having complex structures, and the second would comprise strains in the taxa *Nocardia, Gordona, Corynebacterium* and the *rhodochrous* complex containing simple unsaturated mycolic acids of varying size. A third sub-group would include an assorted collection of species none of which contained mycolic acids. Good examples (Table 7) of this latter sub-group are representatives of *Saccharopolyspora* (Minnikin, Alshamaony & Goodfellow, 1975; Lacey & Goodfellow, 1975), *Streptomyces mediterranei* (Pommier & Michel, 1973), a strain of *Corynebacterium pyogenes*, four *Nocardia* strains (Lechevalier, Horan & Lechevalier, 1971; Lechevalier, Lechevalier & Horan, 1973) and possibly other species of *Nocardia* (*N. aerocologenes*, *N. coeliaca*, *N. orientalis*, *N. rugosa* and *N. tenuis*) and strains of *Micropolyspora* (Mordarska, Mordarski & Goodfellow, 1972). Lipid analyses have highlighted the anomalous taxonomic position of these strains and demonstrated that further general studies on their classification are needed. Lipid studies have already suggested that certain bacteria responsible for cases of bovine farcy, formerly labelled *Nocardia farcinica*, should in fact be assigned to the genus *Mycobacterium* (Asselineau, Lanéelle & Chamoiseau, 1969; Chamoiseau & Asselineau, 1970; Chamoiseau, 1973) and lipid analyses of further isolates will probably be valuable in clarifying the classification of these bacteria.

Lipids have been used mainly as an aid to the classification of nocardioform bacteria but there is no reason why some of the techniques should not be developed into standard methods for the identification of these organisms. Characters such as the mycolic acids are so unique in structure and reliable in occurrence that, if they are positively identified, it is possible to classify bacteria containing them in one of a very limited number of genera. Other lipid characters are less powerful and for their successful standardisation of growth conditions will be of paramount importance.

5. References

Acharya, N. P. V., Senn, M. & Lederer, E. (1967). Sur la présence et la structure de mycolates d'arabinose dans les lipides liés de deux sources de mycobactéries. *C.r. hebd. Séanc. Acad. Sci., Paris* **264C,** 2173.

Adam, A., Senn, M., Vilkas, E. & Lederer, E. (1967). Spectrométrie de masse de glycolipides 2. Diesters de tréhalose naturels et synthétiques. *Eur. J. Biochem.* **2,** 460.

Alshamaony, L., Goodfellow, M. & Minnikin, D. E. (1976). Free mycolic acids in the classification of *Nocardia* and the "*rhodochrous*" complex. *J. gen. Microbiol.* **92,** 188.

Alshamaony, L., Goodfellow, M., Minnikin, D. E. & Mordarska, H. (1976). Free mycolic acids in the classification of *Gordona* and the "*rhodochrous*" complex. *J. gen. Microbiol.* **92,** 183.

Asselineau, J. (1961). Sur la composition des lipides de *Corynebacterium diphtheriae. Biochim. biophys. Acta* **54,** 359.

Asselineau, J. (1966). *The Bacterial Lipids.* Paris: Hermann.

Asselineau, J., Lanéelle, M. A. & Chamoiseau, G. (1969). De l'etiologie du farcin de zébus tchadiens nocardiose ou mycobactériose. II. Composition lipidique. *Rev. Élev. Méd. vét. Pays trop.* **22,** 205.

Asselineau, J. & Lederer, E. (1950). Structure of the mycolic acids of mycobacteria. *Nature, Lond.* **166,** 782.

Azuma, I., Kanetsuna, F., Tanaka, Y., Mera, M., Yanagihara, Y., Mifuchi, I. & Yamamura, Y. (1973). Partial chemical characterization of the cell wall of *Nocardia asteroides* strain 131. *Jap. J. Microbiol.* **17,** 154.

Azuma, I., Kimura, H. & Yamamura, Y. (1965). Isolation of arabinose mycolate form was D fraction of human type tubercle bacillus Aaoyama B strain. *J. Biochem., Tokyo* **57,** 571.

Azuma, I., Ohuchida, A., Taniyama, T., Yamamura, Y., Shoji, K., Hori, M., Tanaka, Y. & Ribi, E. (1974). The mycolic acids of *Mycobacterium rhodochrous* and *Nocardia corallina. Biken's J.* **17,** 1.

Azuma, I. & Yamamura, Y. (1962). Studies in the firmly bound lipids of human tubercle bacillus. 1. Isolation of arabinose mycolate. *J. Biochem., Tokyo* **52,** 200.

Azuma, I. & Yamamura, Y. (1963). Studies on the firmly bound lipids of human tubercle bacillus. *J. Biochem., Tokyo* **53,** 275.

Ballio, A. & Barcellona, S. (1968). Relations chimiques et immunologiques chez les actinomycétales. 1. Les acides gras de 43 souches d'actinomycètes aérobies. *Annls Inst. Pasteur, Paris* **114,** 121.

Ballio, A., Barcellona, S. & Boniforti, L. (1965). The component fatty acids of lipids from some *Streptomyces* spp. *Biochem. J.* **94,** 11c.

Ballio, A., Barcellona, S. & Salvatori, T. (1968). Identification of 9,10-methylenehexadecanoic acid in some aerobic "Actinomycetales" by a combined gas chromatographic—mass spectrometric technique. *J. Chromat.* **35,** 211.

Barber, M., Jolles, P., Vilkas, E. & Lederer, E. (1965*a*). Determination of amino acid sequences in oligopeptides by mass spectrometry. 1. The structure of fortuitine, an acylnonapeptide methyl ester. *Biochem. biophys. Res. Commun.* **18,** 469.

Barber, M., Wolstenholme, W. A., Guinand, M., Michel, G., Das, B. C. & Lederer, E. (1965*b*). Determination of amino acid sequences in oligopeptides by mass spectrometry. II. The structure of peptidolipin NA. *Tetrahedron Lett.* 1331.

Batrakov, S. G., Pilipenko, T. V. & Bergelson, L. D. (1971). A new ornithine-containing lipid from actinomycetes. *Dokl. Akad. Nauk SSSR* **200,** 226.

Batt, R. D., Hodges, R. & Robertson, J. G. (1971). Gas chromatography and mass spectrometry of the trimethylsilyl ether methyl ester derivatives of long chain hydroxy acids from *Nocardia corallina. Biochim. biophys. Acta* **239,** 368.

Beau, S., Azerad, R. & Lederer, E. (1966). Isolement et charactérisation des dihydro-ménaquinones des myco- et corynébactéries. *Bull. Soc. Chim. biol.* **48,** 569.

Behal, V., Prochazkova, V. & Vaněk, Z. (1968). Regulation of biosynthesis of secondary metabolites. II. Fatty acids and chlortetracycline in *Streptomyces aureofaciens*. *Folio microbiol., Praha* **14,** 112.

Bergelson, L. D., Batrakov, S. G. & Pilipenko, T. V. (1970). A new glycolipid from *Streptomyces*. *Chem. Phys. Lipids* **4,** 181.

Bloch, H. (1950). Studies on the virulence of tubercle bacilli. Isolation and biological properties of a constituent of virulent organisms. *J. exp. Med.* **91,** 197.

Boone, C. J. & Pine, L. (1968). Rapid method for the characterisation of actinomycetes by cell wall composition. *Appl. Microbiol.* **16,** 279.

Bordet, C., Etémadi, A. H., Michel, G. & Lederer, E. (1965). Structure des acides nocardiques de *Nocardia asteroides*. *Bull. Soc. chim. Fr.* 234.

Bordet, C., Karahjoli, M., Gateau, O. & Michel, G. (1972). Cell walls of nocardiae and related actinomycetes: identification of the genus *Nocardia* by cell wall analysis. *Int. J. Syst. Bact.* **22,** 251.

Bordet, C. & Michel, G. (1963). Étude des acides gras isolés de plusieurs espèces de *Nocardia*. *Biochim. biophys. Acta* **70,** 613.

Bordet, C. & Michel, G. (1964). Isolement d'un nouvel alcool, le 16-hentriaconatanol a partir des lipides de *Nocardia brasiliensis*. *Bull. Soc. Chim. biol.* **46,** 1101.

Bordet, C. & Michel, G. (1966*a*). Les nocardols, alcools à haut poids moléculaire de *Nocardia asteroides*. *C.r. hebd. Séanc. Acad. Sci., Paris* **262C,** 1294.

Bordet, C. & Michel, G. (1966*b*). Isolement et structure des nocardones, cétones à haut poids moléculaire de *Nocardia asteroides*. *C.r. hebd. Séanc. Acad. Sci., Paris* **262C,** 1810.

Bordet, C. & Michel, G. (1969). Structure et biogenèse des lipides à haut poids moléculaire de *Nocardia asteroides*. *Bull. Soc. Chim. biol.* **51,** 527.

Bouisset, L., Breuilland, J. & Michel, G. (1963). Etude de l'ADN chez les Actinomycetales: comparaison entre les valeurs du rapport A+T/G+C et les caractères bactériologiques des corynebacterium. *Annls Inst. Pasteur, Paris* **104,** 756.

Bousfield, I. J. (1972). A taxonomic study of some coryneform bacteria. *J. gen Microbiol.* **71,** 441.

Bowie, I. S., Grigor, M. R., Dunckley, G. G., Loutit, M. W. & Loutit, J. S. (1972). The DNA base composition and fatty acid constitution of some Gram-positive pleomorphic soil bacteria. *Soil Biol. Biochem.* **4,** 397.

Brennan, P. & Ballou, C. E. (1967). Biosynthesis of mannophosphoinositides by *Mycobacterium phlei*. *J. biol. Chem.* **242,** 3046.

Brennan, P. & Ballou, C. E. (1968). Phosphatidylmyoinositol monomannoside in *Propionibacterium shermanii*. *Biochem. biophys. Res. Commun.* **30,** 69.

Brennan, P. J. & Lehane, D. P. (1971). The phospholipids of corynebacteria. *Lipids* **6,** 401.

Brennan, P. J., Lehane, D. P. & Thomas, D. W. (1970). Acylglucoses of the corynebacteria and mycobacteria. *Eur. J. Biochem.* **13,** 117.

Brennan, P. J., Rooney, S. A. & Winder, F. G. (1970). The lipids of *Mycobacterium tuberculosis* BCG: fractionation, composition, turnover and the effects of isoniazid. *I. J. med. Sci.* **3,** 371.

Campbell, I. M. & Bentley, R. (1968). Inhomogeneity of vitamin K_2 in *Mycobacterium phlei*. *Biochemistry, N.Y.* **7,** 3323.

Campbell, I. M. & Naworal, J. (1969). Composition of the saturated and monounsaturated fatty acids of *Mycobacterium phlei*. *J. Lipid Res.* **10,** 593.

Carroll, K. K., Cutts, J. H. & Murray, E. G. D. (1968). The lipids of *Listeria monocytogenes*. *Can. J. Biochem.* **46,** 899.

Chamoiseau, G. (1969). De l'étiologie du farcin de zébus tchadiens: nocardiose ou mycobacteriose? 1. Étude bactériologique et biochimique. *Rev. Élev. Méd. vét Pays trop.* **22,** 195.

Chamoiseau, G. (1973). *Mycobacterium farcinogenes* agent causal du farcin du boeuf en Afrique. *Ann. Microbiol (Inst. Pasteur)* **124A,** 215.

Chamoiseau, G. & Asselineau, J. (1970). Examen des lipides d'une souche de *Nocardia farcinica*: presence d'acides mycoliques. *C.r. hebd. Séanc. Acad. Sci., Paris* **270D,** 2603.

Cross, T. & Goodfellow, M. (1973). Taxonomy and classification of the actinomycetes. In *Actinomycetes: Characteristics and Practical Importance.* Eds F. A. Skinner & G. Sykes. p. 11. London: Academic Press.

Cummins, C. S. (1962). Chemical composition and antigenic structure of cell walls of *Corynebacterium, Mycobacterium, Actinomyces* and *Arthrobacter*. *J. gen Microbiol.* **28,** 35.

Cummins, C. S. & Harris, H. (1956). The chemical composition of the cell wall in some gram-positive bacteria and its possible value as a taxonomic character. *J. gen. Microbiol.* **14,** 583.

Diara, A. & Pudles, J. (1959). Sur les lipides de *Corynebacterium ovis*. *Bull. Soc. Chim. biol.* **41,** 481.

Dunphy, P. J., Phillips, P. G. & Brodie, A. F. (1971). Separation and identification of menaquinones from microorganisms. *J. Lipid Res.* **12,** 442.

Etémadi, A-H. (1963). Isolement des acides isopentadécanoique et isoheptadécanoique des lipides de *Corynebacterium parvum*. *Bull. Soc. Chim. biol.* **45,** 1423.

Etémadi, A-H. (1964). Techniques microanalytiques d'étude de structure d'esters α-ramifiés β-hydroxylés. Chromatographie en phase vapeur et spectrométrie de masse. *Bull. Soc. chim. Fr.* 1537.

Etémadi, A-H. (1967*a*). The use of pyrolysis gas chromatography and mass spectroscopy in the study of the structure of mycolic acids. *J. Gas Chromat.* **5,** 447.

Etémadi, A-H. (1967*b*). Correlations structurales et biogénétiques des acides mycoliques en rapport avec la phylogenèse de quelques genres d'actinomycétales *Bull. Soc. Chim. biol.* **49,** 695.

Etémadi, A-H. (1967*c*). Les acides mycoliques structure, biogenese et intérêt phylogénétique. *Exposés. a. Biochim. méd.* XXVIII, 77.

Etémadi, A-H., Gasche, J. & Sifferlen, J. (1965). Identification d'homoloques supérieurs des acides corynomycolique et corynomycolénique dans les lipides de *Corynebacterium* 506. *Bull. Soc. Chim. biol.* **47,** 631.

Etémadi, A-H., Markovits, J. & Pinte, F. (1966). Sur les différents types d'acides nocardiques isolés de *Nocardia opaca*. *C.r. hebd. Séanc. Acad. Sci., Paris* **263,** 835.

Etémadi, A-H., Okuda, R. & Lederer, E. (1964). Sur la structure de l'acide α-smegma-mycolique. *Bull. Soc. chim. Fr.* 868.

Farshtchi, D. & McClung, N. M. (1970). Effect of substrate on fatty acid production in *Nocardia asteroides. Can. J. Microbiol.* **16,** 213.

Fujii, K. & Fukui, S. (1969). Relationship between vitamin B_{12} content and ratio of mono-unsaturated fatty acids to methyl-branched fatty acids in *Corynebacterium simplex* cells grown on hydrocarbons. *FEBS Letters* **5,** 343.

Gale, P. H., Arison, B. H., Trenner, N. R., Page, A. C. Jr. & Folkers, K. (1963). Characterization of vitamin K_9 (H) from *Mycobacterium phlei. Biochemistry, N.Y.* **2,** 200.

Gerber, N. N. (1971). Prodigiosin-like pigments from *Actinomadura* (*Nocardia*) *pelletieri. J. Antibiot., Tokyo* **24,** 636.

Gerber, N. N. (1973). Minor prodiginine pigments from *Actinomadura madurae* and *Actinomadura pelletieri. J. Heterocycl. Chem.* **10,** 925.

Goldfine, H. (1972). Comparative aspects of bacterial lipids. *Adv. Microbiol. Physiol.* **8,** 1.

Gomes, N. F., Ioneda, T. & Pudles, J. (1966). Purification and chemical constitution of the phospholipids from *Corynebacterium diphtheriae* PW8. *Nature, Lond.* **211,** 81.

Goodfellow, M. (1971). Numerical taxonomy of some nocardioform bacteria. *J. gen. Microbiol.* **69,** 33.

Goodfellow, M. (1973). Characterisation of *Mycobacterium, Nocardia, Corynebacterium* and related taxa. *Annls Soc. belge Méd. trop.* **53,** 287.

Goodfellow, M., Lind, A., Mordarska, H., Pattyn, S. & Tsukamura, M. (1974). A co-operative numerical analysis of cultures considered to belong to the *rhodochrous* complex. *J. gen. Microbiol.* **85,** 291.

Goodfellow, M., Minnikin, D. E., Patel, P. V. & Mordarska, H. (1973). Free nocardomycolic acids in the classification of nocardiae and strains of the *rhodochrous* complex. *J. gen. Microbiol.* **14,** 185.

Goren, M. B. (1972). Mycobacterial lipids, selected topics. *Bact. Rev.* **36,** 33.

Guinand, M. & Michel, G. (1966). Structure d'un peptidolipide isole de *Nocardia asteroides,* la peptidolipine NA. *Biochim. biophys. Acta* **125,** 75.

Guinand, M., Michel, G. & Lederer, E. (1958). Sur les lipides de *Nocardia asteroides*; isolement de lipopeptides. *C.r. hebd. Séanc. Acad. Sci., Paris* **246,** 848.

Guinand, M., Michel, G. & Lederer, E. (1964). Structure de la peptidolipine NA. *C.r. hebd. Séanc. Acad. Sci., Paris* **259,** 1267.

Guinand, M., Vacheron, M. J. & Michel, G. (1970). Structure de parois cellulaires des *Nocardia.* I. Isolement et composition des parois de *Nocardia kirovani. FEBS Letters* **6,** 37.

Hertzberg, S. & Jensen, S. L. (1967). Bacterial carotenoids. XX. The carotenoids of *Mycobacterium phlei* strain Vera. The structures of the phlei xanthophylls -two novel tertiary glucosides. *Acta chem. scand.* **21,** 15.

Hill, L. R. (1966). An index to deoxyribonucleic acid composition of bacterial species. *J. gen. Microbiol.* **44,** 419.

Hofheinz, W. & Grisebach, H. (1965). Die Fettsäuren von *Streptomyces erythreus* und *Streptomyces halstedii. Z. Naturf.* **20,** 43.

Ioneda, T., Lederer, E. & Rozanis, J. (1970). Sur la structure des diesters de tréhalose ("cord factors") produit par *Nocardia asteroides* et *Nocardia rhodochrous*. *Chem. Phys. Lipids* **4,** 375.

Jensen, S. L. & Weedon, B. C. L. (1964). The structure of leprotene. *Naturwissenschaften* **51,** 482.

Jones, D. (1975). A numerical taxonomic study of coryneform and related bacteria. *J. gen. Microbiol.* **87,** 52.

Kanetsuna, F. & Bartoli, A. (1972). A simple chemical method to differentiate *Mycobacterium* from *Nocardia*. *J. gen. Microbiol.* **70,** 209.

Kanetsuna, F. & San Blas, G. (1970). Chemical analysis of a mycolic acid-arabinogalactan-mucopeptide complex of mycobacterial cell wall. *Biochim. biophys. Acta* **208,** 434.

Kanzaki, T., Sugiyama, Y., Kitano, K., Ashida, Y. & Imada, I. (1974). Quinones of *Brevibacterium*. *Biochim. biophys. Acta* **348,** 162.

Kataoka, T. & Nojima, S. (1967). The phospholipid compositions of some actinomycetes. *Biochem. biophys. Acta* **144,** 681.

Kawanami, J. (1971). Lipids of *Streptomyces toyocaensis* on the structure of siolipin. *Chem. Phys. Lipids* **7,** 159.

Kawanami, J. & Otsuka, H. (1969). Lipids of *Streptomyces sioyaensis*. VI. On the β-hydroxy fatty acids in siolipin. *Chem. Phys. Lipids* **3,** 135.

Khuller, G. K. & Brennan, P. J. (1972*a*). The polar lipids of some species of *Nocardia*. *J. gen. Microbiol.* **73,** 409.

Khuller, G. K. & Brennan, P. J. (1972*b*). Further studies on the lipids of corynebacteria. The mannolipids of *Corynebacterium aquaticum*. *Biochem. J.* **127,** 369.

Kikuchi, M. & Nakao, Y. (1973). Relation between cellular phospholipids and the excretion of L-glutamic acid by a glycerol auxotroph of *Corynebacterium alkanolyticium*. *Agr. Biol. Chem.* **37,** 515.

Kimura, A., Kawanami, J. & Otsuka, H. (1967*a*). Lipid of *Streptomyces sioyaensis*. Part 1. Lipid composition of *Streptomyces sioyaensis* and its comparison with those of various strains of Actinomycetales. *Agr. Biol. Chem.* **31,** 441.

Kimura, A., Kawanami, J. & Otsuka, H. (1967*b*). Lipids of *Streptomyces sioyaensis*. *J. Biochem., Tokyo* **62,** 384.

Kosaric, N. & Carroll, K. K. (1971). Phospholipids of *Listeria monocytogenes* *Biochem. biophys. Acta* **239,** 428.

Kostiw, L. L., Boylen, C. W. & Tyson, B. J. (1972). Lipid composition of growing and starving cells of *Arthrobacter crystallopoietes*. *J. Bact.* **111,** 103.

Krasilnikov, N. A., Koronelli, T. V. & Rozynov, B. V. (1972). Aliphatic and mycolic acids of *Mycobacterium paraffinicum*. *Mikrobiologiya* **41,** 808.

Krasilnikov, N. A., Koronelli, T. V., Rozynov, B. V. & Kalyuzhnaya, T. V. (1973). Mycolic acids of pigmented paraffin-oxidizing mycobacteria. *Mikrobiologiya* **42,** 240.

Krembel, J. & Etémadi, A-H. (1966). Sur la biogenèse des "acides mycoliques en C_{60}" isolés de *Mycobacterium smegmatis*. *Bull. Soc. Chim. biol.* **48,** 67.

Kunesch, G., Ferluga, J. & Etémadi, A-H. (1966). Etude microanalytique de la structure d'acides mycoliques insaturés par chromatographie pyrolytique et spectrométrie de masse: mise au point et applications. *Chem. Phys. Lipids* **1,** 41.

Kusamran, K., Polgar, N. & Minnikin, D. E. (1972). The mycolic acids of *Mycobacterium phlei*. *J. C. S. Chem. Comm.* 111.

Lacave, C., Asselineau, J. & Toubiana, R. (1967). Sur quelques constituants lipidiques de *Corynebacterium ovis*. *Eur. J. Biochem.* **2,** 37.

Lacey, J. & Goodfellow, M. (1975). A novel actinomycete from sugarcane bagasse: *Saccharopolyspora hirsuta* gen. et sp. nov. *J. gen. Microbiol.* **88,** 75.

Lanéelle, G. (1963). Nature des acides mycoliques de *Mycobacterium paratuberculosis*; application de la chromatographie sur couche mince à leur fractionnement. *C.r. hebd. Séanc. Acad. Sci., Paris* **257,** 781.

Lanéelle, G., Asselineau, J. & Chamoiseau, G. (1971). Presence de mycosides C' (formes simplifiées de mycoside C) dans les bacteries isolées de bovins atteints du farcin. *FEBS Letters* **19,** 109.

Lanéelle, G., Asselineau, J., Wolstenholme, W. A. & Lederer, E. (1965). Détermination de séquences d'acides aminés dans des oligopeptides par la spectrométrie de masse. III. Structure d'un peptidolipide de *Mycobacterium johnei*. *Bull. Soc. chim. Fr.* 2133.

Lanéelle, M-A. (1966). Sur la relation entre des cétones à haut poids moléculaire isolées de *Nocardia brasiliensis* et les acides nocardomycolique. *C.r. hebd. Séanc. Acad. Sci., Paris* **263C,** 560.

Lanéelle, M-A. & Asselineau, J. (1968). Sur les lipides de *Propionibacterium freudenreichii*. *C.r. hebd. Séanc. Acad. Sci., Paris* **266D,** 1901.

Lanéelle, M-A. & Asselineau, J. (1970). Caracterisation de glycolipides dans une souche de *Nocardia brasiliensis*. *FEBS Letters* **7,** 64.

Lanéelle, M-A., Asselineau, J. & Castelnuovo, G. (1965). Études sur les mycobactéries et les nocardiae. IV. Composition des lipides de *Mycobacterium rhodochrous, M. pellegrino* sp., et de quelques souches de nocardiae. *Annls Inst. Pasteur, Paris* **108,** 69.

Lanéelle, M-A., Asselineau, J. & Castelnuovo, G. (1968). Relations chimiques et immunologiques chez les Actinomycétales. IV. Composition chimique des lipides de quatre souches de *Streptomyces* et d'une souche *N.* (*Str.*) *gardneri*. *Annls Inst. Pasteur, Paris* **114,** 305.

Lanéelle, M-A., Lanéelle, G. & Asselineau, J. (1963). Sur la presence d'ornithine dans les lipides bactériens. *Biochim. biophys. Acta* **70,** 97.

Lechevalier, H. A., Lechevalier, M. P. & Gerber, N. N. (1971). Chemical composition as a criterion in the classification of actinomycetes. *Adv. appl. Microbiol.* **14,** 47.

Lechevalier, M. P. & Lechevalier, H. (1970). Chemical composition as a criterion in the classification of aerobic actinomycetes. *Int. J. Syst. Bact.* **20,** 435.

Lechevalier, M. P. & Lechevalier, H. (1974). *Nocardia amarae* sp. nov., an actinomycete common in foaming activated sludge. *Int. J. Syst. Bact.* **24,** 278.

Lechevalier, M. P., Horan, A. C. & Lechevalier, H. (1971). Lipid composition in the classification of nocardiae and mycobacteria. *J. Bact.* **105,** 313.

Lechevalier, M. P., Lechevalier, H. A. & Heintz, C. E. (1973). Morphological and chemical nature of the sclerotia of *Chainia olivacea* Thirumalachar and Sukapure of the order Actinomycetales. *Int. J. Syst. Bact.* **23,** 157.

Lechevalier, M. P., Lechevalier, H. & Horan, A. C. (1973). Chemical characteristics and classification of nocardiae. *Can. J. Microbiol.* **19,** 965.

Lederer, E. (1967). Glycolipids of mycobacteria and related microorganisms. *Chem. Phys. Lipids* **1,** 294.

Lederer, E., Pudles, J., Barbezat, S. & Trillat, J. J. (1952). Sur la constitution chimique de l'acide coryno-mycolique de bacille diphtérique. *Bull. Soc. chim. Fr.* 93.

Markovits, J., Pinte, F. & Etémadi, A-H. (1966). Sur la structure des acides mycoliques dicarboxyliques insaturés isolés de *Mycobacterium phlei. C.r. hebd. Séanc. Acad. Sci., Paris* **263C,** 960.

Markovits, J., Vilkas, E. & Lederer, E. (1971). Sur la structure chimique des cires D, peptidoglycolipides macromoléculaires des souches humaines de *Mycobacterium tuberculosis. Eur. J. Biochem.* **18,** 287.

Maurice, M. T., Vacheron, M. J. & Michel, G. (1971). Isolément d'acides nocardiques de plusieurs espèces de *Nocardia. Chem. Phys. Lipids* **7,** 9.

Michel, G., Bordet, C. & Lederer, E. (1960). Isolément d'un nouvel acide mycolique: l'acide nocardique à partir d'une souche de *Nocardia asteroides. C.r. hebd. Séanc. Acad. Sci., Paris* **250,** 3518.

Michel, G. & Lederer, E. (1962). Isolément et constitution chimique des nocardols de *Nocardia asteroides. Bull. Soc. chim. Fr.* 651.

Minnikin, D. E. & Abdolrahimzadeh, H. (1974). The replacement of phosphatidylethanolamine and acidic phospholipids by an ornithine-amide lipid and a minor phosphorus-free lipid in *Pseudomonas fluorescens* NCMB 129. *FEBS Letters* **43,** 257.

Minnikin, D. E., Abdolrahimzadeh, H. & Baddiley, J. (1974). Replacement of acidic phospholipids by acidic glycolipids in *Pseudomonas diminuta. Nature, Lond.* **249,** 268.

Minnikin, D. E., Alshamaony, L. & Goodfellow, M. (1975). Differentiation of *Mycobacterium, Nocardia,* and related taxa by thin-layer chromatographic analysis of whole-cell methanolysates. *J. gen. Microbiol.* **88,** 200.

Minnikin, D. E., Patel, P. V. & Goodfellow, M. (1974). Mycolic acids of representative strains of *Nocardia* and the "*rhodochrous*" complex. *FEBS Letters* **39,** 322.

Minnikin, D. E. & Polgar, N. (1967). The methoxymycolic and keto mycolic acids from human tubercle bacilli. *J.C.S. Chem. Comm.* 1172.

Miquel, A-M., Das, B. C. & Etémadi, A-H. (1966). Sur la structure des α-kansamycolones, cétones à haut poids moléculaire isolées de *Mycobacterium kansasii. Bull. Soc. chim. Fr.* 2342.

Mordarska, H. (1968). A trial of using lipids for the classification of actinomycetes. *Arch. Immun. Ther. exp.* **16,** 45.

Mordarska, H., Guzeva, L. N. & Agre, N. S. (1973). Lipids from the mycelia of thermophilic actinomycetes. *Mikrobiologiya* **42,** 165.

Mordarska, H. & Mordarski, M. (1969). Comparative studies on the occurrence of lipid A, diaminopimelic acid and arabinose in *Nocardia* cells. *Arch. Immun. Ther. exp.* **17,** 739.

Mordarska, H. & Mordarski, M. (1970). Cell lipids of *Nocardia.* In *The Actinomycetales.* Ed. H. Prauser. Jena: Gustav Fischer.

Mordarska, H., Mordarski, M. & Goodfellow, M. (1972). Chemotaxonomic characters and classification of some nocardioform bacteria. *J. gen. Microbiol.* **71,** 77.

Mordarska, H., Mordarski, M. & Pietkiewicz, D. (1973). Chemical analysis of hydrolysates and cell extracts of *Nocardia pellegrino*. *Int. J. Syst. Bact.* **23,** 274.

Mordarska, H. & Réthy, A. (1970). Preliminary studies on the chemical character of the lipid fraction of *Nocardia*. *Arch. Immun. Ther. Exp.* **18,** 455.

Morgan, E. D. & Polgar, N. (1957). Constituents of the lipids of tubercle bacilli. Part VIII. Studies on mycolic acid. *J. chem. Soc.* 3779.

Morris, L. J. (1964). *New Biochemical Separations.* Eds A. T. James & L. J. Morris. London: D. Van Nostrand.

Moss, C. W. & Cherry, W. B. (1968). Characterization of the C_{15} branched-chain fatty acids of *Corynebacterium acnes* by gas chromatography. *J. Bact.* **95,** 241.

Moss, C. W., Dowell, V. R., Lewis, V. J. & Schekter, M. A. (1967). Cultural characteristics and fatty acid composition of *Corynebacterium acnes*. *J. Bact.* **94,** 1300.

Moss, C. W., Dowell, V. R. Jr., Farshtchi, D., Raines, L. J. & Cherry, W. B. (1969). Cultural characteristics and fatty acid composition of propionibacteria. *J. Bact.* **97,** 561.

Nocard, E. (1888). Note sur la maladie de boeufs de la Guadeloupe, connue sous le nom de farcin. *Annls Inst. Pasteur, Paris* **2,** 293.

Noll, H., Bloch, H., Asselineau, J. & Lederer, E. (1956). The chemical structure of the cord factor of *Mycobacterium tuberculosis*. *Biochim. biophys. Acta* **20,** 299.

Norgård, S., Aasen, A. J. & Jensen, S. L. (1970). Bacterial carotenoids. XXXII C_{50}-carotenoids 6. Carotenoids from *Corynebacterium poinsettiae* including four new C_{50}-diols. *Acta chem. scand.* **24,** 2183.

Okabe, S., Shibukawa, M. & Ohsawa, T. (1967). L-glutamic acid fermentation with molasses. Part IX. Relation between the lipid in the cell membrane from *Microbacterium ammoniaphilum* and the extracellular accumulation of L-glutamic acid. *Agr. Biol. Chem.* **31,** 789.

Okazaki, H., Kanzaki, T. & Fukuda, H. (1968). L-glutamic acid fermentation. Part V. Behaviour of oleic acid in an oleic acid-requiring mutant. *Agr. Biol. Chem.* **32,** 1464.

Okazaki, H., Sugino, H., Kanzaki, T. & Fukuda, H. (1969). L-glutamic acid fermentation. Part VI. Structure of a sugar lipid produced by *Brevibacterium thiogenetalis*. *Agr. Biol. Chem.* **33,** 764.

Okuyama, H., Kankura, T. & Nojima, S. (1967). Positional distribution of fatty acids in phospholipids from mycobacteria. *J. Biochem.* **61,** 732.

Otsuka, S. & Shiio, I. (1968). Fatty acid composition of the cell-wall membrane fraction from *Brevibacterium flavum*. *J. gen. appl. Microbiol., Tokyo* **14,** 135.

Pangborn, M. C. (1968). Structure of mycobacterial phosphatides. *Ann. N.Y. Acad. Sci.* **154,** 133.

Pennock, J. F. (1966). Occurrence of vitamins K and related quinones. *Vitam. Horm.* **24,** 307.

Pommier, M-T. & Michel, G. (1972). Isolement et caractéristique d'un nouveau glycolipide de *Nocardia caviae*. *C.r. hebd. Séanc. Acad. Sci., Paris* **275C,** 1323.

Pommier, M-T. & Michel, G. (1973). Phospholipid and acid composition of *Nocardia* and nocardoid bacteria as criteria of classification. *Biochem. Syst.* **1,** 3.

Pridham, T. G. & Lyons, A. J. Jr. (1969). Progress in clarification of the taxonomic and nomenclatural status of some problem actinomycetes. *Devs ind. Microbiol.* **10,** 183.

Promé, J-C., Lacave, C. & Lanéelle, M-A. (1969). Sur les structures de lipides à ornithine de *Brucella melitensis* et de *Mycobacterium bovis. C.r. hebd. Séanc. Acad. Sci., Paris* **269C,** 1664.

Prottey, C. & Ballou, C. E. (1968). Diacyl myoinositol monomannoside from *Propionibacterium shermanii J. biol. chem.* **243,** 6196.

Pudles, J. & Lederer, E. (1954). Sur l'isolement et la constitution chimique de l'acide coryno-mycolénique et de deux cétones des lipides du bacille diphtérique. *Bull. Soc. Chim. biol.* **36,** 775.

Raines, L. J., Moss, C. W., Farshtchi, D. & Pittman, B. (1968). Fatty acids of *Listeria monocytogenes. J. Bact.* **96,** 2175.

Röhrscheidt, E. & Tárnok, I. (1972). Untersuchungen an *Nocardia*-pigmenten. Chromatographische Eigenschaften der Farbstoffe und ihre Bedeutung für die Differenzierung pigmentierter *Nocardia*-stamme. *Zentbl. Bakt. ParasitKde Abt. I.* **221,** 221.

Schleifer, K. H. & Kandler, O. (1972). Peptidoglycan types of bacterial cell walls and their taxonomic implications. *Bact. Rev.* **36,** 407.

Scholes, P. B. & King, H. K. (1965). Electron transport in a Park-Williams strain of *Corynebacterium diphtheriae. Biochem. J.* **97,** 754.

Schubert, K., Rose, G. & Hörhold, C. (1967). Cholesterin in *Streptomyces olivaceus. Biochim. biophys. Acta* **137,** 168.

Schwartz, A. C. (1973). Terpenoid quinones of the anaerobic *Propionibacterium shermanii.* 1. (*II, III*)—Tetrahydromenaquinone—9. *Arch Mikrobiol.* **91,** 273.

Senn, M., Ioneda, T., Pudles, J. & Lederer, E. (1967). Spectrométrie de masse de glycolipides 1. Structure du "cord factor" de *Corynebacterium diphtheriae. Eur. J. Biochem.* **1,** 353.

Shaw, N. (1968). The lipid composition of *Microbacterium lacticum. Biochem. biophys. Acta* **152,** 427.

Shaw, N. (1970). Bacterial glycolipids. *Bact. Rev.* **34,** 365.

Shaw, N. & Dinglinger, F. (1969). The structure of an acylated inositol mannoside in the lipids of propionic acid bacteria. *Biochem. J.* **112,** 769.

Shaw, N. & Stead, D. (1970). A study of the lipid composition of *Microbacterium thermosphactum* as a guide to its taxonomy. *J. appl. Bact.* **33,** 470.

Shaw, N. & Stead, D. (1971). Lipid composition of some species of *Arthrobacter. J. Bact.* **107,** 130.

Shibukawa, M., Kurima, M. & Ohuchi, S. (1970). L-Glutamic acid fermentation with molasses. Part XII. Relationship between the kind of phospholipids and their fatty acid composition in the mechanism of extracellular accumulation of *L*-glutamate. *Agr. Biol. Chem.* **34,** 1136.

Sone, N. (1974). Isolation of a novel menaquinone with a partly hydrogenated side chain from *Propionibacterium arabinosum. J. Biochem., Tokyo* **76,** 133.

Stees, J. L., Brown, O. R. & Grigsby, R. D. (1969). Structural features of the major pigment component of *Nocardia corallina. Microbios* **2,** 199.

Stodola, F. H., Lesuk, A. & Anderson, R. J. (1938). The chemistry of lipids or tubercle bacilli. LIV. The isolation and properties of mycolic acid. *J. biol. Chem.* **126,** 505.

Suzuki, T., Tanaka, K., Matsubara, I. & Kinoshita, S. (1969). Trehalose lipid and α-branched-β-hydroxy fatty acid formed by bacteria grown on n-alkanes. *Agr. Biol. Chem.* **33,** 1619.

Tabaud, H., Tisnovska, H. & Vilkas, E. (1971). Phospholipides et glycolipides d'une souche de *Micromonospora. Biochimie* **53,** 55.

Tadayon, R. A. & Carroll, K. K. (1971). Effect of growth conditions on the fatty acid composition of *Listeria monocytogenes* and comparison with fatty acids of *Erysipelothrix* and *Corynebacterium. Lipids* **6,** 820.

Takinami, K., Yoshii, H., Yamada, Y., Okada, H. & Kinoshita, K. (1968). Control of L-glutamic acid fermentation by biotin and fatty acid. *Amino Acid Nucleic Acid* **18,** 120.

Tárnok, I. & Tárnok, Zs. (1970). Carotenes and xanthophylls in mycobacteria. 1. Technical procedures; thin-layer chromatographic patterns of mycobacterial pigments. *Tubercle, Lond.* **51,** 305.

Tárnok, I. & Tárnok, Zs. (1971). Carotenes and xanthophylls in mycobacteria. 11. Lycopene α- and β-carotene and xanthophyll in mycobacterial pigments. *Tubercle, Lond.* **52,** 127.

Thoen, C. O., Karlson, A. G. & Ellefson, R. D. (1971*a*). Fatty acids of *Mycobacterium kansasii. Appl. Microbiol.* **21,** 628.

Thoen, C. O., Karlson, A. G. & Ellefson, R. D. (1971*b*). Comparison by gas-liquid chromatography of the fatty acids of *Mycobacterium avium* and some other non-photochromogenic mycobacteria. *Appl. Microbiol.* **22,** 560.

Thoen, C. O., Karlson, A. G. & Ellefson, R. D. (1972). Differentiation between *Mycobacterium kansasii* and *Mycobacterium marinum* by gas-liquid chromatographic analysis of cellular fatty acids. *Appl. Microbiol.* **24,** 1009.

Trevisan, V. (1889). *I generi e le specie delle Batteriaceae.* Milano.

Tsukamura, M. (1971). Proposal of a new genus, *Gordona,* for slightly acid-fast organisms occurring on sputa of patients with pulmonary disease and in soil. *J. gen. Microbiol.* **66,** 15.

Vacheron, M-J., Arpin, N. & Michel, G. (1970). Isolement d'esters de phleixanthophylle de *Nocardia kirovani. C.r. hebd. Séanc. Acad. Sci., Paris* **271C,** 881.

Vacheron, M-J. & Michel, G. (1971). Présence d'acides de poids moléculaire élevé dans les triglycérides isolés des parois de *Nocardia kirovani. C.r. hebd. Séanc. Acad. Sci., Paris* **273C,** 1778.

Vilkas, E., Adam, A. & Senn, M. (1968). Isolement d'un nouveau type de diester de trehalose a partir de *Mycobacterium fortuitum. Chem. Phys. Lipids* **2,** 11.

Vilkas, E. & Rojas, A. (1964). Sur les lipides de *Mycobacterium fortuitum. Bull. Soc. Chim. biol.* **46,** 689.

Voiland, A., Bruneteau, M. & Michel, G. (1971). Étude du mycoside C_2 de *Mycobacterium avium.* Determination de la structure. *Eur. J. Biochem.* **21,** 285.

Voss, J. G. (1970). Differentiation of two groups of *Corynebacterium acnes. J. Bact.* **101,** 392.

Walker, R. W., Barakat, H. & Hung, J. G. C. (1970). The positional distribution of fatty acids in the phospholipids and triglycerides of *Mycobacterium smegmatis* and *M. bovis* BCG. *Lipids* **5,** 684.

Walker, R. W. & Bastl, C. P. (1967). The glycolipids of *Arthrobacter globiformis. Carbohyd. Res.* **4,** 49.

Walker, R. W. & Fagerson, I. S. (1965). Studies of the lipids of *Arthrobacter globiformis.* 1. Fatty acid composition. *Can. J. Microbiol.* **11,** 229.

Weeks, O. B. & Andrewes, A. G. (1970). Structure of the glycosidic carotenoid corynexanthin. *Archs Biochem. Biophys.* **137,** 284.

Welby-Gieusse, M., Lanéelle, M-A. & Asselineau, J. (1970). Structure des acides corynomycoliques de *Corynebacterium hofmanii* et leur implication biogénétique. *Eur. J. Biochem.* **13,** 164.

Whiteside, T. L., De Siervo, A. J. & Salton, M. R. J. (1971). Use of antibody to membrane adenosine triphosphatase in the study of bacterial relationships. *J. Bact.* **105,** 957.

Wilkinson, S. G. & Bell, M. G. (1971). The phosphoglucolipid from *Pseudomonas diminuta. Biochim. biophys. Acta* **248,** 293.

Yamada, K. & Komagata, K. (1970). Taxonomic studies on coryneform bacteria. III. DNA base composition of coryneform bacteria. *J. gen. appl. Microbiol., Tokyo* **16,** 215.

Yanagawa, S., Fujii, K., Tanaka, A. & Fukui, S. (1972). Lipid composition and localization of 10-methyl branched-chain fatty acids in *Corynebacterium simplex* grown on n-alkanes. *Agr. Biol. Chem.* **36,** 2123.

Yanagawa, S., Tanaka, A. & Fukui, S. (1972). Fatty acid compositions of *Corynebacterium simplex* grown on 1-alkenes. *Agr. Biol. Chem.* **36,** 2129.

Yano, I., Furukawa, Y. & Kusunose, M. (1968). Incorporation of radioactivity from methionine-methyl-^{14}C into phospholipids by *Nocardia polychromogenes. J. Biochem., Tokyo* **63,** 133.

Yano, I., Furukawa, Y. & Kusunose, M. (1969*a*). Phospholipids of *Nocardia coeliaca. J. Bact.* **98,** 124.

Yano, I., Furukawa, Y. & Kusunose, M. (1969*b*). Occurrence of α-hydroxy fatty acids in Actinomycetales. *FEBS Letters* **4,** 96.

Yano, I., Furukawa, Y. & Kusunose, M. (1970*a*). 2-hydroxy fatty acid-containing phospholipid of *Arthrobacter simplex. Biochem. biophys Acta* **210,** 105.

Yano, I., Furukawa, Y. & Kusunose, M. (1970*b*). α-hydroxy fatty acid-containing phospholipids of *Nocardia leishmanii. Biochem. biophys. Acta* **202,** 189.

Yano, I., Furukawa, Y. & Kusunose, M. (1971*a*). Fatty-acid composition of *Arthrobacter simplex* grown on hydrocarbons. Occurrence of α-hydroxy-fatty acids. *Eur. J. Biochem.* **23,** 220.

Yano, I., Furukawa, Y. & Kusunose, M. (1971*b*). Occurrence of acylated trehaloses in *Nocardia. J. gen. appl. Microbiol., Tokyo* **17,** 329.

Yano, I. & Saito, K. (1972). Gas chromatographic and mass spectrometric analysis of molecular species of corynomycolic acids from *Corynebacterium ulcerans. FEBS Letters* **23,** 352.

Yano, I., Saito, K., Furukawa, Y. & Kusunose, M. (1972). Structural analysis of molecular species of nocardomycolic acids from *Nocardia erythropolis* by the combined system of gas chromatography and mass spectroscopy. *FEBS Letters* **21,** 215.

Yribarren, M., Vilkas, E. & Rozanis, J. (1974). Galactosyl-diglyceride from *Actinomyces viscosus*. *Chem. Phys. Lipids* **12,** 172.

8. Serological Relationships Between *Nocardia*, *Mycobacterium*, *Corynebacterium* and the "*Rhodochrous*" Taxon

A. Lind and M. Ridell

Institute of Medical Microbiology, University of Göteborg, Göteborg, Sweden

Contents

1. Introduction

The antigenic composition of many species of diverse genera have been fairly well characterized. Knowledge of, e.g., the presence of cross-reacting and of species-specific antigens has been a prerequisite for the application of serodiagnostic methods and for the use of serological methods in determinative bacteriology. Serological techniques have also been found to be an important tool in taxonomical studies.

Antigenic mosaics of great complexity have been demonstrated among bacteria of the genera *Nocardia, Mycobacterium, Corynebacterium* and also among those classified in the "*rhodochrous*" complex, a taxon in search of an established generic niche (see chapter 2). The taxonomy of these groups of bacteria has been extensively studied but many problems remain unsolved. Thus, the borderlines between the genera are not yet settled, the delineation between several species within these taxa is incomplete and new species probably still remain to be discovered and established.

An improved and extended knowledge of the serological relationships between the genera *Nocardia, Mycobacterium, Corynebacterium* and the *rhodochrous* taxon would be useful in disentangling a complicated taxonomical situation. Further it would be of clinical benefit in, e.g., studies aiming at more specific reagents for use in diagnostic skin tests, the preparation of new types of vaccines, development of amended methods for serological identification of bacteria as well as for the elaboration of new serodiagnostic methods.

In this review, reports on sero-taxonomic studies carried out on the genera *Nocardia, Mycobacterium, Corynebacterium* and the *rhodochrous* taxon are presented. The emphasis is laid on the occurrence of antigens cross-reacting between these taxa. It should also be mentioned that this review is restricted to serology *sensu stricto* and accordingly studies on cellular immunity are not included.

2. Serological Analyses

(a) *Analyses by means of agglutination techniques*

Schneidau & Shaffer (1960), using a slide-agglutination technique, studied the possible occurrence of serological cross-reactivity between bacteria representing a number of species of mycobacteria and nocardiae. They found that strains designated *N. asteroides, N. brasiliensis, N. corallina* and *N. opaca* exhibited various degrees of cross-reactivity when tested with antimycobacterial sera. Alternatively, mycobacteria such as *M. fortuitum, M. phlei, M. smegmatis* and *M. tuberculosis* did occasionally cross-react with antisera prepared against the *Nocardia* strains mentioned.

Cummins (1962) also used an agglutination technique when analysing the antigenic structure of cell walls of *Nocardia, Mycobacterium, Corynebacterium* and some other genera. He compared the results of the serological tests with those obtained from analyses of the chemical composition of the cell walls of strains classified in these genera. A common antigenic component was identified in all the nocardiae, mycobacteria and corynebacteria strains which had arabinose and galactose as their principle wall sugars. Some strains of corynebacteria and three strains of *N. pelletieri* which had a different pattern of wall components, appeared to lack this common cell wall antigen. Cummins concluded that the results of the serological studies seemed to confirm the relationships suggested by the wall composition data.

The agglutination method has been found to be a useful tool in species and type determination of mycobacteria (Shaefer, 1965, 1967). Further application of this method for taxonomical studies of the taxa treated in this review might contribute to an extended knowledge of their antigenic relationships.

(b) *Analyses by means of hemagglutination techniques*

Thurston, Phillips & Pier (1968) sensitized erythrocytes with antigens from unheated culture filtrates of *N. asteroides* and found that such erythrocytes reacted with antisera prepared against nocardiae and also against antisera prepared against *M. bovis* and *M. kansasii.* Specificities not apparent in the direct test were evident when hemagglutination and hemolytic inhibition tests were used. Inhibition reactions showed the likelihood that the photochromogenic and scotochromogenic strains tested were antigenically more closely related to nocardiae than they were to other mycobacteria. They indicated that this noteworthy finding should be further investigated.

(c) *Analyses by means of complement-fixation techniques*

In 1956 Kwapinski reported on a serological relationship between *M. tuberculosis* and *C. diphtheriae* using the complement-fixation method. Later he studied the serological similarities of the cell walls from a great number of Actinomycetales strains (Kwapinski, 1964). Cell walls of different species of *Mycobacterium* seemed to share some antigens with the walls of *N. asteroides, N. madurae, N. farcinica* (ATCC3318), *N. rubra* and *Jensenia canicruria.* Serological relationships between walls of mycobacteria and those of *N. caprae, N. corallina, N. brasiliensis, N. pelletieri, N. eppingeri* and *N. lutea* were less prononounced. The walls of *M. marinum* (*balnei*), *M. avium* and *M. thamnopheos* exhibited the least serological cross-reactivity with the *Nocardia* cell wall antisera. Practically no serological relationship was found between the walls of mycobacteria and *N. leishmanii, N. erythropolis, N. opaca, N.*

caviae, *N. polychromogenes*, *N. pretoriana*, *N. rangoonensis* and *N. blackwellii*.

Pier, Thurston & Larsen (1968) performed comparative studies on sera from cattle with nocardiosis or mycobacteriosis. Using, among other methods, the complement-fixation test they found that a mycobacterial complement-fixing antigen did not differentiate between sera from cattle infected with either *N. asteroides* or a *Mycobacterium* species. However, a nocardial antigen did not produce cross-reactions with sera of cattle infected with mycobacteria.

(d) *Analyses by means of immunodiffusion techniques*

The most comprehensive studies on the serological relationships between *Nocardia*, *Mycobacterium*, *Corynebacterium* and the *rhodochrous* taxon have been performed with the immunodiffusion and immunoelectrophoresis methods.

Castelnuovo *et al.* (1964) carried out a study on the serological relationships between mycobacteria and nocardiae in which immunodiffusion, as well as immunoelectrophoresis, techniques were employed. They demonstrated that all mycobacterial and nocardial strains studied possessed common antigens. Some strains labelled *M. rhodochrous*, *M. pellegrino*, *N. rubra*, *N. corallina* and *N.* (*Serratia*) sp. were "antigenically identical or closely related," and it was proposed that these taxa be classified within one group. They considered that this taxon was more closely related to the nocardiae than to the mycobacteria. In a later report Castelnuovo *et al.* (1968) concluded that strains of the genera *Streptomyces*, *Nocardia* and *Mycobacterium* have at least one antigen or hapten in common.

Kwapinski (1966*a*) studied the so-called exo-antigens, i.e. culture filtrates, prepared from strains of Actinomycetales including, among others, mycobacteria, nocardiae and corynebacteria. Employing ordinary double-dimension immunodiffusion, as well as immunoelectrophoresis, techniques he demonstrated that culture filtrates of strains belonging to these genera, as well as those prepared from *rhodochrous* strains, cross-reacted with antisera against strains of the heterologous taxa. As a rule only one precipitinogen was found to be in common but in 8% of the analyses 2–4 precipitinogens were registered.

Kwapinski (1966*b*) and Kwapinski *et al.* (1973) also studied the serological relationships between cytoplasmatic antigens from strains of certain Actinomycetales species. In the studies of these antigens *Nocardia* strains were found "mainly to be immunologically related to the saprophytic and scotochromogenic mycobacteria." However, these investigators considered it significant that the cytoplasmic antigens of *N. brasiliensis* were related exclusively to such antigens of pathogenic mycobacteria such as *M. tuberculosis*, *M. kansasii*, *M. avium* etc. Based on the demonstrated antigenic patterns Kwapinski and his co-workers divided the studied strains into a relatively large number of serogroups. Kwapinski has treated the serological relationships between different Actinomycetales species in a comprehensive review (Kwapinski, 1969).

Since 1957 the serological relationships between strains of the genus *Mycobacterium* have been studied at the Institute of Medical Microbiology in Göteborg by means of various immunodiffusion techniques. During the last decade these studies have been extended to include taxa such as *Nocardia*, *Corynebacterium* and the *rhodochrous* complex. Special attention has been paid to the existence of antigens common to these different taxa. These intergeneric serological investigations have so far included some 270 strains carrying 64 different species designations on their arrival at the Institute.

As a temporary aid the species names represented by the studied strains have been placed in one of five artificial groups called "*Nocardia*," "*Mycobacterium*," "*Corynebacterium*," "*Rhodochrous*" and "*Nocardia farcinica*" (see Table 1). This grouping is justified for the presentation of the results and the discussion but should not be considered as a base for revision of the taxonomical status of the species listed.

Comparative immunodiffusion analyses were performed employing in total 16 reference precipitation systems representing different species. The species names of the strains from which the reference systems were prepared are shown in Table 2. A schematic figure showing the pattern obtained in a comparative immunodiffusion analysis using one of the reference systems is shown in Fig. 1. It can be seen that five precipitinogens are common for the test strain and the reference strain. Three of the

Table 1
Species names of studied strains

Nocardia	***Rhodochrous***
Nocardia asteroides	*Bacillus havaniensis*
Nocardia blackwellii	*Bacillus mycoides corallinus*
Nocardia brasiliensis	*Bacillus mycoides roseus*
Nocardia caviae	*Bacillus rubricus*
Nocardia dassonvillei	*Corynebacterium equi*
Nocardia madurae	*Corynebacterium fascians*
	Corynebacterium rubrum
Mycobacterium	*Corynebacterium* sp.
Mycobacterium avium	*Jensenia* sp.
Mycobacterium borstelense	*Micrococcus rhodochrous*
Mycobacterium bovis var. BCG	*Mycobacterium agreste*
Mycobacterium fortuitum	*Mycobacterium crystallophagum*
Mycobacterium gastri	*Mycobacterium eos*
Mycobacterium gordonae	*Mycobacterium lacticola*
Mycobacterium intracellulare	*Mycobacterium pellegrino*
Mycobacterium kansasii	*Mycobacterium rhodochrous*
Mycobacterium marianum	*Mycobacterium rubropertinctum*
Mycobacterium marinum	*Mycobacterium rubrum*
Mycobacterium microti	*Mycobacterium* sp.
Mycobacterium phlei	*Nocardia corallina*
Mycobacterium smegmatis	*Nocardia erythropolis*
Mycobacterium terrae	*Nocardia globerula*
Mycobacterium vaccae	*Nocardia polychromogenes*
Mycobacterium xenopi	*Nocardia restrictus*
	Nocardia rhodnii
Corynebacterium	*Nocardia rubra*
Corynebacterium bovis	*Nocardia rubropertincta*
Corynebacterium diphtheriae	*Nocardia* sp.
Corynebacterium hoffmanii	*Proactinomyces erythropolis*
Corynebacterium minutissimum	*Proactinomyces globerula*
Corynebacterium murium	*Proactinomyces restrictus*
Corynebacterium pseudotuberculosis	*Proactinomyces* sp.
Corynebacterium pyogenes	*Rhodococcus rhodochrous*
Corynebacterium renale	
Corynebacterium ulcerans	***Nocardia farcinica***

precipitinogens of the test strain are identified by means of the reference system. These three precipitinogens correspond to the precipitates designated *bB*, *dD* and *eE* of the reference system. The presence of *b* and *d* in the test strain is demonstrated by the fusion of lines while the presence of *e* is demonstrated only by deviation phenomena.

Table 2
Reference systems

Species	No. of labelled precipi-tinogens	Species	No. of labelled precipi-tinogens
Nocardia		*Mycobacterium bovis* var. BCG	12
Nocardia asteroides	7	*Mycobacterium fortuitum*	15
Nocardia caviae	10	*Mycobacterium kansasii*	12
		Mycobacterium marianum	10
Rhodochrous		*Mycobacterium marinum*	8
Mycobacterium pellegrino	7	*Mycobacterium microti*	15
Nocardia corallina	7	*Mycobacterium phlei*	13
Nocardia polychromogenes	10	*Mycobacterium smegmatis*	11
Nocardia sp. *Serratia*	5	***Nocardia farcinica***	8
Myocobacterium			
Mycobacterium avium	7		

In concordance with studies performed by other investigators using immunodiffusion methods Lind & Norlin (1963), Norlin (1965) and Norlin, Lind & Ouchterlony (1969) showed the presence of several interspecies cross-reacting precipitinogens within the genus *Mycobacterium*. The common mycobacterial precipitinogens found by them were designated α, β, γ etc.

By means of nine mycobacterial reference systems Ridell & Norlin (1973) analysed 56 *Nocardia* and *rhodochrous* strains and

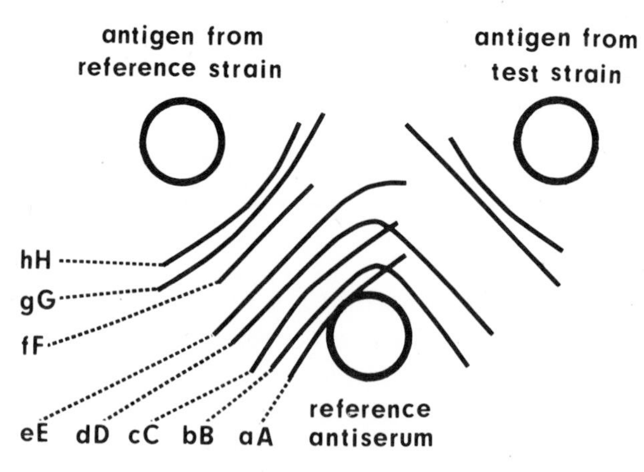

Fig. 1. Schematic illustration of a comparative immunodiffusion analysis.

found that many of the tested strains had one, two or three precipitinogens in common with the mycobacterial reference strains. Two of the common precipitinogens in the tested *Nocardia* and *rhodochrous* strains were identified as the common mycobacterial precipitinogens α and β (see Table 3).

Table 3

Presence of common mycobacterial precipitinogens in the tested Nocardia *and* rhodochrous *strains*

Species	No. of tested strains	No. of strains in which mycobacterial common precipitinogens were identified		
		α	$p\alpha$	β
N. asteroides	10	10		2
N. blackwellii	1	1		
N. brasiliensis	2	2		
N. caviae	3	3		1
N. corallina	7		7	3
N. farcinica	8	8		5
N. polychromogenes	2		2	1
N. rubra	10		10	9
N. rubropertincta	2		2	1
Nocardia sp.	1		1	1
"*M.*" *pellegrino*	6		4	3
"*M.*" *rhodochrous*	4		4	3
Total	56	24	30	29

Two different kinds of the precipitinogen α were found. One was demonstrated in all of the strains designated *N. asteroides*, *N. blackwellii*, *N. brasiliensis*, *N. caviae* and *N. farcinica*. This precipitinogen was identical to the mycobacterial common precipitinogen α. The other one was revealed in strains with the species designations *N. corallina*, *N. polychromogenes*, *N. rubra*, *N. rubropertincta*, "*M.*" *pellegrino* and "*M.*" *rhodochrous*. This precipitinogen, designated $p\alpha$, was closely related but not identical with α. Strains in which the precipitinogen $p\alpha$ was shown represent species currently classified in the *rhodochrous* taxon. Thus, serological differentiation based on the presence of either α or $p\alpha$ is of taxonomical value.

The precipitinogen β was demonstrated in about half of the tested *Nocardia* and *rhodochrous* strains. It has also been found in most of the mycobacterial species examined. Therefore, precipitinogen β cannot be used for inter-generic delineations.

Ridell (1974) analysed 125 *Nocardia, Mycobacterium* and *rhodochrous* strains by means of two reference precipitation systems representing *N. corallina* and "*M.*" *pellegrino*. Most of the tested strains had one, two or three precipitinogens in common with the *N. corallina* and the "*M.*" *pellegrino* reference strains. For some of the strains, however, five, six or seven precipitinogens were revealed indicating a comparatively high degree of serological similarity between these strains and the reference strain. Table 4 gives details on all the tested strains (in total 28) referred to the species *N. corallina, N. rubra,* "*M.*" *pellegrino* and "*M.*" *rhodochrous* when tested by means of the two reference systems. Ten strains, designated *N. rubra,* "*M.*" *pellegrino* or "*M.*" *rhodochrous*, revealed a close serological relationship with the "*M.*" *pellegrino* reference strain. This indicates that these 11 strains (results encircled in Table 4, left column) might belong to one taxonomic group. Eight other strains, designated *N. corallina, N. rubra* or "*M.*" *rhodochrous*, displayed a close serological relationship with the *N. corallina* reference strain and these nine strains (results encircled in Table 4, right column) may therefore belong to a second taxonomic group. Further, eight strains with the epithets *N. corallina, N. rubra* or "*M.*" *rhodochrous* showed little serological similarity to either of the two reference strains and possibly represent additional groups. This study indicates that the serotaxonomical results are not in concordance with the species designations of the studied strains.

The taxonomical status of strains with the designation *N. farcinica* is unclear (see chapter 1). Ridell (1975) has studied strains with this species-designation by means of the immunodiffusion method. Antigen preparations from 34 *N. farcinica* strains were analysed by means of 12 reference systems prepared from various nocardial and mycobacterial strains and from strains of the *rhodochrous* taxon. Based on the number of common precipitinogens two distinct clusters were recognized. The strains of one of the clusters displayed a close serological relationship with the mycobacterial reference strains and should probably be classified in the genus *Mycobacterium*. The strains in the other cluster

Table 4

Number of demonstrated precipitinogens in strains designated N. corallina, N. rubra, "M." pellegrino *and* "M." rhodochrous *revealed by means of the* "M." pellegrino *and the* N. corallina *reference systems*

Strains	Reference system "*M.*" *pellegrino*	*N. corallina*
N. corallina		
N28	3	3
N29	2	3
N30	1	⑥
N31	2	⑦
N32 (reference strain)	1	⑦
N52	3	⑥
N53	4	⑦
N. rubra		
N19	3	2
N20	⑥	3
N21	3	2
N23	2	⑦
N24	⑦	3
N25	1	⑦
N26	⑤	2
N27	3	2
N50	3	2
N51	3	3
"*M.*" *pellegrino*		
N12	⑦	2
N13 (reference strain)	⑦	2
N14	⑥	1
N15	⑤	3
N16	⑦	3
N48	⑥	3
N49	⑥	3
"*M.*" *rhodochrous*		
N35	1	⑦
N36	2	⑦
N45	⑥	2
N46	3	3

The number of precipitinogens ≥5 are encircled.

showed a close serological relationship with the *Nocardia* reference strains and should probably be retained in this genus. However, most of the tested strains irrespective of the cluster they were classified in cross-reacted with the reference strains to some extent.

Studies on mycobacterial ribosomes have been performed by means of the immunodiffusion method (Ridell, Baker, Lind, Norlin & Ouchterlony, to be published). It was found that ribosomal preparations from mycobacterial strains of various species cross-reacted with extracts of disintegrated cells prepared from *Nocardia* as well as from *rhodochrous* strains.

In a cooperative study performed by the International Working Group on Mycobacterial Taxonomy (IWGMT) (Goodfellow *et al.*, 1974) cultures considered to belong to the *rhodochrous* complex were examined using chemical, numerical and immunodiffusion techniques. In this investigation 78 *rhodochrous* cultures, 12 marker cultures with the genus designation *Mycobacterium* and 8 with the designation *Nocardia* were studied. Twelve of the 16 reference systems mentioned earlier were employed in the serological analyses. These systems represented *M. bovis* var. *BCG*, *M. microti*, *M. kansasii*, *M. marianum*, *M. avium*, *M. fortuitum*, *M. phlei*, *N. asteroides*, *N. caviae*, "*M.*" *rhodochrous*, "*M.*" *pellegrino* and *N. polychromogenes*. Only the number of precipitates of the reference systems which showed deviation due to the influence of a corresponding precipitinogen in the culture under study was taken into consideration. Consequently, the number of registered precipitinogens was lower in this study than in the earlier investigations (Ridell & Norlin, 1973; Ridell, 1974). However, this study also demonstrated precipitinogens common for *Mycobacterium*, *Nocardia* and the *rhodochrous* taxon. The analyses carried out in this study resulted in fairly good agreement between the serological, chemical and numerical data.

The question of a generic location for the *rhodochrous* strains cannot be settled until their relationship with other established genera such as *Corynebacterium* have been ascertained. In immunodiffusion analyses we demonstrated that nine tested strains of *Corynebacterium* with different species designations had at least one or two precipitinogens in common with one or more of the *Mycobacterium*, *Nocardia* or *rhodochrous* reference strains.

In summary it can be stated that bacteria belonging to the genera *Nocardia*, *Mycobacterium*, *Corynebacterium* and to the *rhodochrous* taxon have precipitinogens in common. Our results concerning

Table 5
Number of demonstrated common precipitinogens

Taxa	Reference systems		
	Nocardia	"*Rhodochrous*"	*Mycobacterium*
Nocardia	0–total	0–4	0–3
"*Rhodochrous*"	0–4	0–total	0–3
Mycobacterium	0–3	0–3	0–total
Corynebacterium	0–2	0–2	0–2

demonstrated common precipitinogens are summarized in Table 5. It was found that most of the pertinent taxa shared up to three precipitinogens. However, a small number of strains referred to the *rhodochrous* taxon had four precipitinogens in common with some of the *Nocardia* reference strains. Some of the strains referred to *Nocardia* had also four precipitinogens in common with some of the *rhodochrous* reference strains. Only two precipitinogens have hitherto been shown as common for corynebacterial strains and strains of the other taxa. It should, however, be emphasized that the actual numbers of demonstrated common precipitinogens should not be considered as final. By further development of methods for the preparation of antigens and antisera as well as by using more sensitive techniques for analyses it is probable that additional common precipitinogens will be revealed.

3. Immunochemical Analyses

Cummins (1962) found a common antigenic component in strains of corynebacteria, mycobacteria and nocardiae which had arabinose and galactose as their principle cell wall sugars. Further studies on these wall polysaccharides were later performed by Zamora, Bojalil & Bastarrachea (1963) and Estrada-Parra, Zamora & Bojalil (1965). They studied the immunochemistry of polysaccharides of *N. brasiliensis* and *N. asteroides.* Two polysaccharides were isolated from *N. brasiliensis*, one designated "Poly I Nb" and the other "Poly II Nb." Poly I Nb was chemically characterized as a polymer of D-arabofuranose and D-galactopyranose in a molar ratio 3 : 1. This polysaccharide, Poly I Nb, prepared from *N. brasiliensis* was apparently identical to a corresponding polysaccharide of *N. asteroides*—they cross-reacted serologically—and was therefore regarded as group-specific. Poly II Nb, which, in addition to arabinose and galactose, contained mannose was considered to be species-specific. It was also shown that the group-specific Poly I Nb was precipitated with sera from patients with tuberculosis and leprosy giving a single band of identity in immunodiffusion tests. In this connection it should be noted that antibodies against the previously mentioned precipitinogen β (Norlin, 1965) have been demonstrated in sera from patients with leprosy (Norlin *et al.*, 1966). The possible relation or identity between Poly I Nb and the precipitinogen β has not yet been investigated.

In a series of papers Azuma and co-workers have studied the chemical characteristics of cell walls in mycobacteria, nocardiae and corynebacteria (Azuma, Ajisaka & Yamamura, 1970; Yamamura, Misaki & Azuma, 1972; Azuma *et al.*, 1973). They have proposed that the wall structures of these microorganisms appear to have a "mycolic acid-arabinogalactan-mucopeptide" complex in common. Two polysaccharides, arabinomannan and arabinogalactan, were purified from bacteria classified in these three genera and investigated in great detail with regard to their chemical structure. These purified preparations showed a potent serological activity in precipitation, complement-fixation and passive hemagglutination tests with rabbit antisera prepared against heat-killed homologous and heterologous whole cells or cell walls (Yamamura, Misaki & Azuma 1972).

At the Mycobacterial Antigen Workshop held in San Francisco, California, 1971 (Mycobacterial Antigen Workshop, 1972) it was stated that arabinomannan can exist in varying molecular sizes of polymeric forms; it could be separated and isolated in several electrophoretically distinct fractions, all of which were found to be antigenically identical by immunodiffusion. In contrast, arabinogalactan did not exhibit a varying molecular composition. At this workshop Daniel stated that arabinomannan and arabinogalactan are related to Seibert's polysaccharide I. Further, Kubica concluded that at least the arabinomannan antigen is common to mycobacteria, nocardiae and, perhaps, even to corynebacteria.

In comparison with the studies on common antigens of polysaccharide nature there are less investigations published concerning common protein antigens. However, Affronti (1959) showed that a PPD preparation obtained from an *N. asteroides* strain cross-reacted in immunodiffusion analysis with an antiserum against *M. intracellulare*. Kwapinski has reported on serological analyses of cytoplasmatic protein fractions from different strains of the order Actinomycetales (Kwapinski & Snyder, 1961; Kwapinski, 1966*b*; Kwapinski *et al.*, 1973). It was stated that the antigen in these plasms is a polysaccharide-nucleo-protein complex and that cross-reactions occurred between plasm-preparations derived from strains of *Nocardia, Mycobacterium* and *Corynebacterium.*

4. Concluding Remarks

In spite of the considerable amount of serological mapping of the common antigens which has been accomplished several lacunae do exist and it is obvious that supplementary serotaxonomical studies are necessary.

Taking into consideration the present serotaxonomical facts it would be of definitive interest to analyse in greater detail the antigenic determinants from the physio-chemical point of view. Such studies should comprise polysaccharides, proteins as well as ribosomal materials, primarily from the taxa mentioned—*Nocardia, Mycobacterium, Corynebacterium* and the *rhodochrous* complex—but also from other related taxa. Such studies are a prerequisite for a proper immunochemical taxonomy.

When immunochemically well characterized compounds have been isolated they should also be analysed concerning their biological properties and their clinical importance should be elucidated.

5. Acknowledgements

This review was made possible by economical support from the Medical Faculty of the University of Göteborg, the World Health Organization, the Swedish National Association against Heart and Chest Diseases and the Ellen, Walter and Lennart Hesselman's Foundation.

6. References

Affronti, L. F. (1959). Purified protein derivatives (PPD) and other antigens prepared from atypical acid-fast bacilli and *Nocardia asteroides*. *Am. Rev. Tuberc. pulm. Dis.* **79,** 284.

Azuma, I., Ajisaka, M. & Yamamura, Y. (1970). Polysaccharides of *Mycobacterium bovis* Ushi 10, *Mycobacterium smegmatis, Mycobacterium phlei* and atypical *Mycobacterium* Pl. *Infect. Immun.* **2,** 347.

Azuma, I., Kanetsuna, F., Tanaka, Y., Mera, M., Yanagihara, Y., Mifuchi, I. & Yamamura, Y. (1973). Partial chemical characterization of the cell wall of *Nocardia asteroides* strain 131. *Jap. J. Microbiol.* **17,** 154.

Castelnuovo, G., Bellezza, G., Duncan, M. E. & Asselineau, J. (1964). Etudes sur les mycobactéries et les nocardiae. *Annls Inst. Pasteur, Paris* **107,** 828.

Castelnuovo, G., Bellezza, G., Giuliani, H. I. & Asselineau, J. (1968). Relations chimiques et immunologiques chez lez Actinomycetales. *Annls Inst. Pasteur, Paris* **114,** 139.

Cummins, C. S. (1962). Chemical composition and antigenic structures of cell walls of *Corynebacterium, Mycobacterium, Nocardia, Actinomyces* and *Arthrobacter. J. gen. Microbiol.* **28,** 35.

Estrada-Parra, S., Zamora, A. & Bojalil, L. F. (1965). Immunochemistry of the group-specific polysaccharide of *Nocardia brasiliensis. J. Bact.* **90,** 571.

Goodfellow, M., Lind, A., Mordarska, H., Pattyn, S. & Tsukamura, M. (1974). A co-operative numerical analysis of cultures considered to belong to the *rhodochrous* complex. *J. gen. Microbiol.* **85,** 291.

Kwapinski, J. B. (1956). Antigenic relationship between the genera *Mycobacterium* and *Corynebacterium. Bull. Acad. pol. Sci. Cl.* 11 *Sér. Sci. biol.* **4,** 379.

Kwapinski, J. B. (1964). Antigenic structure of the actinomycetales. VII. Chemical and serological similarities of cell walls from 100 actinomycetales strains. *J. Bact.* **88,** 1211.

Kwapinski, J. B. (1966*a*). Antigenic structure of the actinomycetales. X. Spectra of serological activities of the exo-antigens. *Zentrbl. Bakt. ParasitKde Abt.* 1 **200,** 80.

Kwapinski, J. B. (1966*b*). Antigenic structure of actinomycetales. XI. Spectra of the serological activity of the plasm antigens. *Zentrbl. Bakt. ParasitKde Abt* 1 **200,** 380.

Kwapinski, J. B. (1969). *Analytic Serology of Microorganisms*, vols. 1 & 2. New York: Interscience Publishers.

Kwapinski, J. B. G., Kwapinski, E. H., Dowler, J. & Horsman, G. (1973). The phyloantigenic position of nocardiae revealed by examination of cytoplasmic antigens. *Can. J. Microbiol.* **19,** 955.

Kwapinski, J. B. & Snyder, M. L. (1961). Antigenic structure and serological relationships of *Mycobacterium, Actinomyces, Streptococcus* and *Diplococcus. J. Bact.* **82,** 632.

Lind, A. & Norlin, M. (1963). A comparative serological study of *M. avium, M. ulcerans, M. balnei,* and *M. marinum* by means of double diffusion-in-gel methods. A preliminary investigation. *Scand. J. clin. Lab. Invest.* **15,** 152.

Mycobacterial Antigen Workshop (1972). An evaluation of a reference system for antigens of *Mycobacterium tuberculosis*. *Am. Rev. resp. Dis.* **106,** 142.

Norlin, M. (1965). Unclassified mycobacteria: a comparison between a serological and a biochemical classification method. *Bull. Un. int. tuberculose* **36,** 25.

Norlin, M., Lind, A. & Ouchterlony, Ö. (1969). A serologically based taxonomic study of *M. gastri. Z. ImmunForsch. exp. Ther.* **137,** 241.

Norlin, M., Navalkar, R. G., Ouchterlony, Ö. & Lind, A. (1966). Characterization of leprosy sera with various mycobacterial antigens using double diffusion-in-gel analysis-III. *Acta Path. microbiol. scand.* **67,** 555.

Pier, A. C., Thurston, J. R. & Larson, A. B. (1968). A diagnostic antigen for nocardiosis: comparative tests in cattle with nocardiosis and mycobacterosis. *Am. J. vet. Res.* **29,** 397.

Ridell, M. (1974). Serological study of nocardiae and mycobacteria by using "*Mycobacterium*" *pellegrino* and *Nocardia corallina* precipitation reference systems. *Int. J. Syst. Bact.* **24,** 64.

Ridell, M. (1975). A taxonomical study of *Nocardia farcinica* using serological and physiological characters. *Int. J. Syst. Bact.* **25,** 124.

Ridell, M. & Norlin, M. (1973). Serological study of nocardia by using mycobacterial precipitation reference systems. *J. Bact.* **113,** 1.

Schaefer, W. B. (1965). Serologic identification and classification of the atypical mycobacteria by their agglutination. *Am. Rev. resp. Dis.* **92,** 85.

Schaefer, W. B. (1967). Type-specificity of atypical mycobacteria in agglutination and antibody adsorption tests. *Am. Rev. resp. Dis.* **96,** 1165.

Schneidau, J. D. & Shaffer, M. F. (1960). Studies on nocardia and other actinomycetales. II. Antigenic relationships shown by slide agglutination tests. *Am. Rev. resp. Dis.* **82,** 64.

Thurston, J. R., Phillips, M. & Pier, A. C. (1968). Extracellular antigens of *Nocardia asteroides.* III. Immunological relationships demonstrated by erythrocyte-sensitizing antigens. *Am. Rev. resp. Dis.* **97,** 240.

Yamamura, Y., Misaki, A. & Azuma, I. (1972). Chemical and immunological studies on polysaccharide antigens of mycobacteria, nocardia and corynebacteria. *Bull. Un. int. tuberculose* **47,** 181.

Zamora, A., Bojalil, L. F. & Bastarrachea, F. (1963). Immunologically active polysaccharides from *Nocardia asteroides* and *Nocardia brasiliensis*. *J. Bact.* **85,** 549.

9. Sensitin Tests as an Aid in the Taxonomy of *Nocardia* and its Pathogenicity

MOGENS MAGNUSSON

Tuberculin Department, Statens Seruminstitut, DK-2300 Copenhagen, S. Denmark

Contents

1. Introduction

Microbial sensitins are species-specific substances prepared from cultures of micro-organisms and capable of eliciting delayed type (cell-mediated) immune reactions in hosts immunized with the same micro-organisms or with micro-organisms belonging to the same or related species. Sensitins are prepared from the microbial cells or from the liquid culture medium used for the cultivation of the cells. Sensitin tests are now usually carried out by intradermal injection of a small amount of the sensitin and the size of the local reaction measured at the injection site, after 24—48 h.

The studies reported by Zupnik in 1903 (Zupnik, 1903) appear to be the first on cell-mediated immunity to *Nocardia*. Immunological studies of *Nocardia* were reviewed in 1970 (Kurup, Randhawa & Gupta, 1970) and Ortiz-Ortiz *et al.* (see chapter 16) have covered the studies of cell mediated immunity in human beings and animals with nocardiosis and in animals with experimentally induced hypersensitivity to *Nocardia* sensitins. The field is reviewed from the aspect of the method of preparing the sensitins, including the material (culture filtrate or microbial cells) used for the preparation.

In the present review, the application of *Nocardia* sensitin tests to man and animals is evaluated from a taxonomic point of view and as an aid in the study of the pathogenicity of *Nocardia*. The results of some previously unpublished studies are included in the review.

2. Sensitin Tests as Aid in the Taxonomy of *Nocardia*

The methods and materials used are reviewed and thereafter the taxonomic results of the same studies are considered in relation to the orthodox classification of the isolates.

(a) *Materials and methods*

A taxonomic study of *Nocardia* strains by sensitin tests may consist of a direct comparison of two or more isolates. The principle and the main steps in such a study are as follows:

Each isolate, or a suitable fraction of it, is used for the immunization of separate experimental animals in order to induce delayed type hypersensitivity in them. When the hypersensitivity has developed comparative tests are performed in each animal with each of the isolates under study or with suitable fractions (sensitins) of each isolate. The relative sizes of the response of each animal to each of the sensitins are measured. If the relative response to the sensitins is the same in all animals, and thus independent of the isolate used as immunogen, the isolates cannot be distinguished under the conditions of the study. If the relative response to the sensitins varies in the animals, dependent on the isolates used as immunogens, the isolates are distinguishable.

In the following account, the procedure by which the same comparative intradermal tests with sensitins of all the strains studied are carried out in animals immunized with each of the strains under study will be called *comparative reciprocal intradermal sensitin tests*. Other procedures, based on comparative intradermal sensitin tests, in which *all* of the homologous and heterologous tests are not carried out are considered to be less discriminative. Taxonomic conclusions may of course also be drawn from such studies and from investigations performed with other objects in view.

Some details of the materials and methods used in the studies included in this review are shown in Table 1.

Immunogens

In the work performed before 1940–1950 living microorganisms were used for immunization. However in later studies killed microorganisms in combination with adjuvants have generally been preferred.

Experimental animals

In the majority of studies white guinea pigs have been used though rabbits and cattle are occasionally employed. Dyson & Slack (1963) stated that infected rabbits would probably be the best animals for further study and that guinea pigs were often erratic reactors. This conflicts with the reviewer's experience, gained from extensive studies of guinea pigs and limited studies of rabbits. Guinea pigs appear to give better defined delayed type skin reactions than rabbits.

Method of immunization

The immunization route, the dose of immunogen, and in the case of multiple immunizations the time interval between injections vary with the immunogen. The procedures followed often seem to be greatly influenced by local tradition.

Sensitins

The substances used as sensitins are usually different from those employed as immunogens. The statement by Drake & Henrici (1943) that the sensitin of *N. asteroides* is heat-labile and cannot be demonstrated in culture filtrates is not correct (Magnusson & Mariat, 1968).

Purification of the sensitins by chemical or physico-chemical methods has been described in some studies with the aim of reducing the cross reactivity of the preparations. On the other hand, Magnusson (1962) used purified sensitins prepared in the same way as tuberculin purified protein derivative and in later work (Magusson & Mariat, 1968) used dilutions of crude, heat-sterilized culture filtrates as sensitins with equally satisfactory results. There is hardly any convincing evidence to support the assumption that chemically or physico-chemically purified sensitins give smaller cross reactions than crude sensitins, therefore the various purification procedures will not be considered in detail.

Sensitin tests

Zupnik (1903) injected sensitins subcutaneously and measured the thermal response 2–8 h later. In the majority of subsequent investigations small amounts (0·02–5 μg) of sensitins are injected intradermally and the size of the local reaction measured 24–48 h later. *In vitro* tests of cell mediated immunity to *Nocardia* have also been carried out recently, (Ortiz-Ortiz, Bojalil & Contreras, 1972; Ortiz-Ortiz, Contreras & Bojalil, 1972*a,b*; Braendstrup *et al.*, 1975).

Conclusion

When used for taxonomic purposes, methods involving comparative reciprocal sensitin tests appear to give more reliable results than methods involving other sensitin tests. The former methods can be applied to isolates which show a satisfactory growth on a standardized, non-immunogenic medium and which, when properly administered, elicit delayed type hypersensitivity in guinea pigs.

(b) *Contribution of sensitin tests to the delineation of species of* Nocardia *and to the identification of* Nocardia *isolates at the species level*

In recent years there has been considerable uncertainty and disagreement about the delineation of *Nocardia farcinica* Trevisan and *N. asteroides* Eppinger (see chapter 1, Editorial note, 1974), and *N. farcinica* has been considered a *nomen dubium* (Lechevalier, Horan & Lechevalier, 1971; Ridell & Norlin, 1973). In recent investigations, strains designated *N. farcinica* and *N. asteroides* formed heterogeneous assemblages (Berd, 1973; Bradley, 1973; Georg *et al.*, 1961; Goodfellow, 1971; Kwapinski, 1970; Kurup & Schmitt, 1973; Lechevalier, Horan & Lechevalier, 1971; Magnusson & Mariat, 1968; Mariat, 1965; Pier & Fichtner, 1971; Ridell &

Table 1

Methods and materials employed in taxonomic studies of Nocardia *by sensitin tests*

Reference	Microorganisms studied	Immunogen						Animal			Immunization route							Sensitin					Sensitin test				
		Living micro-organisms			Killed micro-organisms																		*in vivo*				*in vitro*
		Vehicle broth-dl	*Vehicle* oil-dl	+ *IFA*	*Vehicle* oil	+ IFA	+ culture filtrate + IFA	Guinea pig	Rabbit	Cow	intraperitoneal	intramuscular	intravenous	intratesticular	intramammary	intradermal	subcutaneous	microorganisms	extracts	culture filtrate	proteins	polysaccharide	intraperitoneal	intradermal	intraoccular	subcutaneous	macrophage migration inhibition
Zupnik, 1903	*Nocardia* sp. *Mycobacterium tuberculosis* etc	+						+			+									+							
Bretey, 1933	*N. asteroides* *N. eppingeri* *M. tuberculosis* etc	+						+			+							+		+				+		+	
Mathieson *et al.*, 1935	*N. asteroides* etc	+						+	+									+								+	
Goyal, 1937	*Nocardia* sp. *M. tuberculosis* etc	+						+	+		+							+		+			+	+	+		

Drake & Henrici, 1943	*N. asteroides* *M. tuberculosis*		+				+						+					+	+	+	+	+
Affronti, 1959	*N. asteroides* *M. tuberculosis* etc			+			+	+												+		+
Magnusson, 1962	*N. farcinica* *N. asteroides* *N. brasiliensis* *M. intracellulare* etc			+			+									+				+		+
Pier & Enright, 1962	*N. farcinica* *N. sebivorans* etc								+											+		+
Dyson & Slack, 1963	*N. asteroides*	+			+			+		+		+				+			+	+	+	+
Bojalil & Magnusson, 1963	*N. asteroides* *N. brasiliensis*			+			+								+					+		+
Pier & Keeler, 1965	*N. farcinica* *N. sebivorans*				+			+	+		+									+		+
Kingsbury & Slack, 1967	*N. asteroides*			+			+	+			+						+					+
Magnusson & Mariat, 1968	*N. farcinica* *N. asteroides* *N. blackwellii* *N. brasiliensis* *Nocardia* sp. *M. farcinogenes* var *senegalense*			+		+	+								+				+	+		+
Pier, Thurston & Larsen, 1968	*N. farcinica* *N. sebivorans*	+							+			+		+						+		+
Kingsbury & Slack, 1969	*N. asteroides* etc			+			+										+					+

Table 1

Reference	Microorganisms studied	Immunogen							Animal			Immunization route							Sensitin					Sensitin test				
		Living micro-organisms			Killed micro-organisms																			*in vivo*				*in vitro*
		Vehicle broth-dl	*Vehicle* oil-dl	+*IFA*	*Vehicle* oil	+IFA	+culture filtrate	+IFA	Guinea pig	Rabbit	Cow	intraperitoneal	intramuscular	intravenous	intratesticular	intramammary	intradermal	subcutaneous	microorganisms	extracts	culture filtrate	proteins	polysaccharide	intraperitoneal	intradermal	intraoccular	subcutaneous	macrophage migration inhibition
Takeya, Nakayama & Muraoka, 1970	*N. asteroides* *N. globerula* *N. polychromogenes* etc			+					+					+								+			+			
Ortiz-Ortiz, Contreras & Bojalil, 1972*a*	*N. asteroides* *N. brasiliensis*			+					+						+								+		+			+
Ortiz-Ortiz, Contreras & Bojalil, 1972*b*	*N. asteroides* *N. brasiliensis*			+					+								+					+			+			+
Braendstrup *et al.*, 1975	*N. farcinica* *M. tuberculosis*				+				+								+					+			+			+

* IFA: Incomplete Freunds adjuvant.

Norlin, 1973; Ridell, 1975; Tsukamura, 1969). A contributory cause of the taxonomic confusion is the uncertainty as to whether type strains of the two species do exist.

Nocard reported the isolation of a microorganism which caused farcy in cattle in Guadeloupe (Nocard, 1888). The following year Trevisan described this organism under the name *Nocardia farcinica* (Trevisan, 1889). On the basis of a study of a culture (ATCC3318) believed to be Nocard's original, and a similar strain isolated and identified by Dr C. P. Fitch of New York, Waksman (1957) gave a more recent description of this species. According to Sneath & Skerman (1966), ATCC3318 is the type strain of *N. farcinica.* The description of *N. farcinica* in the eighth edition of Bergey's *Manual of Determinative Bacteriology* (1974) is based on this strain.

The type strain of *N. farcinica* carries the following culture collection numbers: ATCC3318, NCTC4524, PSA*162, Gordon 3318, Goodfellow N34, Magnusson 752, and 341). However, Chamoiseau & Asselineau (1970) and Lechevalier, Horan & Lechevalier (1971) have found that ATCC3318 and NCTC4524 are chemotaxonomically distinct. The distinctness of these two strains has been confirmed by a serological method (Ridell, 1975), and by comparative reciprocal intradermal sensitin tests (unpublished observations). Strain ATCC3318 contains nocardomycolic acids whereas strain NCTC4524 possesses mycobacterial mycolates (Lechevalier, Horan & Lechevalier, 1971).

Ridell & Norlin (1973) found that strain ATCC3318, and two other strains designated as *N. farcinica,* did not differ from the majority of the *Nocardia* strains examined as they all shared a low number (at least two) of immunoprecipitation lines with a series of mycobacteria. In contrast, five other strains designated *N. farcinica* showed at least five immunoprecipitation lines with one or more of the same series of mycobacteria.

By comparative reciprocal intradermal sensitin tests strain ATCC3318 (Magnusson 752) could be grouped with four other isolates labelled *N. farcinica,* four designated *N. asteroides* and with a strain labelled *N. blackwellii* (Magnusson & Mariat, 1968). Three other strains designated *N. farcinica* were clearly distinct from this

* Progetto Sistematica Actinomiceti, Istituto "P. Stazzi", Milano, Italy.

group of 10 strains. The authors considered that the 10 strains which formed a homogeneous group should be classified as *N. farcinica*, and that the three labelled *N. farcinica* be classified into two other species. By the same method NCTC4524 (Goodfellow N34, Magnusson 341) cannot be distinguished from an authentic strain of *Mycobacterium farcinogenes* var. *senegalense* Chamoiseau (Magnusson, unpublished observations). Thus it is quite evident that ATCC3318 (Gordon 3318, Magnusson 752) is distinct from NCTC4524 (Goodfellow N34, Magnusson 341).

Eppinger described a new pathogenic microorganism (*Cladothrix asteroides*), isolated from a man, which caused "pseudotuberculosis" (Eppinger, 1891). This organism was named *Nocardia asteroides* by Blanchard (1896). A more recent description of *N. asteroides* is given by McClung (1974). As an *ad hoc* reference strain of *N. asteroides* the strain, PSA165, studied by Silvestri *et al.* (1962) and by Dr Ruth Gordon (IMRU 727), was proposed (Sneath & Skerman, 1966). It is the same isolate as ATCC19247, Goodfellow N317, Magnusson 809 and 353.

On the basis of examining more than 600 strains, received as *Streptomyces, Nocardia* or *Mycobacterium* sp., Gordon & Mihm (1957) described some physiological and biochemical characteristics of a taxon containing 79 isolates. Forty-three of these isolates were labelled *N. asteroides* and 21 others carried 17 different specific epithets, the latter included strains bearing the following names which were considered to conform to the original species description: *Nocardia farcinica* Trevisan, *N. asteroides* (Eppinger, 1891) Blanchard, *N. blackwellii* (Eriksson, 1935) Waksman and Henrici, *N. sebivorans* (Gorrill & Heptinstall, 1954), and *N. polychromogenes* (Vallée, 1903) Waksman and Henrici. The authors designated this taxon *Nocardia asteroides* (Eppinger) Blanchard and gave the reasons cited below for using this name instead of the epithet *N. farcinica*, the name that has priority according to the rules of the bacteriological code:

> *Nocardia asteroides* has become firmly established in human medicine and, as reviewed by Bohl, Jones, Farrel, Chamberlain, Cole and Ferguson, 1953, not uncommon in veterinary medicine. *Nocardia farcinica,* on the other hand has nearly disappeared from scientific reports, as evidenced by its being listed only twice since 1937. Only three strains of

Nocardia farcinica, two of which were originally identical, were donated to this collection, indicating that *Nocardia farcinica* has also nearly disappeared from the larger culture collections and the collections of investigators interested in this taxon. In contrast, 43 of the 79 strains representing this species were received as *Nocardia asteroides* or its varieties. There was also, as previously stated, a possibility that the interpretation of the original account of *Nocardia farcinica* was a mistaken one, and that strains W3318, W3399 and 611 did not typify farcinica."

The classification and nomenclature used by Gordon & Mihm have been adopted in several laboratories, and many additional strains have later been identified by cultural, physiological and biochemical tests as *Nocardia asteroides* (Palmer, Harvey & Wheeler, 1974). However, as mentioned above, several reports on the heterogenicity of the taxon (taxa) designated *N. asteroides* by Gordon & Mihm have also been published. In the opinion of the reviewer the taxon (taxa) designated *N. asteroides* by Gordon & Mihm includes several species, among which are *N. farcinica*, *N. asteroides* and *N. sebivorans*.

On the basis of later studies of additional isolates (Gordon & Mihm, 1959; Gordon & Mihm, 1962*b*), further characteristics of the taxon were described. Mariat (1965) found a higher frequency than Gordon & Mihm did of *N. asteroides* strains being able to utilize sodium citrate as carbon source (88%, as against 34%) and a much higher frequency of strains forming acid from inositol (38% as against 3%).

By comparative reciprocal intradermal sensitin tests on guinea pigs the proposed reference strain of *N. asteroides* (PSA165, IMRU727, Magnusson 809) was clearly distinct from the *N. farcinica* strains studied, including ATCC3318, and from a strain of *N. asteroides* (Magnusson 81) used as laboratory reference strain for this species (Magnusson & Mariat, 1968). The authors considered, therefore, that *N. asteroides* seemed to be a separate species (distinct from *N. farcinica*).

By serological typing of 50 strains of *N. asteroides* (in the wider sense of the term used by Gordon & Mihm) and 16 strains of related organisms by immunodiffusion precipitation Pier & Fichtner (1971) distinguished four serotypes and two type mixtures. The biochemical and physiological characteristics of strains belonging to one of the serotypes (Type I) were remarkably consistent.

By application of comparative reciprocal intradermal sensitin tests on 19 of the 50 strains studied by Pier & Fichtner the data shown in Table 2 were obtained. There is good agreement between the results obtained with the two immunological methods. Strains classified by immunodiffusion precipitation as Serotype I were classified as *Nocardia asteroides* (in a narrow sense of the term) by comparative intradermal sensitin tests, the four other strains listed in the table being indistinguishable from the laboratory reference strain of *N. asteroides* (NADL404, Magnusson 81).

Table 2

Relationship between immunodiffusion precipitation serotypes (Pier & Fichtner, 1971) and sensitin types (species) of 19 strains of "N. asteroides"

Strain No. (NADL)	Immunodiffusion precipitation serotype	Sensitin type (species)
404, 40N, 41N 443–1, 443–2	I	*Nocardia asteroides*
19N, 47N 1011, India 8	II II + IV	*Nocardia sebivorans*
34N, India 10 N.f. 3399 503, 2N, 44N ATCC3318	III III + IV	*Nocardia farcinica*
36N, 43N	IV	*Nocardia* sp. A
727	IV	*Nocardia* sp. B

Strains classified by immunodiffusion precipitation as Serotype II or II + IV were identified as *N. sebivorans* by sensitin tests, and strains classified as Serotype III and III + IV as *N. farcinica*. Finally two strains classified by immunodiffusion precipitation as Serotype IV were classified by sensitin tests in the same (unnamed) taxon as seven other *Nocardia* strains (Magnusson 610, 685, 708, 712, 911, 913 and 914). On the other hand the proposed type strain of *N. asteroides*, strain 727, was also classified as Serotype IV, but appeared by sensitin tests to be distinct from the other strains. The fact that the serotype mixtures, II+IV and III+IV, cannot be distinguished by sensitin tests from the corresponding pure serotypes (II and III) respectively may be explained in two ways. One possibility is that the apparent serotype mixtures are not real mixed types, the more complex immunodiffusion precipitation pattern might be due to the corresponding antigen being more concentrated that the antigens of strains with a simpler immunodiffusion precipitation pattern. If so, the immunodiffusion precipitation line corresponding to Serotype IV should disappear on dilution of the antigens before the lines corresponding to Serotypes II and III, respectively, disappear. The other possibility is that immunoprecipitation is a more discriminating method than the comparative sensitin tests.

By comparative intradermal sensitin tests taxa in addition to those listed in Table 2 have been observed within the assemblage of strains classified by Gordon & Mihm as *N. asteroides*. However, at present it is not known whether any of these additional taxa are identical with previously described species.

Kingsbury & Slack (1969) performed comparative intradermal tests, with sensitins prepared from three strains designated *N. asteroides* Ohio, strain 70 and "3672", in guinea-pigs immunized with, amongst other strains, *N. asteroides* 3672 and five similarly labelled strains L-50, ATCC3318, ATCC9504, ATCC6846 and ATCC9970. Comparative tests with the first two preparations were also performed in guinea-pigs immunized with *N. farcinica* ATCC13781. Quantitative differences in the specificity of the three sensitins were found. The variations in cross reactivity which the authors consider as strain specificity are really due to the fact that *N. asteroides* is a complex assemblage of strains which should be classified into a number of species, including *N. farcinica.* Among the six strains designated *N. asteroides* used for immunization were the type strain of *N. farcinica,* (ATCC3318), and *N. blackwellii* ATCC6846 which cannot be distinguished from *N. farcinica* by comparative reciprocal intradermal sensitin tests (Magnusson & Mariat, 1968). On the other hand, the isolate designated *N. farcinica* ATCC13781 which was also included in the study does not belong to *N. farcinica* (Magnusson & Mariat, 1968). Sensitin from strain Ohio gives relatively stronger reactions than sensitins of *N. asteroides* strains 70 and 3672 in guinea-pigs immunized with *N. farcinica* ATCC3318 and ATCC6846 indicating that strain Ohio should probably be classified as *N. farcinica.* The relatively small cross reactions in guinea-pigs immunized with *N. asteroides* 3672, L-50 and ATCC9970 indicate that these isolates are distinct from *N. farcinica.* The cross reactions observed in the guinea-pigs immunized with ATCC9504 indicate that this strain may also be classified as *N. farcinica,* although the same guinea-pigs also showed large reactions to sensitin from "*N. asteroides*" 3672.

By intradermal tests with a dose of 0·2 μg of sensitin prepared from *N. asteroides* strain 3672 Kingsbury & Slack (1967) obtained reactions >9 mm in guinea-pigs immunized with the same and with three other strains of *N. asteroides,* reactions 7–9 mm in guinea-pigs immunized with two strains of *N. farcinica* and two of *N. brasiliensis,* and reactions <7 mm in guinea-pigs immunized with two other strains of *N. brasiliensis.* By intradermal tests with the same dose of sensitin prepared from *N. asteroides* strain Ohio, (which presumably ought to read *N. farcinica* strain Ohio), Kingsbury & Slack in the same study obtained reactions >9 mm in guinea pigs immunized with two strains of *N. farcinica* (ATCC3318 and ATCC6846) and with strain ATCC9504, which may be read as *N. farcinica* ATCC9504, reactions 7–9 mm in guinea pigs immunized with *N. asteroides* strain 3672 and reactions <7 mm in guinea-pigs immunized with two other strains of *N. asteroides* and with four strains of *N. brasiliensis.*

In the numerical taxonomic study carried out by Tsukamura (1969) strains received as *N. asteroides* were divided into two clusters. The cluster which included strain ATCC3318 was named as *N. farcinica* Trevisan and the second cluster *N. asteroides* (Eppinger) Blanchard. The first cluster comprised 38 strains and included four strains identified as *N. farcinica* by comparative reciprocal intradermal sensitin tests. This taxon may therefore be the same as the *N. farcinica* group delineated by Magnusson & Mariat (1968), the acetamidase positive subgroup of *N. asteroides* Berd (1973), the *N. asteroides* group A recovered by Kurup & Schmitt (1973), and cluster B of Ridell (1975).

Two subcultures (N77 and N122) of *N. blackwellii* ATCC6846, identified as *N. farcinica* by comparative reciprocal intradermal sensitin tests (Magnusson & Mariat, 1968), were classified in subgroup IA in the numerical taxonomic study of Goodfellow (1971). However, the same subgroup included two other strains (N317, see p. 244, and *N. phenotolerans* N100) which are both distinct from *N. farcinica* according to comparative reciprocal intradermal sensitin tests (Magnusson unpublished observations). Thus, the subgroup in question seems to be heterogeneous and appears to include *N. farcinica*.

The cluster designated *N. asteroides* (Eppinger) Blanchard by Tsukamura (1969) comprised 21 strains and included two isolates identified as *N. asteroides* by comparative reciprocal intradermal sensitin tests and two others classified by Pier & Fichtner (1971) as immunoprecipitation Serotype 1. According to the observations reported above, the latter two isolates may also be identified as *N. asteroides* (in the narrow sense of the term) by comparative reciprocal intradermal sensitin tests. However, Tsukamura's second group also includes strain Gordon 553, which has been classified by Pier & Fichtner (1971) as Serotype IV, and which, therefore, is likely to belong to the sensitin type *Nocardia* sp. A (Table 2). Strain Gordon 553 would thus appear to be distinct from *N. asteroides* in the narrow sense of the term defined by comparative reciprocal intradermal sensitin tests. Tsukamura's second group (Tsukamura, 1969) also includes *N. polychromogenes* W-3409, a strain which is also distinct from *N. asteroides* according to comparative reciprocal intradermal sensitin tests (Magnusson, unpublished observations). Tsukamura's second group, therefore, comprises isolates belonging to more than one sensitin type (*N. asteroides* in the narrow sense of the term). The same may also be the case with the acetamidase negative subgroup of *N. asteroides* described by Berd (1973).

N. asteroides group D described by Kurup & Schmitt (1973) includes strain NADL-19N, which has been identified as *N. sebivorans* by sensitin tests, see Table 2.

Conclusion

The species name *Nocardia farcinica* has been used in recent publications both for isolates, which cannot be distinguished from *N. farcinica* ATCC3318 by comparative reciprocal intradermal sensitin tests and for strains which, by the same technique, cannot be distinguished from *N. farcinica* NCTC4524 or from an authentic strain (#378) of *Mycobacterium farcinogenes* var. *senegalense*, received from Dr G. Chamoiseau. The two latter strains cannot be distinguished from each other by comparative reciprocal intradermal sensitin tests in guinea-pigs, but they can be distinguished from strain ATCC3318. According to the reviewer the first-mentioned strain (ATCC3318) is the type strain of *N. farcinica*, and only strains which cannot be distinguished from this strain, should be classified (designated) as *N. farcinica*. The use of the name *N. farcinica* for strains which cannot be distinguished from NCTC4524 is misleading and should be discontinued; such strains may be classified as *M. farcinogenes* var. *senegalense* (Chamoiseau, 1973).

The species name *N. asteroides* has also been applied, and is still being used, for strains with varying properties. On the one hand, the name has been used in a wide sense for assemblies of strains with varying sets of biochemical and physiological properties; on the other hand it has been used in a narrow sense for strains which cannot be distinguished from a reference culture of the species by comparative reciprocal intradermal sensitin tests or by other suitable discriminative techniques.

The establishment of a suitable type strain of *N. asteroides* may permit a better delineation of this species and would therefore be useful. Preliminary studies indicate that the strain which has been proposed as a neotype strain of *N. asteroides* is not suitable for this purpose since it can be distinguished by comparative reciprocal intradermal sensitin tests from nearly all other strains of *N. asteroides* (in the wider sense of the term) with which it has so far been compared.

Within the taxon *N. asteroides* (in the wider sense), several homogeneous larger or smaller subclusters of strains can be distinguished by comparative reciprocal intradermal sensitin tests. Two of these are established species (*N. farcinica* and *N. sebivorans*) though other subclusters may be identical with other known species, among which is *N. asteroides* in the narrower sense.

Isolates of *N. farcinica* (see above) may be identified by comparative reciprocal intradermal sensitin tests, by immunodiffusion precipitation tests or by a combination of conventional biochemical and physiological tests. *Nocardia sebivorans* and *N. asteroides* (in the narrow sense) may be identified by comparative reciprocal intradermal sensitin tests or immunodiffusion precipitation tests. Further studies are required before strains belonging to these species can be identified by physiological and biochemical tests. Comparative reciprocal intradermal sensitin tests indicate that *N. polychromogenes* is not a subjective synonym of *N. asteroides.*

Snijders (1924) proposed the name *N. otitidis-caviarum* for an organism he isolated from the middle ears of guinea-pigs in Sumatra. Eriksson (1935) described the same organism using the name *N. caviae,* and Gordon & Mihm (1962*a*) gave an extended description of the species. Gordon & Mihm used the name *N. caviae* Eriksson, and generally this name has subsequently been used. However, Lessel (1962) states that *N. otitidis-caviarum* is the correct name of the species according to the bacterological code. The type strain is ATCC14629 (NCTC1934).

Studies of four strains of *N. otitidis-caviarum* received from, and identified by, Dr Ruth Gordon have shown that isolates belonging to this species can also be identified by comparative reciprocal intradermal sensitin tests in guinea-pigs (Magnusson, unpublished observations). By this method, a total of 13 isolates of *N. otitidis-*caviarum had been identified to date.

A description of *N. brasiliensis* based on studies of 50 strains has been given by Gordon & Mihm (1959). No type strain seems to exist though strain ATCC19296 (IP 337, Gordon 845, Goodfellow N318) has been proposed as a reference strain (Sneath & Skerman, 1966).

Studies of two strains of *N. brasiliensis* received from, and identified by, Dr Ruth Gordon have indicated that isolates belonging to this species may be identified by comparative intradermal reciprocal sensitin tests in guinea-pigs (Magnusson, 1962; Magnusson & Mariat, 1968). With this method a total of six isolates of *N. brasiliensis* had been identified to date.

An isolate labelled *N. brasiliensis* N14, (NCTC10300, ATCC19295) (see Goodfellow, 1971), received from Dr M. Goodfellow, could be distinguished both from the laboratory reference strain (Magnusson 117) and from the proposed reference strain of *N. brasiliensis* by comparative reciprocal intradermal sensitin tests in guinea-pigs. In the numerical taxonomic study reported by Goodfellow (1971) the strain in question formed a cluster with *N. brasiliensis* N318 at a similarity level of 89%, negative matches included. This seems to indicate that this strain is also biochemically distinct from the proposed reference strain.

Dr W. Kaplan has observed (W. Kaplan, C.D.C., Atlanta, Georgia, U.S.A., pers. comm., 1968) that strains of *N. brasiliensis* can be clearly distinguished from some strains of *N. asteroides* by fluorescent antibody techniques but are very difficult to distinguish from other strains of the latter species. This observation could not be confirmed by comparative reciprocal intradermal sensitin tests in guinea-pigs. By such tests, isolates of both categories of *N. asteroides* could clearly be distinguished from strains of *N. brasiliensis.*

By other immunological methods, which also seem to reflect the species specificity of the delayed type hypersensitivity of guinea-pigs, one strain of *N. brasiliensis* could also be distinguished from a strain of *N. asteroides* in the wider sense (Ortiz-Ortiz, Bojalil & Contreras, 1972; Ortiz-Ortiz, Contreras & Bojalil, 1972*a*,*b*).

(c) *Comparative intradermal tests with sensitins from isolates classified in the genus* Nocardia *and in other genera*

By comparative reciprocal intradermal sensitin tests in guinea-pigs an authentic culture of *Nocardia intracellularis* (Cuttino & McCabe, 1949) could not be distinguished from four isolates of "Battey organisms" (*Mycobacterium intracellulare*) though it could be distinguished from two strains of *M. avium* (Magnusson, 1962). This observation confirmed the results of studies (Schneidau & Shaffer, 1957; Shephard, 1958), in which other methods had been used and in which the microorganism in question was classified in the genus *Mycobacterium.*

It appears from the same study (Magnusson, 1962), and from a subsequent one (Magnusson & Mariat, 1968), that isolates of *N. asteroides, N. brasiliensis* and *N. farcinica* can be distinguished by comparative reciprocal intradermal sensitin tests from *Mycobacterium rhodochrous* strain Gordon 482, *Mycobacterium lacticola* NCTC8154 (Gordon: *M. rhodochrous* N8154) and from strains belonging to 19 other species of *Mycobacterium.*

Even in studies in which less discriminating methods have been used, the specificity of delayed type hypersensitivity induced by isolates of *Nocardia* and strains belonging to other genera has been demonstrated. Hereby, a distinction between isolates belonging to species of *Nocardia* and to other genera has also been possible.

Zupnik (1903) found that the minimum reactive dose of tuberculin injected intraperitoneally in guinea-pigs immunized with a strain of actinomycetes from a cow, a strain isolated from a human case of actinomycosis, and a strain isolated from air—presumably all three strains of *Nocardia*—was about one third of that necessary to give a positive reaction in non-immunized guinea-pigs. Guinea-pigs immunized with an acid-fast Birt and Leishman Actinomyces strain gave negative reactions even to double the dose required in animals immunized with the other strains of actinomycetes.

Bretey (1933) demonstrated the specificity of delayed type hypersensitivity in guinea-pigs immunized with a strain of *Nocardia eppingeri*, with two strains of *Streptothrix*, and with a strain of *M. bovis* by skin testing with sensitins prepared from the strains used for immunization and from a strain of *N. asteroides*.

Goyal (1937) demonstrated the specificity of delayed type hypersensitivity in tuberculous guinea-pigs and guinea-pigs immunized with a series of isolates classified as *N. asteroides* or other species of *Nocardia* or *Streptomyces* (*Streptothrix*). He concluded that the various strains of *Nocardia* sp. and *Streptomyces* sp. could not be differentiated on the basis of the intradermal sensitin tests in guinea-pigs. This conclusion, however, is hardly correct. The results would rather seem to show that the three isolates *N. asteroides eppingeri* NCTC3258, *N. asteroides eppingeri* NCTC4519 and *Nocardia* sp. strain Thomoff form a group that can be distinguished from a strain of *N. eppingeri* originating from the Institut Pasteur collection, from *N. madurae* and from strains A21-22, G12 and Laporte. The strain of *N. eppingeri* can apparently also be distinguished from *N. madurae* and from strains G12 and Laporte.

Affronti (1959) demonstrated the specificity of delayed type hypersensitivity induced in rabbits by strains of *N. asteroides*, *M. kansasii* and *M. intracellulare*, and Takeya, Nakayama & Muraoka (1970) demonstrated similarly specificity induced in guinea-pigs by strains of *N. polychromogenes* CBS (M-6) and 10 species of *Mycobacterium*. Sensitins from *N. asteroides* Gordon 399 and *N. globerula* NRRLB1306 (M-75) gave reactions <4 mm in guinea-pigs immunized with *M. rhodochrous*, *M. chitae*, *M. fortuitum*, *M. smegmatis*, *M. phlei* and *M. abscessus* (Takeya, Nakayama & Muraoka, 1970). Likewise, Drake & Henrici (1943) distinguished between *M. tuberculosis* and four isolates of *N. asteroides* by sensitin tests.

Mathieson *et al.* (1935) immunized guinea-pigs and rabbits with living and killed culture of *N. gypsoides* (Henrici & Gardner, 1921) and *N. asteroides* isolated from pleura and diaphragma of a cow which died of actinomycosis (Bishop & Fenstermacher, 1933), using one strain of each species. By skin tests with tuberculin and homologous sensitin no cross sensitization was demonstrated.

Kingsbury & Slack (1967), by intradermal tests with 2 μg of sensitin prepared from *N. asteroides* strain 70, obtained reactions >9 mm in guinea-pigs immunized with the same strain and with *N. asteroides* "Packer", and reactions <7 mm in guinea-pigs immunized with strains of *Streptomyces madurae, S. griseus, S. paraguayensis, S. somaliensis, M. tuberculosis* and with strains belonging to three other species of *Mycobacterium.* By intradermal tests with the same dose of sensitin prepared from *N. asteroides* strain Ohio (which presumably ought to read *N. farcinica* strain Ohio, reactions >7 mm were obtained in guinea-pigs immunized with the two *N. asteroides* strains mentioned above, and reactions <7 mm in guinea-pigs immunized with the same four species of *Streptomyces* and the same four species of *Mycobacterium.*

The specificity of the immune responses to comparative reciprocal intradermal sensitin tests in guinea-pigs may be characterized by the *specificity difference,* or spd value (Magnusson, 1962). The specificity difference is the difference between the sum of the

Table 3
Average inter-*generic and* intra-*generic specificity differences (in mm) of sensitins of* Nocardia *sp. and* Mycobacterium *sp.*

	Nocardia sp.	*Mycobacterium* sp.
Nocardia sp.	13·2 (4 species, 6 isolates)	15·5 (4 *Nocardia* sp., 19 *Mycobacterium* sp., 76 isolates)
Mycobacterium sp.	15·5 (4 *Nocardia* sp., 19 *Mycobacterium* sp., 76 isolates)	12·2 (19 species, 35 isolates)

homologous reactions and the sum of the heterologous reactions. From previously published observations (Magnusson, 1962) the average intra- and inter-generic spd values shown in Table 3 can be calculated. Each species was represented by one strain only. The average inter-generic specificity difference for the genera *Nocardia* and *Mycobacterium* appears to be only very slightly larger than the corresponding average intra-generic specificity differences. If assemblages of four species of *Nocardia* and 19 species of *Mycobacterium* are representative, it would appear that comparative reciprocal intradermal sensitin tests cannot be used for the delineation of genera. This means that, as a rule, it cannot be decided whether a given species belongs to *Nocardia, Mycobacterium* or to a third genus simply on the basis of sensitin tests.

Conclusions

Organisms belonging to the species *N. brasiliensis* and *N. otitidiscaviarium* can be identified by comparative reciprocal intradermal sensitin tests in guinea-pigs.

A delineation between the genera *Nocardia* and *Mycobacterium* cannot be made by comparative reciprocal intradermal sensitin tests in guinea-pigs.

By such tests isolates of *Nocardia* can be identified at the species level, provided that they grow on the usually applied synthetic, non-immunogenic medium and provided that suitable type or reference strains are available.

3. Sensitin Tests as Diagnostic Aids

Sensitin tests have a potential value as aids in the diagnosis of infections caused by *Nocardia* as long as the infecting microorganism elicits cellular immunity in the infected organism, and provided the cellular immunity elicited and the sensitins used are specific. It is evident from what has been said earlier that sensitins from *Nocardia* are species specific and that their cross reactivity depends on the relatedness of the strains used for their preparation.

In the following sections studies of hypersensitivity in man and in cattle to sensitins of *Nocardia* are reviewed in order to determine whether the first two provisions mentioned are fulfilled.

Intradermal tests with *Nocardia* sensitins may be performed with the same technique as Mantoux tuberculin tests. In man 0·1 ml of the sensitin is injected intradermally on the volar side of the forearm. In cattle, 0·2 ml of the sensitin may be injected intradermally into the cervical skin (Pier & Enright, 1962). The size of the local reaction at the injection site should be measured 24–72 h after the injection.

Our present knowledge of the levels of hypersensitivity attained in individuals infected with *Nocardia* is very limited, and it is therefore uncertain which doses of sensitin should be used. Hitherto purified preparations of sensitins have been used in doses of 0·2–2 μg (Bojalil & Magnusson, 1963) or 2–5 μg (Ortiz-Ortiz & Bojalil, 1972).

By comparative intradermal tests, the different preparations should be injected simultaneously, since a skin test may give a change in the level of hypersensitivity. Multiple comparative testing is performed by simultaneous injections of three or more preparations. By such testing interactions between tests may occur, which means that the response to a given preparation depends on what other preparations are given at the same time. A strong reaction to one preparation may depress the reaction to another given at the same time. Our present knowledge in this field is insufficient.

The influence of one skin test reaction on subsequent reactions will be stronger if the level of hypersensitivity is low. In an individual with waning hypersensitivity a skin test even with a non-immunogenic sensitin may re-establish the maximal level of the hypersensitivity reached earlier.

(a) *Tests in patients with nocardiosis*

Fourteen patients with mycetoma caused by *N. brasiliensis* showed large reactions to intradermal tests with 0·2 μg of purified protein derivative of *N. brasiliensis* sensitin, while the same dose of the preparation elicited "negative" reactions (≤6 mm) in healthy adults and in patients with various other diseases, including tuberculosis (Bojalil & Zamora, 1963; Bojalil & Magnusson, 1963). A dose of 2 μg of the same preparation gave positive reactions in a few tuberculin positive persons.

One of the 14 patients mentioned above, and one tuberculin positive person, gave a reaction of 9 mm to an intracutaneous test with 0·2 μg of purified protein derivative of *N. asteroides* sensitin. A dose of 2 μg of the same preparation elicited fairly large reactions in a large number of tuberculin positive persons. No persons known to be infected with *N. asteroides* were examined (Bojalil & Magnusson, 1963).

In 10 patients with mycetoma due to *N. brasiliensis* skin tests with purified cytoplasmic extracts of *N. brasiliensis* (2–5 μg) elicited larger reactions than tests with similar extracts of *N. asteroides* (Ortiz-Ortiz & Bojalil, 1972). Low if any, reactivity, was observed when the same preparations were used in 25 patients with tuberculosis, 16 patients with leprosy and in healthy individuals (Ortiz-Ortiz & Bojalil, 1972). The results of comparative intradermal tests in three additional patients infected with *N. brasiliensis* are shown in Table 4. Only the patient from Yaoundé, Cameroun, gives a considerably stronger reaction to the *N. brasiliensis* sensitin than to the other sensitins applied simultaneously.

Of two patients infected with *N. farcinica* only one gave a significant reaction to the *N. farcinica* sensitin (see Table 4).

One patient infected with *N. asteroides* (in the wider sense of the term) showed no reaction to an intradermal test with 0·1 ml of a heavy suspension of a heat-killed vaccine prepared from her lesions (Benbow, Smith & Grimson, 1944). However, the patient was in a very debilitated state at the time of testing and may have been anergic. A patient from Senegal infected with *N. asteroides* gave no significant reaction to *N. asteroides* sensitin or to any of the other four sensitins applied (see Table 4). A third patient infected with *N. asteroides* showed a reaction appearing as a well demarcated dark strawberry red area of edema, 13 mm in diameter, to an intracutaneous injection of 0·5 ml(!) of a Berkefeld-filtered extract of another strain of *N. asteroides* (*Glover et al.*, 1948). The sensitin was prepared from unheated defatted organisms.

The results of comparative tests in two additional patients are included in Table 4. However, these results are very difficult to evaluate, since a test with sensitin prepared from the infecting microorganism was not carried out in the first patient, and the microorganism infecting the other patient was not identified.

Table 4

Comparative intradermal reactions of patients infected with Nocardia *sp. to various sensitins*

Identification of patient	Reference*	Organism isolated	Dose (μg/ 0·1 ml)	Induration (mm) (48 h)					
				N. brasiliensis sensitin RS117	*N. otitidis-caviarum* sensitin RS606	*N. asteroides* sensitin RS81	*N. farcinica* sensitin RS84	Human tuberculin RT23	*M. avium* sensitin RS10
Yaoundé, Cameroun; Odile. Mycetoma	Gamet *et al.* (1964)	*N. brasiliensis*	0·1	>15	0	0	0	+	—
Mexico; 2517	(1)	*N. brasiliensis*	0·1	0	20	—	0	30	—
Mexico; 2500	(1)	*N. brasiliensis*	0·1	0	0	0	0	0	—
Tunis, Dr Jaminer.	(1)	*N. otitidis-caviarum* 8718 B	0·1	2	0	3·5	0	9	—

Senegal, M. Babacar G; cerebral nocardiosis	(1)	*N. asteroides* 8707 B	0·1	0	0	4	0	0	—
Denmark; P.R.N. 425	(2)	*N. farcinica*	0·1	—	—	—	15	—	7
			0·02	—	—	—	—	18	—
Yugoslavia, Sombor; Dr Panić	(3)	*N. farcinica*	0·1	—	0	5	3·5	—	—
Mexico; 945	(1)	*Nocardia* sp., 672†	0·1	6	0	0	13	0	—
Mexico; 148/64	(1)	Non identified isolate	0·1	8	15	0	0	0	—

* (1) Dr F. Mariat, Institut Pasteur, Paris, pers. comm.
(2) Dr P. Ottsen, Sorø, Denmark, pers. comm.
(3) Dr J. Panić, Sombor, Yugoslavia, pers. comm.
† Distinguishable from *N. brasiliensis*, *N. otitidis-caviarum*, *N. asteroides* and *N. farcinica* by comparative reciprocal intradermal sensitin tests on guinea-pigs.

(b) *Epidemiological studies with* Nocardia *sensitins in man*

Edwards & Krohn (1957) performed comparative intradermal tests, in a study population in South India, with purified protein derivative of human tuberculin and of one of several sensitins made from various mycobacteria other than mammalian tubercle bacilli; the latter included *M. intracellulare* which was referred to as *Nocardia intracellularis.* Later studies (Schneidau & Shaffer, 1957; Shephard, 1958; Magnusson, 1962) however have revealed that this microorganism belongs to the genus *Mycobacterium.* When quoting the work of Edwards & Krohn (1957) Bleiker (1964) refers to the microorganism as *Nocardia.*

Bleiker (1964) performed comparative intracutaneous tests with purified protein derivative of *N. brasiliensis* (0·1 μg) in 259, 6–14 year old school children in Surinam. In three of the children positive reactions (≥8 mm) were obtained with the *N. brasiliensis* sensitin. In similar tests with purified protein derivative of *N. asteroides* sensitin (0·1 μ g) and of human tuberculin (0·02 μ g) in 241 school children of the same age group in the same country, a positive reaction (≥8 mm) was obtained with the *N. asteroides* sensitin in a single child.

The extract from *N. brasiliensis* mentioned above appeared useful in epidemiological studies, since skin reactivity was demonstrated in individuals working or living in areas in which *N. brasiliensis* was isolated from the soil (Ortiz-Ortiz & Bojalil, 1972).

(c) *Tests on cattle*

After having shown *Nocardia asteroides* (in the wider sense) to be a causative agent of mastitis in cattle in California, Pier *et al.*, prepared a series of preparations for the demonstration of delayed type hypersensitivity induced by this microorganism (Pier, Gray & Fosatti, 1957, 1958; Pier, Mejia & Willers, 1961; Pier & Enright, 1962). By skin tests with these preparations delayed type hypersensitivity could be demonstrated four weeks after experimental infection of cattle (Pier & Enright, 1962). Skin tests with the preparations were also used for detection of infected animals in infected herds (Pier & Enright, 1962). Cross reactions with tuberculin in the same animals and reactions of tuberculin positive animals to the *Nocardia* sensitins were not observed. Improved methods for the preparation of a two-week old culture filtrate sensitin of *N. farcinica* and *N. sebivorans* were described by Pier & Keeler (1965). Chemical fractionation accomplished on a diethyl amino ethyl cellulose column evolved two fractions that were active in eliciting cutaneous hypersensitivity in experimentally infected calves. Sensitin activity was not demonstrated when using similar preparations of log-phase culture filtrates.

In later studies (Pier, Thurston & Larsen, 1968) on cattle experimentally or naturally infected with *N. asteroides* (in the wider sense of the term), and with four species of mycobacteria (*M. bovis, M. paratuberculosis, M. fortuitum* and *Mycobacterium* sp. Runyon group III), positive skin reactions of diagnostic value were obtained with *N. asteroides* sensitin 4–5 weeks after the infection. The sensitin in a standard dose, or 10 × the standard dose, did not elicit any cross reactions in skin tests on cattle infected with mycobacterial culture. Tuberculin in standard dose gave no positive skin reactions in cattle infected with *N. asteroides*, while tuberculin in 10 × the standard dose elicited measurable reactions in the same animals.

Conclusion

Sensitins for intradermal tests prepared from *N. farcinica, N. sebivorans, N. asteroides, N. brasiliensis* and *N. otitidis-caviarum* are available for clinical trials.

For revealing delayed type hypersensitivity induced by *N. asteroides* (in a wider sense of the term) several different preparations should be used, due to the varying degrees of cross reactivity among the preparations.

The results obtained with various *N. brasiliensis* sensitins in human populations seem to show that the delayed hypersensitivity in persons infected with this microorganism is species specific, and that such tests may be of diagnostic and epidemiological value.

Very few, if any, results of tests with sensitins of the other species of *Nocardia* are available.

N. asteroides (in a wider sense of the term) has been demonstrated as a casual agent of mastitis in cattle. The results of skin tests on cattle with sensitins of *Nocardia* are encouraging and further studies of the epidemiological and diagnostic value of such tests are required.

4. Acknowledgement

My participation in the First International Conference on the *Biology of the Nocardiae* was supported by a grant (512-3811) from the Danish Medical Research Council.

5. References

Affronti, L. F. (1959). Purified protein derivatives (PPD) and other antigens prepared from atypical acid fast bacilli and *Nocardia asteroides. Am. Rev. Tuberc. pulm. Dis.* **79,** 284.

Benbow, E. R. Jr., Smith, D. T. & Grimson, K. S. (1944). Sulfonamide therapy in actinomycosis. Two cases caused by aerobic partially acid-fast actinomyces. *Am. Rev. Tuberc. pulm. Dis.* **49,** 395.

Berd, D. (1973). *Nocardia asteroides.* A taxonomic study with clinical correlations. *Am. Rev. resp. Dis.* **108,** 909.

Bishop, L. M. & Fenstermacher, R. (1933). An acid-fast actinomyces obtained from lesions resembling bovine tuberculosis. *Cornell Vet.* **23,** 287.

Blanchard, R. (1896). Parasites végétaux a l'exclusion des bactéries. In *Traité de pathologie Générale,* vol. II. Ed. Ch. Bouchard, p. 811. Paris: G. Masson.

Bleiker, M. A. (1964). *Tuberculine-Onderzoek in Suriname II.* Koninklijke Drukkerij. Amsterdam: van Gorcum & Comp.

Bojalil, L. F. & Magnusson, M. (1963). Specificity of skin reactions of humans to Nocardia sensitins. *Am. Rev. resp. Dis.* **88,** 409.

Bojalil, L. F. & Zamora, A. (1963). Precipitin and skin tests in the diagnosis of mycetoma due to *Nocardia brasiliensis. Proc. Soc. exp. Biol. Med.* **113,** 40.

Bradley, S. G. (1973). Relationships among mycobacteria and nocardiae based upon deoxyribonucleic acid reassociation. *J. Bact.* **113,** 645.

Braendstrup, O., Magnusson, M., Maxild, J. & Werdelin, O. (1975). Demonstration of delayed-type hypersensitivity in guinea-pigs by the agarose plate technique. *Acta path. Microbiol. scand.* **83,** 52.

Bretey, J. (1933). Sur les propriétés allergisantes des *Nocardia* et des *Streptothrix* et sur la toxicité de leurs extraits. *C.r. Séanc. Soc. Biol.* **113,** 53.

Buchanan, R. E. & Gibbons, N. E. (Eds). (1974). *Bergey's Manual of Determinative Bacteriology,* 8 edn. Baltimore: Williams and Wilkins.

Chamoiseau, G. (1973). *Mycobacterium farcinogenes* agent causal du farcin du boeuf en Afrique. *Ann Microbiol.* (*Inst. Pasteur*) **124A,** 215.

Chamoiseau, G. & Asselineau, J. (1970). Examen des lipides d'une souche de *Nocardia farcinica*: présence d'acides mycoliques. *C.r. hebd. Séanc. Acad. Sci., Paris* **270,** 2603.

Cuttino, J. & McCabe, A. (1949). Pure granulomatous nocardiosis: a new fungus disease distinguished by intracellular parasitism. A description of a new disease in man due to a hitherto undescribed organism *Nocardia intracellularis,* n. sp., Including a study of the biologic and pathogenic properties of the species. *Am. J. Path.* **25,** 1.

Drake, C. H. & Henrici, A. T. (1943). *Nocardia asteroides.* Its pathogenicity and allergic properties. *Am. Rev. Tuberc. pulm. Dis.* **48,** 184.

Dyson, J. E. Jr. & Slack, J. M. (1963). Improved antigens for skin testing in nocardiosis. I. Alcohol precipitates of culture supernates. *Am. Rev. resp. Dis.* **88,** 80.

Editorial note. (1974). In *Bergey's Manual of Determinative Bacteriology,* 8 edn. Eds Buchanan, R. E. & Gibbons, N. E. p. 730. Baltimore: Williams and Wilkins.

Edwards, L. B., Edwards, P. Q. & Palmer, C. E. (1959). Sources of tuberculin sensitivity in human populations. A summing up of recent epidemiologic research. *Acta. tuberc. scand.* **47,** 77.

Edwards, L. B. & Krohn, E. F. (1957). Skin sensitivity to antigens made from various acid-fast bacteria. *Am. J. Hyg.* **66,** 253.

Eppinger, H. (1891). Ueber eine neue, pathogene *Cladothrix* und eine durch sie hervogerufene Pseudotuberculosis (cladothrichica). *Beitr. path. Anat.* **9,** 287.

Eriksson, D. (1935). The pathogenic aerobic organisms of the *Actinomyces* group. *Medical Research Council Report* No. **203,** London: H.M.S.O.

Gamet, A., Brottes, H. & Essomba, R. (1964). Nouveaux cas de mycétomes dépistés au Cameroun. *Bull. Soc. Path. exot.* **57,** 1191.

Georg, L. K., Ajello, L., McDurmont, C. & Hosty, T. S. (1961). The identification of *Nocardia asteroides* and *Nocardia brasiliensis. Am. Rev. resp. Dis.* **84,** 337.

Glover, R. P., Herrell, W. E., Heilman, F. R. & Pfuetze, K. H. (1948). Nocardiosis. *Nocardia asteroides* infection simulating pulmonary tuberculosis. *J. Am. med. Ass.* **136,** 172.

Goodfellow, M. (1971). Numerical taxonomy of some nocardioform bacteria. *J. gen. Microbiol.* **69,** 33.

Gordon, R. E. & Mihm, J. M. (1957). A comparative study of some strains received as nocardiae. *J. Bact.* **73,** 15.

Gordon, R. E. & Mihm, J. M. (1959). A comparison of four species of mycobacteria. *J. gen. Microbiol.* **21,** 736.

Gordon, R. E. & Mihm, J. M. (1962*a*). Identification of *Nocardiae caviae* (Erikson) *nov. comb. Ann. N.Y. Acad. Sci.* **98,** 628.

Gordon, R. E. & Mihm, J. M. (1962*b*). The type species of the genus *Nocardia. J. gen. Microbiol.* **27,** 1.

Gorrill, R. H. & Heptinstall, R. H. (1954). The animal pathogenicity of *Nocardia sebivorans nov. spec. J. Path. Bact.* **68,** 387.

Goyal, R. K. (1937). Étude microbiologique, experimentale et immunologique de quelques streptothricées. *Annls Inst. Pasteur, Paris* **59,** 94.

Henrici, A. T. & Gardner, E. L. (1921). The acid-fast actinomycetes with a report of a case from which a new species was isolated. *J. infect. Dis.* **28,** 232.

Kingsbury, E. W. & Slack, J. M. (1967). A polypeptide skin test antigen from *Nocardia asteroides.* I. Production, chemical and biological characterization. *Am. Rev. resp. Dis.* **95,** 827.

Kingsbury, E. W. & Slack, J. M. (1969). A polypeptide skin test antigen from *Nocardia asteroides.* II. Further studies on the specificity of a Nocardia active polypeptide. *Sabouraudia* **7,** 85.

Kurup, P. V., Randhawa, H. S. & Gupta, N. P. (1970). Nocardiosis: a review. *Mycopath. Mycol. appl.* **40,** 193.

Kurup, P. V. & Schmitt, J. A. (1973). Numerical taxonomy of *Nocardia. Can. J. Microbiol.* **19,** 1035.

Kwapinski, J. B. G. (1970). Serological taxonomy and relationships of Actinomycetales. In *The Actinomycetales.* Ed. H. Prauser. p. 345. Jena: Gustav Fischer.

Lechevalier, M. P., Horan, A. C. & Lechevalier, H. (1971). Lipid composition in the classification of nocardiae and mycobacteria. *J. Bact.* **105,** 313.

Lessel, E. F. (1962). Status of the name *Nocardia otitidis-caviarum* Snijders 1924. *Int. Bull. bact. Nomencl. Taxon.* **12,** 191.

McClung, N. M. (1974). Family VI. *Nocardiaceae* Castellani and Chalmers 1919. In *Bergey's Manual of Determinative Bacteriology,* 8 edn. pp. 726–746. Baltimore: Williams and Wilkins.

Magnusson, M. (1962). Specificity of sensitins. III. Further studies in guinea pigs with sensitins of various species of *Mycobacterium* and *Nocardia. Am. Rev. resp. Dis.* **86,** 395.

Magnusson, M. & Mariat, F. (1968). Delineation of *Nocardia farcinica* by delayed type skin reactions on guinea pigs. *J. gen. Microbiol.* **51,** 151.

Mariat, F. (1965). Étude comparative de souches de *Nocardia* isolées de mycétomes. *Annls Inst. Pasteur, Paris* **109,** 90.

Mathieson, D. R., Harrison, R., Hammond, C. & Henrici, A. T. (1935). Allergic reactions of actinomycetes. *Am. J. Hyg.* **21,** 405.

Nocard, E. (1888). Note sur la maladie des boeuf de la Guadeloupe connue sous le nom de farcin. *Annls Inst. Pasteur, Paris* **2,** 19.

Ortiz-Ortiz, L. & Bojalil, L. F. (1972). Delayed skin reactions to cytoplasmic extracts of *Nocardia* organisms as a means of diagnosis and epidemiological study of *Nocardia* infection. *Clin. Exp. Immun.* **12,** 225.

Ortiz-Ortiz, L., Bojalil, L. F. & Contreras, M. F. (1972). Delayed hypersensitivity to polysaccharides from *Nocardia. J. Immun.* **108,** 1409.

Ortiz-Ortiz, L., Contreras, M. F. & Bojalil, L. F. (1972*a*). The assay of delayed hypersensitivity to ribosomal proteins from *Nocardia. Sabouraudia* **10,** 147.

Ortiz-Ortiz, L., Contreras, M. F. & Bojalil, L. F. (1972*b*). Cytoplasmic antigens from *Nocardia* eliciting a specific delayed hypersensitivity. *Infect. Immun.* **5,** 879.

Palmer, D. L., Harvey, R. L. & Wheeler, J. K. (1974). Diagnostic and therapeutic considerations in *Nocardia asteroides* infection. *Medicine, Baltimore* **53,** 391.

Pier, A. C. & Enright, J. B. (1962). *Nocardia asteroides* as a mammary pathogen of cattle. III. Immunologic reactions of infected animals. *Am. J. vet. Res.* **23,** 284.

Pier, A. C. & Fichtner, R. E. (1971). Serologic typing of *Nocardia asteroides* by immunodiffusion. *Am. Rev. resp. Dis.* **103,** 698.

Pier, A. C., Gray, D. M. & Fosatti, M. J. (1957). Nocardia infection of the bovine mammary gland—a preliminary report. *J. Am. vet. med. Ass.* **131,** 327.

Pier, A. C., Gray, D. M. & Fosatti, M. J. (1958). *Nocardia asteroides*—a newly recognized pathogen of the mastitis complex. *Am. J. vet. Res.* **19,** 319.

Pier, A. C. & Keeler, R. F. (1965). Extracellular antigens of *Nocardia asteroides.* I. Production and immunologic characterization. *Am. Rev. resp. Dis.* **91,** 391.

Pier, A. C., Mejia, M. J. & Willers, E. H. (1961). *Nocardia asteroides* as a mammary pathogen of cattle. I. The disease in cattle and the comparative virulence of 5 isolates. *Am. J. vet. Res.* **22,** 502.

Pier, A. C., Thurston, J. R. & Larsen, A. B. (1968). A diagnostic antigen for nocardiosis: comparative tests in cattle with nocardiosis and mycobacteriosis. *Am. J. vet. Res.* **29,** 397.

Ridell, M. (1975). Taxonomic study of *Nocardia farcinica* using serological and physiological characters. *Int. J. Syst. Bact.* **25,** 124.

Ridell, M. & Norlin, M. (1973). Serological study of Nocardia by using mycobacterial precipitation reference system. *J. Bact.* **113,** 1.

Schneidau, J. D. Jr. & Shaffer, M. F. (1957). Studies on Nocardia and other Actinomycetales. I. Cultural studies. *Am. Rev. Tuberc. pulm. Dis.* **76,** 770.

Shephard, C. C. (1958). Behaviour of the atypical mycobacteria in He La cells. *Am. Rev. Tuberc. pulm. Dis.* **77,** 968.

Silvestri, L., Turri, M., Hill, L. R. & Gilardi, E. (1962). A quantitative approach to the systematics of actinomycetes based on overall similarity. *Symp. Soc. gen. Microbiol.* **12,** 333.

Sneath, P. H. A. & Skerman, V. B. D. (1966). A list of type and reference strains of bacteria. *Int. J. Syst. Bact.* **16,** 1.

Snijders, E. P. (1924). Cavia-scheefkopperij, een nocardiose. *Geneesk. Tijdschr. Ned.-Indië* **64,** 85.

Takeya, K., Nakayama, Y. & Muraoka, S. (1970). Specificity in skin reaction to tuberculin protein prepared from rapidly growing mycobacteria and some Nocardia. *Am. Rev. resp. Dis.* **102,** 982.

Trevisan, V. (1889). *I generi e la specie dell Batteriacee.* p. 9. Milano: Zanaboni e Gabuzzi.

Tsukamura, M. (1969). Numerical taxonomy of the genus *Nocardia. J. gen. Microbiol.* **56,** 265.

Vallee, H. (1903). Sur un nouveau *Streptothrix* (*Streptothrix polychromogenes*). *Annls Inst. Pasteur, Paris* **17,** 288.

Waksman, S. A. (1957). Family Actinomycetaceae Buchanan. In *Bergey's Manual of Determinative Bacteriology*, 7 edn. Eds Breed *et al.* p. 713. Baltimore: Williams and Wilkins.

Zupnik, L. (1903). Ueber die Tuberkulinreaktion. *Dt. Arch. klin. Med.* **76,** 290.

10. Host-Phage Relationships in Nocardioform Organisms

H. Prauser

Zentralinstitut für Mikrobiologie und Experimentelle Therapie, Forschungszentrum für Molekularbiologie und Medizin, Akademie der Wissenschaften der DDR, Jena, German Democratic Republic

Contents

1. Introduction

A thorough review of the literature has revealed little information regarding host-phage relationships in nocardioform organisms. Nocardioform organisms are defined as branching bacteria which reproduce by fragmentation into irregular bacilli and coccoid elements (Prauser, 1974*a*, 1975). In the author's opinion there are 13 genera which harbour nocardioform actinomycetes. It has been suggested that *Nocardia* Trevisan *sensu* Waksman (1961) be subdivided into at least four genera on the basis of actinophage host ranges, cell wall composition and morphology (Prauser & Falta, 1968). They include: *Nocardia* Trevisan *sensu* McClung (1974), *Oerskovia* (Prauser *et al.*) Lechevalier (1972), *Nocardioides* Prauser (1974*b*, 1976*a*), and *Actinomadura* (Lechevalier & Lechevalier) J. Meyer (in Prauser, 1976*b*). However, *Actinomadura dassonvillei* (Brocq-Rousseu) Lechevalier & Lechevalier (1970) was found to be nocardioform and recently combined in the new genus *Nocardiopsis* J. Meyer (in Prauser 1974*c*, 1976*b*, Meyer 1976). The other nocardioform taxa include *Actinomyces, Rothia, Promicromonospora, Agromyces, Bacterionema, Intrasporangium, Arachnia,* and now, tentatively, *Saccharopolyspora* Lacey & Goodfellow (1974).

Although the genus *Mycobaterium* includes some nocardioform organisms they will not be considered here. A discussion of mycobacteriophage has been reviewed in the proceedings of a symposium on host-virus relationships in *Mycobacterium, Nocardia* and *Actinomyces* by Juhasz & Plummer (1970).

Studies of nocardiophage date back to 1958 when Bradley and his co-workers studied the taxonomic relationships between actinomycetes utilizing actinophage susceptibilities (Bradley & Anderson, 1958). Brownell employed nocardiophage in genetic investigation and succeeded in utilizing phage sensitivity as a genetic marker (Brownell & Adams, 1968). In the course of their studies they obtained the most detailed information regarding nocardiophages, ϕC and ϕEC (Brownell, Adams & Bradley, 1967). Following Bradley's lead, Prauser and his co-workers used actinophage host ranges for the identification and classification of actinomycetes (Prauser, 1967). They extended the studies to nocardioform genera other than *Nocardia* and to many sporoactinomycetes including thermophilic species. In 1972, Andrzejewski & Pietkiewicz started to study lysogeny in *Nocardia asteroides.* Further information has been given by Rautenstein *et al.* (1967); Riverin, Beaudoin & Vezina (1970); Sivcev & Bönicke (1970), and a few others.

2. Isolation and Source of Phages

For the majority of the nocardioform species, phage have not yet been isolated. The actinophages reported to date were isolated from either soil samples or from cultures of actinomycetes, by methods worked out with phages of the *Eubacteriales* (Adams, 1959). Most of the phage have been isolated from samples rich in organic matter, such as compost soils, mire and soils from hotbeds regardless of their geographic origin. Among the phages isolated from soil samples are the nocardiophages ϕC, ϕEC (Brownell, Adams & Bradley, 1967), and R1 (Riverin, Beaudoin & Vezina, 1970), and the mycobacteriophage Bo2 (Sivcev & Bönicke, 1970). All phages employed in Bradley's study which were active against nocardiae and mycobacteria, and all *Nocardia, Mycobacterium, Nocardioides, Oerskovia, Promicromonospora* and *Arthrobacter* phages employed by Prauser were isolated by methods involving the addition of soil. Bradley, Anderson & Jones (1961) employed peptone-yeast extract broth inoculated with a prospective actinomycete host and one teaspoonful of soil. After 24 h the mixture was centrifuged, filter sterilized and the filtrate assayed for phage by spreading 0·05 ml of filtrate with 10^7–10^8 colony-forming units (c.f.u.) of the prospective host onto peptone-yeast extract agar. Single plaques were isolated and purified by serial passages. Prauser & Falta (1968) applying similar techniques inoculated a suspension of a single soil sample with up to 20 preincubated prospective hosts and then spotted the filtrates against the same 20 strains on conventional soft-agar overlay dishes. Mixtures of several soils were also tested for phage with one or more prospective hosts.

Polyvalent *Streptomyces* phage have been isolated most frequently, whereas nocardiophage are rarely polyvalent. Expressed in percentages, the successful isolation trials ranged from 3% for *Nocardia corallina* ATCC4273 to more than 60% for streptomycetes. The relative small number of strains of *Nocardioides, Oerskovia* and *Promicromonospora* isolated in different laboratories seems to be incompatible with the ease of getting the respective phage from soil. Either the isolation techniques have been inappropriate for these actinomycetes or hosts of a taxonomic position different from those described here occur in soil.

Other phage were obtained from cultures of nocardiae without the addition of soil. Andrzejewski & Pietkiewicz (1972) and Pietkiewicz & Andrzejewski (1973*b*), isolated their nocardiophage by sterile filtration from cultures of several *Nocardia asteroides* strains after 9–28 days, in one exceptional case after 2 days. Pulverer, Schütt-Gerowitt & Schaal (1974) also obtained phage from strains of this species.

Phages interfering with the production of antibiotics were frequently reported for streptomycetes and in two cases for nocardiae. Thiemann *et al.* (1964*a,b*) isolated phage highly active in cultures of the rifamycin producer *Nocardia mediterranei*. Rautenstein *et al.* (1967) found temperate phage in the course of the production of ristomycin by strains of *Proactinomyces* (*Nocardia*) *fructiferi* var. *ristomycini*.

3. The Role of Host-Range Studies in the Classification of Nocardioform Organisms

Bradley and his co-workers were the first to try to elucidate phylogenetic relationships between genera of the Actinomycetales by the use of actinophage host ranges (Bradley & Anderson, 1958).

As judged by the ability of *Streptomyces* phages to attack strains of *Nocardia,* the Bradley group concluded a close relationship of *Streptomyces* and *Nocardia* (Bradley & Anderson, 1958; Jones & Bradley, 1961, 1962, 1964; Bradley, Anderson & Jones, 1961). Although Anderson & Bradley (1961) recognized that three strains previously sensitive to *Streptomyces* phage had been mislabeled as streptomycetes (*Nocardia sp.* 52, *N. paraguayensis, Actinomadura madurae*), phage likewise active against strains of different genera were not denied, i.e. *Streptomyces* and *Micromonospora*; and *Streptomyces* and *Nocardia* (Bradley, Anderson & Jones, 1961; Jones & Bradley, 1962).

Rautenstein *et al.* (1967) in their paper on phages of *Proactinomyces* (*Nocardia*) *fructiferi* var. *ristomycini* stated that they isolated from soil actinophage for several strains of *Actinomyces* (*Streptomyces*) and some strains of *Proactinomyces* (*Nocardia*). In a paper dealing with the relationships between mycobacteria and nocardiae Manion *et al.* (1964) summarized their conclusions from the host-range studies on actinomycetes as follows: "Based on viral susceptibility, the actinomycetes have now been arranged so that streptomycetes and mycobacteria are joined by the nocardiae. The nocardiae with persistent mycelia seem to be the intermediate type, whereas those which fragment rapidly are depicted as a branch." Studies of different groups working on taxonomy (Lechevalier & Lechevalier, 1970; Lechevalier, Lechevalier & Gerber, 1971; Cross & Goodfellow, 1973; Prauser, 1970*a,* 1975) did not support the suggestion of a close relationship between *Streptomyces* and *Nocardia. Nocardia gibsonii* reported to be susceptible to *Streptomyces* phages (Gilmour & Ingalsbe, 1959) is now placed in the genus *Streptomyces* (Pridham & Tresner, 1974).

Host range studies by Prauser & Falta (1968) used 19 phages and 80 strains of *Nocardia, Mycobacterium, Nocardioides, Oerskovia, Promicromonospora, Streptomyces, Actinomadura* and other genera of the *Streptomycetaceae.* To date, more than 500 actinomycete strains and over 50 actinophages have been included in the host range studies. The largest set of experiments involved 435 strains and 28 phages. This material represented 23 genera of Actinomycetales and 197 species, 55 of them in 6 nocardioform genera: *Nocardia* (152 strains/42 species including *N. asteroides* and other aerial mycelium-forming nocardiae, strains of the "*rhodochrous*" complex, *Mycobacterium* (14 strains/9 species), *Nocardioides* (32 strains/1 species), *Intrasporangium* (1 strain/1 species), *Oerskovia* (9 strains/1 species), and *Promicromonospora* (4 strains/1 species). The phages employed in the study included: 7 *Nocardia* N-phages, 6 *Nocardioides* X-phages, 3 *Oerskovia* O-phages, and 3 *Promicromonospora* P-phages. Moreover, 6 polyvalent S-phages active against strains of all genera of the Streptomycetaceae, two genus-specific *Micromonospora* phages, and one phage active against *Arthrobacter tumescens* were used. Selected data are provided in Table 1. N-phages are genus-specific for *Nocardia*. They attack nocardiae with extended primary and aerial mycelium, e.g. *Nocardia asteroides*, as well as nocardiae with reduced branching and lacking aerial mycelium, e.g. *Nocardia corallina*. For the X-, O- and P-phages it is uncertain whether they are species- or genus-specific since the first two genera are at present represented by only one species. In the case of *Oerskovia* only strains of *Oerskovia turbata* were available. No strain of the rapidly growing mycobacteria was effected by any of the phages. There was also no case of cross-lysis between the *Streptomyces* and *Micromonospora* phages when tested against strains of the respective taxa. This observation does not agree with the afore-mentioned results and conclusions of Bradley's group. The *Streptomyces* phages were family-specific (Table 1) an observation confirmed by examination of 152 *Streptomyces* strains representing 109 species and 17 strains from other genera of the Streptomycetaceae.

The only exceptions in the taxon-specificity of the phages used in Prauser's study were the phages X4 and X6. These phages were also effective against strains of *Nocardioides albus* and *Arthrobacter simplex*, both of which have the same wall composition. As suggested by Yamada & Komagata (1972) and Schleifer & Kandler (1972) *Arthrobacter simplex* is possibly related to actinomycetes which possess the same wall type. At present, common lysis of microorganisms which are morphologically as different as the motile, single-celled coryneform bacterium, *Arthrobacter simplex*

Table 1
Host ranges of selected strains and phages

	Streptomycetaceae														
	Streptomyces								Other genera						
Phage	ATCC3004	INA2659	NCIB8232	ATCC19773	ATCC8664	NCIB8238	ATCC3356	ATCC25494	RIA563	ISP5601	RIA655	IMET9902	IMRU3857	M2572
S1	•	•	+	+	+	•	+	+	+	•	+	+	•	•
S2	+	+	+	+	+	+	+	+	+	+	+	+	+	•
S3	+	•	•	•	•	•	•	•	•	•	•	•	•	•
S4	•	•	•	+	+	•	+	•	+	•	•	+	•	+
S6	+	•	+	•	+	•	+	+	+	•	+	+	•	+
S7	•	+	+	•	+	+	+	+	•	+	+	+	+	+
X1	•	•	•	•	•	•	•	•	•	•	•	•	•	•
X2	•	•	•	•	•	•	•	•	•	•	•	•	•	•
X3	•	•	•	•	•	•	•	•	•	•	•	•	•	•
X4	•	•	•	•	•	•	•	•	•	•	•	•	•	•
X5	•	•	•	•	•	•	•	•	•	•	•	•	•	•
X6	•	•	•	•	•	•	•	•	•	•	•	•	•	•
N2	•	•	•	•	•	•	•	•	•	•	•	•	•	•
N4	•	•	•	•	•	•	•	•	•	•	•	•	•	•
N5	•	•	•	•	•	•	•	•	•	•	•	•	•	•
N9	•	•	•	•	•	•	•	•	•	•	•	•	•	•
O1	•	•	•	•	•	•	•	•	•	•	•	•	•	•
O2	•	•	•	•	•	•	•	•	•	•	•	•	•	•
O3	•	•	•	•	•	•	•	•	•	•	•	•	•	•
O4	•	•	•	•	•	•	•	•	•	•	•	•	•	•
P1	•	•	•	•	•	•	•	•	•	•	•	•	•	•
P2	•	•	•	•	•	•	•	•	•	•	•	•	•	•
P3	•	•	•	•	•	•	•	•	•	•	•	•	•	•
Y1	•	•	•	•	•	•	•	•	•	•	•	•	•	•
Cell wall type				I							I			

+ = sensitive to phage; • = not sensitive to phage; nt = not tested.
Cell wall type: Number according to Lechevalier & Lechevalier (1967). I = LL-diaminopimelic acid, glycine; IV = *meso* diaminopimelic acid, arabinose, galactose; VI = lysine, aspartic acid.
Phages: see text.
Strains (in the order given in the Table), in brackets: source of the strain.
Streptomycetaceae: ATCC3004 *S. albus* (W. Kurylowicz); INA2659 *S. glaucescens* (G. F. Gause); NCIB8232 *S. griseus* (W. Kurylowicz); ATCC19773 *S. hirsutus* (E. B. Shirling); ATCC8664 *S. lavendulae* (W. Kurylowicz); NCIB8238 *S. olivaceus* (W. Kurylowicz);

Table 1—continued
Host ranges of selected strains and phages

	Nocardioides albus								*Arthrobacter*		*Nocardia*					
Phage	IMET7801	IMET7806	IMET7807	IMET7809	IMET7810	IMET7812	IMET7818	IMET7822	SG1020	NCIB8929	369	673	751	IMET7330	ATCC4273	ATCC3318
S1	·	·	·	·	·	·	·	·	·	·	·	·	·	·	·	·
S2	·	·	·	·	·	·	·	·	·	·	·	·	·	·	·	·
S3	·	·	·	·	·	·	·	·	·	·	·	·	·	·	·	·
S4	·	·	·	·	·	·	·	·	·	·	·	·	·	·	·	·
S6	·	·	·	·	·	·	·	·	·	·	·	·	·	·	·	·
S7	·	·	·	·	·	·	·	·	·	·	·	·	·	·	·	·
X1	+	+	+	·	+	+	·	+	·	·	·	·	·	·	·	·
X2	+	+	+	+	·	+	+	+	·	·	·	·	·	·	·	·
X3	·	+	·	·	+	+	·	+	·	·	·	·	·	·	·	·
X4	·	+	·	·	·	+	·	·	+	·	·	·	·	·	·	·
X5	+	+	+	·	·	+	·	+	·	·	·	·	·	·	·	·
X6	+	+	·	·	+	·	+	+	+	+	·	·	·	·	·	·
N2	·	·	·	·	·	·	·	·	·	·	·	·	+	+	+	·
N4	·	·	·	·	·	·	·	·	·	·	+	+	+	·	·	+
N5	·	·	·	·	·	·	·	·	·	·	+	·	·	·	·	·
N9	·	·	·	·	·	·	·	·	·	·	·	·	·	+	+	·
O1	·	·	·	·	·	·	·	·	·	·	·	·	·	·	·	·
O2	·	·	·	·	·	·	·	·	·	·	·	·	·	·	·	·
O3	·	·	·	·	·	·	·	·	·	·	·	·	·	·	·	·
O4	·	·	·	·	·	·	·	·	·	·	·	·	·	·	·	·
P1	·	·	·	·	·	·	·	·	·	·	·	·	·	·	·	·
P2	·	·	·	·	·	·	·	·	·	·	·	·	·	·	·	·
P3	·	·	·	·	·	·	·	·	·	·	·	·	·	·	·	·
Y1	·	·	·	·	·	·	·	·	·	·	·	·	·	·	·	·
Cell wall type	I								I		IV					

ATCC3356 *S. viridochromogenes* (W. Kurylowicz); ATCC25494 *Streptoverticillium rubrireticuli* (E. B. Shirling); RIA563 *Chainia poonensis* (V. D. Kuznetsov); ISP5601 *Actinopycnidium caeruleum* (E. B. Shirling); RIA655 *Actinosporangium violaceum* (V. D. Kuznetsov); IMET9902 *Microellobosporia cinerea* (H. Prauser); IMRU3857 *Microellobosporia flavea* (H. A. Lechevalier); M2572 *Elytrosporangium brasiliense* (H. A. Lechevalier).

Nocardioides albus: IMET7801–7822 (H. Prauser).

Arthrobacter: IMET SG1020 *Arthrobacter sp.* (Sci. Serv., Ottawa), NCIB8929 *Arthrobacter simplex* (NCIB).

Nocardia: 369 *N. asteroides* (F. Mariat); 673 *N. calcarea* (P. Hirsch); 751 *N. caviae* (F. Mariat); IMET7330 *N. caviae* (H. Mordarska); ATCC4273 *N. corallina* (I. Szabo); ATCC3318 *N. farcinica* (J. E. Thiemann); ATCC4276 *N. opaca* (I. Szabo); SR *N. pellegrino* (H. Mordarska); IMET7026 *N. uniformis* (I. Szabo); CBS247.45 *N. vaccinii* (G. F. Gause).

Table 1—continued

Host ranges of selected strains and phages

				Oerskovia turbata						*Promicromono-spora citrea*					nf act.
ATCC4276	SR	IMET7026	CBS247.45	IMET7003	IMET7006	IMET7011	Örskov 27	IMET7135	IMRU761	RIA562	IMET7085	IMET7260	IMET7261	IMET7262	IMET7005
·	·	·	·	·	·	·	·	·	·	·	·	·	·	·	·
·	·	·	·	·	·	·	·	·	·	·	·	·	·	·	·
·	·	·	·	·	·	·	·	·	·	·	·	·	·	·	·
·	·	·	·	·	·	·	·	·	·	·	·	·	·	·	·
·	·	·	·	·	·	·	·	·	·	·	·	·	·	·	·
·	·	·	·	·	·	·	·	·	·	·	·	·	·	·	·
·	·	·	·	·	·	·	·	·	·	·	·	·	·	·	·
·	·	·	·	·	·	·	·	·	·	·	·	·	·	·	·
·	·	·	·	·	·	·	·	·	·	·	·	·	·	·	·
·	·	·	·	·	·	·	·	·	·	·	·	·	·	·	·
·	·	·	·	·	·	·	·	·	·	·	·	·	·	·	·
·	·	·	·	·	·	·	·	·	·	·	·	·	·	·	·
+	+	·	·	·	·	·	·	·	·	·	·	·	·	·	·
·	·	+	+	·	·	·	·	·	·	·	·	·	·	·	·
·	+	·	·	·	·	·	·	·	·	·	·	·	·	·	·
+	+	·	·	·	·	·	·	·	·	·	·	·	·	·	·
·	·	·	·	nt	+	nt	+	+	+	·	·	·	·	·	·
·	·	·	·	+	+	+	+	+	+	·	·	·	·	·	·
·	·	·	·	+	+	+	+	+	+	·	·	·	·	·	·
·	·	·	·	+	+	+	·	+	+	·	·	·	·	·	·
·	·	·	·	·	·	·	·	·	·	+	+	·	·	·	·
·	·	·	·	·	·	·	·	·	·	+	+	+	+	+	·
·	·	·	·	·	·	·	·	·	·	+	·	+	+	·	·
·	·	·	·	·	·	·	·	·	·	·	·	·	·	·	+
						VI						VI			VI

Oerskovia turbata: IMET7003, 7006, 7011 (H. Prauser); *Ørskov* 27 (H. Lautrop), IMET7135 (H. Prauser); IMRU761 (Fabian, Prague).

Promicromonospora citrea: RIA562 (V. D. Kuznetsov); IMET7085 (L. Muller, Jena); IMET7260, 7261, 7262 (H. Prauser).

Nocardioform actinomycete of unknown genus position: IMET7005 (H. Prauser).

Culture collections: ATCC = American Type Culture Collection, Rockville; CBS = Centraalbureau voor Schimmelcultures, Baarn; IMET = Institut für Mikrobiologie und experimentelle Therapie, Jena; IMRU = Institute of Microbiology, Rutgers, New Brunswick; INA = Institute for New Antibiotics, Moscow; ISP = International Streptomyces Project; NCIB = National Collection of Industrial Bacteria, Aberdeen; RIA = U.S.S.R. Research Institute for Antibiotics, Moscow.

and the typically nocardioform actinomycete, *Nocardioides albus*, by the same phage is unique. Furthermore, it could be shown that some of the *Streptomyces* phages, particularly phage S7, may cause distinct clearing effects without traceable phage multiplication on *Nocardioides albus*, when high-titre phage suspensions were applied.

Prauser (1968, 1970*a*,*b*) has proposed a possible relationship between actinomycetes based in part on phage host ranges. He postulated that the organisms now placed in the order Actinomycetales (Gottlieb, 1974) may have evolved from different origins within cornyeform bacteria (Prauser, 1975). One evolutionary line would be the LL-diaminopimelic acid-containing coryneform bacteria such as *Arthrobacter simplex* by *Nocardioides albus*, *Streptomyces* and the other *Streptomycetaceae*. Another group may have evolved from *meso*-diaminopimelic and mycolic acid-containing coryneform bacteria, namely the mycobacteria, nocardiae and micropolysporas. Actinomycetes possessing other diamino acids (*Actinomyces*, *Oerskovia* and *Promicromonospora*) are not thought to be related to the afore-mentioned taxa.

Phage typing systems for reliable identification of actinomycete species or genera have not yet been established since there are no phages able to lyse all strains of a given taxon, with the exceptions of the *Oerskovia*, and *Promicromonospora* phages (Prauser & Falta, 1968). Nevertheless, most of the phages proved to be very useful for the quick determination of newly isolated actinomycetes and for the elucidation of taxonomic uncertainties concerning the genus position of unknown actinomycetes. In the experience of this author, subsequent cell wall analyses and other taxonomic studies always confirm the taxonomic position indicated by the host ranges.

4. Characteristics of Specific Phages

(a) *Nocardiophage φC and φEC*

Two nocardiophages, φC and φEC, both isolated from soil, were studied to determine their host-phage relationships and for use as genetic markers (Brownell, Adams and Bradley, 1967). Phage φC infected the *Nocardia canicruria* mating type (Mat-Ce) of *Nocardia erythropolis*. Phage φEC infected both the mat-Ce and the mat-cE mating types of *Nocardia erythropolis*. Moreover, phage ϕC attacked one of eight strains of *Nocardia corallina* tested while phage ϕEC infected the latter and one of two *Nocardia opaca* strains tested. Strains of *Nocardia brasiliensis, N. farcinica, N. rubra, N. asteroides, N. rhodnii* and *Mycobacterium rhodochrous* were not lysed by either strain. Phage MJP1 displayed host specificity similar to ϕEC, but also infected *Nocardia farcinica*. Later, one strain each of *Nocardia globerula, N. calcarea* and *N. restrictus* was found sensitive to ϕC and ϕEC (Brownell & Crockett, 1971). In addition, Bradley & Huitron (1973) detected the susceptibility of a *Nocardia gypsoides* strain to both phages.

Brownell, Adams and Bradley (1967) initiated studies to further characterize the phages. Phage ϕC was found to possess a head 52 nm in diameter and a tail 192 nm long; phage ϕEC showed similar size and shape. The phages differed in latent period (ϕC: 25 min, ϕEC: 180 min), burst size (ϕC: 60 p.f.u./infective centre, ϕEC: 20 p.f.u./infective centre), and adsorption (ϕC: 95%, ϕEC: 34%. Both phages were sensitive to NaCl and chloroform, but were not affected by diethyl ether. Optimum growth of phage ϕC ($1{\cdot}8 \times 10^{10}$ p.f.u./ml) was achieved by propagation on peptone yeast-extract broth containing $0{\cdot}01$ M $-$ $Ca(NO_3)_2$ at 30°C, using a heavy suspension of the host from an 18 h culture and by adding the phage after 2 h incubation at a concentration of 10^4 p.f.u./ml. On a chemically defined medium containing glycerol, nicotinamide, salts, and amino acids, 7×10^{10} p.f.u./ml could be obtained. Unique to the nocardiae, ethanol was the only solvent able to extract a phage-inactivating substance from whole cells or isolated walls of all strains to which the phages ϕC and ϕEC could attach (Brownell & Crockett, 1971). This was true even of *Nocardia restrictus* ATCC14807 on which the phages do not propagate. After the ethanol extraction the attachment of phages to cells or isolated walls decreased markedly. The inactivating complex was effective against both ϕC and ϕEC; however, the latter was 10-fold more inactivated. Protein of some complexity could be shown to be one of the constituents of the phage inhibitor. T.l.c. revealed lipoproteins. Both phages ϕC and ϕEC were shown to be temperate (Crockett & Brownell, 1972). Serial transfers of broth cultures of Mat-Ce cells (formerly called *Nocardia canicruria*) infected with phage ϕEC resulted in 99% phage producing colonies after seven serial passages (Crockett & Brownell, 1972). A phage-producing strain, Ce-3(ϕEC), was selected from these colonies and used in further experiments. The plaques of the phages produced by the Ce-3 (ϕEC) strain were clear on strain Ce-3 and resembled those of ϕEC phage. Strain Ce-3(ϕEC) turned out to be immune to the attack of ϕEC. Lysogeny of strain Ce-3(ϕEC) was concluded from a low frequency of spontaneous curing (0·4%), the inability to cure the strain with phage-specific antiserum, normal adsorption of ϕEC on Ce-3(ϕEC) cells with the absence of subsequent plaque formation, a low rate of spontaneous induction ($2{\cdot}5 \times 10^{-4}$), and a somewhat higher frequency of induction by treatment with u.v. irradiation or mitomycin C. Phage ϕC could also attach to Ce-3(ϕEC) cells but failed to propagate, which was explained as lysogenic superinfection immunity.

Strains lysogenic for phage ϕC were obtained by a similar technique using a mutant of Ce-3 (Brownell & Clark, 1974). Phages produced from spontaneously induced populations also propagate in substrains of the Mat-cE mating type strains of *Nocardia erythropolis* which is normally resistant to phage ϕC. However, the progeny from this infection is unable to reinfect the Mat-cE strains and revert to their original specificity. Thus, lysogenization has caused a host-controlled, non-inheritable, phenotypic alteration on the prophage.

Maturation of phage ϕEC in *Nocardia erythropolis* was examined in ultrathin sections of infected cells by Serrano, Serrano & Brownell (1974). Attached phage particles could be shown after 10 min incubation. The first phage-like particles appeared after 30 min. Clusters of them were observed in the less dense areas of the cytoplasm after 50–90 min. Cell lysis, resulting in ghost cells as well as host cell survivors, was demonstrated after 120 min.

Successful use of phage ϕC as a genetic marker in recombination studies between Mat-Ce and Mat-cE mating types was reported by Brownell & Adams (1968).

(b) *Nocardiophage R1*

Phage R1 infects a strain of *Nocardia restrictus*, two strains of *N. corallina* and one strain of *N. convoluta* and *N. globerula*. Phage R1 has a hexagonal head 75 nm in diameter. The flexible, non-contractile tail is 330 nm long and terminates in a round plate without spikes. A collar-like structure near the head has been observed (Riverin, Beaudoin and Vezina, 1970). The latent period was 25–28 min and the burst size 61 p.f.u./infective centre, 80% of the phages attached after 5 min. Acridine orange staining as well as depression of phage production by actinomycin D indicated the occurrence of double stranded DNA. Phage R1 is sensitive to heat, u.v. irradiation, NaCl and deviations from pH 7. Purified phage suspensions gave irregular, large or small, turbid or clear plaques with or without a halo. Phages from different types of plaques yielded similar mixtures of different plaques. Mutants resistant to phage R1 occurred very rarely.

(c) *Nocardia N-phages, Nocardioides X-phages, Oerskovia O-phages and* Promicromonospora *P-phages*

These phages, isolated by Prauser & Falta (1968), were used to study relationships between nocardioform organisms. The phage were cultivated on soft-agar overlays containing M79 medium, a complex organic medium. The overlays were inoculated with host cells and a dilution of phage lysates, which would produce single plaques, and incubated for 24 h at 28°C. Most of the *Nocardia* phages displayed small, more or less turbid plaques indicating small burst sizes (N2, N5, Ng: 0·3 mm in diameter; N3, N8: 0·5-0·8 mm; N4, N6: 1·0–2·0 mm). The plaques of N3 and N6 were clear. The *Nocardioides albus* phages produced clear to slightly turbid, medium-sized plaques (X4: 0·8–1·2 mm; X6: 1·0–2·0 mm; X2: 1·5–2·3 mm; X1: 2·0 mm; X5: 1·8–2·8 mm). The lysate of P1 phage displayed a mixture of clear plaques 0·3–0·5 and 1·0–2·0 mm in diameter on *Promicromonospora citrea* strains. The phages O2 and O4 produced clear plaques of 0·5–2·0 mm on representatives of *Oerskovia turbata.* All of the N-, X-, O- and P-phages studied to date display the type B morphology of Bradley (1967). The *Nocardia* phage N5 has a head with a hexagonal profile, 69 nm in diameter and a flexible tail 240 nm long and 9 nm in width, terminating in a thickened base plate (Fig. 1). The hexagonal head of the *Nocardioides* phage X2 measures 63 nm in diameter, its flexible tail 220 nm in length and 10 nm in width with a base plate. Phage X6 with a hexagonal head 50 nm in diameter, a slender tail 200 nm long and 6·5 nm wide which usually appears

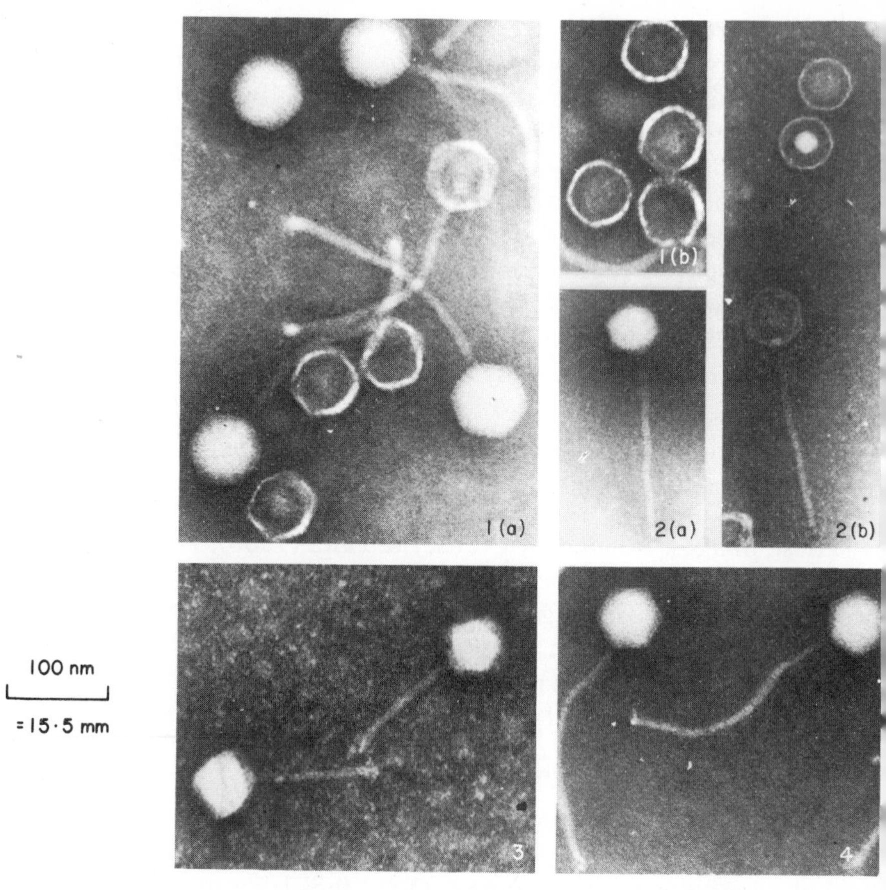

Figs 1–4. Electron micrographs of actinophages. Fig. 1. (a) (b) *Nocardia*-phage N5. (b) "Empty" heads showing a string of pearls-like peripheral structure indicating the capsid. Fig. 2. (a) (b) *Nocardioides*-phage X6. (b) "Empty" heads. Fig. 3. *Oerskovia*-phage O2. Fig. 4. *Promicromonospora*-phage P1. Marker bar denotes 100 nm for all micrographs. (From *diploma biologist Regine Velttermann, Jena*).

nearly straight in the electron micrographs (Fig. 2) and does not possess a base plate. The head of the *Oerskovia* phage O2 is 50 nm in diameter. Its tail, 125 nm long and 9 nm wide, terminates in a base plate (Fig. 3). The *Promicromonospora* phage P1 shows a very distinct hexagonal head, 58 nm in diameter and a tail 230 nm long and 9 nm wide with a basal plate (Fig. 4). Thus, the morphology of these phages is similar to other actinophages studied. The latent period of phage X1 was found to be 110 min and phage X6

150 min. Spotting of *Nocardioides* phage suspensions on strains other than those originally taken as hosts, frequently resulted in turbid to very turbid lytic areas on the test dishes. Phage X2 and strain IMET7807 were chosen to test for the possibility of lysogenization. From a turbid lytic area a strain tentatively labelled IMET7807(X2) was isolated which no longer displayed any sensitivity to X2. Filtrates from broth cultures of 10 sub-strains of IMET7807(X2) were effective against IMET7807, but failed to infect the 10 substrains. However, lysogenization remains to be verified.

(d) *Mycobacteriophage Bo2*

The *Mycobacterium phlei* phage Bo2 is of interest here since it could be successfully adapted to *Nocardia asteroides* and *N. brasiliensis* (Sivcev & Bönicke, 1970). The Bo2 phage was isolated from soil and was unable to infect several species of nocardiae tested, however, it did cause a slight clearing of a strain of *Nocardia asteroides* and *N. brasiliensis* when phage titers of 10^{10} p.f.u./ml were spotted. After three transfers of Bo2 phages on *N. asteroides*, the resulting phage suspensions were no longer active against the host strain *Mycobacterium phlei* F89, but the clearing effect on *N. asteroides* could still be demonstrated. Turbid plaques less than 0·5 nm in diameter were recognized after 20 transfers. Clones resulting from single plaques of the Bo2 phage differed in the number of transfers necessary to produce single plaques on *N. asteroides* and to lose their activity against the original *Mycobacterium* host strain. A strain of *Nocardia brasiliensis* behaved similarly to the *N. asteroides* strain with respect to host range alteration. The original host ranges of the adapted phages could be restored by their propagation on the original host strain; however, the activity against the *Nocardia* strains was not completely lost. The underlying mechanisms remain to be clarified.

(e) *Other phages*

From fermentation cultures of *Proactinomyces* (*Nocardia*) *fructiferi* var. *ristomycini* Rautenstein *et al.* (1967) isolated several phages active against this organism. The phages, virulent mutants of temperate actinophages, lysed six other nocardiae showing a similar antimicrobial spectrum, but were ineffective against 49 other strains of nocardiae tested. Virulent mutants could also be obtained by treatment of free phages or lysogenic cultures with rubomycin at concentrations of 0·25–2·0 and 0·15–100 μg/ml, respectively (Rautenstein *et al.*, 1970). The virulent phages differ in morphology, plaque characteristics, host range and antigenic properties. All five phages examined by electron micrographs possessed hexagonal heads and non-contractile, flexible or straight tails. The largest phage has a head 60 nm in diameter and a tail 150 nm long. The smaller phage ranged in size around 53 nm and 87 nm. All tails are 13 nm wide, however, the basal plates were different in shape and display spike-like appendages.

Filtrates from broth cultures of 12 strains of *Nocardia asteroides* caused phagolysis when tested against the same 12 strains (Andrzejewski & Pietkiewicz, 1972; Pietkiewicz & Andrzejewski, 1973*b*). Phage from eight of the filtrates were purified by three serial single-plaque transfers. These phages displayed identical activity against 8 of the 12 strains. Four of the phages lysed strains from which they were isolated. Spontaneous plaque formation was observed in all susceptible strains and 4 of the 8 sensitive strains. The results have been discussed but not yet explained. Ultraviolet irradiation resulted in lytic agents from 6 of the 12 strains studied by Pietkiewicz and Andrzejewski, (1973*a*). A phage also isolated from *N. asteroides* had a head 42 nm in diameter with a icosahedral shape (Andrzejewski & Müller, 1975). The tail was a flexible, non-contractile structure 140 nm long and 6 nm wide terminating in a conical structure. Several insusceptible substrains were obtained by Andrzejewski & Tarnok (1974), which produced phages even after 10 passages. Curing by phage antiserum failed. Artificial lysogenization of the indicator strain was concluded.

After Azarowicz, Erikson & Armen (1954) had provided the first electron microscopical information on nocardiophages, Anderson & McClung (1961) demonstrated that nocardiophages and other actinophages generally resemble the gross morphology of *Escherichia coli* phages. Among 43 actinophages studied by Bradley & Ritzi (1967), nine were nocardiophages including MNP1, MNP2, MNP7, MJP1, MJP20, ϕC, ϕEC and MHP8. Most of the heads showed somewhat elongated more or less hexagonal profiles ranging from 47×53 nm to 85×88 nm. The length of the tails were 120 nm for a medium-sized phage to 364 nm for the largest

(MNP8). A correlation between size and shape, and generic placement of the host could not be found.

Four nocardiophages were included in comparative studies on the action of ultrasonic vibrations on actinophages by Anderson & Bradley (1964). The sensitivity of the nocardiophages was found to be within the range of the *Streptomyces*, *Mycobacterium* and T2 phage, though there were marked differences. Generally, the large phages proved more sensitive than small ones.

5. Concluding Remarks

For the majority of the nocardioform genera, for most of the nocardiae and for many other *Actinomycetales* genera, phages have not yet been demonstrated. From the observations on the few phages that have been studied, it appears that the phage infecting nocardioform organisms resemble phages of the single-celled bacteria. All phages of the nocardioform organism examined to date show the morphological type B of Bradley (1967). However, the developmental cycle of the actinomycetes which involves the development of multicaryote hyphal districts, results in the loss of physiological susceptibility to phages in the ageing parts of the mycelia. This characteristic may have played a role in discouraging investigators to study the host-phage relationships of these phages. Thus, fundamental questions are lacking and a lot of work remains to be done.

6. References

Adams, M. H. (1959). *Bacteriophages*. London: Interscience Publishers.

Anderson, D. L. & Bradley, S. G. (1961). Susceptibility of nocardiae and mycobacteria to actinophage. *Antimicrob. Ag. Chemother.* 1961, 898.

Anderson, D. L. & Bradley, S. G. (1964). The action of ultrasonic vibrations on actinophages. *J. gen. Microbiol.* **37,** 67.

Anderson, D. L. & McClung, N. M. (1961). Morphology of actinophages. *Bact. Proc.* **1961.**

Andrzejewski, J. & Müller, G. (1975). Über die Morphologie eines *Nocardia asteroides*-Bakteriophagen. *Zentbl. Bakt. ParasitKde Abt.* I. **230,** 379.

Andrzejewski, J. & Pietkiewicz, D. (1972). Über die Isolierung von Bakteriophagen aus lysogenen *Nocardia asteroides*-Stämmen. *Zentbl. Bakt. ParasitKde Abt. I.* **219,** 366

Andrzejewski, J. & Tarnok, I. (1974). Lysogeny and artificial lysogenization in *Nocardia. Proc. Intern. Conf. Nocardiae, Media, Venezuela.* Ed. G. Brownell. p. 88. Augusta: McGowen.

Azarowciz, E. N., Erikson, J. O. & Armen, D. M. (1954). Information from the electronmicrograph. *J. biol. photogr. Ass.* **22,** 49.

Bradley, D. E. (1967). Ultrastructure of bacteriophages and bacteriocins. *Bact. Rev.* **31,** 230.

Bradley, S. G. & Anderson, D. L. (1958). Taxonomic implication of actinophage host-range. *Science, N.Y.* **128,** 413.

Bradley, S. G., Anderson, D. L. & Jones, L. A. (1961). Phylogeny of actinomycetes as revealed by susceptibility to actinophage. *Devs ind. Microbiol.* **2,** 223.

Bradley, S. G. & Huitron, M. E. (1973). Genetic homologies among nocardiae. *Devs ind. Microbiol.* **14,** 189.

Bradley, S. G. & Ritzi, D. (1967). Structure of actinophages for *Streptomyces* and *Nocardia. Devs. ind. Microbiol.* **8,** 206.

Brownell, G. H. & Adams, J. N. (1968). Linkage and segregation of a mating type-specific phage and resistance characters in nocardial recombinants. *Genetics, Princeton* **60,** 437.

Brownell, G. H., Adams, J. N. & Bradley, S. G. (1967). Growth and characterization of nocardiophages for *Nocardia canicruria* and *Nocardia erythropolis* mating types. *J. gen. Microbiol.* **47,** 247.

Brownell, G. H. & Clark, J. E. (1974). Host range alteration of nocardiophage ϕC following lysogenization of *Nocardia erythropolis J. gen. Virol.* **23,** 247.

Brownell, G. H. & Crockett, J. K. (1971). Inactivation of nocardiophages ϕC and ϕEC by extracts of bacteriophage-attachable cells. *J. gen. Virol.* **8,** 894.

Crockett, J. K. & Brownell, G. H. (1972). Isolation and characterization of a lysogenic strain of *Nocardia erythropolis. J. gen. Virol.* **10,** 737.

Cross, T. & Goodfellow, M. (1973). Taxonomy and classification of the actinomycetes. In *Actinomycetales: Characteristics and Practical Importance.* Eds G. Sykes & F. A. Skinner. p. 11. London and New York: Academic Press.

Gilmour, C. M. & Ingalsbe. (1959). Studies on *Streptomyces* phage II. Host range and species differentiation. *J. Bact.* **78,** 193.

Gottlieb, D. (1974). Order I. *Actinomycetales* Buchanan 1917, 162. In *Bergey's Manual of Determinative Bacteriology,* 8 edn. Eds R. E. Buchanan & N. E. Gibbons, p. 657. Baltimore: Williams and Wilkins.

Jones, L. A. & Bradley, S. G. (1961). Susceptibility of actinomycete type species to actinophages. *Bact. Proc.* **1961,** 74.

Jones, L. A. & Bradley, S. G. (1962). Relationship of *Streptoverticillium* and *Jensenia* to other actinomycetes. *Devs ind. Microbiol.* **3,** 257.

Jones, L. A. & Bradley, S. G. (1964). Relationships among streptomycetes, nocardiae, mycobacteria and other actinomycetes. *Mycologia* **56,** 505.

Juhasz, S. E. & Plummer, G. (Eds). (1970). *Host-Virus Relationships* in *Mycobacterium, Nocardia* and *Actinomyces.* Springfield: Charles C. Thomas.

Lacey, J. & Goodfellow, M. (1974). A new genus of nocardioform actinomycetes? *Proc. Intern. Conf. Biol. Nocardiae, Merida. Venezuela.* Ed. G. Brownell. p. 22. Augusta: McGowen.

Lechevalier, H. A. & Lechevalier, M. P. (1967). Biology of actinomycetes. *A. Rev. Microbiol.* **21,** 71.
Lechevalier, H. A. & Lechevalier, M. P. (1970). A critical evaluation of the genera of aerobic actinomycetes. In *The Actinomycetales.* Ed. H. Prauser. p. 393. Jena: Gustav Fischer.
Lechevalier, H. A., Lechevalier, M. P. & Gerber, N. N. (1971). Chemical composition as a criterion in the classification of actinomycetes. *Adv. appl. Microbiol.* **14,** 47.
Lechevalier, M. P. (1972). Description of a new species, *Oerskovia xanthineolytica,* and emendation of *Oerskovia* Prauser *et al. Int. J. Bact.* **22,** 260.
Manion, R. E., Bradley, S. G., Zinneman, H. H. & Hall, W. H. (1964). Interrelationships among mycobacteria and nocardiae. *J. Bact.* **87,** 1056.
McClung, N. M. (1974). Family VI. *Nocardiacea* Castellani and Chalmers 1919, 1040. In Bergey's Manual of Determinative Bacteriology, 8 edn. Eds. R. E. Buchanan & N. E. Gibbons. p. 726. Baltimore: Williams and Wilkins.
Meyer, J. (1976). *Nocardiopsis* gen. nov.—a new genus of the Actinomycetales. *Int. J. Syst. Bact.* **26** (in press).
Pietkiewicz, D. & Andrzejewski, J. (1973*a*). The effects of UV induction on phage release by *Nocardia asteroides. Med. dósw. Mikrobiol.* **25,** 171 (in Polish).
Pietkiewicz, D. & Andrzejewski, J. (1973*b*). Weitere Untersuchungen über *Nocardia asteroides*-Bakteriophagen. *Zentbl. Bakt. ParasitKde Abt.* 1 **224,** 376.
Prauser, H. (1967). Contributions to the taxonomy of the Actinomycetales. *Publ. Fac. Sci. Univ. Brno,* **K40,** 196.
Prauser, H. (1968). State and tendencies in the taxonomy of Actinomycetales. *Publ. Fac. Sci. Univ. Brno,* **K43,** 295.
Prauser, H. (1970*a*). Characters and genera arrangement in the Actinomycetales. In *The Actinomycetales.* p. 407. Ed. H. Prauser. Jena: Gustav Fischer.
Prauser, H. (1970*b*). Application of actinophages on the taxonomy of Actinomycetales. *Publ. Fac. Sci. Univ. Brno* **K47,** 123.
Prauser, H. (1974*a*). Host-phage relationships in nocardioform organisms. *Proc. Intern. Conf. Biol. Nocardiae, Merida, Venezuela.* Ed. G. Brownell. p. 84. Augusta: McGowen.
Prauser, H. (1974*b*). The actinomycete-genus *Nocardioides. Proc. Intern. Conf. Biol. Nocardiae, Merida, Venezuela.* Ed. G. Brownell. p. 20. Augusta: McGowen.
Prauser, H. (1974*c*). *Nocardioides* Prauser, *Nocardiopsis.* J. Meyer, and *Micropolyspora fascifera* Prauser—new taxa of the Actinomycetales, *Actinomycetologist, Tokyo* **24,** 14.
Prauser, H. (1975). The Actinomycetales—an order? In *Proc. Intern. Cong. IAMS, Tokyo,* vol. 1. p. 19. Tokyo.
Prauser, H. (1976a). *Nocardioides*—a new genus of the Actinomycetales. *Int. J. Syst. Bact.* **26,** 58.
Prauser, H. (1976*b*). New nocardioform organisms and their relationships. In *Actinomycetes—the Boundary Microorganisms.* Ed. T. Arai. Tokyo: Toppan (in press).
Prauser, H. & Falta, R. (1968). Phagensensibilität, Zellwandzusammensetzung and Taxonomie von Actinomyceten. *Z. allg. Mikrobiol.* **8,** 39.

Pridham, T. G. & Tresner, H. D. (1974). Family VII. Streptomycetaceae. Waksman and Henrici 1943, 339. In *Bergey's Manual of Determinative Bacteriology*, Eds. R. E. Buchanan & N. E. Gibbons. p. 747. Baltimore: Williams and Wilkins.

Pulverer, G., Schütt-Gerowitt, H. & Schaal, K. P. (1974). Bacteriophages of *Nocardia. Proc. Intern. Conf. Biol. Nocardiae, Merida, Venezuela.* Ed. G. Brownell. p. 82. Augusta: McGowen.

Rautenstein, Ya. I., Tikhonenko, A. S., Solovieva, N. Ya., Belyaeva, N. N. & Filatova, A. D. (1970). Obtaining of and some properties of virulent mutants of temperate phages of a polylysogenic *Proactinomyces* (*Nocardia*) *fructiferi. Izv. Akad. Nauk SSSR* **1970,** 272. (In Russian.)

Rautenstein, Ya. I., Trenina, G. A., Solovieva, N. Ya & Filatova, A. D. (1967). Actinophage of lysogenic culture of *Proactinomyces* (*Nocardia*) *fructiferi*, producer of the antibiotic ristomycin. *Mikrobiologiya* **36,** 482. (In Russian.)

Riverin, M., Beaudoin, J. & Vezina, C. (1970). Characterization of a nocardiophage for *Nocardia restrictus. J. gen. Virol.* **6,** 395.

Schleifer, K. H. & Kandler, O. (1972). Peptidoglycan types of bacteral cell walls and their taxonomic implications. *Bact. Rev.* **1972,** 407.

Serrano, J. A., Serrano, A. A. & Brownell, G. H. (1974). Ultrastructure studies on the maturation of nocardiophages ϕEC. *Proc. Intern. Conf. Biol. Nocardiae, Merida, Venezuela.* Ed. G. Brownell. p. 86. Augusta: McGowen.

Sivcev, J. & Bönicke, R. (1970). Host-controlled changes of phage *phlei* Bo2 by *Nocardia asteroides* and *Nocardia brasiliensis.* In *Host-Virus Relationships in Mycobacterium, Nocardia* and *Actinomyces.* Eds. S. E. Juhasz & G. Plummer. p. 53. Springfield: Charles C. Thomas.

Thiemann, J. E., Hengeller, C., Virgilio, A., Buelli, O. & Licciardello, G. (1964*a*). Rifamycin. 33. Isolation of actinophages active on *Streptomyces mediterranei* and characteristics of phage-resistant strains. *Appl. Microbiol.* **12,** 261.

Thiemann, J. E., Hengeller, C., Virgilio, A., Buelli, O. & Licciardello, G. (1964*b*). Rifamycin. 34. Physicochemical characterization of actinophages active on *Streptomyces mediterranei. Appl. Microbiol.* **12,** 269.

Waksman, S. A. (1961). *The Actinomycetes*, vol. II. Baltimore: Williams & Wilkins.

Yamada, J. & Komagata, K. (1972). Taxonomic studies on coryneform bacteria. V. Classification of coryneform bacteria. *J. gen. appl. Microbiol., Tokyo* **18,** 417.

11. Genetic Studies in *Nocardia erythropolis*

J. N. ADAMS

Department of Microbiology, School of Medicine, University of South Dakota, Vermillion, South Dakota 57069, U.S.A.

and

G. H. BROWNELL

Department of Cell and Molecular Biology, Medical College of Georgia, Augusta, Georgia 30902, U.S.A.

Contents

1. Introduction

As members of the Actinomycetales, the nocardiae have long been noted for their apparent variability and complexity. The complexities of the group led Bisset & Moore (1949) to propose a taxonomic scheme, based on morphological features, to separate soil diphtheroids from actinomycetes. The prime criterion in their proposal depended upon the complexity of cell wall development. In that and a subsequent report, Bisset & Moore (1950) proposed the genus *Jensenia* to include the soil diphtheroids. The genus *Jensenia*, with *J. canicruria* as the type species, was separated from other actinomycetes including the genus *Nocardia*.

The apparent intermediary nature of the bacteria classified in the genus *Nocardia*, led McClung (1949, 1954, 1955) to carry out extensive light microscopic examinations of growing, representative members of the genus. Adams & McClung (1960, 1962*a*,*b*) continued these studies and presented evidence showing that the organisms were not nearly as morphologically complex as previously reported (Adams, 1963). Therefore, on morphological bases (Adams & McClung, 1960, 1962*a*), these authors considered *J. canicruria* as a member of the genus *Nocardia*. Gordon & Mihm (1961) believed *J. canicruria* to be a member of the mycobacteria and included the species in the synonomy of "*Mycobacterium*" *rhodochrous*. In later investigations, Adams, Adams & Brownell (1970) confirmed their previous investigations of *J. canicruria* and felt the name to be a later synonym for *Nocardia erythropolis*. More detailed discussions of the present status of the taxonomy of *N. erythropolis* nee *J. canicruria* will be found elsewhere in this volume (see chapter 2).

From the reports briefly mentioned above, it is apparent that certain confusion surrounds some organisms included in the genus *Nocardia*. The intermediary status of the taxonomic group and the morphological complexity of the nocardiae compelled us to investigate some members of the group on a more basic biological level, the genetic level. With knowledge of the hereditary mechanisms of the nocardiae, we felt the diverse and disparate views of the nocardiae might be reconciled.

From a morphological viewpoint (Adams & McClung, 1962*a*) *N. erythropolis* appeared to be an organism relatively intermediate in complexity when compared with other organisms included in *Nocardia*. Owing to the controversy concerning the taxonomic position of such strains, the intermediary nature of *N. erythropolis* seemed to dictate a starting point for genetic investigations. Should unique hereditary mechanisms of *N. erythropolis* be recognized, then similarities or dissimilarities of genetic mechanisms in other members of the controversial group could be sorted out and a firmer basis of taxonomic understanding of the organisms might be anticipated. For these reasons appropriate mutants of *N. erythropolis* were prepared in the laboratory and mating experiments attempted. Unfortunately, our mutants of *N. erythropolis* did not interact genetically (Adams & Bradley, 1963; Adams, 1964). The mutants were self-sterile. This lack of genetic interaction, obviously, did not point to solutions of the problems alluded to. Therefore, *N. canicruria*, an organism similar to *N. erythropolis* which had been included in our morphological studies was tested for genetic interactions (Adams & Bradley, 1963; Adams, 1964). These investigations, like those with *N. erythropolis*, were unsuccessful.

Studies of our original hypothesis on the basis of genetic interactions seemed to be thwarted. However, if our thesis concerning relatedness of these organisms was even remotely correct, then nothing would be lost in attempts to observe genetic interaction between our mutants of *N. erythropolis* and *N. canicruria*. Consequently, matings were attempted with these mutants. Recombinants were recovered. Heterothallism, in the classical sense, seemed to govern fertility in these nocardiae.

It was obvious that simple answers would not be forthcoming even with the new-found genetic interactions. The nocardiae continued to be complex and diverse. The following exposition is concerned with our accumulated knowledge of genetic interactions between these strains, both originally considered as members of disparate genera.

2. The *Nocardia erythropolis* Mating Types

In the initial investigations of recombination (Adams & Bradley, 1963; Adams, 1964) standard microbial genetic techniques were used for selection and identification of auxotrophic nutritional mutants. After appropriate nutritionally complementary mutants were isolated and identified, matings were attempted. Matings and estimates of spontaneous reversion frequencies at selectable loci were made by cultivating each tested parental type individually and in mixed culture. Following suitable periods of incubation, suspensions were prepared from cultures of the individual parental types and plated on appropriate selective media to estimate reversion frequencies. Similar suspensions were prepared from mixed populations to determine if genetic interactions permitted the production of selectable prototrophic recombinants.

A sizable number of auxotrophic mutants of *Nocardia erythropolis* were prepared. Many matings were attempted (Adams & Bradley, 1963; Adams, 1964) with a variety of mutants serving as parental types. None of these matings produced recombinants. Similar results in attempted matings of nutritionally complementary mutants of *N. canicruria* were observed. When interspecific matings were tried with nutritionally complementary mutants of *N. ery*-

thropolis and *N. canicruria*, prototrophic recombinants were recovered at frequencies greatly exceeding spontaneous mutational reversions at the selected loci.

In these preliminary experiments, Adams (1964) used multiple auxotrophs of both *N. erythropolis* and *N. canicruria*. This allowed selection and identification of recombinants containing auxotrophic loci inherited from each parental type. For example, a mating was carried out using an histidine, arginine requiring strain of *N. erythropolis* (*his-1 arg-1*) and an adenine, alanine requiring strain of *N. canicruria* (*ade-2 ala-1*). Recombinants were selected on medium supplemented with histidine (His) and adenine (Ade). This selection permitted the recovery of recombinant strains requiring both His and Ade. One of the recombinants, R7, with the genotype *ade-2 his-1*, contained the *his-1* locus derived from the *N. erythropolis* parental type and the *ade-2* locus derived from the *N. canicruria* parental type. Consequently, R7 contained one locus in common with each of the parental types from which it was derived, and one disparate locus compared with each of these parental types. This conformation of genes permitted the performance of test crosses with each of the parental types and an examination of the inherited distribution of mating factors among recombinant progeny.

Recombinant R7, was crossed with *N. erythropolis* and recombinants were recovered on His-supplemented selective medium. These recombinants resulted from genetic interactions between the *arg-1* and *ade-2* loci. Upon mating R7 with *N. canicruria* and selecting for recombinant progeny on Ade-supplemented medium, recombinants were also recovered. Identical matings were carried out with recombinant R8, an *ade-2 his-1* sibling of R7. When these test backcrosses were examined it was discovered that R8 produced recombinant progeny if mated with *N. erythropolis*. Recombinants were not recovered when R8 was mated with *N. canicruria*.

Self-sterility of *N. erythropolis* and *N. canicruria* mutants and inter-strain fertility was a major observation (Adams & Bradley, 1963). These findings suggested that these nocardiae were heterothallic. Thus, a single factor could control compatibility between strains. This genetic factor was first considered to be similar to plus or minus compatibility controlling factors of yeast or other fungi. From this hypothesis, it was postulated that recombinants from a cross of *N. erythropolis* by *N. canicruria* would inherit the mating type of either *N. erythropolis* or of *N. canicruria.* Consequently, a recombinant should backcross with *N. erythropolis* or with *N. canicruria,* but not both. This hypothesis did not adequately explain the observations noted above since R7 recombined with both *N. erythropolis* and *N. canicruria.*

The next simplest hypothesis to explain the observed fertility patterns is that two loci, rather than one, control mating. These loci (Adams, 1967, 1970; Brownell & Adams, 1968) were arbitrarily designated the *E* locus and the *C* locus, with their respective alleles, *e* and *c*. By empirically determining mating behaviour of a given strain with a variety of other recombinant and standard parental strains, the effects of the postulated loci on mating compatibility was ascertained (Adams, 1967, 1970; Brownell & Adams, 1968; Brownell & Kelly, 1969). Of the four possible mating factor genotypes, all were observed. The standard strain of *N. erythropolis* bears the *cE* loci while the standard strain of *N. canicruria* has the loci, *Ce.* The *CE* containing recombinants (Adams, 1964) were recognized by their ability to mate with both standard strains, *cE* and *Ce.* The remaining genotype, *ce,* was characteristic of recombinants which were infertile with any other strains under normal mating conditions (Adams, 1967, 1970; Brownell & Kelly, 1969). From these investigations it was apparent that in order to obtain fertile matings of these nocardiae, one strain must contain the *C* allele while the other strain must bear the *E* allele. Strains derived from just one of the original strains of *N. erythropolis* or *N. canicruria* could contain but one of the alleles *E* or *C.* Therefore, intra-strain fertility between nutritionally complementary mutants of homologous origin were not observed.

Some attempts have been made to determine the nature of the mating factors. Strain treatment with acridine dye (acriflavine) known to eliminate plasmids associated with fertility control in *Escherichia coli* (Hirota, 1960), did not affect fertility of these nocardiae (Adams, 1964). In fact, acriflavine resistance was later added to the *N. erythropolis* linkage map (Brownell & Adams, 1967, 1968; Brownell & Kelly, 1969). Brownell & Kelly (1969) confirmed the absence of effects on either compatibility patterns or recombination frequencies when using acriflavine-treated strains. They also established chromosomal locations for the allele pairs *E*/*e* and *C*/*c* on the *N. erythropolis* linkage map. They observed linkage of the *C* locus to *purB2* (*ade-2*) on the left of the linkage map. The *E* locus segregated with *strA1*, a marker to the right of the map. Brownell & Walsh (1972) observed colonial variants which produced recombinants at frequencies approaching 10^{-4}. This increased frequency was approximately 100-fold greater than that observed in earlier investigations. Later reports confirmed these observations (Adams & Adams, 1974) but these investigators were unable to produce mutants bringing about higher frequency recombination. From these reports, plasmids cannot presently be recognized to play a role in nocardial fertility. This is in contrast to the situation observed in both *E. coli* and *Streptomyces coelicolor* (Vivian, 1971).

Folkens (1971) found no differences in cell wall composition of *N. erythropolis* when compared with *N. canicruria*. Similar findings were reported by Komura, Komagata & Mitsugi (1973) and by Kasweck (1974). Differences in wall composition were considered as possible means of differentiating mating types. A loose association of nocardiophage sensitivity and mating characteristics had been reported earlier (Brownell, Adams & Bradley, 1967). With further studies, however, the association of phage sensitivity with mating capacity was found to segregate among recombinants (Brownell & Adams, 1968). We must await future investigations to determine the precise nature of the control of fertility by these mating factors.

Although the means by which chromosomally located nocardial mating factors exert their influence is not presently known, the multi-factor fertility control in these nocardiae appears to be of great importance. Brownell (1974) observed a new nocardial mating type. This type was derived from cultures received as *Mycobacterium rhodochrous* (ATCC184) from the American Type Culture Collection. These strains recombined with *N. erythropolis*, mating type *cE*, but did not recombine with the *Ce* mating type. Recombinants recovered from fertile crosses of "*M.*" *rhodochrous* with *N. erythropolis* were of three types. The first was fertile with both "*M.*" *rhodochrous* and with the *Ce* mating type. The second was fertile only with "*M.*" *rhodochrous* while the third was limited to fertility with the *cE* mating type. These preliminary results do not allow a clear-cut decision as to the nature of this strain of "*M.*" *rhodochrous*, designated the *M* mating type. It may contain a third mating type locus or another alternative allele of either the *E* or *C* locus. However, the discovery of this mating type suggests more wide spread natural occurrence of varying mating types than previously recognized. Further substantiation of the existence of naturally occurring heterothallism among bacteria was recently reported. Using a well established member of the genus *Mycobacterium, M. smegmatis,* Mizuguchi & Tokunaga (1971) reported that two strains, designated Jucho and Lacticola, exhibited fertility patterns similar to those observed in the *N. erythropolis–N. canicruria* system. In later reports (Mizuguchi, 1972; Tokunaga, Mizuguchi & Suga, 1973) four mating types were described among recombinants from fertile matings. A direct analogy of this system was made with the multi-factor *N. erythropolis* mating system described above. Certainly, naturally occurring heterothallism and multiple mating factor control of fertility among these nocardiae and similar organisms have been amply demonstrated. Undoubtedly, we can expect to observe similar mating phenomena in other nocardioform taxa when other organisms have been more adequately investigated.

For the sake of clarity, we have maintained the specific epithets which various authors ascribed in reference to the mating strains we have used. We have done so in hopes of helping to alleviate confusion for the reader and provide a complete understanding of the background of these strains. While there are some differences between these strains, the similarities more than outweigh the differences. These observations led us (Adams, Adams & Brownell, 1970) to propose that *N. canicruria* be included in the synonomy of *N. erythropolis.* While there may be discussion of the generic and species assignments for this organism (see chapter 2), there should be no disputing the idea that there are mating types of the organism bearing the label, *N. erythropolis.*

In order to maintain separate identities for the mating types, we have arbitrarily chosen to use strain number designations for these strains. All strains derived from our original *N. erythropolis* culture bear our original numerical strain designation, "2". Mutants derived from this strain carry the numerical designation 2-*n*; the *n* suffix represents a sequential mutant isolation number. All strains derived from our culture originally designated as *J. canicruria* bear the label *N. erythropolis* with a numerical suffix designation, 3-*n*. The suffix is indicative of our original numerical strain designation for this culture. Where appropriate, recombinant strains will bear numerical identities and their mating type loci, if known. To avoid confusion both a numerical and mating locus designation may be suffixed to the label *N. erythropolis* (Ce3 or cE2). For the present we have accepted the appellation *N. erythropolis* for all of our mating strains. This convention, with appropriate suffixes, will be used throughout the remainder of this chapter.

3. The Linkage Map of *Nocardia erythropolis*

After the recognition of recombination between mutants of *N. erythropolis* strains cE2 and Ce3 (Adams & Bradley, 1963; Adams, 1964), extensive investigations were begun to establish the presence or absence of linkage between observed mutant loci. A variety of mutants were produced with both auxotrophic and inhibitor resistant phenotypes. Analyses were carried out by using one or more characters from each mated parental type as a selective characteristic. Remaining characteristics were examined for their frequency of segregation among the selected recombinants.

Segregation of unselected genes expressing resistance or susceptibility to acriflavine, erythromycin, streptomycin and tetracycline was analyzed in selected prototrophic recombinants produced in matings of *N. erythropolis* cE2 with *N. erythropolis* Ce3 (Brownell & Adams, 1967). On the basis of the segregation of the unselected characteristics, the organisms were shown to be functionally haploid and appeared to contain not more than one chromosome. From the data presented, all observed genes were postulated to be present in a linear linkage array. The ordering of the genes in *N. erythropolis* cE2 mutant substrain was *tetB10 eryB9 his-3 purA1 acr-2 strA1*. Respectively these genes control: resistance to tetracycline and erythromycin, deficiency for histidine and for purine, and resistance to acriflavine and streptomycin. The ordering of genes in the *N. erythropolis* Ce3 mutant substrain was:

purB2 tetA9 eryA7 acr-11 strB2. Excluding the loci for acriflavine resistance, *acr-2* and *acr-11*, resistance loci in *N. erythropolis* cE2 were not allelic to and showed lateral displacement from genes controlling phenotypically similar resistance in *N. erythropolis* Ce3 (Fig. 1). Recombination phenomena between these strains were postulated to occur as a result of formation of a heterogenomic zygote in which new combinations of parental characteristics were produced. Production of selectable, haploid recombinants was ascribed to crossing over between the parental genomes with subsequent haploidization of the zygote and segregation of the recombinant genomes.

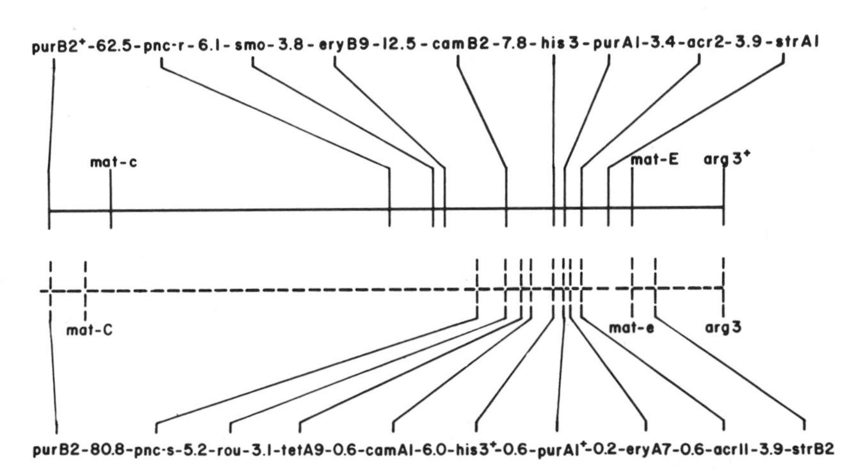

Fig. 1. Linkage maps of some genes in the *Nocardia erythropolis* chromosome. (—) The linkage group of the cE2 mating type with respective loci; (– – – –) the cE3 mating type linkage group with its loci. The genes and their alleles are illustrated by vertical bands and are placed according to their relative distance from each other. Values for the map distances between loci are given in percentage. Description of mutant loci are in conformity with the recommendations of Demerec *et al.* (1966). Symbols: +, synthesized; –, required; r, resistant; s, sensitive; pur, purine; pnc, phage ϕC; rou, rough colony morphology; smo, smooth colony morphology; tet, tetracycline; ery, erythromycin; cam, chloramphenicol; his, histidine; acr, acriflavine; str, streptomycin; mat, mating gene.

Subsequently, nocardiophage ϕC was isolated from soil (Brownell, Adams & Bradley, 1967). This phage infected only *N. erythropolis* Ce3 and suggested a possible relationship between mating types and ϕC sensitivity. Since *N. erythropolis* cE2 was resistant to infection by ϕC, recombinants were obtained from matings of the two strains and analyzed for segregation of ϕC sensitivity and for their fertility in backcrosses to the parental types (Brownell & Adams, 1968). The locus *pcn-r* (resistance to nocardiophage ϕC) of *N. erythropolis* cE2 was found linked between *purB2* and *eryB9*. The alternative allele, *pcn-s* contained in *N. erythropolis* Ce3, was linked between *purB2* and *tetA9*. The genes controlling mating behavior were found to segregate from those expressing phage sensitivity. These results invalidated the use of sensitivity to phage ϕC for scoring mating types among recombinants. Later, Brownell & Kelly (1969) established the location of the *mat-E* locus in association with *strA1* and confirmed the location of the mating type *mat-C* locus near the *purB2* gene. The *pcn-s* allele was shown in these studies to be quite far removed from the *mat-C* mating type locus.

The addition of chloramphenicol resistance markers in these strains, which had previously been mapped, permitted further confirmation and clarification of previously reported linkage relationships (Brownell & Adams, 1968). The locus *camB2* of the cE2 mating type was shown to be linked between *his-3* and *eryB9* while *camA1* of Ce3 mating type was situated between *tetA9* and *his-3*$^+$ These additional loci also exhibited lateral displacement of genes controlling similar phenotypes. However, the general linear array of genes in the single linkage group was reconfirmed. Although serious attention was given the possibility that these loci were present in a circular array, our results were not clarified further by trying to establish a circular, rather than linear, array.

Although it is unlike other bacterial chromosome maps with circular gene arrays, an extensive linear linkage map has been constructed for *N. erythropolis* (Adams, 1974). This map presently contains more than 40 established loci controlling a large variety of characteristics. There exist certain anomalies in this map. The region to the left of *hisA* (*his-3*) seems to comprise 97% of the known map. The map region to the right of *hisA* appears to contain only 3% of the total map. These distances, of course, are relative crossover distances as measured by the degree of segregation of unselected markers. Such figures may not be real approximations of recombinational distances along the linkage map since presently there are no known means of obtaining absolute map distances in these nocardiae.

Certain aberrations of recombination may contribute to inaccurate distance representations between various loci on the linkage map. Adams (1968) observed that loci, situated to the right of *hisA*, originating in *N. erythropolis* mating type cE2 are found infrequently among recombinants of cE2 by Ce3 matings. The phenotype for *argA*1, a gene to the extreme right of *hisA* was found in low numbers of recombinants even though selectable recombinants containing this gene should have resulted from single crossovers. Phenotypes containing the *argAl*$^+$ locus, although ostensibly arising as the result of three separate crossovers, occurred in approximately 4% of the recombinant populations analysed from the mating. A second anomaly of recombination, possibly contributing to inaccuracies of the map, was also observed in these crosses. The majority of recombinants were found (Brownell & Adams, 1967, 1968; Adams, 1968) to contain a predominance of loci contributed by the Ce3 mating type.

These recombinational variations led Adams (1968) to postulate two models to account for the findings. The first envisaged the mechanism of recombination as a polar transfer phenomenon. In this model, the chromosome was considered to be transferred from mating type cE2 to Ce3, with polar orientation. Markers to the left of *hisA*, beginning with *purB2*$^+$ of the cE2 mating type, were considered as the first markers transferred. Consequently, loci to the right of *purB2*$^+$ would appear at lower frequencies among recombinants in proportion to their distance to the right of *purB2*$^+$. In this model, markers to the right of *hisA* would be transferred last. If termination of chromosomal transfer occurred before most markers to the right of *hisA* were introduced into the Ce3 recipient, these markers would appear to be located in a proportionately smaller chromosomal segment. In this instance the markers would generally appear to be "excluded" from recombinants and distances between loci in the "excluded" region would seem disproportionately small.

A second model was also considered to account for disproportionate linkage map distances in respect to the chromosome segments to the left and right of *hisA*. Since mating types cE2 and Ce3 were known not to be derived from an homologous ancestor, it was postulated that the chromosomes of the cE2 and Ce3 mating types were not completely homologous. Some evidence for this concept was presented by Brownell & Adams (1968). Under these conditions, those regions of greatest chromosomal homology would more frequently permit synapsis and crossovers between loci in the region of homology than loci situated in areas of inhomogeneity. Consequently, the area to the left of *hisA* was considered as a homologous area in both mating types. The area to the right of *hisA* appeared to be inhomogeneous. Therefore, crossover distances could be expected to be much reduced between genes located in the region. More discussion on the effects of incomplete chromosome homology is provided in the section on the molecular aspects of the genome.

The data of Adams (1968) did not permit a clear cut choice between the postulated models described above. However, from the results reported by Brownell & Walsh (1970) and by Folkens (1971) it became apparent that the hypothesis of gene exclusion was not absolute. In matings of cE2 by Ce3 Brownell & Walsh (1970) employed contraselective loci in various combinations in order that all mapped characters could be subjected to unselected analysis. The greater majority of the recombinant class types they recovered appeared as stable haploids. Others behaved as heterozygous diploids and segregated new phenotypes. In these investigations of the heterogenotes, all regions of the parental genomes were found to be involved in segregation. However, the majority of recombinants resulted from crossovers in the regions to the left of *hisA*. The recognition of heterogenotes for all areas of the mapped regions implied that the entire mapped region could become merozygotic under standard mating conditions. Folkens (1971) reported similar findings, although he used a different cE2 mating strain in his experiments. In both these reports, the number of heterogenomic recombinants comprised a relatively small fraction of the selected recombinants. However, these results provided clear evidence that marker exclusion was not absolute. Since the stable, haploid recombinants appeared to possess the anomalies of recombination described in detail above, greater clarification of the anomalous linkage effects was not possible.

Presently, it would appear the marker exclusion is a phenomenon encountered among recombinants produced in these nocardial matings. However, a small percentage of zygotes produce heterogenotes with a capacity for further segregation of parental genes (Brownell & Walsh, 1970; Folkens, 1971). In the future, analysis of segregants from these heterogenotes may provide a convenient means for establishing absolute nocardial map distances. A second approach to the solution of these problems encountered in visualizing the linkage map is provided by the existence of EC mating types. These mating types are self-fertile (Brownell & Kelly, 1969; Adams, 1970). Matings of mutants derived from a single EC progenitor type should permit examination of recombinational phenomena in isogenic strains. Many of the problems with respect to chromosomal inhomogeneity of parental types will be alleviated with the use of such strains. Thus, two potent tools now exist for examining these unsolved problems.

4. Plasmid and Prophage Inheritance

The mating type strains of *Nocardia erythropolis* (Ce3 and cE2) are differentially infected by nocardiophage ϕC while both strains are susceptible to infection by a closely related phage, ϕEC. Phage ϕC was distinct from other nocardiophage in its ability to attach quickly and efficiently to its host (Brownell, Adams & Bradley, 1967). Phage-specific antiserum prepared with either phage ϕC or ϕEC was found to cross-react and inactivate both phage (Brownell & Clark, 1974). A substance solubilized by ethanol from the surface of Ce3 and cE2 mating strains, presumably a phospholipid which serves as the phage receptors, was found to inhibit both phage ϕC and ϕEC (Brownell & Crockett, 1972). Both phage served to adsorb out the inhibitory substance in the extracts suggesting that they share a common receptor. The nature of resistance of the cE2 mating type strains to infection by phage ϕC is not known. However, it is known that phage ϕC adsorbs to the cE2 cells and that resistance is not due to super-infection immunity since this phenomenon has been shown to produce immunity to infection by phage ϕEC also (Crockett & Brownell, 1972).

Although both phage ϕC and ϕEC are virulent, a lysogenic strain of the Ce3 mating type carrying the ϕEC-prophage was isolated and characterized by Crockett and Brownell (1972). The lysogen, designated Ce3-ϕEC, was immune to infection by both phage ϕEC and ϕC; has a spontaneous induction frequency of

$2 \cdot 5 \times 10^{-4}$ and a spontaneous cure rate of *c.* 0·4%. Ultraviolet irradiation or treatment with mitomycin C induced the lysogens to produce large numbers of phage. The phage produced by the lysogens retained their virulent nature, that is, they produced clear plaques and lysogenized at extremely low frequencies. The growth rate of the lysogens and other phenotypes studied were comparable to that of the uninfected wild-type Ce3 strain.

Matings between lysogenic and nonlysogenic strains were initiated in an attempt to determine the chromosomal location of the ϕEC-prophage (Brownell, 1972*b*, 1973). Since the nonlysogenic parental types were susceptible to infection by phage produced from spontaneous induction of the lysogenic parental strain, the initial matings were conducted in U-tube apparatus. The U-tube crosses were controlled by separating parental strains with membrane filters which allowed mixing of only free phage. The results of these experiments showed that direct cell contact was required for inheritance of the phage-bearing trait. No recombinants or lysogens which correspond to the nonlysogenic parental phenotypes were recovered from cultures wherein the parental strains were separated by 0·45 μ membrane filters (Table 1). However, mixed cultures produced both lysogens and recombinants. Standard agar mating techniques (Adams, 1964) were later adopted when it was shown that a nonlysogenic parental type strain failed to acquire the phage-bearing trait following mixed growth with a lysogen of the same mating type. For example, when a lysogenic Ce3-ϕEC strain was cultured on the surface of an agar plate with a nonlysogenic Ce3 strain and the nonlysogenic parental types were isolated and tested for the phage-bearing trait, no lysogens were observed. Thus, direct cell contact and mating ability are both required for prophage inheritance and exogenous phage are not responsible for originating new lysogens.

An analysis of the recombinants obtained from matings of the lysogens with nonlysogenic strains showed that the ϕEC-prophage existed as a plasmid in the Ce3-ϕEC mating type lysogens. Mixed growth of a Ce3-53ϕEC lysogen with a nonlysogenic cE2-13 mating strain resulted in $1 \cdot 7 \times 10^{-1}$/ml lysogenic cE2-13 parental type strains (Table 1). In the same crosses recombination involving chromosomal traits occurred at a recovery of $7 \cdot 3 \times 10^{-5}$. While the chromosomal rate was within the normal range, the high incidence of inheritance of the ϕEC-prophage by the nonlysogenic cE2-13

Table 1

Recombinant recovery from crosses of Nocardia erythropolis *lysogens*

	Crosses			
	Ce3-53 (ϕEC) by cE2-13 Recombinant Recovery		cE2-13 (ϕEC) by Ce3-53 Recombinant Recovery	
	(ϕEC)-plasmid*	Chromosomal traits†	(ϕEC)-plasmid*	Chromosomal traits†
U-tube				
Filter separated chambers	$<2{\cdot}0\times10^{-3}$	$<1{\cdot}0\times10^{-8}$	$<2{\cdot}1\times10^{-3}$	$<1{\cdot}0\times10^{-8}$
Mixed cultures	$1{\cdot}0\times10^{-2}$	$6{\cdot}7\times10^{-7}$	$<2{\cdot}5\times10^{-3}$	$7{\cdot}5\times10^{-7}$
PY agar plates	$1{\cdot}7\times10^{-1}$	$7{\cdot}3\times10^{-5}$	$<1{\cdot}2\times10^{-3}$	$7{\cdot}8\times10^{-6}$

Linkage model

cE2-13	+	his-3 ade-1		+
	I		II	
Ce-53	purB2	+	+	arg-3

* The nonlysogenic parental strains were selected by plating the mating mixtures on appropriately supplemented MM. The resulting colonies were then replicated to soft-agar over-lay seeded with a phage-sensitive indicator strain to score for the phage-bearing trait. Recovery is expressed as lysogenic colonies per total colonies replicated.

† Recombinants involving chromosomal traits were recovered by plating the mating mixture on selective media: MM+Ad, MM+Arg or MM. Recovery is expressed as number of recombinants per total cells plated.

Symbols: +, synthesized; −, required; *purB2*, purine; *ade-1*, adenine; *arg-3*, arginine; *his-3*, histidine; PY, peptone yeast-extract; MM, minimal medium.

Linkage model from Brownell (1972*a*).

parental strain clearly points to a plasmid origin of the prophage in Ce3-53ϕEC parental strain.

A lack of reciprocity was observed from mating of the cE2-13ϕEC lysogens with nonlysogenic Ce3-53 strains (Table 1). The Ce3-53 parental type strain failed to inherit the ϕEC-prophage from the cE2-13ϕEC lysogen in any of the colonies tested. The lack of reciprocity or nonplasmid-like inheritance of the ϕEC-prophage from these matings could be the result of either prophage integration into the bacterial chromosome or zygotic induction. It was anticipated that zygotic induction would reduce the recombinant recovery, increase free phage titers, and reduce the number of viable Ce3 cell types. While recombinant recovery involving chromosomal traits was generally 10-fold lower from matings of the cE2-13ϕEC by Ce3-53 (Table 1), there was no significant difference in either phage production or cell mortality. Free phage titers recovered from broth crossed strains all ranged from 4·5 to $6{\cdot}0 \times 10^5$ pfu/ml indicating that if zygotic induction were occurring, it did not lead to significantly higher phage titers. Mortality of the nonlysogenic parental type was found to shift the parental cell ratios in favor of the lysogenic parental strain to the same degree in reciprocal crosses. Matings initiated with 1 : 1 cell ratios showed 1 : 100 (nonlysogens to lysogens) ratios by the time the crosses were harvested in 72 h. Finally, an analysis of the recombinants obtained from matings of the cE2-13ϕEC by Ce3-53 indicated that the ϕEC-prophage was integrated into the chromosome of the cE2-13ϕEC lysogen. Lysogens were found only among those recombinants that originated from crossing over in region II (see linkage model, Table 1). No recombinants which had inherited a portion of the cE2-13ϕEC genophore between the *purB2*$^+$ and *his-3* (cross-overs in region I) acquired the phage-bearing trait. However, 37% of the recombinants which had inherited some portion of the chromosome between the *ade-1* and *arg-3*$^+$ also possessed the ϕEC-prophage. The integration of the ϕEC-prophage to the right of the *his-3 ade-1* is consistent with these observations.

In subsequent studies ϕEC-bearing lysogens of the cE2 mating type strains were isolated which when mated with nonlysogenic Ce3 strains, transferred the ϕEC-prophage as a plasmid rather than an integrated element. Thus, the mating type genes appear not to be a factor affecting the nature of the ϕEC-prophage.

Lysogens of the Ce3 mating type strain were later isolated and found to bear the ϕC-prophage (Brownell & Clark, 1974). The phage produced by the Ce3-ϕC lysogens appeared to have undergone a host-controlled alteration which allowed the newly propagated phage to grow in the normally resistant cE2 mating strains. However, the phage progeny produced by the infected cE2 cells could not re-infect the cE2 mating type strain, that is, the phage produced from a single cycle or burst had "reverted" to their original host range and would propagate only in Ce3 strains. The Ce3-ϕC lysogens were also inducible; u.v. irradiation induced phage production is about 37% of the treated population (Brownell & Clark, 1974).

A more extensive genetic analysis was conducted on the Ce3-ϕC lysogens. The Ce3-ϕC lysogens behaved essentially the same as the cE2-13ϕEC strains since the inheritance of the ϕC-prophage was always accompanied by recombination between chromosomal traits. Matings between Ce3-90ϕC and cE2-13, using the agar plate procedures, produced essentially the same recombinant recovery as did crosses of nonlysogenic strains. Recombinants obtained from either Pur^+His^+ or His^+Arg^+ selection were all phenotypically Try^- (Table 2). The Pur^+His^+ selection produced nonlysogenic (Lyc^+) recombinants while the His^+Arg^+ selection produced lysogenic recombinants (phenotypes 1-3,4; Table 2). The nonlysogenic recombinants all inherited the *purB2*$^+$ allele from the chromosome of the nonlysogenic cE2-13 parental strain by crossing over between the *purB2* and *his-3* characters (see linkage model, Table 2). The lysogenic Try^- recombinants obtained from His^+Arg^+ selection all inherited the *purB2* locus suggesting the linkage of factors which control lysogeny to the *purB2* genes. Selection of the Try^+His^+ phenotype produced both Pur^+ and Pur^- recombinant class types; and again, only the Pur^- class type were lysogenic (phenotypes 5 and 6, Table 2). These recombinants were also Arg^- indicating that the prophage was not associated with the right hand portion of the chromosome as was the ϕEC-prophage in the cE2 mating type lysogens.

Thus, the ϕC and ϕEC prophage have proven to be useful genetic markers and in certain lysogens have been shown to be the

Table 2

The segregation of unselected characters from crosses of lysogenic and nonlysogenic strains

Strains crossed	Selected phenotypes	recombinant* recovery/ml	Class type reference	Recombinant phenotypes	% of recombinant population
3-90(ϕC)	Pur^{+-} His^{+-}	$7{\cdot}0\times10^{-5}$	(1)	Try^{-} Arg^{-} Str-R Tet-R Lyc^{+} Mat-ce	95·1
×			(2)	Try^{-} Arg^{-} Str-R Tet-S Lyc^{+} Mat-ce	2·6
2-13			(3)	Try^{-} Arg^{+} Str-R Tet-R Lyc^{+}†	1·8
	Pur^{+-} His^{+} Arg^{+}	$6{\cdot}5\times10^{-5}$	(4)	Try^{-} Pur^{-} Str-R Tet-R Lyc^{-} Mat-Ce	100·0
	Pur^{+} Try^{+} His^{+-}	$1{\cdot}2\times10^{-5}$	(5)	Pur^{-} Arg^{-} Str-R Tet-R Lyc^{-} Mat-Ce	97·0
			(6)	Pur^{+} Arg^{-} Str-R Tet-R Lyc^{+} Mat-ce	2·1

Linkage model

cE2-13	+	+	mat-c	tet-s	his-3	purA1	+	mat-E	str-s	+
Ce3-90ϕC	purB2	lyc-1	mat-C	tetA9	+	+	try-3	mat-e	strB2	arg-3

Symbols: +, synthesized; −, required; R, resistant; S, sensitive; pur, purine; tet, tetracycline; his, histidine; try, tryptophan; str, streptomycin; mat, mating gene; arg, arginine; lyc, lysogenic.
* Recombinant recovery expressed as the number of recombinants per total cells plated.
† Heterogenotes which will segregate further to produce Arg− class types.

first plasmid traits to exist in the *Nocardia.* Future studies should explore the transducing potentials of these phages or their mutants.

5. Non-Conjugational Inheritance

As with the *Streptomyces,* all attempts to demonstrate phage mediated gene transfers in the genus *Nocardia* have been unsuccessful (Hopwood *et al.,* 1973). Phage induced from genetically marked *N. erythropolis* Ce3 and cE2 lysogens have not been shown to transfer any of the traits analyzed in recipient strains. However, the ϕC and ϕEC bearing lysogens of *N. erythropolis* are not good systems for studying the transducing potentials of nocardiophage because the phage produced by the induction of the lysogens are relatively lytic and re-lysogenize at very low frequencies (Crockett & Brownell, 1972). Whether or not transduction is part of the nocardial genetic armament must await the isolation of more temperate phage or the development of mutant phage from present stocks.

Experiments designated to determine if DNA-mediated transformations occur in the nocardiae have been carried out in several laboratories. Both highly purified as well as crude DNA preparations from the *N. erythropolis* mating strains have been unsuccessful in bringing about genetic transformations in appropriate recipients. While neither transduction or transformation have been observed in these nocardiae, experimentation has been too limited to draw any conclusion as to the role that these genetic mechanisms may play in the genus *Nocardia.*

6. Molecular Nature of the *Nocardia erythropolis* Chromosome

Efforts to construct linkage maps that could provide the molecular tools needed for understanding the nature of recombination between compatible nocardiae have been only partially successful. Nonallelism between characters controlling phenotypically similar traits and aberrant segregation patterns, that is, multiple unselective cross-over events in adjacent regions, have produced linear maps with gene clustering (Brownell & Adams, 1967, 1968). Although the current linkage maps contain almost 40 characters

(Adams, 1974), it is not possible to determine how much of the chromosome is included in the mapped region or effectively to determine map distances.

It was postulated that the nocardial linkage maps were reflecting a lack of chromosome homology and that the synapsis which occurred during zygote formation could not properly aline all alleles. This theory drew some support from DNA : DNA reassociation studies conducted by Clark & Brownell (1972). Their studies showed that the Ce3 strains exhibited essentially complete homology with the cE2 mating type reference DNA. The reference DNA being the radioactively labeled DNA which reassociates with the unlabeled, immobilized test DNA (Table 3). However, when the

Table 3

Relative relatedness of Nocardia erythropolis *mating strains*

	Relative relatedness* to cE2 mating strains		Relative relatedness† to Ce3 mating strains	
Unlabeled DNA source	Relative binding (%)	Tm,e	Relative binding (%)	Tm,e
N. erythropolis cE2	100 (2100)‡	79 ± 2·0	63 ± 2·0	84 ± 0·2
N. erythropolis cE2-13	96 ± 9·0	82 ± 4·0	58 ± 4·0	80 ± 1·0
N. erythropolis Ce3	99 ± 5·0	80 ± 3·0	100 (2900)†	80 ± 2·0
N. erythropolis Ce3-53	101 ± 3·0	84 ± 4·0	100 ± 9·0	79 ± 0·1
Escherichia coli	4·2	77·1	4·2	84
Blank filter	(76·8)‡	—	(163·2)‡	—

* The specific activity of the cE2 reference DNA was 6400 ct/min μg.
† The specific activity of the Ce3 reference DNA was 6800 ct/min μg.
‡ Ct/min μg. The incubation temperatures were 70°C.

reciprocal experiment was performed, the Ce3 strains were used as reference DNA, the cE2-13 strains were able to bind only 58 ± 4% of the Ce3 DNA and the cE2 strains only 63 ± 2·0% (Clark & Brownell, 1972). The data suggested that the genome of the cE2 mating type strains contains nucleotide sequences all of which are also present in the Ce3 mating type chromosome but the Ce3 strains contain DNA with additional sequences. Consequently, the Ce3 DNA could bind the cE2 reference DNA but the reverse would not be true because the base sequences are unique to the Ce3 chromosome.

Because of the significant difference in the genomes of the two parental strains, it was anticipated that DNA reassociation studies between recombinant and parental chromosomes might indicate the contribution of the parental genomes to the zygote. Recombinants that had inherited various areas of the parental chromosomes were tested against parental reference DNA's. All of the recombinants tested exhibited extensive homology to both parental strains, implying that a significant portion of the unique nucleotides were inherited from the Ce3 parental strain. Thus, the unique sequences appear to be conserved in the recombinant genome.

A re-examination of the segregation patterns of the nocardial recombinants revealed that the conservation of specific areas of the chromosome of the Ce3 mating strains could explain the aberrant segregation behaviour observed in certain class types. For example, mating between double auxotrophs may fail to produce recombinants when selected on minimal medium, a result once thought to be due to linkage. However, recombinants selected on supplemented minimal medium frequently produced class types which could have resulted only from multiple cross-overs in closer regions than those required for prototrophy. Table 4 depicts one such cross, the data are selected from Brownell & Walsh (1972). Selection for prototrophy failed to produce recombinants while either arginine or histidine supplemented medium (*purB2*$^+$ *his-3*$^+$ *pur A1*$^+$ or *his-3*$^+$ arg-*3*$^+$ selection) allowed recombinant recovery. The data suggest that while single cross-overs between either *purB2*$^+$ and *his-3*$^+$ or *his-3*$^+$ and *arg-3*$^+$ can occur, the double cross-over event leading to prototrophy is relatively rare (see linkage model, Table 4). However, when recombinant colonies were selected on arginine supplemented plates (PurB2$^+$ His-3$^+$ PurAl$^+$ selection), were replicated and scored for the segregation of unselected traits, 4% of the recombinant population had inherited alleles which required multiple unselective cross-overs (note class type Pnc-S Str-S Arg$^-$, Table 4). If a region of unique nucleotides were to exist in the Ce3 genophore between the *strB2* and *arg-3* loci and if these sequences are conserved in recombinant genomes then the cross-overs in this area would be of a selective nature. The close proximity of the unique sequences to the *arg-3* locus may preclude the inheritance of the *arg-3*$^+$ allele from the cE2 mating types.

Table 4

Unselected gene segregation in Nocardia erythropolis *recombinants*

Selected phenotypes	Recombinant* (recovery/ml)	Major recombinant phenotypes	% of recombinant population	Crossover regions
Prototrophy	$<1{\cdot}0\times10^{-8}$	—	—	—
$PurB2^{+}His\text{-}3^{+}PurA1^{+}$	$3{\cdot}0\times10^{-5}$	Pnc-S Str-R Arg^{-}	88·0	I
		Pnc-S Str-S Arg^{-}	4·0	I, III, IV
		Pnc-S Str-S Arg^{+}†	7·0	—
$His\text{-}3^{+}Arg\text{-}3^{+}$	$4{\cdot}5\times10^{-5}$	Pur^{-} Pnc-S Str-R	97·0	IV
		Pur^{+} Pnc-S Str-R†	2·0	—

Zygote model

cE2-13	+		pnc-r		his-3	purA1		str-s		+
		I		II			III		IV	
Ce3-57	purB2		pnc-s		+	+		strB2		arg-3

* Recombinant recovery is expressed as the number of recombinants per total cell plated.
† Heterogenotes capable of further segregation to produce Arg− phenotypes.

Thus, our current hypothesis predicts that the linkage maps will remain linear with regions of high negative interference due to the existence of unique nucleotide sequences in the Ce3 chromosome.

7. References

Adams, J. N. (1963). Nuclear morphogenesis during the developmental cycles of some members of the genus *Nocardia*. *J. gen. Microbiol.* **33,** 429.

Adams, J. N. (1964). Recombination between *Nocardia erythropolis* and *Nocardia canicruria*. *J. Bact.* **88,** 865.

Adams, J. N. (1967). Evidence for multiple mating factor control of nocardial recombination. *Bact. Proc.* **1967,** 55 (Abstract).

Adams, J. N. (1968). Partial exclusion of the *Nocardia erythropolis* chromosome in nocardial recombinants. *J. Bact.* **96,** 1750.

Adams, J. N. (1970). Recombination and segregation of resistance and auxotrophic characters in crosses of *Nocardia* species (*Mycobacterium rhodochrous* (Overbeck) Gordon & Mihm). Pneumonologie **142,** 164.

Adams, J. N. (1974). Linkage map and list of markers of *Nocardia erythropolis*. In *Handbook of Microbiology*, vol. 4. Eds A. I. Laskin and H. A. Lechevalier. p. 661. Cleveland, Ohio: CRC Press.

Adams, J. N. & Adams, M. M. (1974). Limiting nutrients during mating as a means of increasing recombinant recovery in crosses of *Nocardia erythropolis*. *J. Bact.* **119,** 646.

Adams, J. N. & Bradley, S. G. (1963). Recombination events in the bacterial genus *Nocardia*. *Science, N.Y.* **140,** 1392.

Adams, J. N. & McClung, N. M. (1960). Morphological studies in the genus *Nocardia*. V. Septation in *Nocardia rubra* and *Jensenia canicruria*. *J. Bact.* **80,** 281.

Adams, J. N. & McClung, N. M. (1962*a*). Comparison of the developmental cycles of some members of the genus *Nocardia*. *J. Bact.* **80,** 206.

Adams, J. N. & McClung, N. M. (1962*b*). On the nature of cytoplasmic inclusions of *Nocardia rubra*. *J. gen. Microbiol.* **28,** 231.

Adams, M. M., Adams, J. N. & Brownell, G. H. (1970). The identification of *Jensenia canicruria* Bisset and Moore as a mating type of *Nocardia erythropolis* (Gray and Thornton) Waksman and Henrici. *Int. J. Syst. Bact.* **20,** 133.

Bisset, K. A. & Moore, F. W. (1949). The relationship of certain branched bacterial genera. *J. gen. Microbiol.* **3,** 387.

Bisset, K. A. & Moore, F. W. (1950). *Jensenia*, a new genus of the Actinomycetales. *J. gen. Microbiol.* **4,** 280.

Brownell, G. H. (1972*a*). The compatible nocardiae: A new tool for genetic instruction. *Bull. Georgia Acad. Sci.* **30,** 33.

Brownell, G. H. (1972*b*). Inheritance of factors controlling phage production in prophage bearing strains of *Nocardia erythropolis*. *Am. Soc. Microbiol.*, p. 74 (Abstract).

Brownell, G. H. (1973). Plasmid-like inheritance of the ϕEC prophage from mating of lysogenic nocardiae. *Genetics, Princeton* **74,** 32.

Brownell, G. H. (1974). A new nocardial mating strain. *Am. Soc. Microbiol.*, p. 41 (Abstract).

Brownell, G. H. & Adams, J. N. (1967). Linkage and segregation of unselected markers in matings of *Nocardia erythropolis* with *Nocardia canicruria*. *J. Bact.* **94**, 650.

Brownell, G. H. & Adams, J. N. (1968). Linkage and segregation of a mating type specific phage and resistance characters in nocardial recombinants. *Genetics, Austin* **60,** 437.

Brownell, G. H., Adams, J. N. & Bradley, S. G. (1967). Growth and characterization of nocardiophages for *Nocardia canicruria* and *Nocardia erythropolis* mating types. *J. gen. Microbiol.* **47,** 247.

Brownell, G. H. & Clark, J. E. (1974). Host range alterations of nocardiophage ϕC following lysogenization of *Nocardia erythropolis*. *J. gen. Virol.* **23,** 247.

Brownell, G. H. & Crockett, J. K. (1972). Inactivation of nocardiophage ϕC and ϕEC by extracts of phage attachable cells. *J. Virol.* **8,** 894.

Brownell, G. H. & Kelly, K. L. (1969). Inheritance of mating factors in nocardial recombinants. *J. Bact.* **99,** 25.

Brownell, G. H. & Walsh, R. S., III (1970). Heterogenomic recombinants from compatible nocardiae. *J. Bact.* **104,** 79.

Brownell, G. H. & Walsh, R. S., III (1972). Colony mutants of compatible nocardiae displaying variations in recombining capacity. *Genetics, Austin* **70,** 341.

Clark, J. E. & Brownell, G. H. (1972). Genophore homologies among compatible nocardiae. *J. Bact.* **109,** 720.

Crockett, J. K. & Brownell, G. H. (1972). Isolation and characterization of a lysogenic strain of *Nocardia erythropolis*. *J. Virol.* **10,** 737.

Demerec, M., Adelberg, E. A., Clark, A. J. & Hartman, P. E. (1966). A proposal for a uniform nomenclature in bacterial genetics. *Genetics, Austin* **54,** 61.

Folkens, A. T. (1971). Conditions influencing mating of *Nocardia erythropolis*. Ph.D. Dissertation, University of South Dakota, Vermillion, South Dakota.

Gordon, R. E. & Mihm, J. M. (1961). The specific identity of *Jensenia canicruria*. *Can. J. Microbiol.* **7,** 108.

Hirota, Y. (1960). The effect of acridine dyes on mating type factors in *E. coli*. *Proc. natn. Acad. Sci. U.S.A.* **46,** 57.

Hopwood, D. A., Chater, K. F., Dowding, J. E. & Vivian, A. (1973). Advances in *Streptomyces coelicolor* genetics. *Bact. Rev.* **37,** 371.

Jacob, F. & Wollman, E. L. (1961). *Sexuality and the Genetics of Bacteria*. New York and London: Academic Press.

Kasweck, K. L. (1974). Investigations into mating efficiency, chemical composition and recombination involving a new class of mutations in *Nocardia erythropolis*. Ph.D. Dissertation, University of South Dakota, Vermillion, South Dakota.

Komura, I., Komagata, K. & Mitsugi, K. (1973). A comparison of *Corynebacterium hydrocarboclastus* Iuzuka and Komagata 1964 and *Nocardia erythropolis* (Gray and Thornton) Waksman and Henrici 1948. *J. gen. appl. Microbiol., Tokyo* **19,** 161.

McClung, N. M. (1949). Morphological studies in the genus *Nocardia*. I. Developmental studies. *Lloydia* **12,** 137.

McClung, N. M. (1954). Morphological studies in the genus *Nocardia*. III. The morphology of young colonies. *Ann. N.Y. Acad. Sci.* **60,** 168.

McClung, N. M. (1955). Morphological studies in the genus *Nocardia*. IV. Bright phase contrast observations of living cells. *Trans. Kans. Acad. Sci.* **58,** 50.

Mizuguchi, Y. (1972). Segregation of unselected markers in mycobacterial recombinants. *Jap. J. Microbiol.* **16,** 36.

Mizuguchi, Y. & Tokunaga, T. (1971). Recombination between *Mycobacterium smegmatis* strains Jucho and Lacticola. *Jap. J. Microbiol.* **15,** 359.

Tokunaga, T., Mizuguchi, Y. & Suga, K. (1973). Genetic recombination in mycobacteria. *J. Bact.* **113,** 1104.

Vivian, A. (1971). Genetic control of fertility in *Streptomyces coelicolor* A3(2): plasmid involvement in the interconversion of UF and IF strains. *J. gen. Microbiol.* **64,** 101.

12. Association of Polydeoxyribonucleotides of Deoxyribonucleic Acids from Nocardioform Bacteria

S. G. Bradley

Department of Microbiology, Virginia Commonwealth University, Richmond, Virginia 23298, U.S.A.

and

M. Mordarski

Ludwik Hirszfeld Institute of Immunology and Experimental Therapy, Polish Academy of Sciences, Wrocław, Chałubinskiego ul. 4, Poland

Contents

1. Introduction

Genetic homology in the genus *Nocardia* and related nocardioform bacteria is defined in biological terms as the similarity between sequences of genes or loci but in biophysical terms, deoxyribonucleic acid (DNA) homology is the similarity between nucleotide sequences. Genetic homologies can be determined from data generated by studies using transduction, transformation or syncytic recombination (Bradley, 1966; see chapter 11). It must be noted however that a limited number of genetic determinants can be transferred from one organism to another by episomal elements or plasmids that are not necessarily homologous with the recipient's basic genome (Bradley, 1969). Genetic tests for homology often fail for reasons other than lack of genetic similarity (Brenner *et al.*, 1969); therefore, biophysical methods are more generally useful for comparing two nocardial genomes than recombinational analyses. In this review, the biophysical assay for DNA homology is the capability of single stranded polynucleotides to form duplex molecules.

2. Isolation of DNA

Isolation of DNA suitable for DNA analysis is dependent upon developing satisfactory procedures for rupturing nocardiae and freeing the DNA of contaminating substances. In general, it is necessary to damage the nocardial wall by mechanical abrasion in order to release the intracellular components (Bradley, Brownell & Clark, 1973). Some nocardioform bacteria, however, can be lysed by lysozyme alone or by combinations of lysozyme, pronase and sodium dodecyl sulfate (Tewfik *et al.*, 1968). It is possible to use high concentrations of lysozyme (5 mg/g wet cell mass) for prolonged incubation periods (18 h at 37°C). It is important, however, that the enzymic preparations used to rupture cells or used during purification of DNA be free of, or deprived of, deoxyribonuclease activity. Nocardiae not susceptible to lysis by enzymes or a combination of mechanical abrasion and enzymes may be ruptured by ultrasonic vibrations. Although organic solvents such as ethanol and acetone do not markedly increase the sensitivity of nocardial walls to lysozyme and pronase, treatment with these solvents drastically reduces viability thereby reducing the hazard of infection as a consequence of manipulating gram quantities of pathogenic strains.

In the widely used Marmur technique (Marmur, 1961) protein is dissociated from the nucleic acids by treating the cell extract with 1 M sodium perchlorate. Most of the protein is then denatured and removed by shaking the viscous extract with chloroform and isoamyl alcohol. Ribonucleic acid (RNA) contamination is effectively eliminated by digesting the nucleic acid preparation with ribonuclease; then the polymerized DNA is precipitated with ethanol, leaving the ribonucleotides in solution. The partially purified DNA is dissolved in acetate buffer. Isopropanol, in the presence of acetate anions, selectively precipitates DNA fibers, leaving RNA in solution. Chelating agents such as ethylenediaminetetraacetate and citrate are added to the buffers used during the isolation procedure because they minimize endogenous deoxyribonuclease activity. Despite the utility of Marmur's method, it has some limitations: (a) less than half of the cellular DNA is recovered, (b) the DNA is sheared to fragments of approximately 10^7 daltons or less and (c) contaminating carbohydrate may not be eliminated.

An alternative method for extraction of DNA uses aqueous phenol (Saito & Miura, 1963; Kirby, 1964). Aqueous phenol arrests deoxyribonuclease activity; additionally, extraction of DNA into aqueous phenol requires little shaking, thereby reducing the extent of shearing of the DNA. However, phenol accumulates deleterious peroxides that must be eliminated by distillation or treatment with 8-hydroxyquinoline. Moreover, care must be taken to remove phenol completely from the DNA preparation otherwise the high u.v. absorbance of the phenol will interfere with the routine analysis of DNA. The phenol is eliminated by (a) serial precipitation of the DNA with alcohol and dissolution in an appropriate salt solution, (b) extensive washing of precipitated DNA, wound on a glass rod, with ethanol and/or (c) extraction of DNA solutions with diethyl ether.

When only small samples of cell extracts are available, DNA can be purified by centrifugation in CsCl (Flamm, Bond & Burr, 1966). Because protein tends to float and RNA tends to sediment in dense CsCl, relatively pure DNA can be recovered with minimum manipulation.

Impurities in nocardial DNA preparations may interfere with studies on nucleotide composition or the kinetics and extent of association between single-stranded DNA samples. Such impurities are indicated by (a) opalescence of the DNA solution; (b) an A_{260}/A_{280} ratio of 1·9–2·0; (c) an A_{255}/A_{230} of 2·1–2·2; (d) RNA detected by an orcinol assay (Hatcher & Goldstein, 1969); or (e) protein detected by the Lowry assay (Lowry *et al.*, 1951). Hill, Wayne & Gross (1972) have used agar-gel diffusion of a DNA preparation against concanavalin A to detect contaminating

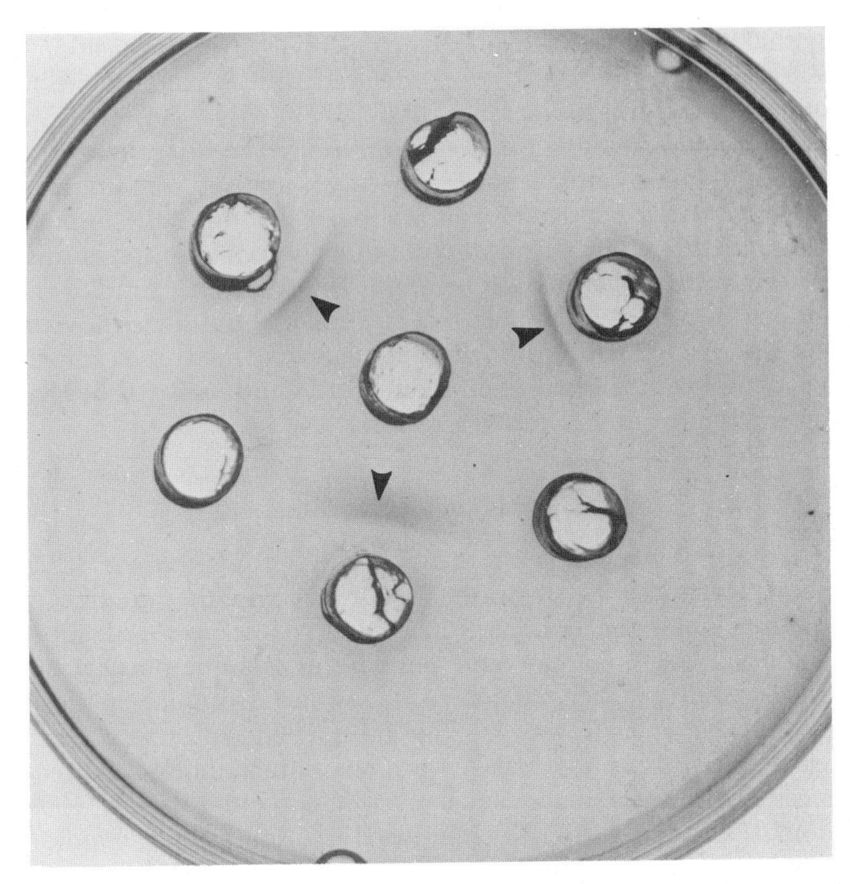

Fig. 1. Demonstration of carbohydrate contaminating nocardial DNA preparations. Bands of precipitation (indicated by arrows) are formed between the center well containing 0·2 ml of 5 mg concanavalin A/ml and certain peripheral wells filled with DNA samples that contain contaminating carbohydrate.

polysaccharides. Precipitation bands indicate the presence of contaminating carbohydrate (Fig. 1). We have detected carbohydrate in DNA samples prepared by the Marmur procedure from *Actinomadura madurae* and *rhodochrous* strains by chemical analysis, by precipitation bands using concanavalin A, and as extraneous bands formed during isopycnic centrifugation in CsCl.

In order to free DNA preparations of contaminating carbohydrate, the samples obtained by the Marmur procedure are treated with α-amylase and β-amylase (100 μg of each enzyme/ml for 1 h at 37°C). Protein in the amylase-treated DNA solution is denatured by shaking with chloroform-isoamyl alcohol; after clarification by centrifugation, the DNA is precipitated with ethanol, harvested and dissolved in saline-citrate diluent. The DNA is precipitated with 2-methoxyethanol, harvested and dissolved in saline-citrate diluent. These added steps usually remove contaminating carbohydrate. Szyba & Mordarski (pers. comm.) have removed contaminating polysaccharides by affinity chromatography using a concanavalin A-Sepharose column. Alternatively, Bradley has added concanavalin A (final concentration 500 μg/ml) to the crude DNA after incubating with ribonuclease. The mixture is agitated for 15 min before the protein is denatured and removed by shaking with chloroform-isoamyl alcohol.

3. Overall Nucleotide Similarity

In double-stranded DNA, the guanine (G) content of the duplex molecule equals the cytosine (C) content and the adenine (A) content equals the thymine (T) content. However, the ratio of (A+T) to (G+C), or the mole per cent of guanine plus cytosine [$\%GC = 100(G+C)/(A+T+G+C)$] may vary from one species to another (Marmur, Rownd & Schildkraut, 1963). In general, closely related organisms have DNA with very similar nucleotide sequences and therefore have very similar GC contents (Mandel, 1969). Similarity in % GC of DNA samples does not necessarily indicate genomic similarity, but dissimilarity in % GC of DNA from two organisms established that they are not identical (Bradley, 1965).

The overall nucleotide composition of DNA can be determined by a variety of techniques. Each method has its virtues and limitations (Bradley & Enquist, 1974). Not only do these analyses of DNA have taxonomic value, but they can also reveal the presence of novel nucleotides (Erikson & Szybalski, 1964). Commonly used methods include (a) buoyant density centrifugation in cesium chloride (Schildkraut, Marmur & Doty, 1962); (b) spectrophotometric measurement of purines released in dilute acid (Huang & Rosenberg, 1966); (c) chromatography of the bases released by hydrolysis (Bendich, 1957); and (d) the temperature corresponding to the midpoint (Tm) of the hyperchromic shift upon heating (Marmur & Doty, 1962). The % GC can be calculated from the Tm according to the formula $2{\cdot}44\,(Tm - 69{\cdot}3 - 15 \log X)$

where $X = 0{\cdot}1$ for $0{\cdot}1 \times$SSC, 1 for $1 \times$SSC and 10 for $10 \times$SSC (Yamada & Komagata, 1970).

It should be noted that DNA of plasmids may have GC contents significantly different from that of the basic genome. Under certain conditions, the amount of plasmid DNA per cell may equal to exceed that of the basic bacterial genome (Kopecko & Punch, 1971). Accordingly, the overall nucleotide composition of DNA from two organisms having identical basic genomes may be significantly different because one strain lacks, or contains little, plasmid DNA whereas the other strain contains large amounts of plasmid DNA. To date, this potential complication has not been knowingly encountered while analyzing DNA from nocardial cultures harvested in the rapid phase of growth or the early stationary phase.

As a general guideline, organisms whose genomic DNA differ by more than 4–5% GC should not be considered members of the same species and organisms whose DNA differ by more than 10% GC should not be considered members of the same genus (Bradley, 1971*a*). Opinions relying upon DNA analyses, however, must be based upon legitimate comparisons (Mandel *et al.*, 1970). It should be noted that many methods for measuring the mole % guanine and cytosine utilize biological reference standards. The % GC for *Escherichia coli*, a frequently used reference, has variously been set at 49–51% GC (Schildkraut, Marmur & Doty, 1962). Moreover, certain methods and formulae for computing GC content show consistent discrepancies, that is, one method may give % GC values higher than another method (Gasser & Mandel, 1968). Uncritical judgements about biological relatedness of organisms based upon comparison of published values may be erroneous (Enquist & Bradley, 1970).

Overall nucleotide composition of DNA samples has been a valuable aid in resolving certain taxonomic dilemmas involving nocardioform organisms (Tewfik & Bradley, 1967; Bradley, 1970; see chapters 1 and 2). The organism previously referred to as "Ørskov's motile Nocardia" contains DNA having 74–76% GC, which is substantially out of the range of other accepted members of the genus *Nocardia* (Table 1). These results support the placement of "Ørskov's motile Nocardia" into a separate genus, *Oerskovia* (Prauser, Lechevalier & Lechevalier, 1970). Overall nucleotide composition of DNA, however, does not help in delineating the genera *Nocardia* and *Mycobacterium* (Bradley, 1971*b*). Data on the overall nucleotide composition of members of the genus *Actinomadura* are consistent with the proposition that these organisms are intermediate between the nocardiae and the streptomycetes.

Table 1
Nucleotide composition of DNA samples from nocardioform bacteria

Taxon	Mole % GC
Actinomadura	68–72
Mycobacteriun	64–69
Nocardia	66–69
Oerskovia	74–76
"*Rhodochrous*" complex	60–69
Streptomyces	69–74

Overall nucleotide composition of DNA has also provided insights into classification at the specific level. Gordon (1966) has referred to the "*rhodochrous*" complex as a good species in search of a genus; however, these organisms fall into two distinct groups with respect to GC content of their DNA (Bradley, 1966). One group, typified by *Nocardia rubra*, possesses DNA having 66–68% GC, whereas the other group, typified by *N. erythropolis*, possesses DNA having 61–63% GC (Bradley & Huitron, 1973). Clearly, the *rhodochrous* complex is not a homogeneous species.

4. DNA/DNA Association

Native nocardial DNA is composed of two complementary polynucleotide strands. The complementary strands can be dissociated by heating or by raising the pH to 12; this process is known as denaturation. Denatured DNA will reassociate when the heated solution is cooled slowly or is maintained at a temperature 20°–30°C below the *T*m or when the alkaline solution is neutralized. Association of a pair of strands of denatured DNA results from a properly aligned collision involving complementary regions. Most collisions between two denatured strands are ineffective but sometimes a collision involving two complementary nucleotide sequences results in the formation of a short duplex segment. Once this duplex segment is formed, further helix formation between the remaining nucleotides of that pair of complementary strands proceeds rapidly (Wetmur & Davidson, 1968).

Nucleotide sequences can be compared by examining DNA/DNA association in free solution or in experiments in which one denatured DNA is immobilized and the other denatured DNA is in solution. Nitrocellulose membrane filters are often used to immobilize one denatured DNA (Warnaar & Cohen, 1966; Monson *et al.*, 1969). Nitrocellulose is preferable to agar for immobilizing single stranded DNA because the DNA-agar complex is not stable at the elevated temperatures needed for association of nocardial DNA (Legault-Demare *et al.*, 1967). In the membrane filter-DNA technique, binding of the DNA in solution can be measured directly (Denhardt, 1966). Alternatively, a competing DNA in solution may also be added, along with the soluble reference DNA (Johnson & Ordal, 1968). The general scheme for DNA association experiments is as follows: (a) one type of DNA is labeled with a radioactive isotope, (b) DNA is isolated, freed of contaminating protein, ribonucleic acid and carbohydrate, (c) the non-radiolabeled DNA is denatured and immobilized on a nitrocellulose membrane filter, (d) the radiolabeled DNA is sheared into lengths having a molecular weight of 200 000 to 500 000 daltons and denatured; and (e) a sample of the sheared, radiolabeled denatured DNA is incubated at 65°–75°C for 15–24 h with the denatured DNA-immobilized on the nitrocellulose membrane. De Ley & Tijtgat (1970) add 30% dimethylsulfoxide to the incubation mixture containing radiolabeled, sheared denatured DNA and the immobilized denatured DNA so that specific association can be achieved at practical incubation temperatures. During the incubation period, complementary nucleotides align, forming duplex molecules of varying stabilities. If the two nucleotide sequences are exactly complementary, the new duplex is as stable as the original duplex (Flavell, Borst & Birfelder, 1974).

Previous investigators have sometimes failed to select incubation temperatures high enough to ensure that the annealed DNA is composed of well matched nucleotide sequences (Yamaguchi, 1967; Okanishi & Gregory, 1970). Duplexes of nocardial DNA formed at 60° and 65°C have low thermal stability whereas those formed at 70° and 75°C are thermally stable. Incubation at lower temperatures has utility, but only in the context of particular experimental goals that are discussed later in this review.

The rate of association of complementary single-stranded DNA molecules is dependent upon a number of experimental conditions: (a) the rate is inversely proportional to the viscosity of the DNA solution; (b) the rate is proportional to the square-root of the molecular weights of the sheared single stranded DNA samples; (c) the rate is slightly faster for DNA samples having a high GC content than for DNA samples having a high AT content; (d) the rate is relatively independent of pH between 5 and 9; (e) the rate is highly dependent upon salt concentration; (f) the rate is inversely proportional to the size and complexity of the genome; (g) the rate is dependent upon the concentration of the DNA; and (h) the rate is dependent upon the purity of the DNA preparation (Brenner, 1970).

In DNA/DNA association experiments in free solution, denaturation and association are monitored optically as changes in absorption at 260 or 270 nm. In free solution techniques, it is not necessary to radiolabel the DNA but both types of DNA are sheared into lengths of 200 000 to 500 000 daltons. The sheared DNA is then incubated at 25°C below the *T*m for *c.* lh in order to allow denatured ends of the duplex strands to reassociate.

For nocardial DNA samples, it is necessary to add formamide to the DNA sample in order to reduce the *T*m to a practical operating temperature. The native DNA is denatured by heating; the absorbance increases about 36% upon thermal denaturation provided that the starting DNA contains negligible contaminating substances and denatured DNA and provided that the maximum temperature is *c.* 10° above the *T*m. The DNA solution is then adjusted to 25° below the *T*m. During the subsequent incubation period, the absorbance decreases, indicating that single-stranded DNA is forming duplex molecules. When 75% of the DNA has associated, the sample is heat denatured to determine the thermal stability of the annealed duplexes (Fig. 2). This serves as an index of the exactness of matching of the nucleotide pairs.

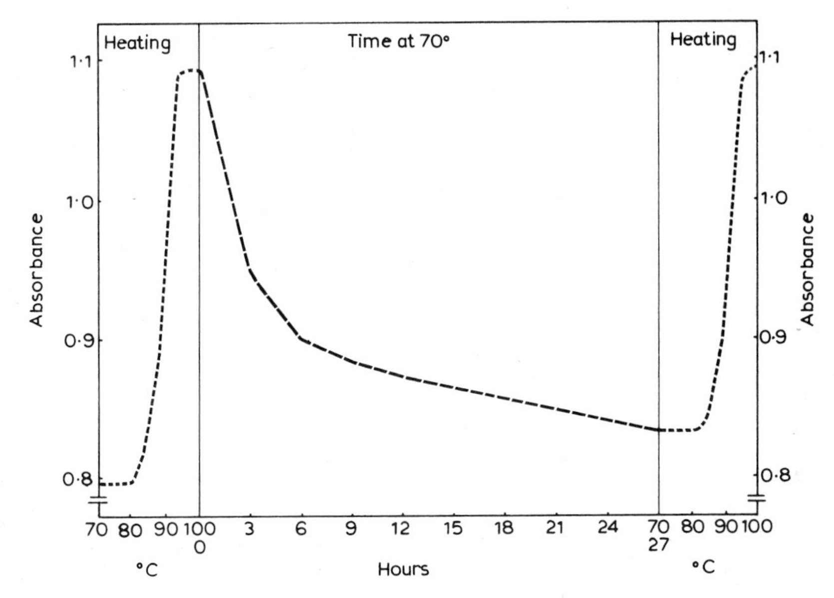

Fig. 2. Denaturation and association of *Nocardia farcinica* 300 DNA. This DNA association experiment was carried out in 25% formamide dissolved in 6×SSSC. The incubation temperature for nucleotide pairing was 70°C. The changes in absorbance were monitored at 270 nm. Note the change in scale of the abscissa from temperature to time and back to temperature.

Nucleotide pairing may be exact, partially matched or may be well matched in one region but unmatched in another region. Optically monitored association measures the proportion of nucleotides matched whereas radioisotopic methods measure the proportion of DNA fragments that associate over a long enough span to hold the pair of DNA fragments together. These differences in analysis account for some of the differences in values obtained by the two methods.

Optically measured DNA association has a number of desirable features for determining relatedness among nocardiae: (a) no radiolabeled DNA is required; (b) high association temperatures can be used without the complication of leaching of the immobilized DNA from the menstruum (agar or nitrocellulose); and (c) absorbance accurately measures associated sequences *per se* but not single stranded loops and unmatched free ends.

Association of a pair of complementary DNA strands results from their collision. Duplex formation involves two steps: (a) the initial pairing of short homologous sequences on the two strands and (b) the fast "zippering" reaction. The rate-determining step appears to be the initial pairing process rather than the subsequent "zippering" process. It is predicted, therefore, that DNA association, adheres to second order kinetics (Wetmur & Davidson, 1968).

5. Kinetics of Association

Denatured DNA from nocardiae consists of equimolar concentrations of the two complementary strands. The velocity of the disappearance of the unassociated nucleotide strands is:

$$-\mathrm{d}C/\mathrm{d}t = kC_1C_2$$

where C_1 and C_2 are the molar concentrations of each of the complementary DNA strands. Because $C_1 = C_2$, the velocity statement becomes:

Equation I
$$-\mathrm{d}C/\mathrm{d}t = kC^2$$

Rearranging:

$$\frac{\mathrm{d}C}{C^2} = -k\,\mathrm{d}t$$

Integrating:

$$\int \frac{\mathrm{d}C}{C^2} = \int -k\,\mathrm{d}t$$

$$-1/C = -(kt + x)$$

or

$$1/C = kt + x$$

Solving for the integration constant X at t (time) = 0:

$$X = 1/C_0$$

Substituting: for X:

$$1/C_t = kt + 1/C_0$$

where C_t is the molar concentration of one of the complementary DNA strands at time "t" and C_0 is the molar concentration of one of the complementary DNA strands at the beginning of experiment ($t = 0$), k is the rate constant and t is time in seconds.

Rearranging:

Equation II $$kt = 1/C_t - 1/C_0$$

Equation III $$kt = (C_0 - C_t)/C_0 C_t$$

$$C_0 C_t kt = C_0 - C_t$$

$$C_0 C_t kt + C_t = C_0$$

$$C_t(C_0 kt + 1) = C_0$$

Equation IV $$C_0 kt + 1 = C_0/C_t$$

Because nucleotide concentration is conveniently measured spectrophotometrically, these expressions can be converted into Absorbance units. In general, 1 g of native DNA/1 has an absorbance at 260 nm of 24 (Britten & Kohne, 1966). The average molecular weight of the nucleotides in DNA is 309 (deoxyadenylic acid, 331; deoxycytidylic acid, 307; deoxyguanylic acid, 347; deoxythymidylate, 322). From the mean of 327, the molecular weight of water (18) is subtracted as a result of formation of the polynucleotide from the individual nucleotides. Accordingly:

Equation V $$C_0 = A_n/309 \times 24 \times 2$$

where A_n is the absorbance of the native DNA at time 0. Note that $A_n/309 \times 24$ is the molar concentration of all nucleotides but $A_n/309 \times 24 \times 2$ is the molar concentration of one of the complementary species of nucleotides.

The proportion of unassociated nucleotides at time "t" can be written as:

Equation VI $$\frac{A_t - A_n}{A_d - A_n}$$

where A_t is the absorbance at 260 nm at time "t" and A_d is the absorbance of the fully denatured DNA. The molar concentration of unassociated DNA at time "t" is:

$$C_t = \frac{A_t - A_n}{A_d - A_n} C_0$$

Substituting into Equation II:

$$kt = 1/C_0 \frac{A_t - A_n}{A_d - A_n} - 1/C_0$$

or

Equation V
$$C_0kt = 1\Big/\left[\frac{A_t - A_n}{A_d - A_n} - 1\right]$$

Equation VIII
$$C_0kt = \frac{A_d - A_n}{A_t - A_n} - 1$$

Substituting *Equation V* for C_0:

$$\frac{A_d - A_n}{A_t - A_n} = 1 + A_nkt/309 \times 24 \times 2$$

Equation IX
$$k = \frac{309 \times 24 \times 2}{A_nt}\left[\frac{A_d - A_n}{A_t - A_n} - 1\right]$$

When half of the denatured DNA has associated:

Equation X
$$k = \frac{1{\cdot}48 \times 10^4}{A_nt}$$

The units of k are liters per mole per second or M^{-1}. seconds^{-1}. If time is expressed in minutes,

Equation XI
$$k = 247/A_nt$$

if time is expressed in hours,

Equation XII
$$k = 4{\cdot}1/A_nt$$

Even when time in minutes or hours is used to compute "k" with the equations XI and XII, the units of "k" remain M^{-1} second^{-1}.

Britten & Kohne (1966) introduced the acronym *Cot*, which is the moles of nucleotide per liter × time in seconds and is readily calculated using the numerically equivalent statement $Cot = \frac{1}{2}(A_{260})$ (time in hours). These workers also introduced the popular "*Cot*-plot" which plots *Cot* (abscissa) versus the fraction of the DNA that has associated (ordinate). The curve generated by an uncomplicated second order reaction is reasonably symmetrical, sigmoid-shaped and makes a relatively straight transition from the completely denatured state to the completely associated state over a 100-fold range in *cot* values (Fig. 3). However it must be stressed that the "*Cot*-plot" does not adequately establish that DNA association is adhering to second order kinetics. The correct second order

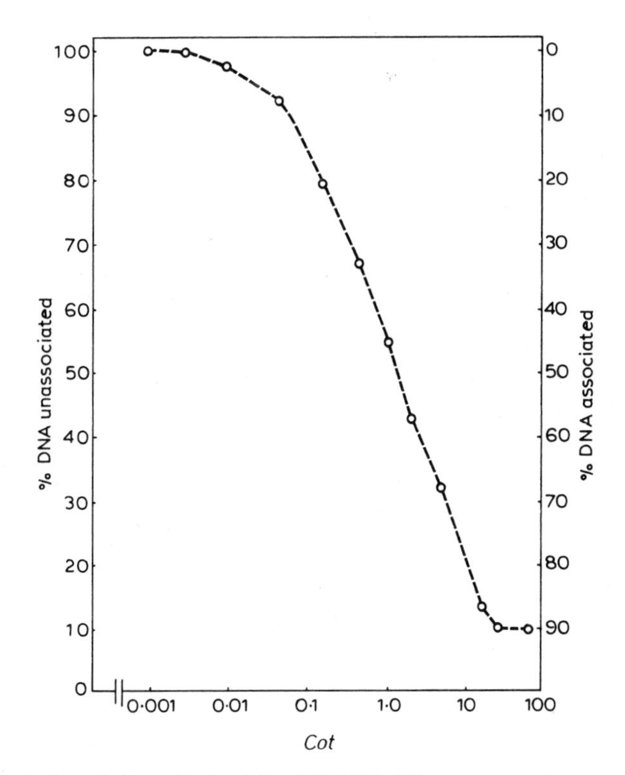

Fig. 3. Association of *Nocardia farcinica* 300 DNA. The per cent of unassociated DNA [$100(A_t - A_n)/(A_d - A_n)$] is plotted against C_0t values [$\frac{1}{2}A_n t$, where t is in hours]. This DNA association experiment was carried out in 25% formamide dissolved in 6×SSC. The incubation temperature for nucleotide pairing was 70°C.

plot for DNA association data consists of time (abscissa) vs $(A_d - A_n)/(A_t - A_n)$ as the ordinate (see *Equation IX*; Fig. 4).

The *Cot* value corresponding to the point at which one-half of the DNA has associated, referred to a *Cot* $\frac{1}{2}$, is characteristic for a given DNA. *Cot* $\frac{1}{2}$, for example, is proportional to genome size provided repeated nucleotide sequences are absent. Accordingly, the size or molecular weight of a nocardial genome can be determined by comparing its *Cot* $\frac{1}{2}n$ to *Cot* $\frac{1}{2}s$ of a standard DNA whose size has been previously established. Gillis, De Ley & DeCleene (1970) have employed the following statement:

$$\frac{Cot\,\frac{1}{2}n}{Cot\,\frac{1}{2}s} = \frac{ks}{kn} = \frac{M_n}{M_s}$$

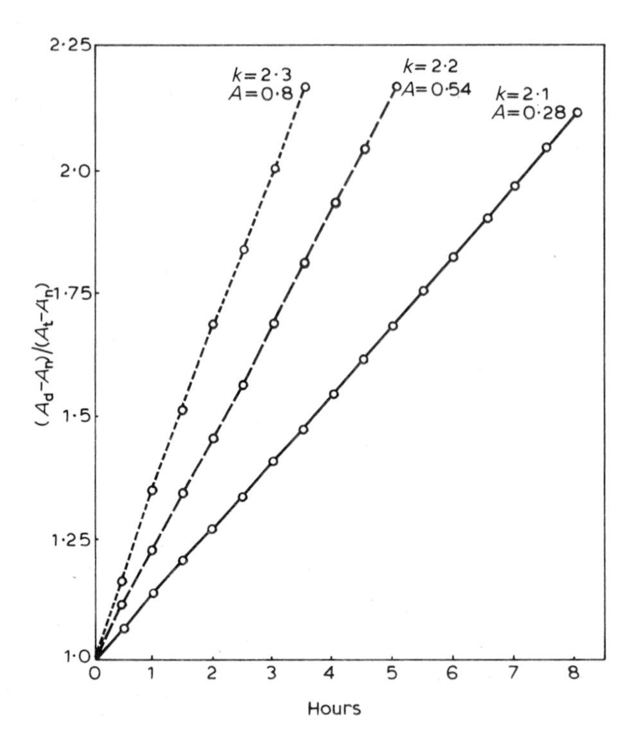

Fig. 4. Kinetics of DNA association. Sheared and denatured DNA of *Nocardia farcinica* 300 renatured according to second-order kinetics. The second-order rate constant (***k***) was $2{\cdot}2 \pm 0{\cdot}11\ M^{-1}\ s^{-1}$ for DNA samples having initial reactant concentrations of $A_{260} = 0{\cdot}8$ (– – – –), 0·54 (— —) and 0·28 (——). A_d, absorbance of denatured DNA; A_n, absorbance of native DNA, A_t, absorbace of reassociating DNA at time T.

where k is the second order rate constant, M is the molecular weight and the subscripts n and s refer to nocardial and standard DNA samples respectively.

The molecular weights of the genomes of diverse species of nocardioform bacteria vary from $2{\cdot}0 \times 10^9$ to $4{\cdot}5 \times 10^9$ daltons. Closely related nocardiae usually have similar genome sizes, although a large deletion may result in a smaller genome and a novel episomal element will result in a larger genome. Accordingly, a difference in molecular weight of DNA does not necessarily preclude relatedness (Brenner, 1970). Clark & Brownell, (1972) and Bradley, Brownell & Clark (1973) have suggested that the genome of the "*canicruria*"-mating type (Mat-Ce) of *N. erythropolis* may contain novel nucleotide sequences that are absent in the genome of the "*erythropolis*"-mating type (Mat-cE). They were unable to discriminate between deletion of nucleotides in the Mat-cE mating type and addition of genetic material to the Mat-Ce mating type.

Cot $\frac{1}{2}$ values can also be used to determine the relatedness between DNA samples from two sources (De Ley, Cattoir & Reynaerts, 1970). The association rates of DNA from each source is determined individually and the rate for an equal mixture of the two DNA samples is determined. The degree of additivity of the individual *Cot* $\frac{1}{2}$ values with respect to the *Cot* $\frac{1}{2}$ value of the mixture reflects the degree of nucleotide homology between the samples. The per cent homology can be calculated from the expression:

$$\% \text{ homology} = 200[Cot\,\tfrac{1}{2}A + Cot\,\tfrac{1}{2}B - Cot\,\tfrac{1}{2}mix]/[Cot\,\tfrac{1}{2}A + Cot\,\tfrac{1}{2}B].$$

This equation was developed by Bradley (1972) and is similar to that of Seidler & Mandel (1971).

6. Taxonomic Inferences

Ridell (1974) has concluded that the specific identification of a substantial number of strains designed as *Nocardia* or *Mycobacterium* must be reconsidered because there are strains which are very different that have been placed in the same species, and there are also strains which are closely related that have been placed in different species. For example, Ridell found that eight strains designated *N. corallina, N. rubra* or "*M.*" *rhodochrous* should be assigned to one taxon whereas 10 other strains designated *N. rubra, M. pellegrino* or "*M.*" *rhodochrous* should be assigned to another distinct taxon (also see chapter 8). Adams & McClung (1962) and Anderson & Bradley (1962) have previously challenged the condensation of a wide variety of nocardioform organisms into a single *rhodochrous* species. Although the generic designation for the members of the *rhodochrous* complex has not been definitively established, many investigators, using a variety of techniques, have established that these organisms do not belong in the genus *Mycobacterium* (Mordarska, Mordarski & Goodfellow, 1972; Goodfellow *et al.*, 1974; see chapter 2). It has been repeatedly demonstrated that mycobacterial walls contain mycolic acids of C_{60} to C_{90} and nocardial walls contain nocardomycolic acids of C_{34} to C_{60}

(Lanéelle, 1966; Lechevalier, Horan & Lechevalier, 1971; Kanetsuna & Bartoli, 1972; Azuma *et al.*, 1974). Tsukamura (1971), Goodfellow, Fleming & Sackin, (1972) and Ridell (1974) have concluded that strains designated *N. rubra*, *N. corallina*, *N. erythropolis*, *M. pellegrino* and *M. rhodochrous* form a taxon clearly distinct from both the genus *Mycobacterium* and the genus *Nocardia*.

Our results with DNA association establish that there is a homogeneous species *N. erythropolis*, which is different from *N. globerula*, *N. rubra* (or *N. corallina*) and *N. pellegrino* (Table 2). *N. globerula* has the same GC content (62%) as *N. erythropolis*.

Table 2
Association of DNA from Nocardia erythropolis 398 *and* 474 *with DNA from selected nocardioform bacteria*

Test DNA from	% homology with 398	474
N. erythropolis 474 (ATCC-4277) (Waksman's 3407)	97	100
N. erythropolis 439 (Gordon's 765A)	100	92
N. erythopolis 398 (McClung's 57)	100	97
N. erythropolis 520 (ATCC17041)	100	100
N. erythropolis 625 (wild type from Michaels)	98	99
N. erythropolis 626 (wild type from McClung) (Waksman's 3407)	100	94
N. erythropolis 305	86	96
N. globerula 472 (ATCC9356)	24	0
N. globerula 631 (ATCC9356)	33	28
N. pellegrino 632 (ATCC15998)	30	21
*N. rubra** 321 (McClung's) R562	0	37
N. rubra 339 (McClung's 107)	24	24
N. thamnopheos 633 (ATCC4445)	23	0

Association of DNA in 6×SSC-25% formamide was monitored optically, as described by Bradley (1973).
* *N. rubra* is probably synonymous with *N. corallina*.

Similarly Mordarski and co-workers have established that there is a homogeneous species *N. pellegrino* (Table 3) which is different from *N. rubra* (or *N. corallina*), *N. erythropolis* and *N. globerula*. *N. pellegrino* has the same GC content (66–68%) as *N. rubra* and *N. corallina*. It should be noted that there seems to be no extant type strains of *N. corallina* or "*Mycobacterium*" *pellegrino* (Sneath & Skerman, 1966). Moreover, there is some question whether the specific epithet *pellegrino* is validly published (Buchanan, Holt & Lessel, 1966).

Table 3

Association of DNA from N. pellegrino P11 and N. asteroides NK20 *with DNA from selected* N. pellegrino *strains*

DNA source	Radiolabeled reference DNA from			
	N. pellegrino P11†		*N. asteroides* NK20‡	
	Relative DNA bound (%)	ΔTm_e* (°C)	Relative DNA bound (%)	ΔTm_e* (°C)
N. pellegrino P11	100	0	13	—
N. pellegrino P1	102	0	16	0
N. pellegrino P2	82	9	12	0
N. pellegrino P3	101	15	15	0
N. pellegrino P4	104	14	15	9
N. pellegrino P5	63	7	10	0
N. pellegrino P6	12	3	15	0
N. pellegrino P7	89	12	15	—
N. pellegrino P8	96	10	11	—
N. pellegrino P9	82	—	10	—
N. pellegrino P10	104	—	100	—
N. asteroides NK20	14	—	17	—

* Temperature at which half of the bound reference DNA is eluted with 0·15 M NaCl–0·0015 M sodium citrate.
† Specific activity 3445 counts min^{-1} μg^{-1}.
‡ Specific activity 6890 counts min^{-1} μg^{-1}.

We have found five strains of the "*N. asteroides*" complex that possess DNA with very similar nucleotides sequences. These strains may belong to *N. asteroides* subtype B, as defined by Schaal (1974). Because this group includes strain ATCC-3318, this species is designated as *N. farcinica* awaiting resolution of the nomenclatural problem concerning this epithet (Table 4, see chapter 1). Several strains considered members of the "*N. asteroides*" complex do not belong to the *N. farcinica* homology group. One of the excluded strains is ATCC-19247 which Sneath & Skerman (1966) recommended as the neotype culture for the taxon *N. asteroides.*

Table 4
Association of DNA from Nocardia farcinica 300 *and* 330 *with DNA from selected nocardiae*

Test DNA from	% homology with 300	330
N. farcinica 300	100	90
N. farcinica 330	90	100
N. farcinica 324	91	95
N. farcinica 333	100	89
N. farcinica 462	96	91
N. asteroides 334	19	15
N. asteroides 323	0	0
N. asteroides 571	22	28
N. brasiliensis 301	0	21
N. brasiliensis 473	13	24
N. caviae 421	0	0
N. erythropolis 458	0	0
N. phenotolerans 514	10	0
Nocardia sp. 304	0	0
N. transvalensis 516	4	0

Association of DNA in 6×SSC-25% formamide was monitored optically, as described by Bradley (1973).

7. Phylogenetic Implications

Taxonomic inferences may be based upon the extents of DNA nucleotide sequences held in common between a test DNA and a reference DNA as determined by DNA/DNA association data. However, DNA samples from related organisms usually contain a spectrum of partially matched sequences as well as identical nucleotide sequences and totally dissimilar sequences. It is not possible, therefore, to arrive at a single value that defines the relatedness of one organism to another. This apparent complication actually provides a basis for inferring phylogenetic relationships among organisms.

The exactness of nucleotide pairing, as well as the extent of association between single-stranded DNA preparations from different strains, can be measured experimentally. In one method, nucleotide pairing is allowed to proceed at two different incubation temperatures. At the higher temperature, only exactly matched sequences should form duplexes. At the lower temperature partially matched and exactly matched sequences should form dup-

Distribution of variation	% Binding NON-EXACTING	EXACTING
identical	100	100
unrelated	0	0
localized	90	90
conserved	10	10
dispersed	100	75
dispersed	100	50
dispersed	75	25
identical/unrelated	50	50
identical/dispersed	100	85
dispersed/unrelated	50	25

Fig. 5. Distribution of mutations and the effects of different patterns of nucleotide heterogeneity on DNA association. The values indicated are typical. Actual experimental values will differ somewhat from these examples.

lexes (Fig. 5). The nonexacting low incubation temperature is usually 30°–35° below the *T*m of the reference DNA and the exacting high temperature is 15°–20° less than the *T*m of the reference DNA. The ratio of (the relative binding at the exacting temperature)/(the relative binding at the nonexacting temperature) is useful for gauging the distribution of nucleotide divergence. We refer to this ratio as the relative binding ratio (RBR) and Brenner (1970) has referred to this value as the thermal binding index (TBI). Because the proportion of nucleotides bound at an exacting incubation temperature must be the same as, or less than the proportion of nucleotides bound at a less exacting temperature, the values of the RBR have a range of 0 to 1.

The evolutionary process can be related to the RBR and the proportion of nucleotides annealed at an exacting incubation temperature. If the reference DNA and test DNA are totally unlike, there will be no exact matching and no partially matched duplexes. If the reference DNA and test DNA are entirely the same, there will be extensive duplex formation at exacting and nonexacting incubation temperatures (Fig. 5). If there is a small localized region of unlike DNA in the test DNA, the RBR is about 1 and the extent of exact binding is high. Biologically this condition could arise by transfer of a small amount of novel genetic information, for example, a plasmid, or by extensive mutation in a small segment of the genome.

If the vast majority of the test DNA is unlike that of the reference DNA, but there is a small conserved segment of homology, the RBR will be about one but the amount of exact binding will be low (Fig. 5). Biologically this condition could arise by transfer of a small amount of novel genetic information or by strong conservation of particular nucleotide sequences, for example those controlling ribosomes.

If random mutation is the basis of genetic diversity, the unmatched nucleotides will be randomly dispersed througout the genome. If genetic divergence, as a consequence of dispersely distributed heterogeneity, has progressed only moderately, then the RBR will be between 0·5 and 1 and the amount of exact binding will be between 50–100%. If divergence by dispersely distributed heterogeneity has progressed extensively the RBR will be low, that is, between 0 and 0·5, and the amount of exact binding will also be low, that is, between 0–50% (Fig. 5). Our results indicate that evolution in the nocardiae has proceeded primarily by random mutation.

8. Neutral Mutations

A fundamental assumption in DNA association analyses is that the ability to form specific DNA duplexes with samples from two nocardiae constitutes evidence for relatedness. Conversely the lack of specific hybrid duplex formation is interpreted as evidence for some degree of unrelatedness. That base mispairing reflects phenotypic dissimilarity, however, is not rigorously proven. It is conceivable that phenetically similar organisms might possess substantial amounts of genomic diversity. In such instances, there would be a significant degree of base mispairing during DNA association.

King & Jukes (1969) have proposed that there may be random neutral genetic mutations that have no effect on the competitive fitness of the organism. Wright (1966) has suggested that neutral mutations can become fixed as evolutionary changes through the action of genetic drift. Kimura & Ohta (1971) went further and advocated that many mutational events are selectively neutral. A molecular basis for neutral mutations can be proposed. It is well established that there is more than one codon for each amino acid. Kimura (1968) pointed out that in the 61 amino acid specifying codons, there are 549 possible single-base substitutions. Of these, 134 substitutions generate synonymous codons. It is conceivable therefore that substantial nucleotide divergence could accumulate within a species with little change in overt phenotype. It is equally conceivable that nucleotide divergence as a result of neutral mutations could proceed rapidly.

9. Concluding Remarks

Methods for isolating highly polymerized nocardial DNA reproducibly and in good yields need to be developed. Nevertheless moderate amounts of DNA, adequately polymerized for most DNA analyses, can be regularly isolated from nearly all readily cultivable nocardioform bacteria. Many nocardial DNA preparations are contaminated with sufficient polysaccharide to interfere with particular analytical procedures. Recently Mordarski and co-workers and Bradley and co-workers have effectively utilized concanavalin A to free nocardial DNA preparations of extraneous carbohydrate. Lack of techniques for isolating covalently closed circular DNA however, limits our capability to detect putative episomal and plasmid DNA in nocardiae.

DNA analyses have definitively resolved many important taxonomic dilemmas about nocardioform bacteria. The hetereogeneity of the taxon "*Mycobacterium*" *rhodochrous* and the homogeneity of the genospecies *Nocardia erythropolis* have been established using both DNA association and genetic recombination (Bradley & Huitron, 1973). Similarly the heterogeneity of the *N. asteroides* complex has been established; however, only one of the DNA groups has been delineated (Bradley, 1973). A number of difficult taxonomic problems remain unresolved; for example, definition of the species of the putative genus *Proactinomyces* (Bradley & Bond, 1974) and of the *N. asteroides* complex.

DNA analyses are able to provide insight into the structure of the nocardial genome. Although the overall genome size has been determined for a number of nocardioform bacteria, deletions and episomal elements have not been definitively detected in this group. Evidence has been presented that there are short conserved DNA segments in the nocardiae but it has not been ascertained whether the same segments are conserved throughout the genus or whether the shared sequences differ in different organisms (Clark & Brownell, 1972; Bradley, Brownell & Clark, 1973). Conceivably some nucleotide sequences, for example, those determining ribosomal RNA structure (Moore & McCarthy, 1967), undergo mutation only rarely, are protected by an effective repair system, or are critical and if altered by mutation are lethal for the organism. Shared nucleotide sequences might also be distributed among the genus as a result of gene transfer or dissemination of episomal or plasmid elements. The extent of gene-flow and its significance in the evolution of the nocardioform bacteria has not been adequately assessed.

It has been suggested, on rather limited data, that genetic diversity in the nocardioform bacteria has arisen primarily through random mutation and selection (Enquist & Bradley, 1970). Additional studies are needed to clarify the significance of the spectrum of thermal stabilities of reassociated and annealed DNA duplexes. It may be possible in the nocardioform bacteria to use DNA association as a tool to determine whether neutral mutations occur in high frequency. Although nucleotide pairing has been measured by a variety of techniques, DNA/RNA hybridization (Gillespie & Spiegelman, 1965), interaction between oligodeoxyribonucleotides and denatured DNA (McConaughy & McCarthy, 1967), DNA/DNA competition (McCarthy, 1967) and chromatography of nucleic acids on hydroxyapatite (Bernardi, 1965) constitute useful techniques that have not been fully exploited in order to resolve the perplexing taxonomic, genetic and physiologic problems presented by the nocardioform bacteria.

10. References

Adams, J. N. & McClung, N. M. (1962). Comparison of the developmental cycles of some members of the genus *Nocardia*. *J. Bact.* **84,** 206

Anderson, D. L. & Bradley, S. G. (1962). Susceptibility of nocardiae and mycobacteria to actinophage. *Antimicrob. Ag. Chemother.* **1961,** 898.

Azuma, I., Ohuchida, A., Taniyama, T., Yamamura, Y., Shoji, K., Hori, M., Tanaka, Y. & Ribi, E. (1974). The mycolic acids of *Mycobacterium rhodochrous* and *Nocardia corallina. Biken's J.* **17,** 1.

Bendich, A. (1957). Methods for characterization of nucleic acids by base composition. In *Methods of Enzymology, vol.* 3. Eds S. P. Colowick & N. O. Kaplan. p. 715. New York: Academic Press.

Bernardi, G. (1965). Chromatography of nucleic acids on hydroxyapatite. *Nature, Lond.* **206,** 779.

Bradley, S. G. (1965). Interspecific genetic homology in actinomycetes. *Int. Bull. bact. Nomencl. Taxon.* **15,** 239.

Bradley, S. G. (1966). Genetics in applied microbiology. *Adv. appl. Microbiol.* **8,** 29.

Bradley, S. G. (1969). The genetic and physiologic bases for microbial variation. *Proc. Sympos. Use of Drugs in Animal Feeds.* p. 287. (Publ. # 1679.) Washington, D.C.: Nat. Acad. Sci.

Bradley, S. G. (1970). Genetic homologies among *Mycobacterium, Nocardia* and *Actinomyces.* In *Host-virus relationships in Mycobacterium, Nocardia and Actinomyces.* Eds S. E. Juhasz & G. Plummer. p. 179. Springfield, Ill.: Charles C. Thomas.

Bradley, S. G. (1971*a*). Criteria for definition of *Mycobacterium, Nocardia* and the *rhodochrous* complex. *Advg. Fronts Pl. Sci.* **28,** 349.

Bradley, S. G. (1971*b*). Phylogenetic relationships among actinomycetes. In *Recent Advances in Microbiology.* Eds A. Pérez-Miravette & D. Pelàez. p. 3. Mexico, D. F.: Assoc. Mex. Micribiol.

Bradley, S. G. (1972). Reassociation of deoxyribonucleic acid from selected mycobacteria with that from *Mycobacterium bovis* and *Mycobacterium farcinica. Am. Rev. resp Dis.* **106,** 122.

Bradley, S. G. (1973). Relationships among mycobacteria and nocardiae based upon deoxyribonucleic acid reassociation. *J. Bact.* **113,** 645.

Bradley, S. G. & Bond, J. S. (1974). Taxonomic criteria for mycobacteria and nocardiae. *Adv. appl. Microbiol.* **18,** 131.

Bradley, S. G., Brownell, G. H. & Clark, J. E. (1973). Genetic homologies among nocardiae and other actinomycetes. *Can. J. Micribiol.* **19,** 1007.

Bradley, S. G. & Enquist, L. W. (1974). Microbial nucleic acids. In *Molecular Microbiology.* Ed. J. B. G. Kwapinski. p. 47. New York: John Wiley & Sons.

Bradley, S. G. & Huitron, M. E. (1973). Genetic homologies among nocardiae. *Devs ind. Microbiol.* **14,** 189.

Brenner, D. J. (1970). Deoxyribonucleic acid divergence in Enterobacteriaceae. *Devs ind. Microbiol.* **11,** 139.

Brenner, D. J., Fanning, G. R., Johnson, K. E., Citarella, R. V. & Falkow, S. (1969). Polynucleiotide sequence relationships among members of Enterobacteriaceae. *J. Bact.* **98,** 637.

Britten, R. J. & Kohne, D. E. (1966). Nucleotide sequence repetition in DNA. *Carnegie Inst.* (*Washington*) *Yearbook* **65,** 78.

Buchanan, R. E., Holt, J. G. & Lessel, E. F. (Eds). (1966). *Index Bergeyana.* Baltimore: Williams and Wilkins.

Clark, J. E. & Brownell, G. H. (1972). Genophore homologies among compatible nocardiae. *J. Bact.* **109,** 720.

De Ley, J., Cattoir, H. & Reynaerts, A. (1970). The quantitative measurement of DNA hybridization from renaturation rates. *Eur. J. Biochem.* **12,** 133.

De Ley, J. & Tijtgat, R. (1970). Evaluation of membrane filter methods for DNA-DNA hybridization. *Antonie van Leeuwenhoek* **36,** 461.

Denhardt, D. T. (1966). A membrane-filter technique for the detection of complementary DNA. *Biochem. biophys. Res. Commun.* **23,** 641.

Enquist, L. W. & Bradley, S. G. (1970). Nucleotide divergence in deoxyribonucleic acids of actinomycetes. *Adv. Fronts Pl. Sci.* **25,** 53.

Erikson, R. L. & Szybalski, W. (1964). The Cs_2SO_4 equilibrium density gradient and its application for the study of T-even phage DNA : glycosylation and replication. *Virology* **22,** 111.

Flamm, W. G., Bond, H. E. & Burr, H. E. (1966). Density-gradient centrifugation of DNA in a fixed angle rotor, a higher order of resolution. *Biochim. biophys. Acta* **129,** 310.

Flavell, R. A., Borst, P. & Birfelder, E.J. (1974). DNA-DNA hybridization on nitrocellulose filters. 2. Concatenation effects. *Eur. J. Biochem.* **47,** 545.

Gasser, F. & Mandel, M. (1968). Deoxyribonucleic acid base composition of the genus *Lactobacillus. J. Bact.* **96,** 580.

Gillespie, D. & Spiegelman, S. (1965). A quantitative assay for DNA/RNA hybrids with DNA immobilized on a membrane. *J. molec. Biol.* **12,** 829.

Gillis, M., De Ley, J. & DeCleene, M. (1970). The determination of molecular weight of bacterial genome DNA from renaturation rates. *Eur. J. Biochem.* **12,** 143.

Goodfellow, M., Fleming, A. & Sackin, M. J. (1972). Numerical classification of "*Mycobacterium*" *rhodochrous* and Runyon's group IV mycobacteria. *Int. J. Syst. Bact.* **22,** 81.

Goodfellow, M., Lind, A., Mordarska, H., Pattyn, S. & Tsukamura, M. (1974). A co-operative numerical analysis of cultures considered to belong to the "*rhodochrous*" taxon. *J. gen. Microbiol.* **85,** 291.

Gordon, R. E. (1966). Some strains in search of a genus—*Corynebacterium, Mycobacterium, Nocardia* or what? *J. gen. Microbiol.* **43,** 329.

Hatcher, D. W. & Goldstein, G. (1969). Improved methods for determination of RNA and DNA. *Anal. Biochem.* **31,** 42.

Hill, E. B., Wayne, L. G. & Gross, W. M. (1972). Purification of mycobacterial deoxyribonucleic acid. *J. Bact.* **112,** 1033.

Huang, P. C. & Rosenberg, E. (1966). Determination of DNA base composition via depurination. *Anal. Biochem.* **16,** 107.

Johnson, J. L. & Ordal, E. J. (1968). Deoxyribonucleic acid homology in bacterial taxonomy: effect of incubation temperature on reaction specificity. *J. Bact.* **95,** 893.

Kanetsuna, F. & Bartoli, A. (1972). A simple chemical method to differentiate *Mycobacterium* from *Nocardia. J. gen. Microbiol.* **70,** 209.

Kimura, M. (1968). Genetic variability maintained in a finite population due to mutational production of neutral and nearly neutral isoalleles. *Genet. Res.* **11,** 247.

Kimura, M. & Ohta, T. (1971). Protein polymorphism as a phase of molecular evolution. *Nature, Lond.* **229,** 467.

King, J. L. & Jukes, T. H. (1969). Non-Darwinian evolution. *Science, N.Y.* **164,** 788.

Kirby, K. S. (1964). Isolation and fractionation of nucleic acids. *Prog. Nucleic Acid Res.* **3,** 1.

Kopecko, D. & Punch, J. D. (1971). Regulation of R-factor replication in *Proteus mirabilis. Ann. N.Y. Acad. Sci.* **182,** 201.

Lanéelle, M. A. (1966). Sur la relation entre des cétones á haut poides moléculaire isolée de *Nocardia brasiliensis* et les acides nocardomycoliques. *C.r. hebd. Séanc. Acad. Sci., Paris* **263,** 506.

Lechevalier, M. P., Horan, A. C. & Lechevalier, H. (1971). Lipid composition in the classification of nocardiae and mycobacteria. *J. Bact.* **105,** 313.

Legault-Demare, J., Desseaux, B., Heyman, T., Seror, S. & Ress, G. P. (1967). Studies on hybrid molecules of nucleic acids. I. DNA-RNA hybrids on nitrocellulose filters. *Biochem. biophys. Res. Commun.* **28,** 550.

Lowry, O. H., Rosebrough, N. J., Farr, A. L. & Randall, R. J. (1951). Protein measurement with Folin phenol reagent. *J. biol. Chem.* **193,** 265.

Mandel, M. (1969). New approaches to bacterial taxonomy. *A. Rev. Microbiol.* **23,** 239.

Mandel, M., Igambi, J., Bergendahl, J., Dodson, M. L., Jr. & Scheltgen, E. (1970). Correlation of melting temperature and cesium chloride buoyant density of bacterial deoxyribonucleic acid. *J. Bact.* **101,** 333.

Marmur, J. (1961). A procedure for the isolation of deoxyribonucleic acid from microorganisms. *J. molec. Biol.* **3,** 208.

Marmur, J. & Doty, P. (1962). Determination of base composition of deoxyribonucleic acid from its thermal denaturation temperature. *J. molec. Biol.* **5,** 109.

Marmur, J., Rownd, R. & Schildkraut, C. L. (1963). Denaturation and renaturation of deoxyribonucleic acid. *Prog. Nucleic. Acid Res.* **1,** 231.

McCarthy, B. J. (1967). Arrangement of base sequences in deoxyribonucleic acid. *Bact. Rev.* **31,** 215.

McConaughy, B. L. & McCarthy, B. J. (1967). The interaction of oligodeoxyribonucleotides with denatured DNA. *Biochem. biophys. Acta* **149,** 180.

Monson, A. M., Bradley, S. G., Enquist, L. W. & Cruces, G. (1969). Genetic homologies among *Streptomyces violaceoruber* strains. *J. Bact.* **99,** 702.

Moore, R. L. & McCarthy, B. J. (1967). Comparative study of ribosomal nucleic acid cistrons in enterobacteria and myxobacteria. *J. Bact.* **94,** 1066.

Mordarska, H., Mordarski, M. & Goodfellow, M. (1972). Chemotaxonomic characters and classification of some nocardioform bacteria. *J. gen. Microbiol.* **71,** 77.

Okanishi, M. & Gregory, K. F. (1970). Methods for the determination of deoxyribonucleic acid homologies in *Streptomyces. J. Bact.* **104,** 1086.

Prauser, J., Lechevalier, M. P. & Lechevalier, H. (1970). Description of *Oerskovia* gen. n. to harbor Ørskov's motile *Nocardia. Appl. Microbiol.* **19,** 534.

Ridell, M. (1974). Serological study of nocardiae and mycobacteria by using "*Mycobacterium*" *pellegrino* and *Nocardia corallina* precipitation reference systems. *Int. J. Ssyt. Bact.* **24,** 64.

Saito, H. & Miura, K. I. (1963). Preparation of transforming deoxyribonucleic acid by phenol. *Biochim. biophys. Acta* **72,** 619.

Schaal, K. P. (1974). Differentiation of strains thought to belong to the species *Nocardia asteroides* by biochemical and serological methods. *Proc. Intern. Conf. Biol. Nocardiae, Merida, Venezuela* p. 12.

Schildkraut, C. L., Marmur, J. & Doty, P. (1962). Determination of the base composition of deoxyribonucleic acid from its buoyant density in CsCl. *J. molec. Biol.* **4,** 430.

Seidler, R. J. & Mandel, M. (1971). Quantitative aspects of deoxyribonucleic acid renaturation: base composition, state of chromosome replication and polynucleotide homologies. *J. Bact.* **106,** 608.

Sneath, P. H. A. & Skerman, V. B. D. (1966). A list of type and reference strains of bacteria. *Int. J. Syst. Bact.* **16,** 1.

Tewfik, E. M. & Bradley, S. G. (1967). Characterization of deoxyribonucleic acids from streptomycetes and nocardiae. *J. Bact.* **94,** 1994.

Tewfik, E., Bradley, S. G., Kuroda S. & Wu, R. Y. (1968). Studies on deoxyribonucleic acids from streptomyces and nocardiae. *Devs ind. Microbiol.* **9,** 242.

Tsukamura, M. (1971). Proposal of a new genus, *Gordona,* for slightly acid-fast organisms occurring in sputa of patient with pulmonary disease and in soil. *J. gen. Microbiol.* **68,** 15.

Warnaar, S. O. & Cohen, J. A. (1966). A quantitative assay for DNA-DNA hybrids using membrane filters. *Biochem. biophys. Res. Commun.* **24,** 554.

Wetmur, J. G. & Davidson, N. (1968). Kinetics of renaturation of DNA. *J. molec. Biol.* **31,** 349.

Wright, S. (1966). Polyallelic random drift in relation to evolution. *Proc. natn. Acad. Sci. U.S.A.* **55,** 1074.

Yamada, K. & Komagata, K. (1970). Taxonomic studies on coryneform bacteria III. DNA base composition of coryneform bacteria. *J. gen. appl. Microbiol, Tokyo* **16,** 215.

Yamaguchi, T. (1967). Similarity in DNA of various morphologically distinct actinomycetes. *J. gen. appl. Microbiol., Tokyo* **13,** 63.

13. The Ecology of Nocardioform Actinomycetes

T. CROSS and T. J. ROWBOTHAM

Postgraduate School of Studies in Biological Sciences, University of Bradford, Bradford BD7 1DP, England

E. N. MISHUSTIN and E. Z. TEPPER

Institute of Microbiology, USSR Academy of Sciences, Moscow, USSR

F. ANTOINE-PORTAELS

Prince Leopold Institute of Tropical Medicine, B-2000 Antwerp, Belgium

and

K. P. SCHAAL and H. BICKENBACH

Institute of Hygiene, University of Cologne, Cologne, Federal Republic of Germany

Contents

1. Introduction

Ecology is the study of the relationships of organisms to their environment. Thus the microbial ecologist must study environmental factors and their effects on the inter-relationships between all organisms, considering not only the microorganisms but also the influence of associated plants and animals. In order to simplify the situation, workers have often studied single species in a few well defined habitats or concentrated on particular substrates and attempted an investigation of the microbes found on or within them. The complete picture relating all organisms and their environments within an ecosystem is undoubtedly extremely complex and can be likened to a series of overlapping jigsaw puzzles composed of many parts fitting together to reveal the relationships, activities and roles of their minutiae. The ecologist must retrieve and decipher the individual pieces in order to place them into the context of a picture. The nocardioform actinomycetes are details within larger pictures or systems and those workers interested in the ecology of these organisms have to provide information on their numbers, distribution, activities, growth, survival and dissemination before their role in microbial processes within ecosystems can be elucidated.

Progress in understanding the ecology of nocardioform actinomycetes has been painfully slow and there have been few detailed studies. The work has been hampered by the confused classification of the organisms included within this very heterogeneous group, and there has been a predeliction among actinomycetologists for studying the ecology of streptomycetes.

Brock (1966) has clearly summarized the areas which require study to provide the information that is necessary for some insight into the properties and factors that ultimately determine the survival and ability of species to grow in natural habitats. One must attempt to define the range of environments in which they live and the size of natural populations. Pure culture studies will then give information on their growth rates, their ability to utilize a variety of nutrients and survive exposure to normal and extreme physical environments. The introduction of isolated organisms into natural environments can be used to determine whether growth occurs or whether numbers decline as a result of unsuitable conditions, competition from other microorganisms, or the lack of a normal food chain involving both associative microbes and a supply of primary substrate. Finally one must discover the methods of dispersal, the conditions necessary for the reactivation of dormant propagules and the susceptibility of a host to a potential pathogen. It might then be possible to predict the role of such organisms in natural habitats or understand at least some of their properties which make them important organisms in food chains, in the biodegradation of useful or discarded manufactured materials, or as pathogens.

We are only just beginning to determine the species and numbers of nocardiae in soil and water. The pathogen *Nocardia asteroides* has been studied extensively in pure culture, but while the information available on its properties gives some insight into its ecology, we have still to demonstrate that growth does occur in soil. We have most information on the numbers, properties and growth cycle of the unnamed nocardioform, Lspi, a member of the "*rhodochrous*" complex. There is little information on the majority of other nocardioform organisms. We do now have useful classifications, some novel isolation methods and considerable information on the ecology and properties of other bacteria and fungi which might be applicable to the nocardioforms. Although the present situation is unsatisfactory, we anticipate several years of exciting new developments in this field.

Much of the information on numbers and distribution has to be abstracted from the results of surveys which have attempted to count the total number of actinomycetes in a given substrate, or were primarily concerned with the streptomycete component. This emphasis on the streptomycetes is understandable for these organisms have proved to be the source of many valuable antibiotics and many microbiologists have searched for media and methods which would facilitate their isolation. The successful methods have come to be regarded as suitable for enumerating all actinomycetes present in a substrate. Hence most of the published lists of genera and species isolated result from the use of media such as starch-casein, egg albumin, glycerol-asparagine and tap-water agars which were designed to isolate streptomycetes.

Reference to such lists would suggest that the nocardioform actinomycetes are a minor component of the actinomycete flora in soil and water samples. In one such survey (Lechevalier & Lechevalier, 1967), 5000 actinomycetes were isolated and classified into genera on the basis of colony morphology. 95·34% proved to be streptomycetes and only 1·98% were recognized as *Nocardia* species; the few other isolates were included in the genera *Micromonospora* and *Antinomadura.* McClung (1949) had earlier attempted to isolate *Nocardia* species from 23 soil samples collected in the United States. Although the total number of actinomycetes in the soils ranged from 300 to 4600 000/g of dry soil he was only able to isolate 61 *Nocardia* and the maximum recovery from any one sample was nine strains. Szabo (1974) isolated 3000 actinomycetes from the A horizon of a forest rendzina soil using principally casein-glucose or glucose-asparagine agars. With the exception of a single strain identified as *Nocardia asteroides*, all were considered to be members of the genus *Streptomyces*.

It has become increasingly evident in recent years that existing methods favour the isolation of streptomycetes and even discourage the growth of nocardioform actinomycetes. *Nocardia* and members of the *rhodochrous* complex do not hydrolyse chitin and few utilize starch and casein (Goodfellow, 1971) or egg albumin (Mishustin & Tepper). Agar media containing such constituents will give streptomycete colonies several millimetres in diameter after 7–14 days incubation but the pin point nocardioform colonies are only visible with the aid of a microscope. These small colonies of branching and fragmenting mycelium appear to depend upon trace amounts of nutrients in the isolation medium, the inoculum and metabolites of adjacent bacterial colonies. Such small colonies may also be obscured by surface films of motile bacteria which often appear on plates inoculated with water samples (Willoughby, 1969).

Recently, a number of methods favouring the isolation of nocardioform actinomycetes have been described. These involve one or more of the following: (a) pretreatment of samples to reduce the numbers of bacteria, fungi and associated actinomycetes, (b) media which discourage the growth of non-nocardioform organisms, and (c) the use of media constituents which favour the growth of nocardioforms so giving easily recognizable colonies. Such methods have already given valuable information on the occurrence and numbers of nocardioforms in various habitats and will be discussed as a preliminary to their properties and possible roles.

2. Nocardioform Actinomycetes Associated with Plants

The numbers, species and activity of nocardioform actinomycetes in soil depends on many factors such as soil type, depth, water content, aeration, pH, organic matter content, climate, season and agricultural treatments etc. These factors are so closely related that they are difficult to separate. The conditions in the microenvironments where actinomycete growth occurs may also differ from the soil considered as a whole. Relatively little information on the soil habitat is given in those papers describing the isolation of nocardioforms from soil.

An important site for microbial activity in soil is the rhizosphere where many microbial/plant interactions occur. Webley, Eastwood & Gimingham (1952) found there were more nocardiae in the rhizosphere of plants stabilizing sand dunes than in adjacent sand. Watson & Williams (1974) also found that certain *Nocardia* strains were associated with the roots of dune plants such as *Ammophila arenaria* and *Agropyron junceiforme.* These nocardioforms were virtually absent from non-rhizosphere soil and soil easily dislodged from the roots by shaking. One of the strains isolated, A78, was included by Goodfellow (1971) in his numerical taxonomic study of nocardioforms and was recovered in the *rhodochrous* cluster, subgroup 14G. A second strain C25, was able to colonise sterile pieces of old *Ammophila* roots placed in dune sand. Bacteria, fungi and other actinomycetes were competing organisms but C25 was abundant after 3 days. Numbers decreased after 14 days but the pattern suggested that this strain and possibly other nocardioforms may be active in the colonisation and breakdown of plant root material.

A second important area of terrestrial microbial activity is leaf litter but the role of actinomycetes in litter breakdown has received little attention (Goodfellow & Cross, 1974). A blueberry pathogen, *Nocardia vaccinii,* is the only reported plant pathogenic species and is thought to persist in leaf mould (Denmaree & Smith, 1952). Most nocardioforms cannot hydrolyse cellulose or starch and would thus seem unlikely to be active in litter breakdown. However, some species degrade humic acids and many species can attack hydrocarbons suggesting that they may assist in the degradation of certain plant materials such as leaf surface waxes.

An unidentified mesophilic *Nocardia* species was isolated in high numbers (1×10^6/g dry wt) from samples of self-heated bagasse, the squashed, chopped fibre left after sugar has been extracted from sugar cane and used for the manufacture of pulp, paper and particle board (Lacey 1974). These isolates have now been included in the new genus *Saccharopolyspora* (Lacey & Goodfellow, 1974).

3. Symbiotic Relationships Between Nocardiae and Insects

A wide variety of blood sucking arthropods has been reported to contain symbiotic microorganisms (Wigglesworth, 1952). Members of the strictly haematophagous reduviid sub-family Triatominae, in particular the species *Rhodnius prolixus* Stål, have been studied in considerable detail and there is now ample evidence for a mutualistic relationship between various species and bacteria found within their alimentary tracts.

The first observation of a microorganism in *R. prolixus* was reported by Duncan (1926) who found a Gram-positive bacillus in the gut contents of all the insects he examined. Erikson (1935) cultured this microorganism, described its growth on various media and named it *Actinomyces rhodnii*: later to be reclassified as *Nocardia rhodnii* by Waksman & Henrici (1943). Wigglesworth (1936) found that the organism consistently occurred within the midgut of *R. prolixus* nymphs being restricted to intercellular crypts formed by epithelial folds. Following a blood meal the microorganisms were expelled into the gut lumen where they became mixed with ingested blood and later expelled in faecal material. The newly hatched nymphs acquire the bacterium from their contaminated egg chorion or the excreta of older confrères (Brecher & Wigglesworth, 1944).

A number of investigators have now cultured the intestinal flora of *R. prolixus* and have, without exception, isolated organisms which they identified as *Nocardia* sp. or *Nocardia rhodnii*. Goodchild (1955), Baines (1956), Mühlpfordt (1959) and Harington (1960*a*) isolated *N. rhodnii* in pure culture from insects reared in their laboratories. Bewig & Schwartz (1955, 1956) isolated *N. rhodnii* from all of the *R. prolixus* specimens they tested, but also found a Gram-positive coccus in one half of their bugs. Gumpert & Schwartz (1962) isolated *N. rhodnii* and *Streptococcus faecalis*, and recovered a *Corynebacterium* from a few specimens. Lake & Friend (1968*a*) cultured *N. rhodnii* from the gut of 197 *R. prolixus* specimens; two bugs also gave a green fluorescent bacterium tentatively identified as a *Pseudomonas* species. In contrast, Cavanagh & Marsden (1969) were unable to find a consistent bacterial flora in *R. prolixus* or the other species of Triatominae examined. Of the 17 specimens of *R. prolixus* cultured, 10 yielded a *Nocardia* species, 4 gave *Mycobacterium fortuitum*, 2 a *Corynebacterium* species, 5 yielded a *Streptococcus/Micrococcus species*, 3 yielded *Streptococcus faecalis* and 2 gave *Escherichia coli*. This situation of a consistent association of *N. rhodnii* with *R. prolixus* reported by most investigators does not appear to hold true for other triatomine species. Bacteria resembling *N. rhodnii* have been isolated occasionally but the majority of insects contained a wide variety of bacteria.

The consistency with which Wigglesworth (1936) found *N. rhodnii* within the alimentary tract of *R. prolixus* led him to postulate a mutualistic ("symbiotic") relationship between the two organisms. This hypothesis has been substantiated by a number of investigations which demonstrated that germ-free *R. prolixus* would not develop beyond the 3rd, 4th or 5th instar stage and rarely reached the adult stage when the host animal was a rabbit or guinea pig (Brecher & Wigglesworth, 1944; Baines, 1956; Harington 1960*a*; Lake & Friend, 1968*b*; Nyirady, 1973; Auden, 1974). The reinfection of *Nocardia* free *R. prolixus* with *N.rhodnii* at any stage of development allowed the bugs to resume their normal developmental pattern and become sexually mature (Brecher & Wigglesworth, 1944; Bewig & Schwartz, 1955; Baines, 1956). A number of other bacteria, not normally associated with *R. prolixus*, were able to perform the same function as *N. rhodnii*. Gumpert (1962*a,b*) reported these bacteria to be *Nocardia asteroides, N. opaca, N. corallina, N. flava, N. rubra* as well as species of *Corynebacterium, Mycobacterium* and a strain of *Pseudomonas*. In contrast, *R. prolixus* reared germ-free and fed on germ-free mice completed their normal developmental stages and the adults laid fertile eggs (Nyirady, 1973; Auden, 1974).

The blood of some animals does not contain sufficient concentrations of certain nutrients required by *R. prolixus* nymphs to allow normal development and the critical metabolites are supplied by the symbiont *N. rhodnii*. It appears that the usual association of *N. rhodnii* with *R. prolixus* is related to the low degree of host specificity in the bug. Opinion is divided on the nature of the critical metabolites supplied; the suggested possibilities include folic acid, pantothenic acid, pyridoxine, thiamine, nicotinic acid and riboflavin (Harington, 1960*b*; Lake & Friend, 1968*b*).

The various identifications of the symbiont as *N. rhodnii* do not appear to have been based upon critical taxonomic criteria or used current classifications. For example, Lake and Friend (1968*a*) employed media such as Dorset's Egg Medium, blood and Sabouraud's agars described by Waksman (1919) and obtained similar growths and pigmentation to these described by Erikson (1935). Gram-strained smears "revealed only a single species of Gram-positive micro-organism" which exhibited branching when young and spherical forms in older cultures. They did, however, notice some variation in the pigmentation of isolates derived from the gut contents of different insects when grown on dextrose agar and glycerin agar. A strain of *N. rhodnii* included by Goodfellow (1971) in his numerical taxonomic study of nocardiae was recovered in the *rhodochrous* complex and one must now question the existence of *N. rhodnii* as a distinct species. It could be argued that many soil inhabiting members of the *rhodochrous* complex would also show the morphology and pigmentation described by Erikson for *N. rhodnii* when grown on media such as nutrient, blood or Czapek's agars. Various investigators must also have been tempted to name a nocardioform actinomycete isolated from the alimentary tract of *R. prolixus* as *N. rhodnii.* The observations of Gumpert (1962*a,b*), that other species of nocardiae including those currently included in the *rhodochrous* complex could perform the same functions as *N. rhodnii* in germ-free insects, would support the view that the taxon *N. rhodnii* is an aggregate of strains (species?) capable of forming a symbiotic and stable relationship with one particular insect. There must be conditions in the alimentary tract which selects for these particular strains and favours those capable of supplying the necessary nutrients. Further detailed taxonomic and ecological studies on the symbiont are obviously required to supplement the extensive work largely directed towards the insect and its developmental stages.

4. The Distribution and Number of Saprophytic Nocardioform Actinomycetes in Soil

That nocardioforms may be far more numerous in soils than had previously been suspected has been emphatically demonstrated by

Mishustin & Tepper. They found that Winogradsky's nitrite agar enhanced the nocardioform character of the colonies, so aiding recognition, and inhibited the growth of bacteria and other actinomycetes which would have made counting difficult, even with the aid of a microscope. Nocardioforms were present in all soil and

Table 1

Number of nocardioforms recovered from different soil types collected in the USSR (Mishustin & Tepper)

Zone	Soil type and condition	Number of nocardioforms/g wt dry Soil ($\times 10^3$)	Number of nocardioforms/g wt dry humus ($\times 10^6$)	% of nocardioforms to total bacteria recovered
Taiga	Peat-bog:			
	virgin	192	0·3	1·9
	cultivated	904	2·9	11·9
	Podzolic:			
	forest	12	0·2	13·0
	Soddy-podzolic:			
	meadow	79	4·3	9·4
	ploughland	196	12·9	7·5
Steppe	Chernozem:			
	virgin	1490	13·8	6·3
	forest	612	6·1	13·2
	ploughland	600	7·2	2·9
Dry steppe	Light chestnut:			
	ploughland	1424	72·0	16·0
	Meadow cinnamonic:			
	irrigated ploughland	3060	117·6	25·0
Subtropic	Krasnozem:			
	forest	91	1·5	11·4
	tea plantation	42	0·9	10·8

Plates of Winogradsky's nitrite medium spread with diluted soil suspensions, incubated at 25°C and counted after 7 days with the aid of a microscope. Soil samples taken from A horizon in virgin soils and from the 0–20 cm region in cultivated soils.

humus samples examined (Table 1) but the proportion of nocardioforms to the other bacteria varied in the different soils. Seasonal changes in numbers were recorded; maximum numbers were detected in spring and lower numbers in summer when soil moisture levels declined to 9·1%. Southern soils, apart from acidic krasnozem soils, contained higher numbers of all microorganisms.

The survey of soils also suggested that the nocardioforms increased where there was active decomposition of humus and where irrigation during the dry summer months maintained microbiological activity (Table 2). Numbers were lower in peat bog

Table 2

Percentage of nocardioforms to other bacteria recovered from various soil types (Mishustin & Tepper)

Soil type	% of nocardioforms
Virgin peat bog	1·9
Cultivated peat bog	11·9
Non-irrigated chestnut soil	16·0
Irrigated meadow cinnamic soil	25·0
Soddy podzolic soil—fallow	28·0
Soddy podzolic soil—continuous planting	11·0

and chernozem soils where there was humus accumulation. Unfertilized soil of the Timiryazev Agricultural Academy near Moscow, which had remained fallow for over 50 years, contained a higher percentage of nocardioforms than soil from adjacent plots which had been continuously planted with rye. In continuously fallow soil the only source of organic matter was humus and Mishustin & Tepper assert that intensive decomposition of humus in soil is correlated with the growth of nocardioforms, in particular the species *N. corallina* and *N. rubra.* This survey also showed that some species were more numerous in certain soils. The soddy podzolic soils were richer in *N. mucosum* and *N. corallina,* chernozems in *N. symbiotica,* and the chestnut soils contained *N. citrea* and *N. symbiotica* in higher numbers.

The very high numbers of nocardioforms recorded by Mishustin and Tepper for soils of the Soviet Union are surprising. Few workers have recorded numbers of this order in surveys of other soils which suggests that we may have grossly underestimated this component of the soil flora, probably as a result of using unsuitable isolation methods. The nitrite isolation medium must now be evaluated by other workers to confirm their findings and the isolated species compared with species described by other workers. The confused taxonomy of species within this group has already been mentioned and it seems likely that many of their strains could be classified within the *rhodochrous* complex and some may prove to be similar to the nocardioform Lspi which will be discussed later in this review.

Oerskovia turbata, a nocardioform actinomycete which fragments into motile elements, was isolated by Øerskov from Danish soil and later named *Nocardia turbata* by Erikson (1954). The species was later transferred to the then monotypic genus *Oerskovia* by Prauser, Lechevalier & Lechevalier (1970). A second species *O. xanthineolytica* has now been described together with a group of non-motile *Oerskovia*-like organisms (NMO's) which were isolated from soil and decaying organic matter by Lechevalier (1972). *Nocardia cellulans* isolated from chalk downland soil (Metcalf & Brown, 1957) is similar to *Oerskovia turbata* (Goodfellow, 1971). *N.cellulans* and the NMO's are cellulolytic and may assist in the degradation of plant material in soil.

In 1969, Gledhill & Casida described a new genus *Agromyces* to accommodate the species *A. ramosus*, a nocardioform previously isolated from various soils in the U.S.A. by Casida (1965). In these soils *A. ramosus* was the numerically dominant bacterium and, by use of a modified dilution-to-extinction isolation procedure in an enriched heart infusion broth, they estimated that the organism could number approximately 10^9/g of soil. The species can also be isolated in lower numbers by conventional dilution plating methods on nutritionally minimal media with shorter incubation times. Two media can be used, either water agar containing 0·2 ml/plate of culture filtrate of *Arthrobacter globiformis*, or a modified Burk *Azotobacter* medium that lacks a carbon source and contains only a minimal nitrogen level (Labeda, Hunt & Casida, 1974). *Agromyces* is a catalase-negative organism producing a true branching mycelium that fragments into coccoid and diphtheroid forms. The colonies on agar are very small, their size usually ranges from 0·1 to 0·5 mm in diameter, and have to be counted with the aid of a microscope. It is surprising that we know so little about the role of this organism in soil when it would appear to be present in such high numbers.

Species belonging to the genus *Actinomadura* have been isolated in low numbers from Japanese soils by Nonomura & Ohara (1971) who first dried the soil in a hot air oven to reduce the numbers of competing bacteria on the isolation plates. Higher numbers have now been recorded in soils from the Soviet Union by the use of selective media containing antibiotics (Lavrova, Preobrazhenskaya & Sveshnikova, 1972; Preobrazhenskaya & Lavrova, 1974). Almost 100% of the actinomycetes isolated on the Gause No. 2 organic medium proved to be streptomycetes but when the antibiotic rubomycin was added as a selective agent, 24·1% of the actinomycetes recovered were classified in the genus *Actinomadura.* When streptomycin was included in the same isolation medium, 11·6% of the actinomycetes isolated were identified as *Actinomadura* and 20·7% belonged to other nocardioform genera. These findings again support the suggestion that many of the currently used isolation media favour the growth of streptomycetes and do not reveal the nocardioform component of soils.

5. The Distribution and Number of Pathogenic Nocardiae in Soil

The distribution of the pathogenic species *Nocardia asteroides, N. caviae* and *N. brasiliensis* has attracted considerable attention over the years. It is generally thought that soil is the reservoir of these organisms because most cutaneous infections in tropical countries occur on the feet (actinomycetoma pedis) or on the back through carrying soil-contaminated sacks (Gonzàlez-Ochoa, 1962). The ability of these organisms to attack *n*-paraffins forms the basis of the long established method of paraffin wax baiting for their isolation from soil. Söhngen (1913) had used wax coated pebbles, half submerged in an organic carbon-free mineral salts medium to isolate acid-fast bacteria from soil, animal excreta and stagnant water. Using this technique coupled with the higher incubation temperature of 47·5°C, Gordon & Hagan (1937) were able to isolate *Nocardia asteroides* from soil. McClung (1960) used an alternative mineral salts medium and wax coated glass rods instead of pebbles and demonstrated the presence of this species in soil samples from three States of the U.S.A. His method has subsequently been used by several other workers to isolate *N. asteroides* from soil, and less frequently the species of *N. brasiliensis* and *N. caviae* (Table 3).

Table 3
Isolations of pathogenic Nocardia *spp. from soil with the paraffin wax baiting technique and direct plating methods*

Species isolated	Soil	Frequency	References
N. asteroides and others	U.S.A.	64 strains from 102 samples	McClung, 1960
N. asteroides	India	4 strains from 27 samples	Kurup, Sandhu & Damodaran, 1964
N. asteroides and *N. caviae*	India	32 strains from 180 soil samples	Kurup & Sandhu, 1965
N. brasiliensis	India	1 strain	Kurup *et al.*, 1967
N. asteroides *N. brasiliensis* *N. caviae* & others	India	104 strains from 134 soil samples	Kumar & Mohapatra, 1968
N. brasiliensis	Mexico	6 strains from 21 soil samples	González-Ochoa, 1962
Nocardia sp.	Philippines	ND	Reyes, 1963
N. asteroides	U.S.A.	ND	Pier, Willers & Mejia, 1961
Nocardia spp.	Australia	ND	Rodda, 1964
N. asteroides	U.S.A.	21 strains from 72 soil samples	Kurup & Schmitt, 1971

ND = not determined.

Portaels has recently conducted an extensive survey of the nocardiae in soil, mud, grass and water samples collected in the Bas-Zaire region of Central Africa. Water samples (50 ml) were centrifuged to give a sediment which was resuspended in a few ml of distilled water for testing. Grass samples (2–3 g) were pulverized with sand and a few ml of distilled water to give a supernatant for testing. These samples of soil and mud were added directly to bottles of McClung's mineral salts medium containing wax coated glass rods, and incubated at 28°C. Growth on the waxed rods was homogenized by shaking with glass beads before streaking on 1% (w/v) glucose-nutrient agar. Colonies bearing aerial mycelium were then transferred to slopes of Lowenstein-Jensen medium for identification.

The 400 samples collected from different ecological areas yielded 93 nocardiae (Table 4). Sixty-four strains were identified as *N. asteroides* although three strains proved to be sensitive to lysozyme. The three strains identified as *N. caviae* originated from

Table 4
Nocardia *strains isolated from the different ecological areas of Bas-Zaire (Portaels)*

Site	Number of strains isolated from samples* of: soil	mud	grass	water	Totals
Kinshasa	10	2	1	1	14
Sugar plantation Kwilu-Ngongo	24	NS	1	NS	25
Savanna between Kinshasa and Matadi	15	10	0	2	27
Forest Mayumbe	8	13	1	4	26
Coast Moanda	1	NS	NS	0	1
Totals	58	25	3	7	93

* 25 samples of soil, mud, grass or water were tested from each site except where indicated: NS = no samples.

savanna soil and a single strain of *N. brasiliensis* came from coastal sand. Four strains were placed in the *rhodochrous* complex (three from water and mud samples) and the remaining strains could not be identified with certainty. The sugar plantation of Kwilu-Ngongo gave the greatest number of nocardioforms and the coastal environments least. Mud and soil samples also appeared to contain more nocardioforms than the water and grass samples. No nocardiae were isolated from sea water.

Schaal and Bickenbach are conducting a similar survey of 800 soil samples collected in various regions of West Germany. They have now reported their results on the first 300 samples using the paraffin wax baiting technique. Growth of nocardiae appearing on the bait, mostly at the surface of the broth, was scraped off and suspended in sterile tap water. The suspension was then streaked on the surface of Diagnostic Sensitivity Test agar (DST, Oxoid) and plates of a selective agar containing tellurite (Schaal, 1972). After 7 and 14 days incubation at 22° and 35°C the plates were examined for typical colonies which were subcultured on DST agar plates and identified.

Actinomycetes with a cell wall Type IV were recovered from 83·3% of the samples tested (Table 5). No striking differences in

Table 5

Prevalence of actinomycetes with a Type IV cell wall in soil samples collected in various regions of West Germany (Schaal & Bickenbach)

Region	Number of samples examined	Number of isolates
Aachen	9	4
Köln	71	59
Bergisches Land	13	10
Münster	30	25
Trier	30	28
Kassel	40	41
München	7	9
Hamburg	40	36
Ruhr river	42	35
Frankfurt	18	12
Total	300	259

the frequency of isolation from various parts of Germany could be observed. The majority of the strains that could be identified with certainty proved to be *N. asteroides*, a large group of strains could be included in the *rhodochrous* complex but the majority could not be identified to the species level (Table 6). *N. brasiliensis* was found

Table 6

Classification of the 259 nocardiae isolated from West German soils (Schaal & Bickenbach)

Species	Number of strains	Percentage
N. asteroides, subgroup A	23	8·9
N. asteroides, subgroup B	12	4·6
N. asteroides, subgroup C	0	0
N. brasiliensis	3	1·2
N. caviae	0	0
N. rubra	10	3·9
rhodochrous group	64	24·7
Unidentified isolates	147	56·7
Total	259	100

rarely in German soils. This is in good agreement with the findings of other workers (Gonzáles-Ochoa, 1962; Reyes, 1963; Kurup *et al.*, 1967; Kurup, Randhawa & Sandhu, 1968). Karup and co-workers detected *N. brasiliensis* only once during the examination of 504 soils from India. The widespread occurrence of strains belonging to the *rhodochrous* complex in German soils is in accordance with the view that this group of organisms is widely distributed in soils all over the world.

Schaal and Bickenbach noticed some correlation between soil type and the distribution of *N. asteroides* (Table 7). Forest soils

Table 7

Prevalence of Nocardia asteroides *(subgroups A and B) and* N. brasiliensis *in various soils of West Germany (Schaal & Bickenbach)*

Soil type	Number of soil samples examined	*N. asteroides* strains		*N. brasiliensis* strains
		Subgroup A	Subgroup B	
Field soils	132	13(9·8%)*	5(3·8%)	2(1·5%)
Forest soils	62	1(1·6%)	1(1·6%)	0
Garden soils	30	3(10·0%)	1(3·3%)	1(3·3%)
Grassland soils	28	2(7·1%)	2(7·1%)	0
Barren soils	48	4(8·3%)	3(6·3%)	0
Total	300	23(7·7%)	12(4·0%)	3(1·0%)

* Percentage of *Nocardia asteroides* subgroup A strains to total nocardiae isolated (Table 6).

contained few pathogenic nocardiae but only slight differences were noticed between the soils from other sources. There is a surprising similarity in the occurrence of *N. asteroides* in the soils of India and West Germany. In India, *N. asteroides* was detected in 10·7% of the soils investigated (Kurup, Randhawa & Sandhu, 1968) while in Germany the recovery was 11·7%. There would appear to be a higher incidence of *Nocardia asteroides* in African and North American soils because Portaels isolated 40 strains of this species from 125 Bas-Zaire soil samples and McClung (1960) isolated 48 strains from 102 soil samples collected in the U.S.A. There appeared to be no relationship between the incidence of the pathogenic nocardiae and soil pH, water content, organic matter or total nitrogen content, supporting the conclusions of McClung (1960). This was in contrast to the pH values of soils containing *N. rubra* which exhibited, without any exception, alkaline reactions of pH 7·5 or more.

The majority of these extensive studies on the distribution of pathogenic nocardiae in soils have relied almost exclusively on the paraffin baiting technique. This technique is of little value for quantitative studies and merely indicates that the organisms are present in a sample. Many other bacteria and fungi can attack paraffin wax and they often grow in such profusion on the coated glass rods as to make the isolation of the poorly competitive nocardioforms almost impossible. Goodfellow & Orchard (1974) tested laboratory strains of *Nocardia* for their *in vitro* susceptibility to 52 antimicrobial agents in an attempt to highlight those compounds which might be useful for the selective isolation of nocardiae from natural habitats. Their survey enabled them to propose a selective medium containing the antifungal antibiotics cycloheximide and nystatin together with chlortetracycline, demethyl chlortetracycline or methacycline, alone or in combinations, to inhibit other bacteria (Orchard & Goodfellow, 1974). Nocardiae were isolated from all of the six soil samples tested (Table 8) and dilution plate counts showed that they contained over 1000 of these organisms per gram dry wt of soil. Preliminary taxonomic studies have identified many of their isolates as *N. asteroides*, suggesting that all previous studies have grossly underestimated the numbers of this opportunistic pathogen in soils.

6. Nocardioform Actinomycetes in Aquatic Habitats

Natural waters contain agarlytic and spreading bacteria which can grow on actinomycete isolation media, making the enumeration of nocardioforms very difficult. Hydrocarbon-utilizing pseudomonads and micrococci abound, rendering the paraffin wax baiting technique almost useless. It is probably as a result of such technical difficulties that the nocardioforms in water have been almost ignored by bacteriologists. The information available is again found as relatively minor parts of much wider microbiological studies.

Table 8
Dilution plate counts of nocardiae in six soils (Orchard & Goodfellow, 1974)

Source of soil sample	Growth of nocardiae ($10^{-3} \times$ no/g dry wt) on isolation plates containing:			
	Demethylchlor-tetracycline (5 μg/ml)	Methacycline (10 μg/ml)	Demethylchlor-tetracycline (5 μg/ml) + Chlortetracycline (45 μg/ml)	Methacycline (10 μg/ml) + Chlortetracycline (45 μg/ml)
Ghana	1·0	1·0	0·1	0·2
Ghana (sandy soil)	2·0	0·5	1·0	0·4
Mexico	4·6	0·9	3·6	1·8
Mexico	2·1	0·5	1·0	0·1
Thailand	3·0	5·0	4·3	4·6
Venezuela	4·1	32·0	4·5	2·7

(a) *Freshwaters*

Water samples from streams, rivers and lakes contain few actinomycetes relative to the total microbial numbers. On isolation plates of colloidal chitin agar or starch-casein-nitrate agar, the typical colonies of *Streptomyces* and *Micromonospora* are evident but the much smaller colonies of the nocardiae are difficult to see and are often obscured by films of spreading Gram-negative bacteria. Mild chlorination or heat treatments to reduce the numbers of competing bacteria in water samples prior to plating out on colloidal chitin agar have been developed (Burman, Oliver & Stevens, 1969; Windle-Taylor, 1969–1970; Burman 1973). Heat-treating water samples at 44°C for 1–2 h prior to plating was found to be very effective and heating inoculated plates or membrane filters at 44°C before incubation had a similar effect. These methods enabled Burman and co-workers to frequently isolate nocardioforms from water.

Eak-Hour & Leclerc (1973) investigated various pretreatment procedures and media for the isolation of actinomycetes from clean and polluted waters, and sediments. A sodium caseinate-asparagine-propionate medium (Olson, 1968) gave the highest recovery of actinomycetes from polluted waters and sediments, whilst chitin agar proved best for relatively clean waters. They found prior heat treatments and chlorination useful for reducing the bacterial population of polluted samples. However, the total numbers of actinomycetes, including *Nocardia*, isolated by Eak-Hour & Leclerc from clean river water (1–5/ml) and polluted river water (2–6/ml) were very low. Willoughby (1971) had previously recorded from 65 to 270 actinomycetes/ml in samples taken from above and below a sewage farm outfall on the River Kent, Cumbria.

Three very interesting nocardioform types were isolated and described by Willoughby (1969) during studies on the actinomycete flora of Blelham Tarn and its streams in the English Lake District (Table 9). His descriptions with photographs and drawings are clear but he preferred at that stage to refer to them under code names rather than attempt an identification. Willoughby found high numbers (10^3–10^4 c.f.u./ml mud) of the most numerous type, Lspi, in the mud of the profundal and littoral regions of the tarn and its inflowing streams. The organism was also recovered from two waterlogged soil samples out of six terrestrial sites examined

Table 9

Colonial appearance of Lspi, Sspi and Nt on colloidal chitin agar isolation plates after 3 weeks incubation at 25°C (Willoughby, 1969)

Lspi (Large spored pink irregular)
Small effuse colonies, usually 0·5–3·0 mm diam. with an arborescent margin and a pale pink to pale orange central papilla. Surface mycelium radiating from the colony centre in curling strands, fragmenting and dividing into spherical elements, 1·0 to 1·4 μm diam. No aerial mycelium. Chitin not hydrolysed.

SSpi (Small spored irregular)
Colonies up to 6 mm diam., creamy white, colour deeper at colony centre. The mycelium branches irregularly to produce a close network and fragments into spherical elements, 0·7 mm in diam. No aerial mycelium. Chitin may be hydrolysed.

Nt (Nocardia type)
Colonies 2–5 mm diam. resembling white or pale yellow miniature chimney-sweep brushes seen end on embedded in the agar. Thin mycelial strands, resembling barbed wire, radiate from the colony centre and fragment into cylindrical elements (0·7 to 3·0×0·7 μm). No aerial mycelium. Chitin not hydrolysed.

near the tarn, and Willoughby concluded that Lspi was a truly aquatic actinomycete. Previously, this actinomycete had been encountered in small numbers on starch-casein-nitrate plates used by Cross & Collins (1966) for the isolation of *Micromonospora* species from the water of Blelham Tarn. Dr J. Lacey of the Rothamsted Experimental Station (pers. comm.) had isolated a very similar organism from Andersen sampler plates taken during studies on the microflora in dust from hay bales opened outdoors. Johnston (1972) had also found Lspi to be the most numerous nocardioform present in his samples of water and mud taken from 14 lakes in the English Lake District. He observed that the numbers of nocardioforms appeared to be related to the nutrient status of the lake, eutrophic lakes such as Blelham Tarn showing higher numbers than oligotrophic lakes such as Ennerdale and Wastwater. From homogenization studies on the surface lake mud of profundal cores he inferred that the nocardioforms (mainly Lspi) were present as either short pieces of mycelium or as fragmentation spores. Both Willoughby and Johnston used a colloidal chitin agar isolation medium.

A second nocardioform isolated by Willoughby and referred to as Sspi was not studied in detail and Johnston does not record its isolation. The third type, Nt, was recovered by Willoughby from stream water (0–58 c.f.u./ml), the higher numbers were isolated during winter. Few recoveries were made from lake water or mud. Nt was isolated in high numbers from stream mud (10^3 c.f.u./ml) and some soils, particularly those under a limed pasture (10^5 c.f.u./g). Willoughby concluded that it was a soil organism being washed into the streams. Johnston (1972) recorded the presence of high numbers of Nt in stream water following periods of heavy rainfall, further evidence that Nt was terrestrial.

These pioneering ecological studies by Willoughby are interesting and significant for they record not only the numbers and descriptions of nocardioform species found in habitats previously studied by many freshwater biologists, but also attempt to distinguish between truly aquatic and terrestrial nocardioforms. The isolation of an actinomycete from water does not prove that it is an aquatic organism or imply that it has a role in that habitat.

Cross and Rowbotham have studied the distribution and properties of the organism Lspi in detail. The species was initially isolated on plates of colloidal chitin agar and counted with the aid of a microscope. Spreading bacterial films, that often obscured the actinomycetes growing on the isolation plates, were eliminated by prior heating 2 ml water samples or homogenates for 6 min at 55°C. A selective medium M3, was devised to encourage the growth and characteristic appearance of the thiamine-requiring species Lspi on the isolation plates and to reduce the radial growth of streptomycete and micromonospora colonies. M3 agar contains KH_2PO_4, 0·466; $Na_2H\,PO_4$, 0·732; KNO_3, 0·1; NaCl, 0·29; $MgSO_3$ 7 H_2O, 0·1; $CaCO_3$, 0·02; Na propionate, 0·2; $FeSO_4$ 7 H_2O, 0·0002; $ZnSO_4$ 7 H_2O, 0·00018; $MnSO_4$ 4 H_2O, 0·00002; cycloheximide 0·05; thiamine HCl, 0·004; agar 18·0 g/l distilled water. The phosphates were autoclaved separately; the vitamin and antibiotic solutions were sterilized by filtration and added to the molten agar before pouring. Plates were incubated at 30°C; Lpsi colonies were counted after 7 days and total actinomycetes after 21 days. M3 and colloidal chitin agar gave equal Lspi counts but M3 gave prominent orange Lspi and smaller micromonospora and streptomycete colonies.

With the aid of the heat pretreatment method in combination with colloidal chitin or M3 agars, Cross and Rowbotham have surveyed Blelham Tarn and its associated streams, and a number of streams flowing into the River Wharfe in Yorkshire for their actinomycete content (Table 10). The majority of streams had counts of between 100 to 500 actinomycetes/ml with *Micromonospora* the predominant genus (over 50% of actinomycete isolates) and nocardioforms not more than 30%. The maority of the nocardioforms (80–90%) could be identified as Lspi. Two streams with actinomycete counts of over 1000/ml contained high numbers of

Table 10

Distribution of the nocardioform Lspi in lake, stream and farm samples (Cross & Rowbotham)

Sample	Lspi c.f.u./ml water, or /g soil, dung etc.
Lake and stream samples, Cumbria	
Wray Beck flowing into Blelham Tarn	18
Tock How Beck flowing into Blelham Tarn (carries some farm effluent)	326
Blelham Tarn, surface water	78
Blelham Tarn, outflow stream	37
Blelham Tarn, surface mud	140×10^5
Stream water samples, Yorkshire	
Hundwith Beck, trout stream	10
Dean Beck, stream water above farm	10
Dean Beck, stream water below farm	1025
Dean Beck, sample from farm drain	1150
Farm samples	
Cow rumen contents	9×10^4
Fresh cow dung from field	4×10^5
Old cow dung from field	$1{\cdot}8\times10^8$
Cow dung incubated in laboratory	
initial count	$1{\cdot}4\times10^7$
count after 3 weeks	$3{\cdot}4\times10^9$
Pasture grass	1×10^5
Soil beneath pasture grass	6×10^5
Grass from ungrazed and non-manured grassland	<25
Soil beneath ungrazed and nonmanured grassland	<100
Grass from ornamental lawn	<25
Silage	25
Animal feed concentrates	40

nocardiaforms representing 70% of the total actinomycetes. Both streams were enriched by dairy farm effluents and showed localized growths of *Sphaerotilus*. In the search for the source of Lspi in streams, attention was directed to the farm environment. Fresh cow dung samples contained 4×10^5 Lspi/g and similar dung samples incubated in the laboratory subsequently gave higher counts ($3{\cdot}4 \times 10^9$ Lspi/g). Grass from grazing pastures or manured meadows carried Lspi (1×10^5/g) but no growth could be demonstrated on grass in controlled experiments in the laboratory and greenhouse.

These surveys suggested that the predominant nocardioform occurring in stream waters, lakes and sediments is a wash-in organism, growing initially on herbivore dung and cycling via contaminated fodder through the animals. A portion of these organisms are washed into streams from dung and manured fields, and probably exhibit little further metabolic activity. This work would also support the view that the majority of actinomycetes in aquatic habitats are wash-in forms with little activity in these habitats. The possible exceptions may be the *Actinoplanes* and the *Micromonospora* which occur in such high numbers in lake muds. Taxonomic studies would place Lspi in the *rhodochrous* complex but as it differs from other members of this group at a 70% similarity level it will shortly be described as a new species.

An alternative assumption was made by Donderski & Strzelczyk (1974) in their studies on the genera and nutritional requirements of bacteria isolated from the water and muds of three lakes in Poland. The *Arthrobacter-Corynebacterium* group were the most numerous bacteria in mud during the summer months whereas *Nocardia* were most numerous during spring and autumn. For example, in the mesotrophic Lake Jasne, *Nocardia* comprised 60% of the isolates in samples taken in May and 51% in January, lower numbers were isolated during the summer months. They found that vitamins were more often required by the benthic organisms and suggested that the distribution of bacteria in a lake system reflected their nutritional requirements and indicated the presence of these special nutrients in the environment.

Mishustin and Tepper have recorded higher numbers of *Nocardia* in soils during winter and spring than in summer months, and these are the months when the higher rainfall would be expected to wash more soil particles and organisms into streams and lakes. Donderski & Strzelczyk (1974) also recorded higher numbers of *Nocardia* in the mud from Polish lakes during the winter and spring months. Many of their aquatic nocardiae would appear to require vitamins for growth and could prove very similar to the thiamine-requiring species Lspi studied by Cross and Rowbotham. If collaborative studies eventually show that the two groups have been isolating and counting similar organisms it seems more plausible to conclude that dung is the growth substrate for these nutritionally demanding organisms. Lspi is strongly aerobic, has an optimum growth temperature of 31°C and can utilize acetate, succinate and propionate as a carbon source when thiamine is available. We feel that it is better to assume that all actinomycetes isolated from aquatic habitats originate from terrestrial sources until conclusive evidence is presented to prove that they are able to grow and not merely survive in an aquatic environment.

Evidence suggesting that *Actinomadura* strains resembling *A. madurae* might be indigenous in streams was presented by Lawson & Davey (1972). They were able to isolate *Actinomadura* strains regularly from a South African stream during a nine month sampling period: the numbers were low (2–20/ml) but did not increase during periods of high rainfall or when both Biochemical Oxygen Demand and total viable bacterial counts for the water were high.

There are also many references in the literature describing cases of nocardiosis in fish. The pathogen implicated in several outbreaks has been identified as *Nocardia asteroides* or a closely related organism and it is hard to decide whether the nocardiae are truly indigenous or contaminants behaving as facultative parasites when presented with susceptible fish under stress or damaged during the intensive rearing conditions. However, the species *N. kampachi* causing nocardiosis in Yellowtails (*Seriola* spp.) (Kariya *et al.*, 1968) can be distinguished from *Nocardia asteroides* (Kubota *et al.*, 1968; Kusuda & Taki, 1973) and has not yet been isolated from any other habitat.

Nocardia amarae has been recovered in high numbers from thick foams which occasionally form on the surface of secondary aeration and settling tanks in sewage-treatment plants of the activated sludge type in the U.S.A. (Lechevalier & Lechevalier, 1974). In most cases this abnormal foam consisted of a mass of *N. amarae* hyphae in which air bubbles were entrapped, but *N. asteroides*, *N. caviae* and species of the *rhodochrous* complex could also be isolated. *N. amarae* seems well adapted to the activated sludge environment. It is aerobic, rich in lipids (11%), and floats to the surface. The species is reported to utilize compounds found in low concentrations in activated sludge and to produce surface active compounds which probably enhance the formation of foam. The natural habitat of this species is not known. It may be present in low numbers in either soil or water but the operating conditions in certain plants select for and then encourage abundant growth.

(b) *Marine habitats*

Nocardioforms have been isolated, usually in more general microbiological studies, from beaches, water and sediments taken both near the coast and far from land. Whether these isolates are washed-in terrestrial strains or belong to the marine environment, and whether they are active in that environment is not known. ZoBell (1946) considered that actinomycetes were widely distributed in marine habitats and was able to isolate nocardiae (proactinomycetes) from dead seaweed. Siebert & Schwartz (1956) also isolated one nocardial strain from air-dried seaweed. Freitas & Bhat (1954) isolated cellulolytic nocardiae and streptomycetes from deteriorating cotton fishing nets and cordage discarded by the Worli fishing village near Bombay but this material could have been contaminated with soil. They identified one cellulolytic strain as *Nocardia corallina* but this determination is doubtful as strains belonging to the *rhodochrous* complex, to which *N. corallina* belongs, do not normally hydrolyse cellulose (Goodfellow, 1971). Other nocardiae isolated by Freitas & Bhat (1954) were able to attack cellulose, agar, dextrin and starch (unusual substrates for nocardiae) and grew best at salt concentrations (2·0–2·5% w/v) near to that of salt water. Humm & Shepard (1946) described two nonmotile nocardioforms under the names *Proactinomyces flavus* (now *N. marina*) from intertidal muddy land, and *P. atlanticus* from seaweed collected on the Atlantic coast of the U.S.A. Both species could attack chitin, alginic acid and agar, and grow in saline media.

Grein & Meyers (1958) investigated the occurrence of actinomycetes including nocardiae in seawater and littoral zones around the Gulf of St. Lawrence, along the west coast of Florida, the Gulf of Mexico (3–5 miles off shore) and near the mouth of the Shark River in New Jersey. *Streptomyces, Micromonospora* and nocardiae were represented in all their samples, and the latter comprised about one tenth of their actinomycete isolates.

All the above isolates originated from areas where there could have been soil contamination. Zobell (1946) had suggested that one would only find true marine organisms in sites far removed from possible terrestrial contamination. Weyland (1969) was able to find *Nocardia* in sediment samples taken at depths of 3400 m and 175 miles off the coast of West Africa. This evidence is convincing but proof of the existence of true marine nocardioforms must await further studies on marine bacteria, their taxonomy, salt requirements and ability to grow in seawater.

Members of the genus *Nocardia* and the *rhodochrous* complex readily utilize a wide variety of hydrocarbons as sole sources of carbon and energy. The role of nocardioforms in the biodegradation of oil slicks appears to be a distinct possibility but more information is needed. The early work of Zobell, Grant & Haas (1943) showed that *Nocardia* formed 5–10% of the total microbial flora of marine sediments enriched with oil.

7. The Properties of Nocardioform Actinomycetes in Pure Culture

We have surveyed the published information on the numbers and distribution of nocardioform actinomycetes in various natural habitats. It is an area of microbial ecology that has attracted few workers and the useful information is relatively meagre. We are only just beginning to appreciate that these organisms are probably more numerous and widespread than previously suspected. The majority of strains that have been isolated came from the soil and this leads one to suspect that what growth and sporulation does occur will take place in the soil. However, soil is a habitat containing many micro-organisms in intense competition. Work on streptomycetes in soil would suggest that most actinomycetes have very limited periods of growth and then persist for periods in the form of dormant or relatively inactive propagules. How can the nocardiae compete for the limited nutrients available in soil in the face of such intense competition from other rapidly growing bacteria and fungi? Can they then survive short or even long periods without a continuing food supply? A consideration of the properties of these organisms in pure culture might indicate some of their attributes which allow them to grow and persist.

The species *Nocardia asteroides* has been studied most extensively and its properties are summarized in Table 11. The results of these

Table 11

Properties of Nocardia asteroides *as determined in various pure culture studies*

Property	Frequency (%)	Reference*
Mycelium fragmenting into rods and cocci	100	a
Aerial mycelium	98–100	a, b
Aerial mycelium segmenting into spores	30	b
Decomposition of adenine	0	c
casein	0	c
hypoxanthine	4	c
tyrosine	0–1	a, c
urea	96	c
xanthine	0	a, c
Growth at 50°C	21–31	a, b
45°C	42–64	a, b
40°C	88–100	a, b
35°C	100	a, b
28°C	100	b
10°C	16–26	a, b, c
minimum growth temperature 0°C		d
Survival at 50°C for 8 h on agar	96	c
Survival at 115°C for 15 min dried on glass		d
Hydrolysis of aesculin	93–100	a, c
cellulose	0	a
chitin	0	a, d
DNA	0	a
gelatin	0	a, d
starch	41–69	a, c
dextrin	50	a
keratin	0	a
inulin	0	a
xylan	0	a
Tween 20	100	a
Tween 40	97	a

Table 11—*contd.*

Property	Frequency (%)	Reference*
Tween 60	84	a
Tween 80	10	a
Resistance to lysozyme	100	a, c
Nitrate reduction	91–98	a, c
Growth at pH 5	2	a
pH 6	98	a
pH 8	100	a
pH 9	98	a
pH 10	62	a
Growth on compounds as sole carbon source:		
paraffin wax	100	a
cellobiose	5	a
glycogen	0	a
L-arabinose	0	a
D-fructose	91	a
D-glucose	100	a
mannose	86	a
lactose	0	a
sucrose	7	a
maltose	74	a
salicin	10	a
sodium acetate	100	a, b
sodium propionate	100	a, b
sodium n-butyrate	100	a
sodium benzoate	0–12	a, c
sodium succinate	95–100	a, c
sodium malate	96–98	a, c
adipic acid	53	a
sebacic acid	84	a
Survives 15 months suspended in distilled and tap water		f, h
High resistance to u.v. light		e
Survives 24 months dried on cover-slips		g
Survives 20 months in sewage		h
Survives 5 to 8 months in unsterile soil		h
Survives up to 6 months on straw		h
Little evidence of antibiotic production		d
Relatively resistant to antibiotics produced by other actinomycetes		d
Opportunistic pathogen in man, animals and fish		(many)

* References.

a. Goodfellow, 1971.
b. Gordon & Mihm, 1962.
c. Gordon *et al.*, 1974.
d. Szabo, 1974.
e. Schaal, 1969.
f. Schaal, 1970*a*.
g. Schaal, 1970*b*.
h. Lindner, 1973.

studies give very few clues to the sites where growth might occur in soil or the substrates which might be preferentially utilized by the species. The one possible lead is suggested by the ability of all strains tested to grow on paraffin wax as the sole carbon source and it would now be interesting to explore the ability of the species to grow on the relatively resistant waxes and hydrocarbon compounds of both plant and animal origin. The organism appears to be able to survive for relatively long periods in soil and water once viable propagules have been formed.

A possible substrate for the growth of other nocardiae species was suggested during the surveys of Mishustin and Tepper when they recorded high numbers of nocardioforms in soil where there was active humus decomposition. Most of the red pigmented strains isolated were able to assimilate benzoic, salicyclic and protocatechuic acids and vanillin. The species *Nocardia rubra*, *N. corallina* and *N. symbiotica* were able to utilize humic acid when tested *in vitro* (Table 12).

Table 12

Decomposition of sodium humate by Nocardia rubra *(Mishustin & Tepper)*

Medium*	% humate decomposed after 4 months incubation at 28°C†
Na humate	22·1
Na humate + yeast autolysate	26·1
Na humate + glucose	18·0
Na humate + KNO_3 + glucose	7·5

* Medium: Winogradsky's mineral medium, lacking an inorganic nitrogen source, containing sodium humate (8 mg/100 ml). Additions: yeast extract or KNO_3 (14 mg N/100 ml) or glucose (0·25 mg/100 ml).

† Humate decomposition estimated from carbon content after humus precipitation by the Tyurin technique.

The studies on Lspi in pure culture coupled with its distribution in natural habitats strongly suggest that herbivore dung is the primary site for growth. The surface of such substrates would supply the oxygen and higher temperatures necessary for growth together with nutrients such as fatty acids and thiamine. The primary mycelium fragments to form coccoid elements which may persist in the soil or contaminate fodder. These coccoid "fragmentation spores" deserve investigating further to determine their resistance to environmental extremes and their longevity. They form when nutrients are limiting and morphologically resemble the coccal forms of *Arthrobacter* spp. which have been suggested to be the survival stage of these species (Luscombe & Gray, 1974).

When isolating Lspi from dung samples, Cross and Rowbotham encountered several other nocardioform species, sometimes in high numbers. It is possible that this substrate may also support the growth of other nutritionally demanding nocardioform species which then survive in soil as spores or mycelial fragments.

The availability of selective media which would allow the isolation of particular species from soil containing a mixed microbial flora should enable us to proceed with soil inoculation studies. The growth of a species in sterilized soil amended with nutrients in the laboratory gives little useful information on the growth of that species in natural habitats. One must demonstrate growth in non-sterile soil before positively stating that soil is the natural habitat of that organism. The nocardioforms, however, pose a technical problem when undertaking such studies. Mycelium added to soil may result in an apparent increase in numbers (c.f.u.) because of fragmentation rather than growth. It will therefore be necessary to use fluorescent antibody techniques in conjunction with viable counts when attempting such back inoculation studies.

8. Role of Nocardioform Actinomycetes in Natural Ecosystems

We feel that it is premature to attribute any specific roles to the nocardioform actinomycetes in natural ecosystems at this time. That we continue to isolate them in high numbers from soil must suggest that they play some part in the complex series of degradative reactions occurring there in competition with many other microorganisms but it is now necessary to concentrate on individual species and explore their specific properties before postulating a role. The heterogeneous collection of organisms accommodated beneath the umbrella term "nocardioform actinomycete" tempts workers to look for common activities but we suggest that they must be regarded as a collection of individual genera or species with specific activities. For example, not all nocardioforms are pathogenic and the known pathogens appear to be opportunistic with growth occurring in natural substrates such as soil and perhaps more rarely in animal tissues. It would be quite wrong to infer that all nocardioforms behaved the same as *Nocardia asteroides*, *N. rubra* or even Lspi. For convenience it has been useful to attempt a survey of the present state of our knowledge of the nocardioform actinomycetes in this chapter. The fragments of information available indicate not only our incomplete knowledge, but also the different avenues such studies must follow in the future.

9. References

Auden, D. T. (1974). Studies on the development of *Rhodnius prolixus* and the effects of its symbiote *Nocardia rhodnii*. *J. med. Ent.* **11,** 63.

Baines, S. (1956). The role of the symbiotic bacteria in the nutrition of *Rhodnius prolixus* (Hemiptera). *J. exp. Biol.* **33,** 533.

Bewig, F. & Schwartz, W. (1955). Über die symbiose von triatomiden mit *Nocardia rhodnii* (Erikson) Waksman and Henrici. *Naturwissenschaften* **42,** 423.

Bewig, F. & Schwartz, W. (1956). Untersuchungen über die symbiose von tieren mit pilzen und bakterien. VIII. Über die physiologieder symbiose bei einigen blutsaugenden insekten. *Arch. Mikrobiol.* **24,** 174.

Brecher, G. & Wigglesworth, V. B. (1944). The transmission of *Actinomyces rhodnii* and its influence on the growth of the host. *Parasitology* **35,** 220.

Brock, T. D. (1966). *Principles of Microbial Ecology*. Englewood Cliffs: Prentice-Hall Inc.

Burman, N. P. (1973). The occurrence and significance of actinomycetes in water supplies. In *The Actinomycetales: Characteristics and Practical Importance*. Eds G. Sykes & F. A. Skinner. London and New York: Academic Press.

Burman, N. P., Oliver, C. W. & Stevens, J. K. (1969). Isolation of actinomycetes. In *Isolation Methods for Microbiologists*. Eds D. A. Shapton & G. W. Gould. London and New York: Academic Press.

Casida, L. E. (1965). Abundant micro-organisms in soil. *Appl. Microbiol.* **13,** 327.

Cavanagh, P. & Marsden, P. D. (1969). Bacteria isolated from the gut of some reduviid bugs. *Trans. R. Soc. trop. Med. Hyg.* **63,** 415.

Cross, T. & Collins, V. G. (1966). *Micromonospora* in an inland lake. *IX Intern. Congr. Microbiol, Moscow* C5, p. 335.

Demaree, J. B. & Smith, N. R. (1952). *Nocardia vaccinii* n.sp. causing galls on blueberry plants. *Phytopathology* **42,** 249.

Donderski, W. & Strzelczyk, E. (1974). Generic composition and nutritional requirements of bacteria isolated from lakes. *Acta. microbiol. pol.* **6,** 67.

Duncan, J. T. (1926). On a bactericidal principle present in the alimentary canal of insects and arachnids. *Parasitology* **18,** 238.

Eak-Hour, C. & Leclerc, H. (1973). Dénombrement des actinomycètes aérobies de l'eau. *Annls Microbiol. (Inst. Pasteur)* **124B,** 533.

Erikson, D. (1935). The pathogenic aerobic organisms of the Actinomyces group. *Spec. Rep. Ser. med. Res. Coun.* No. 203.

Erikson, D. (1954). Factors promoting cell division in a 'soft' mycelial type of *Nocardia*: *Nocardia turbata n.sp. J. gen. Microbiol.* **11,** 198.

Freitas, Y. M. & Bhat, J. V. (1954). Micro-organisms associated with the deterioration of fish-nets and cordage. *J. Univ. Bombay* **23,** 53.

Gledhill, W. E. & Casida, L. E. (1969). Predominant catalase-negative soil bacteria. III. *Agromyces* gen. n., micro-organisms intermediary to *Actinomyces* and *Nocardia. Appl. Microbiol.* **18,** 340.

González-Ochoa, A. (1962). Mycetomas caused by *Nocardia brasiliensis,* with a note on the isolation of the causative organism from soil. *Lab. Invest.* **11,** 1118.

Goodchild, A. J. P. (1955). The bacteria associated with *Triatoma infestans* and some other species of Reduviidae. *Parasitology* **45,** 441.

Goodfellow, M. (1971). Numerical taxonomy of some nocardioform bacteria. *J. gen. Microbiol.* **69,** 33.

Goodfellow, M. & Cross, T. (1974). Actinomycetes. In *Biology of Plant Litter Decomposition.* Eds C. H. Dickenson & G. J. F. Pugh. p. 269. Academic Press: London and New York.

Goodfellow, M. & Orchard, V. A. (1974). Antibiotic sensitivity of some nocardioform bacteria and its value as a criterion for taxonomy. *J. gen. Microbiol.* **83,** 1.

Gordon, R. E., Barnett, D. A., Handerhan, J. E. & Hor-Nay Pang, C. (1974). *Nocardia coeliaca, Nocardia autotrophica* and the nocardin strain. *Int. J. Syst. Bact.* **24,** 54.

Gordon, R. E. & Hagan, W. A. (1937). The isolation of acid-fast bacteria from the soil. *Am. Rev. Tuberc. pulm. Dis.* **36,** 549.

Gordon, R. E. & Mihm, J. M. (1962). The type species of the genus *Nocardia. J. gen. Microbiol.* **27,** 1.

Grein, A. & Meyers, S. P. (1958). Growth characteristics and antibiotic production of actinomycetes isolated from littoral sediments and materials suspended in sea water. *J. Bact.* **76,** 457.

Gumpert, J. (1962*a*). Die Funktion der symbiontischen Bakterien in dem Triatomien. *Zentbl. Bakt. ParasitKde Abt* 1 **184,** 315.

Gumpert, J. (1962*b*). Untersuchungen über die Symbiose von Tieren mit Pilzen und Bakterien. X. Die Symbiose der Triatominen. 2. Infektion symbiotenfreier Triatominen mit symbiontischen und saprophytischen Mikroorganismen und gemeinsame eigenschaften der symbiontischen Stamme. *Z. allg. Mikrobiol.* **2,** 290.

Gumpert, J. & Schwartz, W. (1962). Untersuchungen über die Symbiose von Tieren mit Pilzen und Bakterien. X. Die Symbiose der Triatominen. 1. Aufzucht symbiontenhaltiger und symbiontenfreier Triatominen und eigenschaften der bei Triatominen verkommenden Mikroorganismen. *Z. allg. Mikrobiol* **2,** 209.

Harington, J. S. (1960*a*). Studies on *Rhodnius prolixus*: growth and development of normal and sterile bugs, and the symbiotic relationship. *Parasitology* **50,** 279.

Harington, J. S. (1960*b*). Synthesis of thiamine and folic acid by *Nocardia rhodnii*, the micro-symbiont of *Rhodnius prolixus. Nature, Lond.* **188,** 1027.

Humm, H. J. & Shepard, K. S. (1946). Three new agar-digesting actinomycetes. *Bull. Duke Univ. mar. Stn* **3,** 76.

Johnston, D. W. (1972). *Actinomycetes in Aquatic Habitats.* Ph.D. Thesis. University of Bradford.

Kariya, T., Kubota, S., Nakamura, Y. & Kira, K. (1968). Nocardial infection in cultured yellow tails. (*Seriola quinqueradiata* and *S. purpurascens*). I. Bacteriological study. *Fish Path., Tokyo* **3,** 16. (In Japanese.)

Kubota, S., Kariya, T., Nakamura, Y. & Kira, K. (1968). Nocardial infection in cultured yellow tails (*Seriola quinqueradiata* and *S. purpurascens*). II. Histological study. *Fish Path., Tokyo* **3,** 24. (In Japanese.)

Kumar, R. & Mohapatra, L. N. (1968). Studies on aerobic actinomycetes isolated from soil. I. Isolation and identification of strains. *Sabouraudia* **6,** 140.

Kurup, P. V., Mishra, S. K., Randhawa, H. S. & Abraham, S. (1967). Isolation of *Nocardia brasiliensis* from soil. *Curr. Sci.* **36,** 493.

Kurup, P. V., Randhawa, H. S., & Sandhu, R. S. (1968). A survey of *Nocardia asteroides, N. caviae* and *N. brasiliensis* occurring in soil in India. *Sabouraudia* **6,** 260.

Kurup, P. V. & Sandhu, R. S. (1965). Isolation of *Nocardia caviae* from soil and its pathogenicity for laboratory animals. *J. Bact.* **90,** 822.

Kurup, P. V., Sandhu, R. S., & Damodaran, V. N. (1964). Occurrence of *Nocardia asteroides* in Dehli soil. *Indian J. med. Res.* **52,** 1057.

Kurup, P. V. & Schmitt, J. A. (1971). Isolation of nocardia from soil by a modified paraffin bait method. *Mycologia* **63,** 175.

Kusuda, R. & Taki, H. (1973). Studies on a nocardial infecton of cultured yellow tail. I. Morphological and biochemical characteristics of Nocardia isolated from diseased fishes. *Bull. Jap. Soc. scient Fish.* **39,** 937.

Labeda, D. P., Hunt, C. M. & Casida, L. E. (1974). Plating isolation of various catalase-negative micro-organisms from soil. *Appl. Microbiol.* **27,** 432.

Lacey, J. (1974). Moulding of sugar-cane bagasse and its prevention. *Ann. appl. Biol.* **76,** 63.

Lacey, J. & Goodfellow, M. (1974). A new genus of nocardioform actinomycetes? In *Proc.* 1 *Intern. Conf. Biol. Nocardiae, Merida, Venezuela*, Ed. G. H. Brownell.

Lake, P. & Friend, W. G. (1968*a*). A monoxenic relationship, *Nocardia rhodnii* Erikson in the gut of *Rhodnius prolixus* Ståhl. (Hemiptera: Reduviidae). *Proc. ent. Soc. Ont.* **98,** 53.

Lake, P. & Friend, W. G. (1968*b*). The use of artificial diets to determine some of the effects of *Nocardia rhodnii* on the development of *Rhodnius prolixus. J. Insect Physiol.* **14,** 543.

Lavrova, N. V., Preobrazhenskaya, T. P. & Sveshnikova, M. A. (1972). Isolation of soil actinomycetes on selective media with rubomycin. *Antibiotiki* **17,** 965. (In Russian.)

Lawson, E. N. & Davey, L. M. (1972). A waterborne actinomycete resembling strains causing mycetoma. *J. appl. Bact.* **35,** 389.

Lechevalier, M. P. (1972). Description of a new species, *Oerskovia xanthineolytica,* and emendation of *Oerskovia* Prauser et al. *Int. J. Syst. Bact.* **22,** 260.

Lechevalier, H. A. & Lechevalier, M. P. (1967). Biology of actinomycetes. *Ann. rev. Microbiol.* **21,** 71.

Lechevalier, M. P. & Lechevalier, H. A. (1974). *Nocardia amarae* new species, an actinomycete common in foaming activated sludge. *Int. J. Syst. Bact.* **24,** 278.

Lindner, K. E. (1973). Zur Tenazitat von Nocardien in der Aussenwelt. *Arch. exp. Vet. Med.* **27,** 139.

Luscombe, B. M. & Gray, T. R. G. (1974). Characteristics of Arthrobacter grown in continuous culture. *J. gen Microbiol.* **82,** 213.

McClung, N. M. (1949). Morphological studies in the genus *Nocardia.* I. Developmental studies. *Lloydia* **12,** 137.

McClung, N. M. (1960). Isolation of *Nocardia asteroides* from soil. *Mycologia* **52,** 154.

Metcalf, G. & Brown, M. E. (1957). Nitrogen fixation by new species of *Nocardia. J. gen. Microbiol.* **17,** 567.

Mülpfordt, H. (1959). Der Einfluss der Darmsymbionten von *Rhodnius prolixus* auf *Trypanosoma cruzi. Z. Tropenmed. Parasit.* **10,** 314.

Nonomura, H. & Ohara, Y. (1971). Distribution of actinomycetes in soil. XI. Some new species of the genus *Actinomadura* Lechevalier *et al. J. Ferment. Technol., Osaka* **40,** 904.

Nyirady, S. A. (1973). The germfree culture of three species of Triatominae: *Triatoma protracta* (Uhler), *Triatoma rubida* (Uhler) and *Rhodnius prolixus* (Stål). *J. Med. Ent.* **10,** 417.

Olson, E. H. (1968). Actinomycetes isolation agar. In *Difco: Supplementary Literature.* Detroit (Mich.): Difco Labs.

Orchard, V. A. & Goodfellow, M. (1974). The selective isolation of Nocardia from soil using antibiotics. *J. gen. Microbiol.* **85,** 160.

Pier, A. C., Willers, E. H. & Mejia, M. J. (1961). *Nocardia asteroides* as a mammary pathogen of cattle. II. The sources of nocardial infection and experimental reproduction of the disease. *Am. J. vet. Res.* **22,** 698.

Prauser, H., Lechevalier, M. P. & Lechevalier, H. (1970). Description of *Oerskovia* gen. n. to harbor Ørskov's motile *Nocardia. Appl. Microbiol.* **19,** 534.

Preobrazhenskaya, T. P. & Lavrova, N. V. (1974). Isolation of nocardioform actinomycetes from soil. In *Proc.* 1 *Intern. Conf. Biol. Nocardiae, Merida, Venezuela.* Ed. G. H. Brownell.

Reyes, A. C. (1963). The isolation of Nocardia from Philippine soil. *Acta med. phillipp.* **19,** 103.

Rodda, G. M. A. (1964). Nocardia strains in the environment. *Med. J. Aust.* **2,** 13.

Schaal, K. P. (1969). Die Resistenz der Erreger des Aktinomykosen-Klassische Aktinomykose und Nocardiose-gegen Umwelteinflusse. I. Die Resistenz gegen ultraviolette Strahlen. *Zentbl. Bakt. ParasitKde Abt. I* **211,** 550.

Schaal, K. P. (1970*a*). Die Resistenz der Erreger der Aktinomykosen-Klassische Aktinomykose und Nocardiose-gegen Umwelteinflusse. II. Die Uberlebensfahigkeit in Wasser und in einfachen Wasserigen Losungen. *Zentbl. Bakt. ParasitKde Abt. I.* **215,** 234.

Schaal, K. P. (1970*b*). Die Resistenz der Erreger der Aktinomykosen-Klassische Aktinomykose und Nocardiose-gegen Umwelteinflusse. III. Die Resistenz gegen Austrocknung. *Zentbl. Bakt. ParasitKde Abt. I.* **215,** 483.

Schaal, K. P. (1972). Zur mikrobiologischen Diagnostik der Nocardiose. *Zentbl. Bakt. Hyg. I Abt. Orig* **220,** 242.

Siebert, U. G. & Schwartz, W. (1956). Untersuchungen über das Vorkommen von Mikro-organismen in entstehenden Sedimenten. *Arch. Hydrobiol.* **52,** 321.

Söhngen, N. L. (1913). Benzin, Petroleum, Paraffinol und Paraffin als Kohlenstoff und Energiequell fur Mikroben. *Zentbl. Bakt. ParasitKde. Abt II* **37,** 595.

Szabo, I. M. (1974). *Microbial Communities in a Forest-Rendzina Ecosystem.* Budapest: Akademiai Kiado.

Waksman, S. A. (1919). Cultural studies on species of Actinomyces. *Soil Sci.* **8,** 71.

Waksman, S. A. & Henrici, A. T. (1943). The nomenclature and classification of the actinomycetes. *J. Bact.* **46,** 337.

Watson, E. T. & Williams, S. T. (1974). Studies on the ecology of actinomycetes in soil. VIII. Actinomycetes in a coastal sand belt. *Soil Biol. Biochem.* **6,** 43.

Webley, D. M., Eastwood, D. J. & Gimingham, C. H. (1952). Development of a soil microflora in relation to plant succession on sand dunes, including the "rhizosphere" flora associated with colonising species. *J. Ecol.* **40,** 168.

Weyland, H. (1969). Actinomycetes in North Sea and Atlantic Ocean sediments. *Nature, Lond.* **223,** 858.

Wigglesworth, V. B. (1936). Symbiotic bacteria in a blood sucking insect, *Rhodnius prolixus* Stål (Hemiptera, Triatomidae). *Parasitology* **28,** 284.

Wigglesworth, V. B. (1952). Symbiosis in blood sucking insects. *Tijdschr. Ent.* **95,** 63.

Willoughby, L. G. (1969). A study of the aquatic actinomycetes of Blelham Tarn. *Hydrobiologia* **34,** 465.

Willoughby, L. G. (1971). Observations on some aquatic actinomycetes of streams and rivers. *Freshwat. Biol.* **1,** 23.

Windle-Taylor, E. (1969–1970). Aerobic sporing bacilli, fungi and actinomycetes in slow sand filters. *Rep. Res. bact. chem. biol. Exam. Lond. Waters* **44,** 16.

Zobell, C. E. (1946). *Marine Microbiology.* Waltham, Mass: Chronica Botanica Co.

Zobell, C. E., Grant, C. W. & Haas, H. F. (1943). Marine micro-organisms which oxidise petroleum hydrocarbons. *Bull. Am. Ass. Petrol. Geol.* **27,** 1175.

14. Lipid-Soluble, Iron-Binding Compounds in *Nocardia* and Related Organisms

C. RATLEDGE and P. V. PATEL
Department of Biochemistry, The University of Hull, Hull HU6 7RX, England

Contents

1. Introduction

It is now well established that most, if not all, micro-organisms have evolved specific mechanisms for the uptake of iron from their environment (Lankford, 1973). This applies not only to organisms which are grown in the laboratory but also to those which grow as pathogens either in animals or plants. The exact mechanisms whereby iron is transported into the cell varies from genus to genus but in all cases they follow a similar pattern. A chelator or sequestering agent is secreted by the micro-organism into the environment. This solubilizes the otherwise highly insoluble ferric ion and is then re-absorbed by the cell. The iron is released from its carrier and taken into the cell cytoplasm and the carrier returns into the medium to sequester further amounts of iron.

Iron is, of course, needed for many essential processes once it is within the cell, e.g. for both heme and non-heme iron containing enzymes. Should there be an inadequate supply of iron to the cell then the activities of many of these enzymes decrease with the result that the cell grows slowly and changes in its rate of respiration and pathways of metabolism often occur. There is good reason to believe that the supply of iron limits the growth of micro-organisms when in animal tissue as several experimenters have shown increased growth of bacteria following injections of iron into an infected animal (Bullen, Rogers & Griffiths, 1972; Szabo, 1971; see chapter 15).

The mechanism by which mycobacteria assimilate iron appears to differ from other bacteria in that more than one iron chelator is synthesized by the cell. Extracellular water-soluble chelators are produced (Ratledge *et al.*, 1974; Macham & Ratledge, 1975) as well as lipid-soluble molecules which do not appear to leave the cell. The latter are the mycobactins which were first isolated by Snow and co-workers from *Mycobacterium phlei* in 1949 and have since been found in all mycobacteria with the exception of a few species such as *M. paratuberculosis* (syn. *johnei*) which has a specific requirement for this material (Snow, 1970). The structures of the principal mycobactins are given in Fig. 1. The minor variations in the substituents of the various mycobactins has enabled these materials to be used as chemotaxonomic markers (Snow & White, 1969) although their detection requires the use of sophisticated equipment such as nuclear magnetic resonance spectrometers. Some mycobactins (e.g. S and T) are distinguished only with the greatest difficulty as the nuclei differ only in configuration at an asymmetric centre (not shown in Fig. 1).

The mycobactins form very stable red complexes with Fe^{3+} which are insoluble in water but will dissolve in chloroform, ethanol, methanol and other organic solvents. They are formed maximally when the mycobacteria are grown with insufficient iron in the medium (Antoine & Morrison, 1968; Snow, 1970; Ratledge & Marshall, 1972), and this represents an attempt by the cell to abstract as much iron as possible from the environment in order to sustain growth.

Although most of our work has been concerned with an understanding of how mycobactin functions, we have also been curious to know if the molecules extended outside the genus *Mycobacterium* and if so whether it possible to use their presence, and even structure, for diagnostic and taxonomic purposes (Patel & Ratledge, 1973). We have examined therefore representative strains of the major taxonomic clusters of the nocardiae recovered by Goodfellow (1971) for the presence of these compounds, and also

(a)

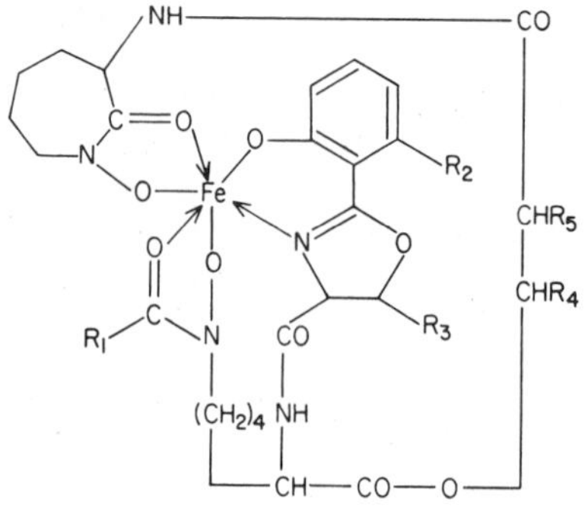

(b)

Organism	Mycobactin isolated	Substituents				
		R_1	R_2	R_3	R_4	R_5
Mycobacterium fortuitum	F*	17, 11 Δ	H	CH_3	CH_3	H
M. thermoresistible	H	19, 17 Δ	CH_3	CH_3	CH_3	H
M. marinum	M	1	H	CH_3	$C_{17}H_{35}$	CH_3
M. phlei	P	17 *cis* Δ	CH_3	H	C_2H_5	CH_3
M. smegmatis	S	17, 15 *cis* Δ	H	H	CH_3	H
M. tuberculosis	T	19 Δ	H	H	CH_3	H

* Occurs as mixture with mycobactin H.

Fig. 1. The mycobactins. (a) General structure of the ferric mycobactins. (b) Substituents of selected mycobactins. Side chains R_1 are alkyl groups having the number of carbon atoms shown (double bonds are indicated where known), alkyl groups with different numbers of carbon atoms have been identified in trace quantities. (From Snow, 1970.)

have looked at other related organisms such as those from the "*rhodochrous*" complex and a species of *Gordona.*

2. Isolation of Mycobactin-Like Compounds from Nocardiae

So far we have found the presence of mycobactin-like materials in all the nocardiae examined and have also found a similar material in *Gordona rubra* (Table 1). Lipid-soluble iron-binding material could not be detected in organisms from the *rhodochrous* complex thereby indicating a distinct division between these organisms and the nocardiae and mycobacteria. As we have only been

Table 1
Organisms examined for lipid-soluble, iron-binding compounds

	Maximum yield of material found (mg/g dry wt)
Nocardia	
N. asteroides (N13; N216; ATCC3318)	72–109
N. brasiliensis (N14; N318; N465; N557)	42–95
N. caviae (N36; N231)	49–89
N. paraffinae (N102)	10
N. phenotolerans (N100)	87
N. sylvodorifera (N217)	58
N. uniformis (N3)	59
Organisms of the ***rhodochrous*** *complex*	
Jensenia canicruria (N53)	No material found
Mycobacterium rhodochrous (N108)	No material found
Nocardia opaca (N38; N124)	No material found
N. pellegrino (N324; N325)	No material found
N. restrictus (N7)	No material found
N. salmonicolor (N5)	No material found
Others	
Gordona rubra	17

The numbering system of strains is that of Goodfellow (1971).

able to examine one species of *Gordona* it would be premature to make any conclusion concerning a distinction between this organism and the *rhodochrous* organisms.

The next problem was to see if the materials from the nocardiae, which we have termed nocobactins, were different in structure to the mycobactins and, if so, whether they could be easily distinguished from them by simple physical procedures rather than having to resort to a complete determination of chemical structure on each occasion. As a preliminary step in this work it was necessary to devise a streamlined method for isolating and purifying these materials as the original procedures used by Snow (1965) involved repeated chromatography on large columns of alumina and took about a week to accomplish purification. Figure 2 summarizes the procedure now used. Once cells have been extracted in ethanol, the iron-binding material can be purified in the course of a single day.

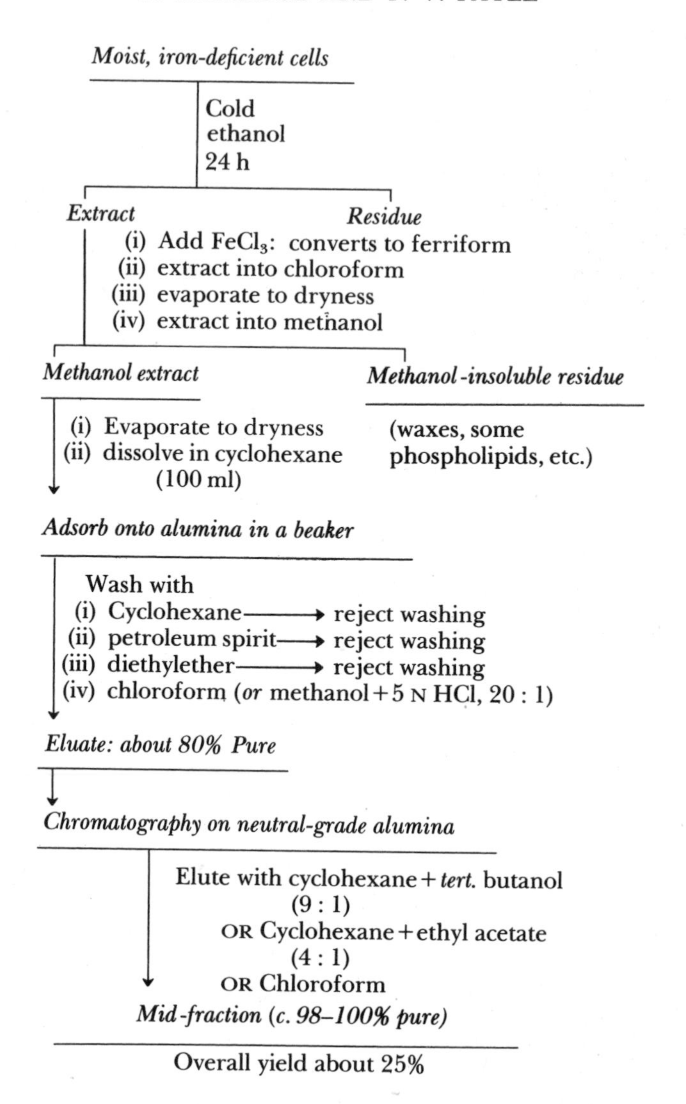

Fig. 2. Scheme for isolating and purifying nocobactins.

Like the mycobactins, the nocobactins are produced in greatly increased amounts during iron-deficient growth and have similar general properties to the mycobactins (Table 2). The maximum amounts of nocobactin produced are around 90 mg/g cell dry wt at 0·05 μg iron/ml of medium. The function of the mycobactins, and probably of the nocobactins also, is to transport iron across the

Table 2

Summary of properties of the nocobactins and mycobactins

Greatly increased formation with iron-deficient growth
Intracellularly located
Extractable with cold ethanol
Insoluble in water; soluble in chloroform, ethanol etc.
Binds with trivalent ions only
Red complexes formed with Fe^{3+}
Requires iron to be presented in a soluble form; will not complex with insoluble or colloidal iron
Functions in the transport of iron across the lipid-rich cell envelope
Iron released into cell by enzymic reduction with NADH

lipid-rich cell envelope (Ratledge & Marshall, 1972). Iron is released from the carrier by enzymic reduction with NAD(P)H; the Fe^{2+}-mycobactin complex readily dissociates as the Fe^{2+} ion has no affinity for the mycobactin. The Fe^{2+} ion is then probably inserted into both heme and non-heme iron proteins.

3. Properties of the Nocobactins

Thin-layer chromatography (t.l.c.) using silica gel G or H with petroleum spirit (b.p. 60°–80°C)-butan-l-ol-ethyl acetate (2 : 3 : 3 by vol.) has been used to separate the nocobactins into three groups. These are the nocobactins of *Nocardia asteroides*, *N. brasiliensis* and *N. caviae* (Table 3). In this particular system there is, however, difficulty in distinguishing the materials from *N. brasiliensis* from some of the mycobactins although preliminary work with other solvent systems and absorbents has indicated that better chromatography systems can probably be devised to separate these two materials. Interestingly, the material from the strain received as *N. asteroides* N216 showed itself to clearly belong to the caviae nocobactins and this has been confirmed in other analyses (see below). The identity of this strain therefore must be questioned and it should be re-examined using conventional taxonomic methods. The nocobactin from *N. uniformis* N3 gave three distinct spots of R_F values 0·25, 0·31 and 0·50 upon t.l.c. in the system described in Table 3. Further examination of each of these three components has not yet been accomplished in detail. The material

Table 3
Thin-layer chromatographic separation of the nocobactins and mycobactins (silica gel G + petroleum spirit-n-butanol-ethyl acetate, 2 : 3 : 3, by vol.)

Organism	R_F of nocobactin or mycobactin	R_R*	Proposed grouping
Nocardia asteroides N13	0·23	0·53	A
N. asteroides ATCC3318	0·23	0·53	A
N. paraffinae N102	0·22	0·51	A
N. sylvodorifera N217	0·24	0·55	A
N. brasiliensis (4 strains)	0·43–0·46	1·0–1·06	B
N. asteroides N216	0·37	0·86	C
N. caviae N36	0·33	0·77	C
N. caviae N231	0·35	0·81	C
N. phenotolerans N100	0·38	0·88	C
Mycobacterium avium	0·44	1·02	
M. smegmatis	0·43	1·00	
M. tuberculosis	0·44	1·02	

* R_R = relative to R_F of mycobactin S.

from *G. rubra* chromatographed with the same mobility as the materials from *N. caviae*.

The organisms known as *N. paraffinae* (N102) and *N. sylvodorifera* (N217) are clearly of the *N. asteroides* type judging from the properties and structures of their nocobactins. This conclusion reinforces the current view, based on other taxonomic determinants, that these names should be discontinued in favour of *Nocardia asteroides*. (We have, however, retained these names here to help demonstrate the usefulness of the nocobactins as chemotaxonic characters but we do not subscribe to their continued usage.)

Another property of mycobactins and nocobactins which may help to distinguish them is their ability to dissociate from the ferric atom at an acidic pH. The typical dissociation curves of the materials from *N. asteroides* and *N. caviae* are shown in Fig. 3. Nocobactins of the *b*(= *brasiliensis*) type precipitate in the acidic solution and their dissociation cannot be determined. However, this may be

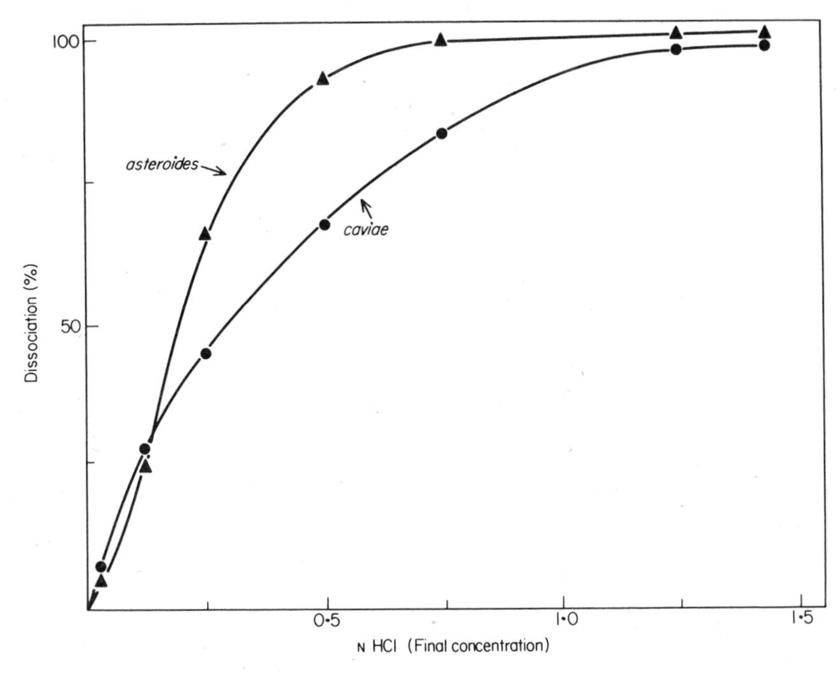

Fig. 3. Dissociation of ferri-nocobactins from *Nocardia asteroides* and *N. caviae*. Ferri-nocobactin dissolved in 3 ml methanol and 1 ml of HCl of varying normalities added; the extinction of the solution at 450 nm was then determined against a reference containing ferri-nocobactin in 5·0 N HCl (i.e. completely dissociated). The normality of HCl given is its final concentration; i.e. $\frac{1}{4}$ of that added.

a useful characteristic to distinguish them from the mycobactins although more strains belonging to this species will have to be examined before this is firmly established. The dissociation values, summarized in Table 4, again permit three types of nocobactin to be recognized. The inclusion of the nocobactin from the dubious strain of *N. asteroides* in the *c*(= *caviae*) group is not as certain as with the chromatography analysis. However, the u.v. spectra of this material clearly distinguishes it from all the other nocobactins of the *a*(= *asteroides*) type (Table 5).

Nocobactins of the *a* type give a unique u.v. absorption spectrum when in the desferri form (Fig. 4). In the ferriform the fine structure is lost and the material shows a λ_{max} at 460 nm. Values for the various extinction coefficients have been given elsewhere (Ratledge & Snow, 1974). In comparison the u.v. spectra of the other nocobactins of *b* and *c* types are not so detailed and are close to

Table 4

Dissociation of nocobactins and mycobactins (Fe^{3+} nocobactin+3HCl$\rightleftharpoons$nocobactin+$FeCl_3$)

Organism	Nomality of HCl required for 50% dissociation*	Proposed groups
Nocardia asteroides N13	0·062	A
N. asteroides ATCC3318	0·192	A
N. sylvodorifera N217	0·175	A
N. brasiliensis (4 strains)	All insoluble in MeOH/HCl	B
N. asteroides N216	0·19	C?
N. caviae N36	0·325	C
N. caviae N231	0·312	C
N. phenotolerans N100	0·25	C
Mycobacterium avium	0·350	
M. smegmatis	0·312	

* Dissociation determined as given in Fig. 3; normality of HCl is its final concentration.

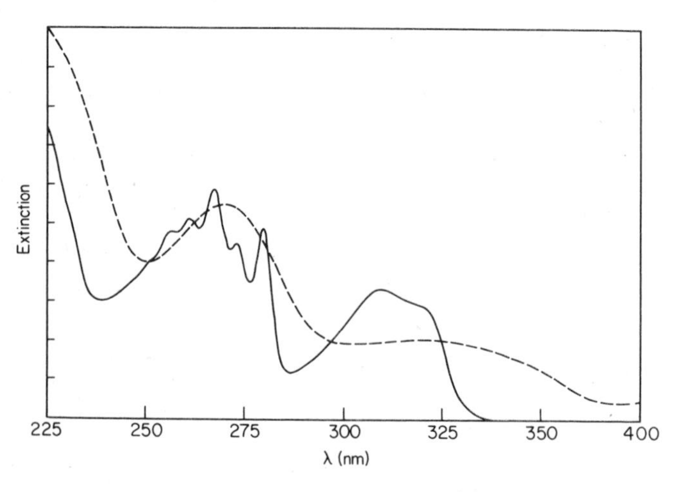

Fig. 4. u.v. spectra of nocobactin *a* type. (———) Desferri-nocobactin; (– – –) ferri-nocobactin. Both determined as solutions in ethanol.

Table 5
Summary of u.v. spectra of desferri nocobactins (in ethanol)

Organism	Spectrum ($u.v._{max}$)	Proposed grouping
Nocardia asteroides N13	multipeak type (see Fig. 4)	A
N. asteroides ATCC3318	multipeak type	A
N. paraffinae N102	multipeak type	A
N. sylvodorifera N217	multipeak type	A
N. uniformis N3	multipeak type	A
Mycobacterium phlei	252, 267, 313	Methyl gp on benzene ring
M. smegmatis	242, 248, 258, 304	Not resolvable
M. tuberculosis	242, 249, 258, 304	Not resolvable
N. asteroides N216	240, 248, 259, 305	Not resolvable
N. brasiliensis (all 4 strains)	240, 247, 260, 305	Not resolvable
N. caviae N36 and N231	242, 249, 260, 303	Not resolvable

those given by the mycobactins indicating a probable equivalence of the aromatic moieties of the molecules. This data is summarized in Table 5. Molecules having a methyl group at the 6 position of the salicylic acid moiety (e.g. mycobactin P) show a slight change in u.v. absorption but this variation has not yet been detected in any of the nocobactins. The nocobactin from *N. uniformis* N3, although exhibiting a mixture of three materials (see above) clearly falls into the nocobactin *a* grouping by its u.v. spectrum and there was no suggestion that its spectrum involved a mixture of two types.

4. Structure of the Nocobactins

Although simple analytical procedures (t.l.c., dissociation values and u.v. spectroscopy) can resolve the three types of nocobactin and distinguish each of them from the mycobactins, for these to have any definitive value the structure of each new material must be determined. If there is no major innovation in the molecule, NMR spectroscopy coupled with mass spectroscopy of the aluminium derivative is usually sufficient for this purpose (Snow &

White, 1969; Greatbanks & Bedford, 1969). However with nocobactins of the *a* type a novel feature of the molecule was indicated by its unique u.v. spectra and the structure of this material could only be elucidated following degradation and synthesis of the aromatic part of the molecule (Ratledge & Snow, 1974).

We have found that the three types of nocobactins are chemically distinct from each other as well as from the mycobactins (Fig. 5). Nocobactins of the *a* type are characterized by an oxazole ring

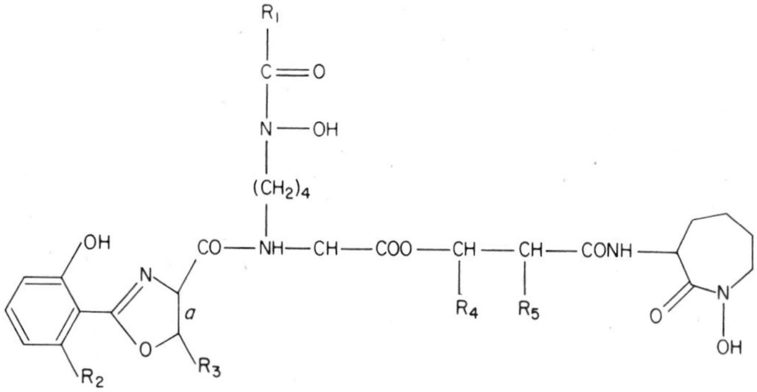

			Substituents			
	a	R_1	R_2	R_3	R_4	R_5
Nocobactin *a* type	Double bond	CH_3	H	CH_3	$C_{11}H_{23}$	CH_3
Nocobactin *b* type	Single bond	alkyl (saturated)	H	CH_3	CH_3	H
Nocobactin *c* type	Single bond	alkyl (saturated)	H	H	CH_3	H
Mycobactin P, S and T types	Single bond	alkyl (unsaturated)	H or CH_3	H or CH_3	CH_3 or C_2H_5	H or CH_3
Mycobactin M type	Single bond	CH_3	H	CH_3	$C_{17}H_{35}$	CH_3

Fig. 5. Structures of the three main types of nocobactins compared with mycobactin structures.

instead of an oxazoline ring and have the aliphatic chain, which is somewhat shorter than usual, in R_4 position (≡ mycobactin M).

Nocobactins of the *b* and *c* type each have long saturated alkyl chains instead of unsaturated ones found in the mycobactins. Nocobactins *b* type is resolvable from the *c* type by possession of a methyl group at R_3. Nocobactins of these two types correspond to the majority of the mycobactins in having the long alkyl side chains at position R_1.

The nocobactin from *N. sylvodorifera*, although falling into the general *a* type, showed a slight shift in u.v. spectra to shorter wavelengths indicating the substitution of a H atom for the methyl group at R_3. Similar minor structural changes might well be found when other molecules from additional species have been examined.

5. For the Future

Mycobactins, and related materials such as the nocobactins, have not yet been extensively used for identification and classification although the occurrence of these materials is undoubtedly of fairly limited distribution. It will be of interest to learn if representatives of the genus *Corynebacterium* possess related materials but it is already clear that most genera within the Actinomycetales, except for *Mycobacterium* and *Nocardia*, do not contain such materials (Ratledge & Chaudhry, 1971). At this stage, based on our fairly limited survey, organisms of the *rhodochrous* complex do not elaborate such a compound and can therefore be distinguished from the mycobacteria-nocardiae-gordonae group. The possibility of distinguishing between these three taxa as well as being able to identify individual species within them now seems entirely feasible using the variations in substituents found in the various mycobactins. Clearly, however, a much larger number of organisms need to be studied to establish whether variations do not occur with different strains belonging to the same species.

The principal difficulty which must be overcome before this method of classification and identification can be adopted is the ability to cause maximum production of these compounds under normal laboratory conditions so that a test for their presence may be carried out without resorting to an elaborate purification of medium and cleansing of glassware. Unfortunately this has not yet been possible. The addition of several iron chelators to the medium have been tried in an attempt to decrease the amount of iron available to the cells and thus derepress the biosynthesis of mycobactin but without success. The only expedient which has so far been found of slight value is the cultivation of the organisms under stationary conditions. This technique however only works when a substantial pellicle of cells is formed as this leads to a heterogeneous population of bacteria wherein the bacteria at the top of the pellicle appear deprived of nutrients, including iron, and synthesize mycobactin in reasonable quantities (Ratledge & Hall, 1971). However, work in our laboratories on this aspect is continuing.

An elucidation of the structure of a mycobactin is probably beyond the scope of many diagnostic laboratories but provided adequate alternative procedures can be found this should not be a handicap. The most useful techniques found to date in our laboratory are t.l.c. and u.v. spectroscopy. The latter technique is however only able to distinguish nocobactins of the *asteroides* type from the others. It may not be unreasonable to expect that further techniques, or improvements in the established ones, will be found which will assist in the identification of the various lipid-soluble, iron-binding compounds which are found in the nocardiae and related organisms.

6. Acknowledgements

The Wellcome Trust is thanked for a research studentship to P.V.P. We are indebted to Dr M. Goodfellow for selecting and supplying all cultures examined in this work.

7. References

Antoine, A. D. & Morrison, N. E. (1968). Effect of iron nutrition on the bound hydroxylamine content of *Mycobacterium phlei*. *J. Bact.* **95,** 245.

Bullen, J. J., Rogers, H. J. & Griffiths, E. (1972). Iron-binding proteins and infection. *Br. J. Haemat.* **23,** 389.

Goodfellow, M. (1971). Numerical taxonomy of some nocardioform bacteria. *J. gen. Microbiol.* **69,** 33.

Greatbanks, D. & Bedford, G. R. (1969). Identification of mycobactins by nuclear-magnetic-resonance spectroscopy. *Biochem. J.* **115,** 1047.

Lankford, C. E. (1973). Bacterial assimilation of iron. *CRC Crit. Rev. Microbiol.* **2,** 273.

Macham, L. P. & Ratledge, C. (1975). Detection of a new group of water-soluble, iron-binding compounds from *Mycobacterium smegmatis*: the exochelins. *J. gen. Microbiol.* **89,** 379.

Patel, P. V. & Ratledge, C. (1973). Isolaton of lipid-soluble compounds that bind ferric ions from *Nocardia* species. *Biochem. Soc. Trans.* **1,** 886.

Ratledge, C. & Chaudhry, M. A. (1971). Accumulation of iron-binding phenolic acids by Actinomycetales and other organisms related to the mycobacteria. *J. gen. Microbiol.* **66,** 71.

Ratledge, C. & Hall, M. J. (1971). Influence of metal ions on the formation of mycobactin and salicylic acid in *Mycobacterium smegmatis* grown in static culture. *J. Bact.* **108,** 314.

Ratledge, C., Macham, L. P., Brown, K. A. & Marshall, B. J. (1974). Iron transport in *Mycobacterium smegmatis*: a restricted role for salicylic acid in the extracellular environment. *Biochim. biophys. Acta* **372,** 39.

Ratledge, C. & Marshall, B. J. (1972). Iron transport in *Mycobacterium smegmatis*: the role of mycobactin. *Biochim. biophys. Acta* **279,** 58.

Ratledge, C. & Snow, G. A. (1974). Isolation and structure of nocobactin NA, lipid-soluble iron-binding compound from *Nocardia asteroides*. *Biochim. J.* **139,** 407.

Snow, G. A. (1965). Isolation and structure of mycobactin T, a growth factor from *Mycobacterium tuberculosis*. *Biochem. J.* **97,** 166.

Snow, G. A. (1970). Mycobactins: iron-chelating growth factors from mycobacteria. *Bact. Rev.* **34,** 99.

Snow, G. A. & White, A. J. (1969). Chemical and biological properties of mycobactins isolated from various mycobacteria. *Biochem. J.* **115,** 1031.

Szabo, I. (1971). The effect of iron and iron-8-oxyquinoline chelate on the virulence of bacteria. *Zentbl. Bakt. ParasitKde Abt. I* **218,** 365.

15. Possible Mechanisms of Nocardial Pathogenesis

B. L. BEAMAN

Department of Medical Microbiology, University of California School of Medicine, Davis, California 95616, U.S.A.

Contents

1. Introduction

The pathogenic nocardiae are variable in their host-parasite response. Generally two basic types of disease are recognized in man (González-Ochoa, 1962). Nocardiosis is primarily a pulmonary infection which frequently disseminates by way of the blood stream and lymphatics to other parts of the body (Aron & Gordon, 1972; Bates & Rifkind, 1971; Freese *et al.*, 1963; Presant, Wiernick & Serpick, 1973; Webster, 1956; Wilhite & Cole, 1966). However, it may also be a primary disease of the central nervous system, kidneys, heart, eyes or other organs (Bach, Monaco & Finland, 1973; Berd, 1973*a*; Benedict & Iverson, 1944; Burpee & Starke, 1971; Cross & Binford, 1962; Failes & Posney, 1965; Hathaway & Mason, 1962; Henderson, Wellman & Weed, 1960; Hoeprich, Brandt & Parker, 1968; Meyer, Font & Shaver, 1970). Clinically and pathologically nocardiosis may be extremely variable presenting as pyogenic infections, encapsulated abscesses, or even granulomas (Langevin & Katz, 1964; Palmer, Harvey & Wheeler, 1974; Presant, Wiernick & Serpick, 1970; Richter *et al.*, 1968;

Susens *et al.*, 1967). Typically nocardial infections are chronic and progressive, however, acute, fulminating nocardiosis have been reported (Carile, Holley & Logan, 1963; Freese *et al.*, 1963; Langeven & Katz, 1964; Palmer, Harvey & Wheeler, 1974; Presant, Wiernick & Serpick, 1970, 1973; Richter *et al.*, 1968; Susens *et al.*, 1967; Whyte & Kaplan, 1969). Nocardial mycetomas are characterized as being chronic granulomatous infections that progressively worsen over a period of several months or years. These mycetomas usually remain localized with spread being by direct extension through the tissues (González-Ochoa, 1962). Further, suppuration and sinus tracts are frequently observed with colonies of bacteria (granules) found within the pus similar to *Actinomyces* infections (Bronner & Bronner, 1971; Greer, 1974; González-Ochoa, 1962).

Three species of *Nocardia* are most frequently recognized in human infections. Nocardiosis is usually caused by *N. asteroides*, but *N. brasiliensis* and *N. caviae* have also been recognized (Bates & Rifkind, 1971; Causey, Arvail & Brinker, 1974; Causey & Sieger, 1974; Diamond & Bennett, 1973). Nocardial mycetoma on the other hand is generally caused by *N. brasiliensis* (Berd, 1973*b*; González-Ochoa, 1962; Green & Adams, 1964; Hogshead & Stein, 1970). There are other actinomycetes that cause mycetomas in man. These include species of the genus *Actinomadura* and will not be discussed here.

2. Experimental Nocardial Infections in Animals

Reports using mice, guinea pigs and rabbits as animal models for studying nocardial pathogenicity have often had contradictory interpretations. For example, González-Ochoa (1973) studied the virulence of *N. asteroides, N. brasiliensis* and *N. caviae* in the foot pads of mice. He concluded that *N. brasiliensis* was more virulent for mice than *N. asteroides* and *N. caviae.* In contrast, Uesaka *et al.* (1971) concluded that *N. asteroides* strains were more virulent for mice when injected into the peritoneal cavity. Other investigators have stated that nocardiae are not pathogenic for mice and that guinea pigs or rabbits must be used. Further, many reports have stressed that adjuvants such as oil or hog gastric mucin enhanced virulence (Gezuele, 1972; Mason & Hathaway, 1969;

Macotela-Ruiz & Mariat, 1963; Newmark, Polack & Ellison, 1971; Strauss & Kligman, 1951; Uesaka *et al.*, 1971). Beaman found that by injecting nocardiae suspended in either saline or mineral oil into the peritoneal cavity of mice, strains of organisms which were highly virulent, less virulent, lowly virulent and non-virulent could be determined. It was found that different isolates of *N. asteroides* varied significantly in their ability to cause disease in young mice. For example, some strains would produce fatal, progressive disease when as few as 10^6 organisms suspended in saline were injected into the peritoneal cavity. In contrast, other strains of *N. asteroides* would not establish progressive, fatal infections even when 10^9 organisms were injected (Beaman, 1973, 1974; Beaman, unpublished data).

Beaman (1973) noted that during the establishment of nocardiae into mouse tissues certain properties of the organisms became altered. It was observed that organisms which were less virulent for mice became more greatly altered during adaptation to the parasitic state than those strains which were more aggressive.

It has been shown that many strains of *Nocardia* as well as other bacteria can undergo an altered morphological development when grown *in vitro* on certain media (Beaman, 1968; Beaman & Shankel, 1969; Duguid & Wilkinson, 1961; Howell & Jordan, 1965; Webley, 1960). Ultrastructural analysis demonstrated that variations in cell wall appearance, cytoplasmic membrane organization and other cytoplasmic structures occurred concomitantly with the morphological modification induced *in vitro* (Beaman, 1968; Beaman & Shankel, 1969; Beaman, Burnside & O'Donnell, 1974). It is reasonable to assume that these structural changes are accompanied by variations in the chemical composition of the affected cell components. As a consequence, the invasiveness of Nocardia as well as the host-parasite relationship could be greatly effected if the same types of modifications observed *in vitro* occurred *in vivo.*

In order to survive within the host and cause disease the nocardiae must be able to neutralize or survive the many defense mechanisms of the host. At the same time they must be able to use the body material as a growth medium. No exotoxins have been described for nocardiae; they do not multiply fast enough within host tissues to give an overwhelming acute infection (nocardial infections are characteristically chronic and progressive); nor is there any evidence that they evade or resist engulfment by phagocytic cells. Most data indicate that nocardiae are readily phagocytized by both polymorphonuclear cells and macrophages (Beaman, 1973; Beaman, 1974; Bourgeois & Beaman, 1974; Uesaka *et al.*, 1971). Further it appears that the pathogenic nocardiae can persist and grow in the intracellular environment, especially within macrophages (Bourgeois & Beaman, 1974). Therefore *Nocardia* should be considered as a facultative intracellular parasite.

It has already been pointed out that all strains of *Nocardia asteroides* are not equally invasive for mice. Secondly, structural and histochemical changes occur as the nocardial cells adapt to growth *in vivo*. Finally, it appears that the less virulent organisms must undergo a greater adaptation than the more invasive strains (Beaman, 1973; Beaman, 1974). These data suggest that different strains of *N. asteroides* are chemically different, and that the chemical nature of the nocardial cells grown in host tissues differ from those grown on such artificial media as Brain Heart Infusion (BHI) and Sauton's agar.

D'Arcy Hart (1968) pointed out that mycobacteria differ in their *in vitro* activities from those *in vivo*. Bekierkunst (1968) noted that mycobacteria grown *in vivo* were not as hydrophobic as the same organisms grown on artificial media. In addition, the *in vivo* grown bacilli were more resistant to 5% NaOH, were more readily stained by Sudan black B, did not readily bind neutral red, and were more virulent than those grown *in vitro* (Bekierkunst, 1968). These observations strongly suggested that changes in lipid content occurred in cells of mycobacteria during growth *in vivo*. Segal & Miller (1965) demonstrated that there was a 52% increase in ether-alcohol extractable lipids as well as a 31% increase in acidified ether-alcohol extractable lipids in mycobacterial cells grown *in vivo* over those grown *in vitro*. Segal (1965) found that tubercle bacilli grown *in vivo* possessed a lower immunizing capacity than those grown *in vitro*. Earlier, Raffel (1953) suggested that possibly the tubercle bacillus during the infection process formed antigens not produced during growth *in vitro*. Kondo & Kanai (1972) reported that the ethanol-ether extractable lipids (fats, phosphatides and wax A) represented 43·7% of the cell weight of *Mycobacterium bovis* grown *in vivo* as compared to 21·3% of the total weight of cells grown *in vitro*. In contrast they found that the chloroform soluble "waxes" and the firmly bound lipids were significantly reduced in *M. bovis* cells grown *in vivo* (Kondo & Kanai, 1972). However, Kanai & Kondo (1974) point out that even though differences exist between *in vitro* and *in vivo* grown mycobacteria, most of the basic chemical components (i.e. tuberculostearic acid, mycolic acid, etc.) of mycobacteria grown *in vitro* can also be detected in cells grown *in vivo*. However, it is known that organisms grown *in vivo* absorb components such as cholesterol from the host. Therefore some of the differences observed between *in vitro* and *in vivo* grown bacteria probably represent absorption of host material (Bekierkunst, 1968).

The staining and histochemical properties of nocardiae grown in animals may give some indication as to what types of changes occur during *in vivo* growth. For example Beaman (1973, 1974) and Beaman & Burnside (1973) have noted that most strains of pathogenic nocardiae are not acid alcohol fast when grown in BHI, but when these same organisms were inoculated into mice they became strongly acid-alcohol-fast after 48–72 h of growth in the host tissues. Although the exact mechanism of the acid-fast stain is not clear, it seems likely that the accumulation of high molecular weight lipids (especially the mycolic acids) bound to the cell wall is somehow responsible for the acid fast staining (Asselineau, 1966). Further, the way this material is bound to the wall appears to be responsible for the degree of acid-fastness (Beaman & Burnside, 1973; Fisher & Barksdale, 1973; Convit & Pinardi, 1972). Therefore these observations concerning the *in vivo* acid-fast staining properties of nocardiae strongly suggest that changes in the "cell wall associated lipids" had occurred. Additional evidence suggesting cell wall changes during *in vivo* growth of nocardiae was obtained using the Gram stain. Organisms which were uniformly Gram-positive *in vitro* became Gram-variable, with beading becoming prominent in many strains. In addition some strains were shown to form pleomorphic, Gram-negative cells when grown in animals without apparent loss of nocardial viability (Beaman, 1973, 1974; Uesaka *et al.*, 1971).

Electron microscopy showed that changes in the cell walls of nocardiae occurred during growth *in vivo* (Beaman, 1973). For example it was found that cells of *N. asteroides* 10905 possessed an enlarged, granular outer layer that appeared to be membrane bound (Beaman, 1973). The exact chemical nature of this material has not yet been established, however, a similar zone surrounding cells of *Actinomyces* and *Nocardia brasiliensis* was thought to be calcium deposits of host origin (Macotela-Ruiz & González-Angulo, 1966; Overman & Pine, 1963; Pine & Overman, 1963). This appears not to be the case with *Nocardia* because a similar zone surrounding nocardial cells was observed during early lag phase of growth *in vitro* (Beaman, Burnside & O'Donnell, 1974*a*), and it was present on altered cells of *N. asteroides* 10905 isolated from macrophages and grown in BHI (Bourgeois & Beaman, 1974). These outer layers did not stain positively for calcium, and preliminary data indicated that this material was composed of lipid, polysaccharide, and peptide of bacterial origin (Beaman, Burnside & O'Donnell, 1974; Beaman, 1974). Additional evidence was obtained *in vivo* that this material was not calcium by the inability to detect calcium histochemically (Beaman, 1974).

Intracellular mycobacteria possess an enlarged outer zone surrounding the organisms. Generally this area appears to be electron lucent and empty which is in contrast to the granular or membranous zone surrounding intracellular nocardiae (Draper & Rees, 1973; Beaman, 1973, 1974; Chapman, Hanks & Wallace, 1959; Edwards, 1970; Imaeda & Convit, 1962; Imaeda, 1965). Draper & Rees (1973) studied the nature of this material around *M. lepraemurium* isolated from livers and spleens of mice. They observed that this zone consisted of parallel fibers wrapped longitudinally around the mycobacterial cell. The purified material was composed of a peptidoglycolipid. They suggested that this substance formed a capsule around the bacterium thus protecting it from the host cell lysosomal enzymes (Draper & Rees, 1973).

3. Host Cell-Parasite Interactions in Nocardial Infections

Macrophages constitute a diversity of reticuloendothelial cells that have their origin in the bone marrow. They are present in all organs of the body and their morphology and metabolism varies depending upon their specific anatomical location and their state of activity within the host. However, all macrophages are mononuclear cells that are capable of phagocytizing particles; storing intra-vital dyes; their cytoplasm contains lysosomes, mitochondria, endoplasmic reticulum and golgi; they are frequently defined as being able to adhere to glass; and their prime role within the host appears to be phagocytosis and intracellular digestion or storage of foreign particles, including dead host cells (Carr, 1973; Gordon & Cohn, 1973; Pearsall & Weiser, 1970; Van Furth, 1970; Vernon-Roberts, 1972; Weiss, 1972).

Polymorphonuclear leukocytes (PMN leukocytes) also develop in the bone marrow. The mature PMN is about 10–15 μm in diameter; has a multilobed nucleus; and has a cytoplasm filled with granules rich in lytic enzymes, little or no endoplasmic reticulum and golgi, and a few small mitochondria. The PMN's primary function within the host is phagocytosis, and they are designed to ingest, neutralize and if possible destroy ingested particles. Therefore both macrophages and PMN's represent a major defense mechanism of the body against infectious agents (Cohn, 1963; Movat, 1971; Simmons & Karnovsky, 1973).

Macrophages and PMN leukocytes can function both nonspecifically and specifically in bacterial infections. That is, they readily eliminate bacteria from contaminated tissue through nonselective phagocytosis of particulate material. In addition both cell types are under the control of humoral and cell mediated immune mechanisms which act specifically in controlling the disease process (Carr, 1973; Cohn, 1963; Mavot, 1971; Simmons & Karnovsky, 1973; Vernon-Roberts, 1972; Weiss, 1972).

Once an infectious agent, such as Nocardia, is phagocytized it may be killed and digested; it may remain viable and possibly replicate or persist within the cell in an altered form; or the antigens of the organism may initiate a cellular and humoral immune response. Many factors influence the fate of the ingested organisms. For example, exposure of macrophages to stimuli such as bacteria or certain bacterial products results in ultrastructural and physiological alterations of the cell. As a consequence of this "macrophage activation" the cells become metabolically more active with an accompanying increase in lysosomal elements. If the macrophage survives exposure to this ingested organism it is said to be "immune," and it becomes a very potent bacteriocidal agent

(Carr, 1973; Pearsall & Weiser, 1970; Solotorovsky & Soderberg, 1972, Weiss, 1972). The route of the inoculation; the metabolic state of both the macrophages and PMN's; the availability, types and concentrations of the lysosomal enzymes within these cell types; the presence of opsonins; the presence of non-specific factors such as complement; the state of the host; and the state of the bacterial invader are additional factors that determine the fate of the organisms entering the body (Carr, 1973; Cohn, 1963; Gordon & Cohn, 1973; Movat, 1971; Pearsall & Weiser, 1970; Simmons & Karnovsky, 1973; Vernon-Roberts, 1972; Solotorovsky & Soderberg, 1972; Weiss, 1972). The specific mechanisms of macrophage and PMN leukocyte interactions in inflamation and body defence is beyond the purview of this review, therefore the readers may wish to consult some of the more recent reviews and books concerning this important topic (Bowden, 1971; Carr, 1973; Dumont, 1972; Gordon & Cohen, 1973; Laskin & Lechevalier, 1972; McCluskey & Cohen, 1974; Movat, 1971; Page, Davies & Allison, 1974; Vernon-Roberts, 1972; Weiss, 1972).

Chronic microbial infections exemplified by tuberculosis result from the organism's ability to resist destruction within the host's macrophages (Solotorovsky & Soderberg, 1972). The organisms can persist within these cells for extended periods of time and may replicate. These types of intracellular parasites are usually limited to macrophages because the macrophages live long enough to permit the parasite to grow. In contrast PMN leukocytes appear not to harbor intracellular parasites since the PMNs are too short lived to permit intracellular growth of the invading organism. Also PMNs may be more effective bacterial killers than macrophages (Carr, 1973; Cohn, 1963; Dumont, 1972; Gordon & Cohen, 1973; Movat, 1971; Page, Davies & Allison, 1974; Solotorovsky & Soderberg, 1972; Vernon-Roberts, 1972).

In pulmonary nocardiosis the pathogenic nocardiae usually enter the host by way of inhalation. The primary mechanisms of removal of these organisms are by ciliary action of the epithelium and by phagocytosis by alveolar macrophages. At this initial stage of contact, PMN leukocytes are probably less important since it has been shown that there are very few PMNs in uninfected lungs (Bowden, 1971; Green, 1973; Truitt & Mackaness, 1971). Therefore, the early events of pulmonary nocardiosis probably involve a macrophage-nocardial interaction. If the macrophage successfully removes the nocardiae, then progressive infection does not occur. On the other hand, if the macrophage cannot destroy the invading nocardiae, an inflammatory lesion ensues and a pneumonitis becomes established (Truitt & Mackaness, 1971; Freese *et al.*, 1963; Palmer, Harvey & Wheeler, 1974; Presant, Wiernick & Serpick, 1973; Langevin & Katz, 1964; Webster, 1956; Wilhite & Cole, 1966). Once the inflammatory response begins, there is an infiltration of PMN leukocytes with the resultant formation of purulent exudate. A necrotic abscess usually results which, with time, may become surrounded by granulation tissue. Macrophages once again become prominant. Typical granulomas composed of macrophages, lymphocytes, epitheloïd cells and giant cells have been reported to occur in man as a result of nocardial infection; however, this type of lesion is most frequently observed in experimental animals (Langevin & Katz, 1964; Richter *et al.*, 1968; Uesaka *et al.*, 1971).

One of the basic hypotheses explaining microbial invasion of compromised individuals by opportunistic organisms such as *Nocardia* is that a deficiency or an alteration in macrophage function occurs in association with the disease or in response to corticosteroid treatment of the host (Pearsall & Weiser, 1970; Vernon-Roberts, 1972; Mishra *et al.*, 1973). It has been suggested, for example, that corticosteroid treatment alters the ability of the lysosomes to fuse with the phagosome, thus the lysosomal enzymes would not reach and kill the ingested organism (Epstein, Verney & Miale, 1967). Further it has been shown the cyclophosphamide (an immunosuppressive drug) renders the macrophage less able to kill by affecting the lysosomal enzymes (Lochard *et al.*, 1971).

Certain microorganisms and their products may have similar effects as corticosteroids and immunosuppressive drugs on macrophage function. It was found that mycobacteria phagocytized by macrophages caused a decrease in the ability of lysosomes to fuse with the phagocytic vesicle. It was postulated that the organisms would then be able to escape the digestive enzymes (Armstrong & D'Arcy Hart, 1973). In addition, it has been found that certain intracellular mycobacteria were surrounded by a layer of peptidoglycolipid that seemed to act as a protective capsule (Draper & Rees, 1973). Because of the close biochemical and biological relationship between the mycobacteria and nocardiae, it seems reasonable to suggest that nocardiae behave similarly to the mycobacteria in their ability to grow intracellularly in macrophages.

The role of the humoral immune response to nocardial infections have not been adequately investigated, however, it is clear that cell mediated immunity is induced by nocardiae (Bojalil & Jamora, 1963; Bojalil & Magnusson, 1963; Kingsbury & Slack, 1969; Ortiz-Ortiz & Bojalil, 1972; Ortiz-Ortiz, Bojalil & Contreras, 1973; Ortiz-Ortiz, Contreras & Bojalil, 1973*a*, Pier, Thurston & Larsen, 1968). Cell mediated immunity is frequently stimulated by organisms which are protected from humoral factors by being intracellular, usually within macrophages; and by possessing cell components (such as the complex lipids of mycobacteria) not readily digested by the host cells (Movat, 1971; Truitt & Mackaness, 1971). Delayed hypersensitivity is a manifestation of cell mediated immunity, and it can be used to measure a cell mediated response (Movat, 1971). Several investigators have studied the delayed hypersensitivity response to nocardial antigens (Bojalil & Jamora, 1963; Bojalil & Magnusson, 1963; Kingsbury & Slack, 1969; Ortiz-Ortiz & Bojalil, 1972; Ortiz-Ortiz, Contreras & Bojalil, 1972*b*); Pier, Thurston & Larsen, 1968). It has been shown that proteins and polysaccharides from nocardiae are capable of eliciting both a delayed hypersensitivity response and inhibit macrophage migration which is an *in vitro* test for cell mediated immunity (Movat, 1971; Ortiz-Ortiz & Bojalil, 1972; Ortiz-Ortiz, Contreras & Bojalil, 1972*a*). These data suggest that the mechanisms of disease caused by nocardiae and the role of intracellular parasitism may be similar to the disease process produced by *Mycobacterium tuberculosis* and similar organisms.

In order to determine how *Nocardia asteroides* adapted to *in vivo* growth, several strains were used to infect *in vitro* cultured mouse peritoneal, guinea pig peritoneal and rabbit alveolar macrophages. The nocardial growth, ultrastructural appearance, and bacteriocidal activities of the macrophages were determined (Bourgeois & Beaman, 1974; Beaman, 1974; Bourgeois, 1975). It was found that with *N. asteroides* 10905, 90% of the viable colony forming units were destroyed during the first 48 h in peritoneal macrophages. After 3 days viable, recoverable organisms reached an equilibrium and remained constant for about 2 weeks. Then quite dramatically the number of organisms recovered from the macrophages increased; however, typical, whole cells of nocardiae could not be found within these macrophages during the period of increase. Nocardial antigens were shown to be present within these cells by using indirect immunofluorescent staining methods. Electron microscopy demonstrated the presence of protoplast and spheroplast-like organisms. By using an osmotically stabilized medium, Gram-negative variants of *N. asteroides* 10905 were isolated from these macrophages (Bourgeois & Beaman, 1974). In sharp contrast, other strains of *N. asteroides* (*N. asteroides* 14759 and a fresh clinical isolate, GUH-2) did not follow the same growth pattern in macrophages as the 10905 strain. *Nocardia asteroides* 14759 and GUH-2 were taken up by the macrophage, but the viable colony forming units per macrophage did not rapidly decrease. Instead, after a few hours nocardial filaments started being formed within the cultured macrophages, and after 24–48 h long, branching filaments were growing out of the host cells. After 48–72 h, microcolonies of nocardiae were all that were observed. The nocardial filaments growing out of the macrophages were characteristically beaded, with many being Gram-negative (Bourgeois & Beaman, 1974; Beaman, 1974; Bourgeois, 1975).

4. L-Forms and Nocardial Pathogenesis

Greer (1974) discussed a case of mycetoma caused by *Nocardia asteroides*. The patient had suffered a gunshot wound in his right foot in 1961. The bullet was removed, the wound healed, and he had no clinical manifestations for 4 years when he suddenly began developing nodules and discharging sinus tracts in the previously injured foot. New lesions appeared while the older ones became latent or inactive. This process continued and progressed until 1971 (10 years after the gunshot wound) when it was found to involve the popliteal fossa and lymph nodes in the groin. No granules or organisms were observed by staining preparations of the exudate obtained from lesions on the foot, in the popliteal fossa, or the inguinal lymph nodes by silver methenamine, Gram, or acid-fast procedures. However, from this material positive cultures of *N. asteroides* were obtained (Greer, 1974). This case represents one of many published reports describing a slow, progressive nocardial disease that appeared to develop after a prolonged latency (in the above case perhaps 4 years). In addition there are numerous reports of failure to microscopically demonstrate nocardiae in infected tissue even though cultures could be obtained from the lesions. Further there are several accounts in which nocardia-like organisms could be seen within the infected tissue, yet positive cultures were not obtained using standard culture conditions (Beckmeyer, 1959; Freese *et al.*, 1963; Kurup *et al.*, 1968; Maniar & Anderson, 1966; Murray *et al.*, 1961; Neu *et al.*, 1967; Stanton & Wright, 1967; Whyte & Kaplan, 1969). González-Ochoa (1973) reported on strains of nocardiae inoculated into the foot pads of mice. He noted that a large number of animals receiving strains of *N. asteroides* developed traumatic tumorous lesions during the first week, but these lesions then disappeared. After 3 weeks the apparently healed lesions once again showed signs of inflamation. This process progressed so that after 2 months the mice had well-established mycetomas (González-Ochoa, 1973). Uesaka *et al.* (1971) studied several species of *Nocardia* inoculated intraperitoneally in mice in the presence of hog gastric mucin. They often found that nocardial organisms could be isolated from tissues in which they could not be detected microscopically. These authors thought that perhaps Gram-negative variants of Nocardia were residing within these tissues. Beaman (1973) investigated the ultrastructural and histochemical properties of nocardial infections in mice, and found that certain strains of nocardiae became Gram-negative in tissues without apparent loss of viability. Further it was observed that these organisms had an altered cell morphology and modified cell wall.

These observations concerning human and animal infections strongly suggest the presence of an altered variant of Nocardia being induced *in vivo*. Further it appears that these variants play an active role in the chronic, progressive nature of nocardial infections. Cell variants of this type could provide a possible mechanism by which recurrent infections or relapses might occur. Since cell wall defective variants frequently have different antibiotic sensitivities than the parent organism, this type of conversion of the nocardiae could explain the observed differences between *in vitro* and *in vivo* drug sensitivities (Palmer, Harvey & Wheeler, 1974;

Peabody & Seabury, 1960; Runyon, 1951; Saltzman, Chick & Conant, 1962; Stites & Glezen, 1967).

Wall defective microbial variants capable of growth as non-rigid cells due to the absence or lack of a rigid cell wall are called L-forms or L-phase variants (Dienes, 1967; McGee *et al.*, 1971; Weibull, 1967). L-forms are capable of continuous growth in their altered form, producing characteristic colonies that have a "fried egg appearance" (Dienes, 1967; McGhee *et al.*, 1971). Since the organisms are Gram-negative and pleomorphic, it is very difficult to distinguish these forms within lesions by conventional light microscopy. Electron microscopy does not readily reveal their presence and based on morphologic criteria alone L-forms cannot, with certainty, be distinguished from mycoplasma. Indeed, L-forms look and behave very differently from their normal bacterial predecessor (Mattman, 1974).

Two basic types of L-forms are generally recognized. Type A L-forms have no apparent cell wall and individual cells are bounded only by a unit membrane. In contrast, type B L-forms retain the outermost layers of their cell wall, but lack components necessary for cell rigidity (Dienes, 1967). True L-forms do not easily revert to the bacterial cell type from which they were derived while transitional phase variants or L-phase variants may revert to the parental organism once the inducing agent is removed (Dienes, 1967; Mattman, 1974).

Variants of *N. asteroides* 10905 were isolated from mouse peritoneal macrophages grown *in vitro*. The cellular and colonial morphology, the Gram reaction and the osmotic sensitivity indicated that cell wall defective variants were induced by the cultured macrophages as shown in Figs 1 and 2 as compared with Fig. 3. Many of the initial isolates appeared to be transitional phase variants, since attempts to transfer them to fresh medium resulted in either no growth, or reversion to normal nocardial cells (Fig. 3). Continued cultivation and transfer of large numbers of these cells to appropriate media, however, eventually gave rise to stable L-phase cells (Fig. 4) which did not revert and could be transferred *in vitro* several times (Bourgeois & Beaman, 1974; Beaman, 1974).

Ultrastructural studies of the altered nocardial cells isolated from macrophages showed that they occurred as both type A and B cell wall variants (Figs 1 and 2). Altered forms that were of the type

B ultrastructure were very large, pleomorphic cells that had obviously lost most of their cell wall rigidity, yet they retained significant portions of the outer layers of the cell envelope (Fig. 2). After continued *in vitro* transfer the L-form colonies appeared to become composed entirely of the type A L-form (Fig. 1; Bourgeois & Beaman, 1974; Beaman, 1974; Bourgeois, 1975). Therefore the work of Bourgeois & Beaman has conclusively shown that transitional phase variants and L-forms of certain strains of *Nocardia asteroides* were induced intracellularly by mouse peritoneal macrophages.

Further studies have shown that L-forms could be isolated from a variety of strains of *Nocardia in vitro* by first inducing protoplasts and spheroplasts with glycine plus lysozyme. In addition, it was found that growing *N. asteroides* 10905 on glycine plus cycloserine and plating the cell wall damaged organisms on L-form medium resulted in L-form colonies similar to that shown in Fig. 4. Some nocardial strains appeared to be more susceptible to L-form conversion than others, thus suggesting chemical differences in the cell wall structure between the various strains tested (Bourgeois, 1975; Bourgeois & Beaman, 1975).

It is not possible to review here the voluminous amount of information that has been published concerning bacterial L-forms and their role, if any, in the disease process. However, there is substantial evidence that L-forms of various pathogenic organisms do indeed play a role in the disease process. For example, it has been suggested that L-forms of *Brucella abortus* play a vital role in the chronic recurring attacks characteristic of brucellosis by a persistence of these altered forms within the cells of the host (Hatten & Sulkin, 1966; McGhee & Freeman, 1970; Mattman, 1974). Cell wall defective variants of mycobacteria have been studied, and it seems likely that they play a significant function in the chronicity and latency characteristic of mycobacterial infections (Mattman, 1974; McCune, Feldman & McDermott, 1966; Thacore & Willett, 1966). L-forms of *Streptococcus* have been implicated in rheumatic heart disease and glomerulonephritis (Mattman, 1974). In addition L-forms have been implicated in syphilis, gonorrhea, urinary tract infections, arthritis, leprosy, osteomyelitis, meningitis, listeriosis, septicemia and endocarditis, pneumonia and other lung diseases, as well as other disease states

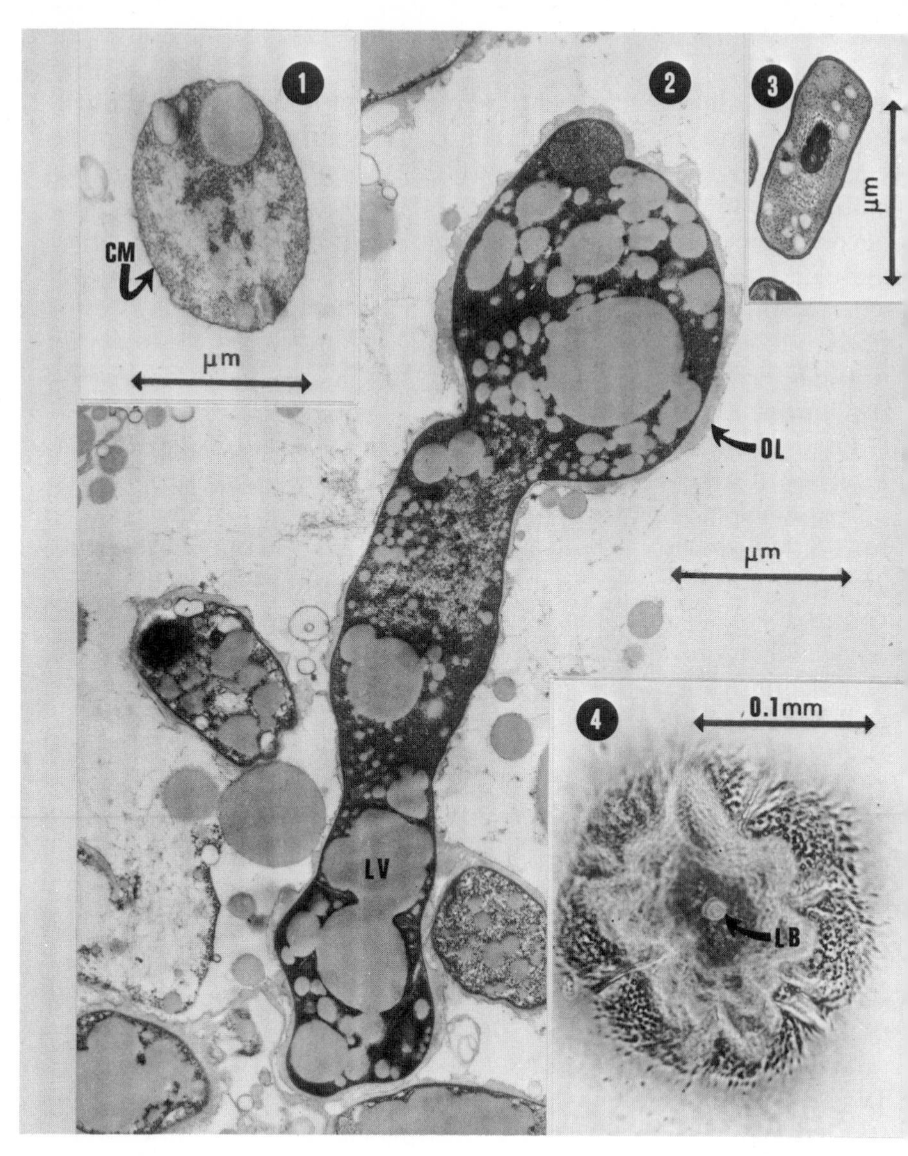

Fig. 1. Thin section of *Nocardia asteroides* 10905 type A L-form grown on BYE-L agar (Bourgeois & Beaman, 1974). Note the absence of the cell wall (arrow): (CM) cytoplasmic membrane: bar is 1 micrometer.

Fig. 2. A section of a type B L-form of *N. asteroides* grown on BYE-L agar. This cell is very large and appears to have lost much of its rigidity. It has an altered cell wall. Note the presence of an outer layer surrounding the cell (arrow): (OL) outer layer of cell envelope; (LV) probable lipid inclusions.

Fig. 3. A section of a "normal" cell of *N. asteroides* 10905 grown on BHI agar for 7 days. Compare the structure and size of the cells shown in Figs 1 and 2 with that in Fig. 3. All micrographs are enlarged approximately the same.

(Mattman, 1974). There are strong arguments in the literature both supporting and discrediting the role of L-forms in disease (Clasener, 1972; Mattman, 1974).

In order to determine whether nocardial L-forms were pathogenic, Bourgeois (1975) suspended in mineral oil stable *N. asteroides* 10905 L-forms that had been maintained *in vitro* for five transfers. This suspension was injected into the peritoneal cavity of several young mice. The oil emulsion was used with the L-form of this strain because it had previously been shown that *N. asteroides* 10905 produced a slowly progressive but fatal infection in mice only when injected in the presence of adjuvants such as oil (Beaman, 1973). After 4 and 6 weeks the mice were sacrificed, and the peritoneal cavity was opened. Every mouse had developed nodular lesions on the spleen and in the peritoneal cavity. In addition there was a macrophage-rich exudate. From both the exudate and the splenic lesions large numbers of nocardial L-forms were isolated. In addition a single, typical *N. asteroides* revertant was isolated from the spleen of one of the mice. Gram stains of the lesions and peritoneal exudate revealed numerous Gram-variable pleomorphic structures with a distinct tendency toward Gram positive beading. Acid fast stains revealed some strongly acid-alcohol fast beads, but normal organisms were not observed. The L-forms recovered from the mice were shown to be derived from *N. asteroides* by specific immunofluorescent staining (Bourgeois, 1975).

Beaman (1975) inoculated several strains of *N. asteroides* and *N. caviae* suspended in either BHI or saline into the peritoneal cavity of mice and followed the organism's growth in the spleen. It was found that the normal organisms recovered from the spleen initially descreased over a period of several days, and then continued to increase until the mice died. While the number of normal organisms recovered from the spleen decreased, there was a dramatic increase in the number of altered forms recovered. These

---

Fig. 4. A phase contrast micrograph of an L-form colony of *N. asteroides* 10905 grown on BYE-L agar for one month. This colony was cultured from *in vitro* grown mouse peritoneal macrophages that had been infected with *N. asteroides* 10905 3 weeks earlier (Bourgeois & Beaman, 1974). Figs 1 and 2 are cells found within sectioned L-form colonies such as shown here. Arrow (LB) points out the large body that is characteristic of bacterial L-form colonies. Note the granular appearance of the periphery of the colony.

altered variants recovered from the animals represented both transition phase variants, since the colonies reverted to typical nocardiae after 4 to 6 weeks incubation, and stable L-forms which could be transferred *in vitro* (Beaman, 1974; Beaman, 1975).

These data clearly establish that cell wall defective variants of Nocardia can be readily induced in cultured macrophages and in mice. It has been shown that at least one strain of *N. asteroides* L-form was, under appropriate conditions, pathogenic for mice and it could be isolated from spleens 6 weeks after inoculation. Since reports of attempts to recover L-forms from patients infected with *Nocardia* have not appeared in the literature, the role of nocardial L-forms in human disease is unknown. However, based on the observations just discussed it seems likely that nocardial L-forms or transitional phase variants play a significant part in certain nocardial infections in man.

5. Chemical Basis for Nocardial Pathogenicity

The nocardiae, mycobacteria, and corynebacteria have a chemically similar and complex cell envelope composed of several classes of free and bound lipids, peptides and polysaccharides. The cell walls of Nocardia have a peptidoglycan backbone composed of *N*-glycolyl-muramic acid, *N*-acetyl glucosamine, L-alanine, D-alanine, D-glutamic acid, and *meso*-diaminopimelic acid (Adam *et al.*, 1969; Azuma *et al.*, 1973; Azuma *et al.*, 1970; Cummins, 1965; Cummins & Harris, 1958; Beaman *et al.*, 1974; Bordet & Michel, 1969; Michel, 1974; Serrano *et al.*, 1972; Vacheron *et al.*, 1972). Vacheron *et al.*, (1972) reported that *Nocardia kirovani* cell walls contained 17·5% free lipids composed of fatty acids, nocardomycolic acids, nocardones, triglycerides and pigment. They reported that 20% of the cell wall weight was composed of nocardomycolic acid and arabinogalactan linked to cell wall. The peptidoglycan represented only 40% of the cell wall weight (Vacheron *et al.*, 1972). Beaman *et al.* (1971) demonstrated that cell walls of *Nocardia rubra* contained peptide or protein material closely associated to the firmly bound lipids. Similar peptide or protein components were shown to be present in cell walls of *N. asteroides* and *N. kirovani* (Michel, 1974; Beaman *et al.*, 1974). In addition the cell envelope of *Nocardia* has been shown to contain trehalose-6,6′-dimycolate and nocobactin (Ioneda, Lederer & Rozanis, 1970; Ratledge & Snow, 1974).

It appears that in mycobacteria many of the cell envelope components play a role in the pathogenicity of these organisms. Most notable of these are wax D, "cord factor," sulfolipids, phospholipids, mycobactin and, to a lesser extent, the mycosides. It seems that some of these substances such as "cord factor" are toxic while others are a significant portion of the organism's antigenic properties that appear to play a role in delayed hypersensitivity as well as possibly protecting the parasite from the defense mechanisms of the host (Anacker *et al.*, 1969; Asselineau, 1966; Azuma, Kimura & Yamamura, 1968; Azuma *et al.*, 1969; Azuma & Yamamura, 1962; Azuma *et al.*, 1970; Barclay *et al.*, 1967; Cummins *et al.*, 1967; Cunto, Kanetsuna & Imaeda, 1969; Golden, Kochen & Spriggs, 1974; Goren, 1972; Goren, Brokl & Schaefer, 1974; Kanetsuna, 1968; Kanetsuna & Blas, 1970; Kochan, 1973; Larson *et al.*, 1963; Maeda, 1972; Misaka *et al.*, 1966; Reid & Mackay, 1967; Ribi *et al.*, 1968).

In 1950, Bloch demonstrated that the surface of virulent tubercle bacilli possessed a toxic factor that was soluble in petroleum ether. He noted that virulent cells extracted with petroleum ether remained viable, but they became less virulent and at the same time lost their ability to grow in the characteristic "serpentine cords" (Bloch, 1950). It was later shown that this toxic lipid, called "cord factor," was trehalose-6,6′-dimycolate (Noll, 1956; Noll *et al.*, 1956). Because the chemical structure of mycolic acids obtained from various strains of mycobacteria are different there are numerous variations of trehalose-6,6′-dimycolate (Asselineau, 1966).

Bekierkunst *et al.* (1969) found that small amounts of trehalose-6,6′-dimycolate isolated from the tubercle bacillus could induce granulomas in the lungs of mice. Furthermore, these lesions appeared to have the same host cell composition as granulomas induced by infection with live *Mycobacterium bovis.* In contrast, Moore, Myrvik & Kato (1972) indicated that the granulomas induced by "cord factor" were of the foreign body type and not due to an allergic or immune mechanism. They found that the purified "cord factor" did not induce granulomas to the same extent as whole BCG, and they concluded that trehalose-6,6″-dimycolate was not the only compound within mycobacteria responsible for the formation of chronic, allergic granulomas characteristic of mycobacterial infections. Kato and colleagues have extensively studied the effect of "cord factors" isolated from mycobacteria and corynebacteria on mitochondria. They noted that mycolic acid or trehalose alone did not cause mitochondrial swelling and disrupton, but when chemically combined to form trehalose-6,6′-dimycolate, mitochondrial structure and physiological activity was greatly affected (Kato, 1966, 1967, 1968, 1969, 1970, 1971; Kato & Fukushi, 1969; Kato & Tanaka, 1967). Kato & Maeda (1974) isolated a trehalose-6-monomycolate from the wax D fraction of virulent human tubercle bacilli. They found that this compound affected mitochondria, but to a lesser extent than "cord factor". From these data, it has been suggested that trehalose mycolates play a role in determining the types of tissues that are susceptible to infection by *M. tuberculosis*, and they appear to be related in some way to the virulence of this pathogen however, the exact role "cord factor" plays in pathogenesis is neither clearly established nor understood. (Goren, 1972; Kato, 1969; Moore, Myrvik & Kato, 1972).

Ioneda, Lederer & Rozanis (1970) isolated a diester of trehalose mycolate from *N. asteroides*. However, the *in vivo* activity of this nocardial "cord factor" has not been determined. Since the "cord factors" of mycobacteria and corynebacteria are toxic, and in *M. tuberculosis* appears to be associated with virulence, it seems likely that "cord factors" from nocardiae are also toxic and may play some role in nocardial pathogenicity.

As indicated earlier, the ultrastructural appearance of the cell walls of nocardiae changes during the growth cycle, and it is further influenced by *in vitro* and *in vivo* growth conditions (Beaman, 1968, 1973; Beaman & Shankel, 1969). In order to determine the chemical changes accompanying ultrastructural alterations, Beaman, Burnside & O'Donnell (1974*a*) studied cell wall composition during the growth cycle of *N. asteroides* 14759 grown in BHI broth. It was found that the peptidoglycan and lipid components were affected most. In lag phase cells, the walls were composed of 36% lipid, 15% peptidoglycan, 29% carbohydrate and 10% nonpeptidoglycan amino acids. In contrast, one week, stationary phase cells had a cell wall composed of 7% lipid, 43%

peptidoglycan, 30% carbohydrate, and 17% amino acids. In addition, the classes of lipids were greatly affected during the growth cycle as demonstrated by gas liquid chromatography (Beaman, Burnside & O'Donnell, 1974). It is probable that such dramatic and basic chemical changes within the wall structure have a significant effect on the host-parasite interaction and may also have an effect on the pathogenic properties of nocardiae.

Kochan (1973) pointed out that the essential requirement for an organism to be pathogenic is that it must be able to grow within the host utilizing components either in the body fluids or in the phagosome as nutrients. Furthermore, it is known that bacterial growth cannot occur without adequate levels of iron. Therefore, the amount of iron available to the pathogen determines it fate within the host (Kochan, 1973; Kochan & Golden, 1974; Weinberg, 1966, 1971).

It was shown that the growth of the virulent strain of *M. tuberculosis* (H37RV) was inhibited in mammalian sera, and that the amount of inhibition was inversely related to the degree of saturation of transferrin (Kochan *et al.*, 1963; Kochan, 1969). By adding amounts of iron greater than the iron binding capacity of transferrin it was found that the tubercle bacillus would then grow (Kochan, 1969).

It has been demonstrated that the ability of many pathogens to grow in experimental animals could be greatly enhanced by increasing the amounts of iron within the host. Sword (1966) found that by adding ferrous sulfate or ferric ammonium citrate to mice the LD_{50} was greatly reduced for virulent strains of *Listeria monocytogenes.* He also found that by adding iron to mice, 10^4 organisms of an avirulent strain of *Listeria* was able to infect and kill these animals (Sword, 1966). Kaye, Gill & Hook (1967) demonstrated that by adding iron to mice, the LD_{50} for *Salmonella typhimurium* was reduced from 10^4 to less than 10 organisms. Similarly it has been shown that the addition of iron to animals greatly enhanced the host's susceptibility to *Klebsiella pneumoniae, Pseudomonas aeruginosa, Pasteurella septica* and *Escherichia coli* (Kochan, 1973; Bullen, Rogers & Griffiths, 1974). Conversely, treating the mice with agents such as iron chelators that reduced the level of iron in the host brought about an increase in host resistance to bacterial infection. Therefore, hyperferremia favors growth of many pathogens within the host while hypoferremia enhances the host's ability to resist invasion by these parasites (Kochan, 1973; Bullen, Rogers & Griffiths, 1974).

It has been shown that mycobacteria synthesize complex secondary hydroxamates that are strong iron-chelating agents (Snow, 1970). These "mycobactins" appear to be closely associated to the lipoidal components of the wall, and they play a major role in iron transport into the cell (Kochan, 1973; Snow, 1970). Further, it was shown that mycobactin promoted mycobacterial growth in tuberculostatic serum by taking iron from the transferrin-iron complex and making it available to the tubercle bacillus (Kochan, 1973). It appears that mycobactin synthesized by these organisms provides a mechanism by which they can obtain adequate amounts of iron from the host, therefore, enabling the mycobacterial cell to more successfully grow within the host's tissues. As a consequence, mycobactin possibly plays a significant role in the overall host-parasite interaction. It may even be considered a virulence factor in that it is firmly associated to the mycobacterial cell, and a loss of this compound would render the organism less able to grow within the host (Kochan, 1973; Snow, 1970).

Chemically similar lipid-soluble, iron-binding compounds have been isolated from *Nocardia* (Ratledge & Snow, 1974; see chapter 14). It was shown that these components increased in cells grown in low-iron environments and decreased in amount when the cells were grown in iron-rich media (Ratledge & Snow, 1974). Ratledge & Snow (1974) called these compounds "nocobactins" to indicate that they were chemically similar to the mycobactins isolated from mycobacteria. The role of iron and nocobactin in nocardial pathogenesis is not established, however, it is logical to assume that iron availability within the host as well as the nocardial cell's synthesis of nocobactin play a role in the host-parasite interaction. Some evidence to support this view was obtained when it is was observed that whole serum was nocardostatic, and nocardiae either failed to grow or the cells was markedly altered when grown in iron-poor media or serum (Goldman & Beaman, pers. comm.; Webley, 1960). Additional evidence that iron is important to nocardial pathogenesis may be based on the observations that mucin greatly enhances nocardial virulence (Strauss & Kligman, 1951; Uesaka *et al.*, 1971). It has been suggested by Fusillo *et al.* (1974)

that mucin generally enhanced bacterial virulence by making iron more readily available. Also, guinea pigs have been shown to be more susceptible to various strains of *Nocardia* than mice or other laboratory animals. It is important to note that guinea pig serum has a higher level of transferrin-bound-iron than most other animals studied (Kochan, 1973).

6. Conclusions

Although the experimental data concerning the mechanisms of nocardial pathogenesis is fragmentary and incomplete, we have attempted to consider most of the possibilities presently available in the literature. As more data are obtained, the factors controlling the pathogenicity of *Nocardia* should become more clearly established, and new factors may come to light. Presently, it appears that chemical components in the cell envelope play a major role in the host-parasite response. Some of these seem to be "cord factor," nocobactin, arabino-galactan-mycolate complexes, perhaps cell wall peptidolipid, and other lipid and polysaccharide components. The fact that major chemical and ultrastrutural changes occur in nocardial cells during their growth cycle, and in response to environmental stimuli, should have profound and basic implications to the overall host-parasite response. In addition the fact that certain cytoplasmic components are capable of eliciting a delayed hypersensitivity within the host cannot be ignored. Furthermore, nocardiae vary considerably in their potential pathogenicity. It should be possible to establish the chemical basis for this variation. Electron microscopy and histochemistry have established that certain strains of *Nocardia* become altered within the host. These observations led to the discovery that nocardiae can function as facultative intracellular parasites, and that altered nocardial forms appear to play a role in nocardial pathogenesis. L-forms of *Nocardia* were isolated from both animals and tissue cultures, and they could be readily induced *in vitro*. It was demonstrated that the L-forms of *N. asteroides* 10905 were pathogenic however, the role of L-forms in human disease remains to be established. The role of iron and nocobactin synthesis appears to be intimately involved in the ability of nocardiae to grow within the host cells and tissues. All of the aspects just mentioned require additional study before answers to the numerous questions regarding the mechanisms of nocardial pathogenicity can be given.

7. Acknowledgements

I am grateful for the help and comments of the many individuals who have patiently discussed the various aspects of this manuscript during its preparation. Much of the work discussed in this review was supported by a Public Health Service research grant from the National Institutes of Health, NIAID, Grant No. RO1-Al-10542.

8. References

Adam, A., Petit, J. R., Wieterbin, J., Sinay, P., Thomas, D. W. & Lederer, E. (1969). L'acide *N*-glycolylmuramimique, constituant des parois de *Mycobacterium smegmatis*: Identification par spectrometrie de masse. *FEBS Letters* **4,** 87.

Anacker, R. L., Ribi, E., Tarmina, D. F., Fadness, L. & Mann, R. E. (1969). Relationship of foot pad sensitivity to purified protein derivatives and resistence to air-borne infections with *Mycobacterium tuberculosis* of mice, vaccinated with mycobacterial cell walls. *J. Bact.* **100,** 51.

Armstrong, J. A. & D'Arcy Hart, P. (1973). Response of cultured macrophages to *Mycobacterium tuberculosis,* with observations on fusion of lysosomes with phagosomes. *J. exp. Med.* **134,** 713.

Aron, R. & Gordon, W. (1972). Pulmonary nocardiosis. *S. Afr. med. J.* **46,** 29.

Asselineau, J. (1966). *The Bacterial Lipids.* Paris: Herman.

Azuma, I., Ajisaka, M. & Yamamura, Y. (1970). Polysaccharides of *Mycobacterium bovis* Ushi 10, *Mycobacterium smegmatis, Mycobacterium phlei* and a typical *Mycobacterium* P1. *Infect. Immun.* **2,** 347.

Azuma, I., Kanetsuna, F., Tanaka, Y., Mera, M., Yanagihara, Y., Mifuchi, I. & Yamamura, Y. (1973). Partial chemical characterization of the cell wall of *Nocardia asteroides* strain 131. *Jap. J. Microbiol.* **17,** 154.

Azuma, I., Kimura, H. & Yamamura, Y. (1968). Chemical and immunological properties of polysaccharides of Wax D extracted from *Mycobacterium tuberculosis* strain Aoyama B. *J. Bact.* **96,** 567.

Azuma, I., Thomas, D. W., Adam, A., Ghuysen, J. M., Bonaly, R., Petit, J. F. & Lederer, E. (1970). Occurrence of N-glycolylmuramic acid in bacterial cell walls. A preliminary survey. *Biochim. biophys. Acta* **208,** 444.

Azuma, I. & Yamamura, Y. (1962). Studies of the firmly bound lipids of human tubercle bacillus. *J. Biochem., Tokyo* **52,** 200.

Azuma, I., Yamamura, Y. & Fukushi, K. (1968). Fractionation of mycobacterial cell walls. Isolation of arabinose mycolate and arabinogalactan from the cell wall fraction of *Mycobacterium tuberculosis* strain Aoyama B. *J. Bact.* **96,** 1885.

Azuma, I., Yamamura, Y., Tahora, T., Onoue, K. & Fukushi, K. (1969). Isolation of tuberculin active peptides from the cell wall fraction of human tubercle bacilli strain Aoyama B. *Jap. J. Microbiol.* **13,** 220.

Bach, M. C., Monaco, A. P. & Finland, M. (1973). Pulmonary nocardiosis: therapy. *J. Am. med. Ass.* **224,** 1378.

Barclay, W. R., Anacker, R., Brehmer, W. & Ribi, E. (1967). Effects of oil-treated mycobacterial cell walls on the organs of mice. *J. Bact.* **94,** 1736.

Bates, R. R. & Rifkind, D. (1971). *Nocardia brasiliensis* lymphocutaneous syndrome. *Am. J. Dis. Child.* **121,** 246.

Beaman, B. L. (1968). An analysis of the biological and ultrastructural properties of *Nocardia* grown on defined and complex media. Ph.D. Thesis. University of Kansas, Lawrence, Kansas, U.S.A.

Beaman, B. L. (1973). An ultrastructural analysis of Nocardia during experimental infections in mice. *Infect. Immun.* **8,** 828.

Beaman, B. L. (1974). Ultrastructural and biochemical approach to nocardial pathogenicity. *Proc. 1 Intern. Conf. Biol. Nocardiae, Merida, Venezuela.* p. 98.

Beaman, B. L. (1975). *In vivo* induction of nocardial L-forms in mice. *Abs. Ann. Meeting Am. Soc. Microbiol.*

Beaman, B. L. & Burnside, J. (1973). Pyridine extraction of nocardial acid-fastness. *Appl. Microbiol.* **26,** 426.

Beaman, B. L., Burnside, J. & O'Donnell, B. (1974*a*). Ultrastructural and biochemical analysis of *Nocardia asteroides* 14759 cell walls during it's growth cycle. *Proc. 1 Intern. Conf. Biol. Nocardiae, Merida, Venezuela,* p. 54.

Beaman, B. L., Kim, K. S., Lanéelle, M. A. & Barksdale, L. (1974). Chemical characterization of organisms isolated from leprosy patients. *J. Bact.* **117,** 1320.

Beaman, B. L., Kim, K. S., Salton, M. R. J. & Barksdale, L. (1971). Amino acids of the cell wall of *Nocardia rubra. J. Bact.* **108,** 941.

Beaman, B. L. & Shankel, D. M. (1969). Ultrastructure of *Nocardia* cell growth and development on defined and complex agar media. *J. Bact.* **99,** 876.

Beckmeyer, W. J. (1959). Nocardiosis: report of a successfully treated case of cutaneous granuloma. *Pediatrics, Springfield* **23,** 33.

Bekierkunst, A. (1968). Biology of mycobacterioses: tubercle bacilli grown *in vivo. Ann. N.Y. Acad. Sci.* **154,** 79.

Bekierkunst, A., Levy, I. S., Yarkoni, E., Vilkas, E., Adam, A. & Lederer, E. (1969). Granuloma formation induced in mice by chemically defined mycobacterial fractions. *J. Bact.* **100,** 95.

Benedict, W. L. & Iverson, H. A. (1944). Chronic keratoconjunctivities associated with *Nocardia. Archs Ophal, N.Y.,* **32,** 89.

Berd, D. (1973*a*). *Nocardia asteroides.* A taxonomic study with clinical correlations. *Am. Rev. resp. Dis.* **108,** 909.

Berd, D. (1973*b*). *Nocardia brasiliensis* infection in the United States. A report of nine cases and a review of the literature. *Am. J. clin. Path.* **60,** 254.

Bloch, H. (1950). Studies on the virulence of tubercle bacilli: isolation and biological properties of a constituent of virulent organisms. *J. exp. Med.* **91,** 197.

Bojalil, L. F. & Jamora, A. (1963). Precipitin and skin tests in the diagnosis of mycetoma due to *Nocardia brasiliensis. Proc. Soc. exp. Biol. Med.* **113m** 40.

Bojalil, L. F. & Magnusson, M. (1963). Specificity of skin reactions of humans to *Nocardia* sensitins. *Amer. Rev. resp. Dis.* **88,** 409.

Bordet, C. & Michel, G. (1969). Structure et bioggenese des lipides a haut poids moleculaire de *Nocardia asteroides. Bull. Soc. Chim. biol.* **51,** 527.

Bourgeois, L. (1975). Structural and biological characterization of macrophage and chemically induced cell wall variants of *N. asteroides* Ph.D. thesis, Georgetown University, Washington, D.C.

Bourgeois, L. & Beaman, B. L. (1974). Probable L-forms of *Nocardia asteroides* induced in cultured mouse peritoneal macrophages. *Infect. Immun.* **9,** 576.

Bourgeois, L. & Beaman, B. L. (1975). *In vitro* induction of spheroplasts and L-forms in the pathogenic nocardiae. *Abs. Ann. Meeting Am. Soc. Microbiol.*

Bowden, D. H. (1971). The alveolar macrophage. *Current Topics in Pathology* **55,** 1.

Bronner, M. & Bronner, M. (Ed.). (1971). *Actinomycosis,* 2 edn. Bristol: John Wright and Sons Ltd.

Bullen, J. J., Rogers, H. J. & Griffiths, G. (1974). Bacterial iron metabolism in infection and immunity. In *Microbial Iron Metabolism,* Ed. J. B. Neilands, p. 517. London: Academic Press.

Burpee, J. C. & Sarke, W. R. (1971). Bilateral metastatic intraocular nocardiosis. *Archs Ophthal, N.Y.* **86,** 666.

Carile, W. K., Holley, K. E. & Logan, G. B. (1963). Fatal acute disseminated nocardiosis in a child. *J. Am. med. Assoc.* **184,** 477.

Carr, I. (1973). *The Macrophage: A Review of Ultrastucture and Function.* London: Academic Press.

Causey, W. A., Arvell, P. & Brinker, J. (1974). Systemic *Nocardia caviae* infection. *Chest* **63,** 360.

Causey, W. A. & Sieger B. (1974). Systemic nocardiosis caused by *Nocardia brasiliensis. Am. Rev. resp. Dis.* **109,** 134.

Chapman, G. B., Hanks, J. H. & Wallace, J. H. (1959). An electron microscope study of the disposition and fine structure of *Mycobacterium lepraemurium* in mouse spleen. *J. Bact.* **77,** 205.

Clasener, H. (1972). Pathogenicity of the L-phase of bacteria. *A. Rev. Microbiol.* **26,** 55.

Cohn, Z. A. (1963). The fate of bacteria within phagocytic cells: the degradation of isotopically labeled bacteria by polymorphonuclear leucocytes and macrophages. *J. exp. Med.* **117,** 27.

Convit, J. & Pinardi, M. E. (1972). A simple method for the differentiation of *Mycobacterium leprae* from other mycobacteria through routine staining techniques. *Int. J. Lepr.* **40,** 130.

Cross, R. M. & Binford, C. H. (1962). Infections by fungi that are commonly primary pathogens: is *Nocardia asteroides* an opportunist? *Lab. Invest.* **11,** 1103.

Cummins, C. J. (1965). Chemical and antigenic studies on cell walls of *Mycobacterium, Corynebacterium* and *Nocardia. Amer. Rev. resp. Dis.* **92,** 63.

Cummins, C. J., Atfield, G., Reese, R. J. W. & Valentine, R. C. (1967). Cell wall composition in *Mycobacterium lepraemurium. J. gen. Microbiol.* **49,** 377.

Cummins, C. J. & Harris, H. (1958). Studies on the cell wall composition and taxonomy of Actinomycetales and related groups. *J. gen. Microbiol.* **18,** 173.

Cunto, G., Kanetsuna, F. & Imaeda, T. (1969). Chemical analysis of the mucopeptide of *Mycobacterium smegmatis*. *Biochem. biophys. Acta* **192,** 358.

D'Arcy Hart, P. (1968). Biology of mycobacterioses: statement of questions. *Ann. N.Y. Acad. Sci.* **154,** 3.

Diamond, R. D. & Bennett, J. E. (1973). Disseminated *Nocardia brasiliensis* infection. *Archs intern. Med.* **131,** 735.

Dienes, L. (1967). Morphology and reproductive processes of bacteria with defective cell walls. In *Microbial Protoplasts, Spheroplasts and L-Forms.* Ed. L. B. Guze. p. 74. Baltimore: Williams and Wilkins.

Draper, P. & Rees, R. J. W. (1973). The nature of the electron-transparent zone that surrounds *Mycobacterium lepraemurium* inside host cells. *J. gen. Microbiol.* **77,** 79.

Duguid, J. P. & Wilkinson, J. F. (1961). Environmentally induced changes in bacterial morphology. *Symp. Soc. gen. Micribiol.* **43,** 326.

Dumont, A. (1972). Ultrastrutural aspects of phagocytosis of facultative intracellular parasites by hamster peritoneal macrophages. *J. Reticuloendothelial Soc.* **11,** 469.

Edwards, R. P. (1970). Electron-microscope illustrations of division in *Mycobacterium leprae. J. med. Microbiol.* **3,** 493.

Epstein, J. M., Verney, E. V. & Miale, T. D. (1967). Studies on the pathogensis of experimental pulmonary aspergillosis. *Am. J. Path.* **51,** 769.

Failes, D. & Posney, K. (1965). Systemic nocardiosis presenting as a breast abscess. *Med. J. Aust.* **1,** 342.

Fisher, C. A. & Barksdale, L. (1973). Cytochemical reactions of human leprosy bacilli and mycobacteria: ultrastructural implications. *J. Bact.* 113, 1389.

Freese, J. W., Young, W. G. Sealy, W. C. & Corant, N. F. (1963). Pulmonary infection by *Nocardia asteroides*: findings in eleven clinical cases. *J. Thorac Surg.* **46,** 537.

Fuzillo, M., Smith, J. C. & Reednick, J. A. (1974). *Abst. Ann. Meeting. Am. Soc. Micro. biol.* **74,** 73 (M44).

Gezuelle, E. (1972). Fatal infection by *Nocardia brasiliensis* in an armadillo. *Sabouraudia* **10,** 63.

Golden, C. A., Kochan, I. & Spriggs, D. R. (1974). Role of mycobactin in the growth and virulence of tubercle bacilli. *Infect. Immun.* **9,** 34.

González-Ochoa, A. (1962). Mycetomas caused by *Nocardia brasiliensis* with a note on the isolation of the causative organism from the soil. *Lab. Invest.* **11,** 1118.

González-Ochoa, A. (1973). Virulence of nocardiae. *Can. J. Microbiol.* **19,** 901.

Gordon, S. & Cohn, Z. A. (1973). The macrophage. *Int. Rev. Cytol.* **36,** 171.

Goren, M. B. (1972). Mycobacterial lipids: selected topics. *Bact. Rev.* **36,** 33.

Goren, M. B., Brokl. O. & Schaefer, W. B. (1974). Lipids of putative relevance to virulence in *Mycobacterium tuberculosis*: phthicerol dimycocerostate and the attenuation indicator lipid. *Infect. Immun.* **9,** 150.

Goren, M. B., Brokl, O. & Schaefer, W. B. (1974). Lipids of putative relevance to virulence in *Mycobacterium tuberculosis*: correlation of virulence with elaboration of sulfatides and strongly acidic lipids. *Infect. Immun.* **9,** 142.

Green, G. M. (1973). Lung defense mechanisms. *Med. Clins N. Am.* **57,** 547.

Green, W. O. & Adams, T. E. (1964). Mycetoma in the United States: a review and report of seven additional cases. *Am. J. clin. Path.* **42,** 75.

Greer, K. E. (1974). Nocardial mycetoma. Va. med. Mon. **101,** 193.

Hathaway, B. M. & Mason, K. N. (1962). Nocardiosis: study of fourteen cases. *Am. J. Med.* **32,** 903.

Hatten, B. A. & Sulkin, S. E. (1966). Intracellular production of *Brucella* L-forms I. Recovery of L-forms from tissue culture cells infected with *Brucella abortus. J. Bact.* **91,** 285.

Henderson, J. W., Wellman, W. E. & Weed, L. A. (1960). Nocardiosis of the eye; report of cases. *Proc. Staff Meet. Mayo Clin.* **35,** 614.

Hoeprich, P. D., Brandt, D. & Parker, R. H. (1968). Nocardial brain abscess cured with cycloserine and sulfonamides. *Am. J. med. Sci.* **255,** 208.

Hogshead, H. P. & Stein, G. H. (1970). Mycetoma due to *Nocardia brasiliensis. J. Bone Jt. Surg.* **52,** 1229.

Howell, A. & Jordan, H. V. (1965). Nutritional control of cellular morphology in an aerobic actinomycete from the hampster. *J. gen. Microbiol.* **38,** 125.

Imaeda, T. (1965). Electron microscopy. A. Products of leprosy research. *Int. J. Lepr.* **33,** 669.

Imaeda, T. & Convit, J. (1962). Electron microscope study of *Mycobacterium leprae* and its environment in a vesicular leprous lesion. *J. Bact.* **83,** 43.

Ioneda, T., Lederer, E. & Rozanis, J. (1970). Sur la structure des diesters de trehalose ("cord factors") produits par *Nocardia asteroides* et *Nocardia rhodochrous. Chem. Phys. Lipids* **4,** 375.

Kanai, K. & Kondo, E. (1974). Chemistry and biology of mycobacteria grown *in vivo. Jap. J. med. Sci. Biol.* **27,** 135.

Kanetsuna, F. (1968). Chemical analyses of mycobacterial cell walls. *Biochim. biophys. Acta* **158,** 130.

Kanetsuna, F. & Blas, G. S. (1970). Chemical analysis of mycolic acid-arabinogalactan-mucopeptide complex of mycobacterial cell wall. *Biochim. biophys. Acta* **208,** 444.

Kato, M. (1966). Studies on the biochemical lesion in experimental tuberculosis in mice. II. Bacterial constituents responsible for the inhibition of the succinate-neotetrazolium reductase system. *Am. Rev. resp. Dis.* **93,** 421.

Kato, M. (1967). Studies of a biochemical lesion in experimental tuberculosis in mice. VI. Effect of toxic bacterial constituents of tubercle bacilli on oxidative phosphorylation in host cell. *Am. Rev. resp. Dis.* **96,** 998.

Kato, M. (1968*a*). Studies of a biochemical lesion in experimental tuberculosis in mice. VII. Structural and functional damage in mouse liver mitochondria under the toxic action of cord factor. *Am. Rev. resp. Dis.* **98,** 260.

Kato, M. (1968*b*). Studies of a biochemical lesion in experimental tuberculosis in mice. VIII. Effect of derivatives and chemical analogues of cord factor on structure and function of mouse liver mitochondria. *Am. Rev. Resp. Dis.* **98,** 668.

Kato, M. (1969*a*). Studies of a biochemical lesion in experimental tuberculosis in mice. IX. Response of respiratory enzymes in the immune mouse liver to cord factor and tuberculin PPD. *Am. Rev. resp. Dis.* **99,** 112.

Kato, M. (1969*b*). Studies of a biochemical lesion in experimental tuberculosis in mice. XI. Mitochondrial swelling induced by cord factor *in vitro*. *Am. Rev. resp. Dis.* **100,** 47.

Kato, M. (1970). Molecular configuration and toxicity of cord factor. *Jap. J. med. Sci. Biol.* **23,** 267.

Kato, M. (1971). Site of action of the cord factor of *Corynebacterium diphtheriae* in mitochondria. *J. Bact.* **107,** 746.

Kato, M. & Fukushi, K. (1969). Studies of a biochemical lesion in experimental tuberculosis in mice. X. Mitochondrial swelling induced by cord factor *in vivo* and accompanying biochemical change. *Am. Rev. resp. Dis.* **100,** 42.

Kato, M. & Maeda, J. (1974). Isolation and biochemical activities of trehalose-6-monomycolate of *Mycobacterium tuberculosis*. *Infect. Immun.* **9,** 8.

Kato, M. & Tanaka, A. (1967). Studies of a biochemical lesion in experimental tuberculosis in mice. V. Further study of a toxic lipid fraction in firmly bound lipids. *Am. Rev. resp. Dis.* **96,** 460.

Kaye, D., Gill, F. A. & Hook, E. W. (1967). Factors influencing host resistance to Salmonella infections: The effects of hemolysis and erythrophagocytosis. *Am. J. med. Sci.* **254,** 205.

Kingsbury, E. W. & Slack, J. M. (1969). A polypeptide skin test antigen from *Nocardia asteroides*. Further studies on the specificity of a nocardin active polypeptide. *Sabouraudia* **7,** 85.

Kochan, I. (1969). Mechanism of tuberculostasis in mammalian serum. I. Role of transferrin in human serum tuberculostasis. *J. infect. Dis.* **119,** 11.

Kochan, I. (1973). The role of iron in bacterial infections with specific considerations of host-tubercle bacillus interaction. *Current Topics in Microbiol. Immun.* **60,** 1.

Kochan, I. & Golden, C. A. (1974). Immunological nature of antimycobacterial phenomenon in macrophages. *Infect. Immun.* **9,** 249.

Kochan, I., Ishak, K., Said, M. & Stotts, J. (1963). Study on the tuberculostatic factor of mammalian serum. *Am. Rev. resp. Dis.* **88,** 818.

Kondo, E. & Kanai, K. (1972). Further demonstration of bacterial lipids in *Mycobacterium bovis* harvested from mouse lungs. *Jap. J. med. Sci. Biol.* **25,** 249.

Kurup, P. V., Sharma, V. N., Viswanatha, R., Sandhu, R. S., Randhawa, H. S. & Damodaran, V. N. (1968). Pulmonary fungal ball due to *Nocardia* species. *Scand. J. resp. Dis.* **49,** 9.

Langevin, R. W. & Katz, S. (1964). Fulminating pulmonary nocardiosis. *Dist. Chest.* **46,** 310.

Larson, C. L., Bell, J. F., List, R. H., Ribi, E. & Wicht, W. C. (1963). Symposium on the relationship of structure of microorganisms to their immunological properties. II. Host-reactive properties of cell walls and protoplasm from *Mycobacterium*. *Bact. Rev.* **27,** 341.

Laskin, A. & Lechevalier, H. (Eds) (1972). *Macrophages and Cellular Immunity.* Cleveland: CRC Press.

Lochard, V. G., Sharbaugh, R. J., Arhelger, R. B. & Grogan, J. B. (1971). Ultrastructural alterations in phagocytic functions of alveolar macrophages after cyclophosphamide administration. *J. Reticuloendothelial Soc.* **9,** 97.

Macotela-Ruiz, E. & González-Angulo, A. (1966). Electron microscopic studies on granules of *Nocardia brasiliensis* in man. *Sabouraudia*, **5,** 92.
Macotela-Ruiz, E. & Mariat, F. (1963). Sur la production de mycetomes experimentaux par *Nocardia brasiliensis* et *Nocardia asteroides. Bull. Soc. Path. exot.* **56,** 46.
Maeda, J. (1972). Isolation and characterization of toxic glycolipid from the firmly bound lipids of human tubercle bacilli. *Jap. J. Bact.* **27,** 467.
Maniar, S. H. & Anderson, P. C. (1966). Mycetoma caused by *Nocardia brasiliensis* in Missouri. *Missouri Med.* **63,** 273.
Mason, K. N. & Hathaway, B. M. (1969). A study of *Nocardia asteroides*: white mice used as test animals. *Archs Path.* **87,** 389.
Mattman, L. (1974). *Cell Wall Deficient Forms.* Cleveland: CRC Press.
McCluskey, R. T. & Cohen, S. (Ed.). (1974). *Mechanisms of Cell Mediated Immunity.* New York: John Wiley and Sons Inc.
McCune, R. M., Feldman, F. M. & McDermott, W. (1966). Microbial persistence. II. Characteristics of the sterile state of tubercle bacilli. *J. exp. Med.* **123,** 469.
McGhee, J. A., Wittler, R. G. Gooder, H. & Charache, P. (1971). Wall-defective variants: terminology and experimental design. *J. infect. Dis.* **123,** 433.
McGhee, J. R. & Freeman, B. (1970). Osmotically sensitive *Brucella* in infected normal and immune macrophages. *Infect. Immun.* **1,** 146.
Meyer, S. L., Font, R. L. & Shaver, R. P. (1970). Intraocular nocardiosis. *Arch. Ophthal., N.Y.*, **83,** 536.
Michel, G. (1974). Cell envelope composition of nocardiae and its interest in taxonomy. *Proc. 1 Intern. Conf. Biol. Nocardiae, Merida, Venezuela.* p. 24.
Misaka, A., Yukawa, S., Tsuchiya, K. & Yamasaki, T. (1966). Studies on cell walls of mycobacteria. I. Chemical and biological properties of the cell walls and mucopeptide of BCG. *J. Biochem. Tokyo*, **59,** 388.
Mishra, S. K., Sandhu, R. S., Randhawa, H. S., Damodaran, V. N. & Abraham, S. (1973). Effect of cortisone administration on experimental nocardiosis. *Infect. Immun.* **7,** 123.
Moore, V. L., Myrvik, Q. N. & Kato, M. (1972). Role of cord factor (trehalose-6,6′-dimycolate) in allergic granuloma formation in rabbits. *Infect. Immun.* **6,** 5.
Movat, H. Z. (Ed.) (1971). *Inflamation, Immunity and Hypersensitivity.* New York: Harper and Row.
Murray, J. F., Finegold, S. M., Frowan, S. & Will, D. W. (1961). The changing spectrum of nocardiosis: A review and presentation of nine cases. *Am. Rev. resp. Dis.* **83,** 315.
Neu, H. C., Silva, M., Hazen, E. & Rosenheim, S. H. (1967). Necrotizing nocardial pneumonitis. *Ann. intern. Med.* **66,** 274.
Newmark, E., Polack, F. M. & Ellison, A. C. (1971). Report of a case of *Nocardia asteroides* keratitis. *Am. J. Ophthal.* **72,** 813.
Noll, H. (1956). The chemistry of cord factor, a toxic glycolipid of *M. tuberculosis. Fortschr. Tuberkforsch.* **7,** 149.
Noll, H., Boch, H., Asselineau, J. & Lederer, E. (1956). The chemical structure of the cord factor of *Mycobacterium tuberculosis Biochim. biophys. Acta* **20,** 299.

Ortiz-Ortiz, L. & Bojalil, L. F. (1972). Delayed skin reactions to cytoplasmic extracts of *Nocardia* organisms as a means of diagnosis and epidemiological study of *Nocardia* infection. *Clin. Exp. Immun.* **12,** 225.

Ortiz-Ortiz, L., Bojalil, L. F. & Contreras, M. F. (1972). Delayed hypersensitivity to polysaccharides from *Nocardia. J. Immun.* **108,** 1409.

Ortiz-Ortiz, L., Contreras, M. F. & Bojalil, L. F. (1972*a*). The assay of delayed hypersensitivity to ribosomal proteins from *Nocardia. Sabouraudia* **10,** 147.

Ortiz-Ortiz, L., Contreras, M. F. & Bojalil, L. F. (1972*b*). Cytoplasmic antigens from *Nocardia* eliciting a specific delayed hypersenstivity. *Infect. Immun.* **5,** 819.

Overman, J. R. & Pine, L. (1963). Electron microscopy of cytoplasmic structures in facultative and anerobic *Actinomyces J. Bact.* **86,** 656.

Page, R. C., Davies, P. & Allison, A. C. (1974). Participation of mononuclear phagocytes in chronic inflamatory diseases. *J. Reticuloendothelial Soc.* **15,** 413.

Palmer, D. L., Harvey, R. L. & Wheeler, J. K. (1974). Diagnostic and therapeutic considerations in *Nocardia asteroides* infection. *Medicine, Baltimore* **53,** 391.

Peabody, J. & Seabury, J. (1960). Actinomycosis and nocardiosis: review of basic differences in therapy. *Am. J. Med.* **28,** 99.

Pearsall, N. N. & Weiser, R. S. (1970). *The Macrophage.* Philadelphia: Lea and Febriger.

Pier, A. C., Thurston, J. R. & Larsen, A. B. (1968). A diagnostic antigen for nocardiosis: comparative tests in cattle with nocardiosis and mycobacteriosis. *Am. J. vet. Res.* **29,** 397.

Pine, L. & Overman, J. R. (1963). Determination of the structure and composition of the 'sulphur granules' of *Actinomyces bovis. J. gen. Microbiol.* **32,** 209.

Presant, C. A., Wiernik, P. H. & Serpick, A. A. (1970). Disseminated extrapulmonary nocardiosis presenting as a renal abscess. *Archs Path.* **89,** 560.

Presant, C. A., Wiernick, P. H. & Serpick, A. A. (1973). Factors affecting survival in nocardiosis. *Am. Rev. resp. Dis.* **108,** 1444.

Raffel, S. (1953). *Immunity.* p. 338. New York: Appleton-Century-Crofts Inc.

Ratledge, C. & Snow, G. A. (1974). Isolation and structure of nocobactin NA, a lipid-soluble iron-binding compound from *Nocardia asteroides. Biochem. J.* **139,** 407.

Reid, J. D. & Mackay, J. B. (1967). The role of delayed hypersensitivity in granulomatous reactions to mycobacteria: II. Reactions to intradermal injections of intact and disintegrated organisms. *Tubercle, Lond.* **48,** 109.

Ribi, E., Anacker, R. L., Barclay, W. R., Brehmer, W., Middlebrook, G., Milner, K. C. & Tarmina, D. F. (1968). Biology of mycobacteriosis: structure and biological functions of mycobacteria. *Ann. N.Y. Acad. Sci.* **154,** 41.

Richter, R. W., Silva, M., Neu, H. C. & Silverstein, P. M. (1968). The neurological aspects of *Nocardia asteroides* infection. *Assoc. Res. Publs Ass. Res. nerv. ment. Dis.* **44,** 424.

Runyon, E. H. (1951). *Nocardia asteroides*: studies of its pathogenicity and drug sensitivities *J. Lab. clin. Med.* **37,** 713.

Saltzman, H. A., Chick, E. W. & Conant, N. F. (1962). Nocardiosis as complication of other diseases. *Lab. Invest.* **11,** 1110.

Segal, W. (1965). Comparative study of *Mycobacterium* grown *in vivo* and *in vitro*. V. Differences in staining properties. *Am. Rev. resp. Dis.* **91,** 285.

Segal, W. & Miller, W. T. (1965). Comparative study of *in vivo* and *in vitro* grown *Mycobacterium tuberculosis*. III. Lipid composition. *Proc. Soc. exp. Biol. Med.* **118,** 613.

Serrano, J. A., Tablante, R. V., de Serrano, A. A., de San Blas, G. C. & Imaeda, T. (1972). Physiological, chemical and ultrastuctural characteristics of *Corynebacterium rubrum*. *J. gen. Microbiol.* **70,** 339.

Simmons, S. R. & Karnovsky, M. L. (1973). Iodinating ability of various leukocytes and their bacteriocidal activity. *J. exp. Med.* **138,** 44.

Snow, G. A. (1970). Mycobactins: iron-chelating growth factors from mycobacteria. *Bact. Rev.* **34,** 99.

Solotorovsky, M. & Soderberg, L. S. (1972). In *Macrophages and Cellular Immunity*. Eds A. Laskin & H. Lechevalier. Cleveland: CRC Press.

Stanton, R. J. & Wright, I. S. (1967). Systemic lupus erythematosus associated with pulmonary nocardiosis. *Archs intern. Med.* **119,** 202.

Stites, D. P. & Glezen, P. (1967). Pulmonary nocardiosis in childhood. *Am. J. Dis. Child.* **114,** 101.

Strauss, R. E. & Kligman, A. M. (1951). The use of gastric mucin to lower resistance of laboratory animals to systemic fungus infections. *J. infect. Dis.* **88,** 151.

Susens, G. P., Al-Shamana, A., Rowe, J. C., Herbert, C. C., Bassis, M. G. & Coggs, G. C. (1967). Purulent constrictive pericarditis caused by *Nocardia asteroides*. *Ann. intern. Med.* **67,** 1021.

Sword, C. P. (1966). Mechanisms of pathogensis in *Listeria monocytogenes* infection. I. Influence of iron. *J. Bact.* **92,** 536.

Thacore, H. & Willett, H. P. (1966). The formation of spheroplasts of *Mycobacterium tuberculosis* in tissue culture cells. *Am. Rev. resp. Dis.* **93,** 786.

Truitt, G. L. & Mackaness, G. B. (1971). Cell-mediated resistance to aerogenic infection of the lung. *Am. Rev. resp. Dis.* **104,** 829.

Uesaka, I., Oiwa, K., Yasuhira, K., Kobara, Y. & McClung, N. M. (1971). Studies on the pathogenicity of *Nocardia* isolates for mice. *Jap. J. exp. Med.* **41,** 443.

Vacheron, M.-J., Guinard, M., Michel, G. & Ghuysen, J.-M. (1972). Structural investigations on cell walls of *Nocardia* sp.: the wall lipid and peptidoglycan moieties of *Nocardia kirovani*. *Eur. J. Biochem.* **29,** 156.

Van Furth, R. (Ed.) (1970). *Mononuclear Phagocytes*. Philadelphia: F. A. Davis Co.

Vernon-Roberts, B. (1972). *The Macrophage: Biological Structure and Function*. Cambridge: Cambridge University Press.

Webley, D. M. (1960). The effect of deficiency in iron, zinc and manganese on the growth and morphology of *Nocardia opaca*. *J. gen. Microbiol.* **23,** 87.

Webster, B. H. (1956). Pulmonary nocardiosis: a review with a report of seven cases. *Am. Rev. Tuberc. publ. Dis.* **73,** 485.

Weibull, C. (1967). The morphology of protoplasts, spheroplasts and L-forms. In *Microbial Protoplasts, Spheroplasts and L-forms*. Ed. L. B. Guze. p. 62. Baltimore: Williams and Wilkins.

Weinberg, E. D. (1966). Roles of metallic ions in host-parasite interactions. *Bact. Rev.* **30,** 136.

Weinberg, E. D. (1971). Roles of iron in host-parasite interaction. *J. infect. Dis.* **124,** 401.

Weiss, L. (1972). *The Cells and Tissues of the Immune System: Structure, Function and Interactions.* Foundations of Immunol. Series., Englewood Cliffs: Prentice Hall, Inc.

Whyte, H. J. & Kaplan, W. (1969). Nocardial mycetoma resembling granuloma faciale. *Archs Derm. Syph.* **100,** 720.

Wilhite, J. L. & Cole, F. H. (1966). Invasion of pulmonary cavities by *Nocardia asteroides. Am. Surg.* **32,** 107.

16. Delayed Hypersensitivity to *Nocardia* Antigens

L. ORTIZ-ORTIZ, MAGDALENA F. CONTRERAS
and L. F. BOJALIL

Departamento de Ecología Humana, Facultad de Medicina, U.N.A.M., México 20, D.F., Mexico

Contents

1. Introduction

Delayed hypersensitivity to nocardiae antigens has been used as a diagnostic device in man and animals infected with *Nocardia* (Kwapinski, 1969). The antigens were obtained in most cases from culture filtrates and contained mainly nucleoprotein-polysaccharides complexes. Previous studies on the specificity of antigens obtained from culture filtrate preparations of various nocardiae (sensitins) have shown that they possess a certain type of species specificity. This is manifested by the fact that humans or animals infected with a given *Nocardia* usually show stronger reactions to antigens produced from homologous nocardiae than to heterologous antigens produced from other species. The problems of nonspecific reactions with these antigen preparations, presumably cross-reactions, and the increasing interest in the pathogenic significance of unclassified acid-fast bacteria form the basis for some broader experimental studies carried out in man or animals to examine the specificity of purified antigens prepared from various nocardiae strains (Magnusson, 1961).

2. Skin Test Antigens from *Nocardia*

(a) *Isolated from culture filtrates*

The first report was that of Arêa Leão (1928) who skin tested a patient with mycetoma by using a culture filtrate from *Actinomyces bovis*, which according to the description could have been a species of the genus *Nocardia*. The sensitin elicited a specific reaction in this patient. Rabbits sensitized by using killed *Nocardia asteroides* emulsified in a water-in-oil mixture showed sensitization when they were intradermally tested with a culture filtrate obtained from the homologous strain as well as to tuberculin. Similar preparations were used later by Drake & Henrici (1943) and by Freund & Lipton (1948) with the same results. They all stated, however, that antigens elicited greater skin test reactions in individuals sensitized with the homologous strain than in those sensitized with the heterologous one. Drake & Henrici (1943) using guinea-pigs and rabbits infected with *N. asteroides* found no evidence of sensitization to tuberculin. On the other hand, Freund's and Lipton's studies showed that, whereas dilutions lower than 1 : 100 of old tuberculin gave a positive reaction in guinea-pigs sensitized with *N. asteroides*, higher dilutions were negative. De Almeida & Lacaz (1941) using a culture filtrate from *N. brasiliensis* did not find evidence of delayed hypersensitivity in patients with mycetoma caused by this actinomycete. Later, Lacaz da Silva (1945) prepared an antigen derived from 26 species of actinomycetes and tested it in a patient with mycetoma, produced by *N. brasiliensis*, with negative results. He also skin tested four more patients with mycetoma produced by other actinomycetes. The reactions were one positive and three negative. Skin tests by McQuown (1955) also failed to cause significant reactivity, which they observed might have been due to anergy. However, other reports describing similar preparations, indicate delayed skin reaction in patients with nocardiosis (Glover *et al.*, 1948).

Attempts at purification were reported by Dyson & Slack (1963), who used a fraction, precipitated with ethyl alcohol, from a culture filtrate of *N. asteroides* as a potent allergen in rabbits sensitized with killed or viable *N. asteroides*. With ion-exchange chromatography culture fluids were freed from gross contamination by cell-wall or intracellular materials (Pier & Keeler, 1965). The isolated antigen was shown to be specific since it produced skin reactions only in cattle experimentally infected with *Nocardia*, but not in those infected with mycobacteria. The purified fraction was composed of 50–60% protein, 7–14% hexose, and up to 1·5% pentose; neither hexosamine nor hexuronic acid were detected. Nucleic acids were also present (Pier & Enright, 1962; Keeler & Pier, 1965; Pier, Thurston & Larsen, 1968).

Magnusson & Mariat (1968) prepared sensitins from culture filtrates of *Nocardia* strains similar to the PPD obtained from mycobacteria (Magnusson, 1961) and studied the differences in specificity between the two. These differences were measured by means of intradermal tests in guinea-pigs sensitized with the strain used for preparation of the sensitins. In the majority of their cultures, classification of the strains based on sensitin specificity was in agreement with those from other methods. The same type of antigens were skin tested in patients with mycetoma caused by *N. brasiliensis.* The inflammatory reaction was usually greatest in these patients when the homologous antigen was used (Bojalil & Magnusson, 1963). On the other hand, Affronti (1959) using culture supernatants of *N. asteroides* fractionated with half-saturated ammonium sulfate solutions found cross-reactivity when this antigen was skin tested in either guinea-pigs sensitized with tubercle bacilli or rabbits sensitized with BCG. However, the largest reactions were elicited in the *Nocardia*-sensitized animals of both species. It is not at all surprising to find cross-reactivity between *Nocardia* and *Mycobacterium,* since as we know, some chemical similarities exist between them (Kwapinski, 1969).

(b) *Isolated from somatic material*

The first attempt to isolate from somatic material of *Nocardia* substances with the capacity to elicit delayed-type hypersensitivity reactions was that of Drake & Henrici (1943). They obtained a polysaccharide from fractionated *N. asteroides* organisms by boiling them in hydrochloric acid, followed by treatment with trichloroacetic acid and precipitation with ethanol. The resulting polysaccharide material produced erythema with no induration in animals sensitized with the homologous strain. A protein material was also obtained by centrifugation of a suspension of ground cells. The supernatant was treated with trichloroacetic acid and the precipitated protein isolated. The protein fraction was injected into the sensitized animals in whom erythema, induration and a small area of necrosis developed. Injection into uninfected control animals was negative.

Later, González-Ochoa and his group (González-Ochoa & Vázquez-Hoyos 1953; González-Ochoa & Baranda, 1953) isolated a crude polysaccharide from *N. brasiliensis* which elicited a specific delayed hypersensitivity reaction in 16 patients infected with *N. brasiliensis.* The mycelial material used for this purpose was extracted with phenol and an ether-acetone mixture in a Soxhlet apparatus. The extracts were treated with ethanol to precipitate the antigen. The authors also reported that individuals with actinomycotic mycetoma due to *N. brasiliensis* and having poor prognoses lacked cutaneous reactivity to the extract. This polysaccharide produced no reactivity in a patient with advanced pulmonary tuberculosis nor in two patients with actinomycosis (*A. bovis*). However, the *N. asteroides* crude polysaccharide did evoke a reaction in patients with actinomycotic mycetoma caused by *N. brasiliensis,* although the reaction was less intense than the one produced by the homologous antigen (González-Ochoa, *et al.*, 1962).

Kingsbury & Slack (1967) isolated an active polypeptide from *N. asteroides.* This was extracted with diluted hydrochloric acid from lipid-free cells. The antigen was precipitated with picric acid and reprecipitated with acetone. The procedure was repeated after dialysis. The polypeptide antigen obtained from three different strains of *N. asteroides* did not show cross-reactions with genera other than *Nocardia.* Cross-reactions occurred with *N. caviae* and one strain of *N. brasiliensis.* When the authors combined the antigen derived from three strains of *N. asteroides,* their results were more specific although occasional cross-reactions were observed with *N. brasiliensis*-sensitized animals (Kingsbury & Slack, 1969).

In our laboratory, for the purpose of studying different somatic materials, we grew *N. asteroides* and *N. brasiliensis* in a modified Proskauer and Beck medium (Youmans & Karlson, 1947) and defatted the culture by repeated treatment with an ethyl alcohol-diethyl ether (1 : 1) solution. The bacteria were suspended in a 0·01 M Tris (hydroxy-methyl)-aminomethane buffer, pH 7·4, containing 0·01 M magnesium acetate and disrupted in a Sorvall-Ribi fractionator at 15 000–30 000 psi. The material was then treated as shown in Fig. 1.

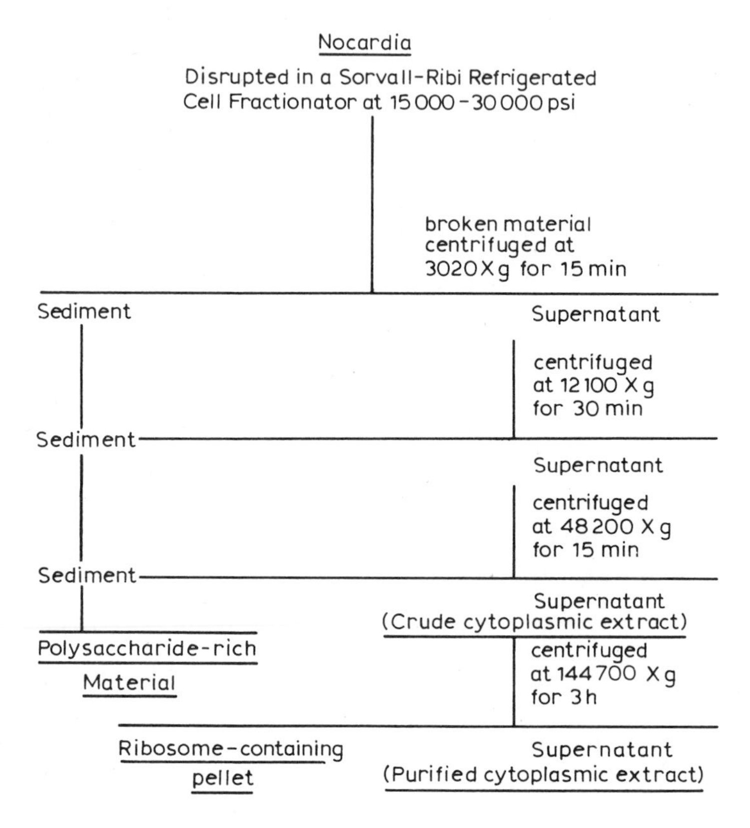

Fig. 1. Method of isolating different somatic components from *Nocardia*.

From the sediment containing most of the cell wall material, a highly purified polysaccharide was isolated (Ortiz-Ortiz, Bojalil & Contreras, 1972*a*). The polysaccharide was able to elicit, not only immediate—but also delayed-type hypersensitivity reactions as well. Although the immediate reactions were of similar intensity in animals sensitized with both types of *Nocardia*, the delayed reaction was greater in the guinea-pigs sensitized with the homologous strain. The protein-free carbohydrate was also able to inhibit the migration of macrophages from *Nocardia*-sensitized animals. The cross-reactivity observed in the anaphylactic response was assumed to be due to a polysaccharide found in cell extracts of both *Nocardia* and *Mycobacterium*, and known as polysaccharide I (Bojalil & Zamora, 1963). On the other hand, the carbohydrate responsible for the delayed reaction seemed to be different from polysaccharide I, since no delayed hypersensitivity was evoked in *Mycobacterium*-sensitized guinea-pigs (Ortiz-Ortiz, Bojalil & Contreras, 1972*a*).

The crude extract from these two *Nocardia* species contained protein, carbohydrate, RNA and DNA materials like that reported by Kwapinski (1969). Skin tests with this crude extract in guinea pigs sensitized against *N. asteroides* and *N. brasiliensis* showed cross-reactivity. Skin reactions, however, were more intense in the homologously sensitized group (Table 1). The inflammatory reactions in all cases were both immediate- and delayed-type hypersensitivities as corroborated by histological studies. By Ouchterlony analysis these two antigenic materials showed several precipitation lines. They also showed lines of identity with sera obtained from tuberculous patients.

Differential centrifugation of the crude extract at 144 700 × **g** for 3 h, followed by separation of the pellet containing ribosomal material produced a supernatant hereafter referred to as purified cytoplasmic extract. This material contained protein, carbohydrate and nucleic acid. Skin tests with these cytoplasmic extracts permitted the differentiation between animals sensitized with *N. asteroides* or *N. brasiliensis* by measuring the magnitude of the indurated skin area. At the doses tested, they evoked a delayed-type hypersensitivity reaction in homologously sensitized animals without showing early inflammatory reactions (Ortiz-Ortiz, Contreras & Bojalil, 1972*b*). This delayed reactivity was further confirmed by *in vitro* experiments in which macrophages obtained from the peritoneal cavities of guinea-pigs sensitized by *Nocardia* were inhibited from migrating after addition of the homologous antigen. Patients with mycetoma due to *N. brasiliensis* have also been skin tested with these antigens which were found to be specific, since they produced no inflammatory reactions in patients with leprosy or tuberculosis or in healthy individuals (Ortiz-Ortiz & Bojalil, 1972).

The ribosomal proteins were obtained from both *Nocardia* species by the methods described previously for mycobacteria (Ortiz-Ortiz, Solarolo & Bojalil, 1971) and their inflammatory

Table 1
Specificity differences of somatic components from Nocardia*

Somatic	Crude cytoplasmic extract		Purified cytoplasmic extract		Ribosomal protein		Polysaccharide material	
Component	*N.a.*†	*N.b.*†	*N.a.*	*N.b.*	*N.a.*	*N.b.*	*N.a.*	*N.b.*
N. asteroides								
crude cytoplasmic extract	—	4	—	5	—	7	—	5
purified cytoplasmic extract	—	7	—	7	—	10	—	7
ribosomal protein	—	12	—	13	—	15	—	12
polysaccharide fraction	—	4	—	4	—	7	—	4
N. brasiliensis								
crude cytoplasmic extract	4	—	7	—	12	—	4	—
purified cytoplasmic extract	5	—	7	—	13	—	4	—
ribosomal protein	7	—	10	—	15	—	7	—
polysaccharide fraction	5	—	7	—	12	—	4	—

* The values in the table are differences (in mm) between homologous and heterologous reactions to skin tests in guinea-pigs sensitized with the strains used for preparation of the somatic components. Values are shown twice.

† N.a. = *Nocardia asteroides*; N.b. = *Nocardia brasiliensis.*

properties studied. Ribosomal proteins from both species elicited delayed-type hypersensitivity in homologously as well as the heterologously sensitized guinea-pigs. However, the reactions were greater in the former group of animals permitting the differentiation between them. The difference in reactivity to the two proteins was even greater when tested by inhibition of macrophage migration tests (Ortiz-Ortiz, Contreras & Bojalil, 1972*c*).

3. Discussion

The work just reviewed, makes it clear that experimental animals or humans sensitized with nocardiae show delayed-type hypersensitivity reactions when skin tested with nocardiae derived from either culture filtrate preparations or somatic materials.

When the antigen preparations were tested for specificity on animals or patients sensitized to other actinomycetes or biologically similar microorganisms, such as mycobacteria, the degree of reactivity varied from nonreciprocal cross-reactivity (Bretey, 1933) to reciprocal cross-reactivity (Goyal, 1937; Freund & Lipton, 1948). In all cases, however, the homologous reaction was greater than and distinguishable from the heterologous skin tests.

It is difficult to evaluate the degree of cross-reactivity or specificity among the preparations discussed, since different tuberculin mixtures and concentrations were used. Purification of culture filtrate preparations increased the specificity of the skin reaction (Drake & Henrici, 1943; Pier & Enright, 1962). The antigen isolated by Pier, Thurston & Larsen (1968) provoked an especially specific response in cattle. Unfortunately, no trials have been performed with this antigen in man.

The isolation of substances from the soma of nocardiae represented the first attempts to look into the biological properties of these components. All fractions so far isolated elicited delayed-type hypersensitivity reactions in *Nocardia*-sensitized animals or patients. They did, however, cross-react to a lesser degree in those animals sensitized to mycobacteria. As we expected, cross-reactivity was greater with the less purified materials. The polypeptide isolated by Kingsbury & Slack (1967) is very promising for diagnosis of nocardiae infections, and if further studies show that it is as specific as the one obtained from mycobacteria (Okada *et al.*, 1963), it could be one of the best choices for skin tests of humans along with the ribosomal protein from nocardiae (Ortiz-Ortiz, Contreras & Bojalil, 1972*c*). The ribosomal material behaved very specifically, and was less cross-reactive than the other materials. Work performed with another acid-fast microorganism, specifically BCG, showed that ribosomal protein had more activity and specificity than conventional PPD preparations (Ortiz-Ortiz, Solarolo & Bojalil, 1971). These purified ribosomal proteins also represent an advantage in standardization, since no contaminants were present.

The ribosome-free supernatant from *Nocardia* somatic material also contained a specific skin test antigen. These purified cytoplasmic extracts inhibited migration only of peritoneal macrophages obtained from the homologous *Nocardia*-sensitized guinea-pigs (Ortiz-Ortiz, Contreras & Bojalil, 1972*b*). It is interesting that the purified antigen from *N. brasiliensis* incited delayed-type hypersensitivity in patients with mycetoma caused by *N. brasiliensis*, but not in those with tuberculosis, or leprosy or in healthy individuals (Ortiz-Ortiz & Bojalil, 1972). The use of these antigens also offers an advantage in epidemiological studies, since positive reactions were obtained in individuals working and living in areas where *Nocardia* was isolated from the soil. The specificity of the purified cytoplasmic extract from *N. asteroides* was not assessed in humans, because of the impossibility of finding individuals infected with these microorganisms. The latter antigen, however, seems to be specific, since it elicited specific delayed reactivity in guinea-pigs sensitized with the homologous microorganisms as well as *in vitro* (Ortiz-Ortiz, Contreras & Bojalil, 1972*b*).

The fact that nitrogen-free polysaccharides obtained from either *N. asteroides* or *N. brasiliensis* could elicit delayed type hypersensitivity in *Nocardia*-sensitized animals was confirmed *in vitro* (Ortiz-Ortiz, Bojalil & Contreras, 1972*a*). Similar types of antigens obtained from *Nocardia* were previously shown to be able to elicit delayed reactivity (Drake & Henrici, 1943; González-Ochoa & Baranda, 1953; González-Ochoa *et al.*, 1962). However, in all cases, doubts arose regarding the purity of the materials used (Dyson & Slack, 1963), since particularly protein or peptide components have been considered as the substances responsible for the sensitizing or eliciting of the responses observed. From the practical point of view, isolation of purified polysaccharides is not advantageous, since other preparations, i.e., ribosomal or cytoplasmic antigens, offer more specificity (Table 1) and are simpler to obtain.

The different antigenic substances herein reported show specificity for the detection of sensitization against *Nocardia.* They can be of importance in studying the role of cell-mediated immunity in the pathogenesis of this disease, since both delayed-hypersensitivity as well as cell-mediated immunity seem to require the presence of specifically sensitized thymus-derived lymphocytes.

4. References

Affronti, L. F. (1959). Purified protein derivatives (PPD) and other antigens prepared from atypical acid-fast bacilli and *Nocardia asteroides. Am. Rev. resp. Dis.* **79,** 284.

Almeida, F. de & Lacaz, S. S. (1941). Estudos sobre o *Actinomyces brasiliensis* (Lindenber, 1909). *Annali Fac. Med. Univ. Sao Paolo* **17,** 577.

Arêa Leão, A. E. (1928). L'intradermo-réaction dans L'actinomycose. Réaction spécifique de la peau avec le filtrat de culture d'*Actinomyces bovis. C. r. Séanc. Soc. Biol.* **98,** 1575.

Bojalil, L. F. & Magnusson, M. (1963). Specificity of skin reactions of humans to *Nocardia* sensitins. *Am. Rev. resp. Dis.* **88,** 409.

Bojalil, L. F. & Zamora, A. (1963). Precipitin and skin tests in the diagnosis of mycetoma due to *Nocardia brasiliensis. Proc. Soc. exp. Biol. Med.* **113,** 40.

Bretey, J. (1933). Sur les propriétés allergisantes des *Nocardia* et des *Streptothrix* et sur la toxicité de leurs extraits. *C. r. Séanc. Soc. Biol.* **113,** 53.

Drake, C. H. & Henrici, A. T. (1943). *Nocardia asteroides*: Its pathogenicity and allergic properties. *Am. Rev. Tuberc. pulm. Dis.* **48,** 184.

Dyson, J. E. & Slack, J. M. (1963). Improved antigen for skin testing in nocardiosis. *Am. Rev. resp. Dis.* **88,** 80.

Freund, J. & Lipton, M. M. (1948). Potentiating effect of *Nocardia asteroides* on sensitization to picryl chloride and on the production of isoallergic encephalomyelitis. *Proc. Soc. exp. Biol. Med.* **68,** 373.

Glover, R. P., Herrell, W. E., Heilman, F. R. & Pfuetze, K. H. (1948). Nocardiosis. *Nocardia asteroides* infection simulating pulmonary tuberculosis. *J. Am. med. Ass.* **136,** 172.

González-Ochoa, A. & Baranda, F. (1953). Una prueba cutánea para el diagnóstico del micetoma actinomicósico por *Nocardia brasiliensis. Revta Inst. Salubr. Enferm. trop., Mexico* **13,** 189.

González-Ochoa, A. & Vázquez-Hoyos, F. (1953). Relaciones serológicas de los principales actinomicetes patógenos, *Revta Inst. Salubr. Enferm. trop., Mexico* **33,** 177.

Gonzálex-Ochoa, A., Shibayama, H., Félix, D. & Anaya, M. (1962). Immunological aspects of actinomycotic mycetoma and nocardiosis. *XII Inter. Congr. Dermatol. Excerp. Med., International Congress Series* No. **55,** 542.

Goyal, R. K. (1937). Étude microbiologique, expérimentale et immunologique de quelques streptothricées. *Annls Inst. Pasteur, Paris* **59,** 94.

Keeler, R. F. & Pier, A. C. (1965). Extracellular antigens of *Nocardia asteroides.* II. Fractionation and chemical characterization. *Am. Rev. resp. Dis.* **91,** 400.

Kingsbury, E. W. & Slack, J. M. (1967). A polypeptide skin test antigen from *Nocardia asteroides.* I. Production, chemical and biologic characterization. *Am. Rev. resp. Dis.* **95,** 827.

Kingsbury, E. W. & Slack, J. M. (1969). A polypeptide skin test antigen from *Nocardia asteroides.* II. Further studies on the specificity of a nocardin active polypeptide. *Sabouraudia* **7,** 85.

Kwapinski, J. B. G. (1969). Analytical serology of actinomycetales. In *Analytical Serology of Microorganisms,* vol. 1. Ed. J. B. G. Kwapinski, p. 1. New York: John Wiley.

Lacaz da Silva, C. (1945). Contribuicao para o estudio dos Actinomicetos productores de micetomas. *Tesis Rec. Med. Univ. Sao Paolo.*

Magnusson, M. (1961). Specificity of mycobacterial sensitins. I. Studies in guinea pigs with purified "tuberculin" prepared from mammalian and avian tubercle bacilli *Mycobacterium balnei* and other acid-fast bacilli. *Am. Rev. resp. Dis.* **83,** 57.

Magnusson, M. & Mariat, F. (1968). Delineation of *Nocardia farcinica* by delayed type skin reactions on guinea pigs. *J. gen. Microbiol.* **51,** 151.

McQuown, A. L. (1955). Actinomycosis and nocardiosis. *Am. J. clin. Path.* **25,** 2.

Okada, Y., Morisawa, S., Syojima, K., Kitagawa, M., Nakashima, S. & Yamamura, Y. (1963). Improved method for the isolation and properties of tuberculine active peptides. *J. Biochem., Tokyo* **54,** 484.

Ortiz-Ortiz, L. & Bojalil, L. F. (1972). Delayed skin reactions to cytoplasmic extracts of *Nocardia* organisms as a means of diagnosis and epidemiological study of *Nocardia* infection. *Clin. Exp. Immun.* **12,** 225.

Ortiz-Ortiz, L., Bojalil, L. F. & Contreras, M. F. (1972*a*). Delayed hypersensitivity to polysaccharides from *Nocardia. J. Immun.* **108,** 1409.

Ortiz-Ortiz, L., Contreras, M. F. & Bojalil, L. F. (1972*b*). Cytoplasmic antigens from *Nocardia* eliciting a specific delayed hypersensitivity. *Infect. Immun.* **5,** 879.

Ortiz-Ortiz, L., Contreras, M. F. & Bojalil, L. F. (1972*c*). The assay of delayed hypersensitivity to ribosomal protein from *Nocardia. Sabouraudia* **10,** 147.

Ortiz-Ortiz, L., Solarolo, E. B. & Bojalil, L. F. (1971). Delayed hypersensitivity to ribosomal protein from BCG. *J. Immun.* **107,** 1022.

Pier, A. C. & Enright, J. B. (1962). *Nocardia asteroides* as a mammary pathogen of cattle. II. Immunologic reactions of infected animals. *Am. J. vet. Res.* **23,** 284.

Pier, A. C. & Keeler, R. F. (1965). *Extracellular antigens of Nocardia asteroides.* I. Production and immunological characterization. *Am. Rev. resp. Dis.* **91,** 391.

Pier, A. C., Thurston, J. R. & Larsen, A. B. (1968). A diagnostic antigen for nocardiosis: comparative tests in cattle with nocardiosis and mycobacteriosis. *Am. J. vet. Res.* **29,** 397.

Youmans, G. P. & Karlson, A. G. (1947). Streptomycin sensitivity of tubercle bacilli. Studies on recently isolated tubercle bacilli and the development of resistance to streptomycin *in vivo. Am. Rev. Tuberc. pulm. Dis.* **55,** 529.

17. Nocardiae and Chemotherapy

A. González-Ochoa

Mycology Laboratory, Institute of Public Health and Tropical Diseases, Mexico

Contents

1. Introduction

Three nocardiae unanimously recognized as being pathogenic: *Nocardia asteroides* (Eppinger, 1891) Blanchard, 1895, *N. brasiliensis* (Lindenberg, 1909) Pinoy, 1913, and *N. caviae* (Erikson, 1935) Gordon & Mihm, 1962, mentioned in the order in which they were discovered, have become increasingly important clinically.

These species whose pathogenic, taxonomic and immunological relationships are so evident, also present great similarities in the chemotherapy of the two diseases which they cause: nocardiosis and actinomycotic mycetoma. Nocardiosis, usually caused by *N. asteroides* and exceptionally by *N. brasiliensis*, consists of chronic infections, suppuration or granulomatosis of the lung, the central nervous system and, less frequently, of other organs. The bacteria live in the tissues either as short acid-fast filaments or as bacillary or coccoid fragments. Actinomycotic mycetoma, caused by nocardiae, is usually due to *N. brasiliensis* though in rare instances *N. caviae* or *N. asteroides* may be responsible. This mycosis appears as fistulous tumours in which the nocardiae, which are able to cause mycetoma, take the form of aggregations of mycelium (microcolonies) known as actinomycotic granules. This disease can also be caused by actinomycetes classified in the genera *Actinomadura* and *Streptomyces*.

Although nocardiosis can be detected in apparently healthy individuals, it usually corresponds to a mycosis caused by opportunistic fungi. The incidence of nocardiosis is rising due to changes in the factors that promote mycoses caused by opportunistic organisms. The factors responsible include rising life expectancies which result in a greater number of persons suffering debilitating ailments such as diabetes, leukemias, lymphomas etc.; the use of immuno-suppresants and antimetabolites, so much in vogue nowadays; the indiscriminate or necessary use of large quantities, or for long periods, of corticosteroids and antibiotics; and the greater number of operations on the heart, kidneys, brain etc. Actinomycotic mycetoma, is diagnosed more and more frequently due to the greater diffusion of knowledge on mycology which increases in parallel with the development of medical knowledge. Consequently the chemotherapy of infections caused by nocardiae is a matter of increasing concern, both to countries with high living standards where cases of nocardiosis are rising, and to countries with low socio-economic levels where more cases of mycetoma are being diagnosed.

It would be useless to undertake a review of chemotherapeutics, antibiotics and other different drugs which have been found to be ineffective in the treatment of nocardiosis and actinomycotic mycetoma caused by nocardiae. This contribution, therefore, will consider only those drugs that have been shown to be effective in the treatment of these mycoses, and which are still in use.

2. The drug,4,4′-diaminodiphenyl sulphone

The drug 4,4′-diaminodiphenyl sulphone (DDS) was synthetized by Fromm & Wittmann (1908). The bacteriostatic action of the drug against streptococci in mice was demonstrated independently by Fourneau *et al.* (1937) and Buttle *et al.* (1937). Rist (1939) found that DDS was active against *Mycobacterium tuberculosis in vitro* and against tuberculosis in guinea pigs. At that time DDS was thought to be highly toxic and the mono- and di-substituted derivatives were used therapeutically in pulmonary tuberculosis, and in leprosy (Faget *et al.*, 1943). However, since the results were discouraging in tuberculosis but highly favourable in leprosy the use of the substituents was restricted to treating the latter.

In a personal communication Latapi in 1946 suggested the use of a di-substituted sulphonic derivative (Diasone*) for the treatment of actinomycotic mycetoma, and Strauss, Kligman and Pillsbury (1951) pointed out the efficacy of sulphones and sulphonamides on nocardiosis. A year later González-Ochoa *et al.* (1952) obtained good results with Diasone in the treatment of two cases of actinomycotic mycetoma caused by *Nocardia brasiliensis.* Halde & Newstrand (1955) published extensive studies on the susceptibility of 23 strains of *N. asteroides* and four strains of *N. brasiliensis* to Diasone, Promizole and Promin, and found that the first two were the most effective. However, they found that different strains showed a wide range of *in vitro* susceptibility; inhibition being caused by concentrations ranging from 1 μg/ml to 10 000 μg/ml. Studies conducted *in vivo* in which mice were inoculated with a strain of *N. asteroides* and treated with Diazone, were inconclusive.

In spite of the supposed toxicity of DDS it was used extensively in the treatment of leprosy, and the incidence of side effects was not higher than those produced by the sulphonic derivatives. In a study of the use of DDS and other therapeutic agents in patients with leprosy, González-Ochoa *et al.* (1952) came to the conclusion that the active parent DDS was superior to its derivatives, because smaller daily doses were required, it could be administered orally, its toxicity was no greater than that of the sulphonic derivatives, and because it cost less. As a result of these studies González-Ochoa and his colleagues started to use DDS, instead of its sulphonic

*Abbott Laboratories.

derivatives, for actinomycotic mycetoma caused by *N. brasiliensis*. They performed *in vitro* studies against some of the aerobic pathogenic actinomycetes and found that *N. brasiliensis* was the most sensitive, its growth partially inhibited at 1 : 50 000 and completely inhibited at 1 : 10 000. González-Ochoa *et al.* (1952) reported their findings of treatment with DDS in experimental infections of mice caused by *N. brasiliensis*, and in eight cases of actinomycotic mycetoma in man produced by the same *Nocardia*. In the experimental infections the results were inconclusive, but with the eight patients very encouraging results were obtained as six of them were cured. In a subsequent paper González-Ochoa (1953) reported his good results with DDS in the treatment of patients having hitherto fatal thoracic mycetomas with pulmonary involvement, and was thereby able to reconfirm the effectiveness of this therapeutic agent (Figs 1 and 2).

In 1946 González-Ochoa and his colleagues used the disubstituted sulphonic derivative Diazone for treating mycetomas but since 1950 DDS has been used for the reasons described above. Between 1950 and 1961 more than 100 patients suffering from actinomycotic mycetoma caused by *N. brasiliensis*, and three from nocardiosis caused by *N. asteroides*, have been treated. In general terms there was a 28% cure rate in the mycetoma cases after 2–4 years of treatment, and depending on the localization, extension of the lesions, bone involvement, and the duration of the ailment, marked improvement was achieved in a further 35% of the cases. One of the cases of pulmonary nocardiosis was cured, but the other two patients died as a result of the disease that had favoured the appearance of nocardiosis.

(a) *Dosage*

The total dose of 200 mg a day, one 100 mg tablet after breakfast and the other after dinner, is recommended. This average dose was proposed after trying doses varying from 100 to 400 mg daily. The dosage used is purely empirical and there is no correlation of dosage with the blood levels necessary to inhibit the *Nocardia*. With 200 mg daily the blood concentration varies from 0·26 mg/ml to 0·88 mg/ml which is about 10 times less than the amount required to partially inhibit *N. brasiliensis in vitro*.

(b) *Clinical results*

Both clinical cure and absolute or radical cure must be considered. Clinical cure occurs early. After one year of treatment the ulcers heal, the fistulae close leaving hyperchromic depressed scars, the nodules disappear, and the tumours are reduced in size. During this time the improvement is also reflected in the laboratory findings. The high sedimentation rate and the neutrophil leucocytosis which are present in the active stage of the disease also return to normal. There is no way of determining whether the mycetoma has been actually cured or whether a latent focus exists. However, if treatment is suspended early, relapses may occur, and to provide a margin of safety DDS should be administered for 2–3 years after clinical recovery is obtained. If treatment is suspended prematurely, the patient may not show improvement again when the drug is reinstituted at a subsequent date. In such cases amputation is indicated if the mycetoma is localized to a limb, and if surgical measures are not possible death occurs.

(c) *Development of resistance*

In addition to the resistance that appears as a result of an early suspension of treatment, González-Ochoa and co-workers have observed that when *Nocardia* infects the lung it continues to spread until it produces an impairment of vital functions and death ensues even when treatment has been continued without interruption. When both of these conditions occur simultaneously—that is, early suspension of treatment and pulmonary invasion—subsequent resistance becomes even more striking. Similar degrees of sensitivity have been observed in the same organism isolated when the patient is improving under DDS treatment, and in those obtained when the lesions did not respond to treatment.

(d) *Side effects*

In spite of the long duration of DDS administration, the serious toxic phenomena described in the literature have not been observed. The commonest side effect is a moderate normocytic iron-deficiency anemia, which is not difficult to control, or to prevent by the administration of iron. In very rare instances patients develop dermatologic eruptions, as well as conjunctivitis,

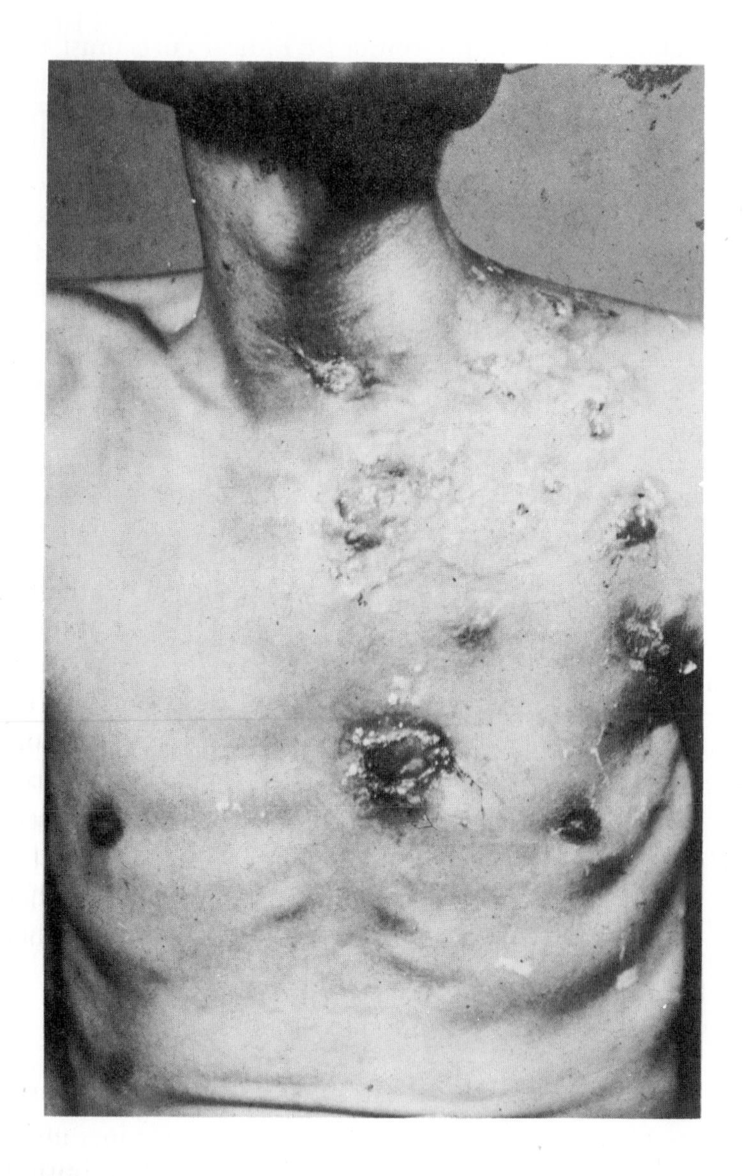

Fig. 1. Thoracic mycetoma with pulmonary involvement before treatment.

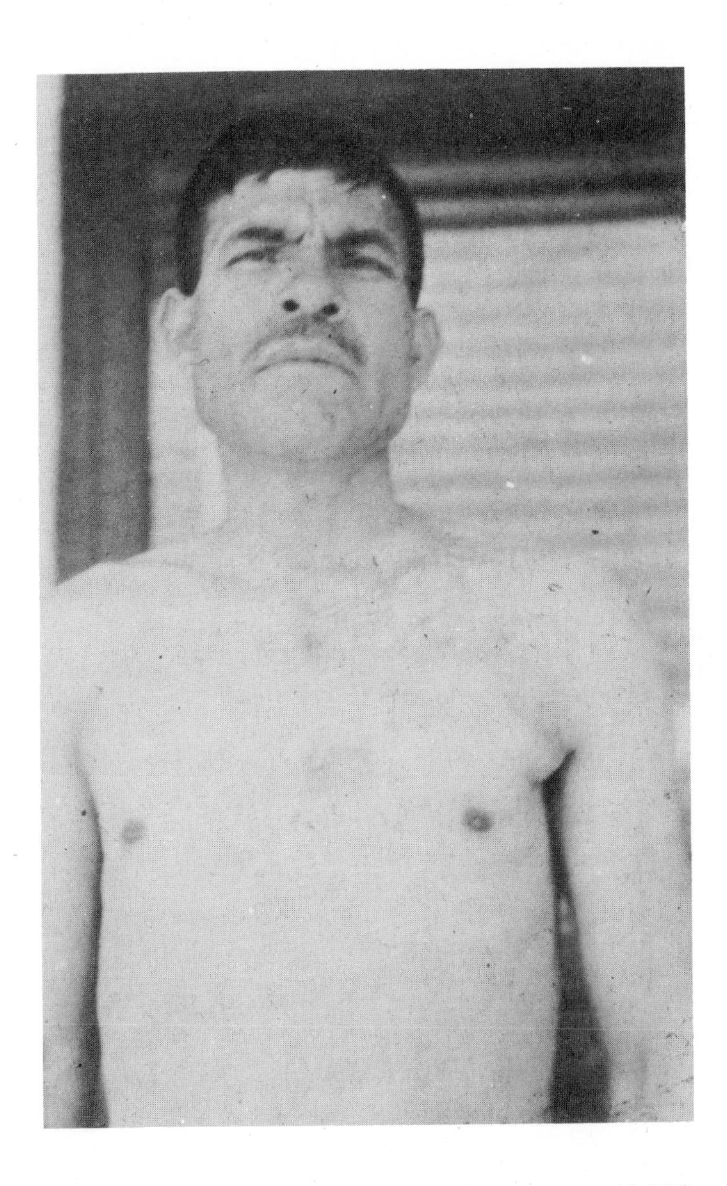

Fig. 2. The same case of thoracic mycetoma after treatment with DDS.

dizziness, and nausea. In some patients the eruptions are polymorphous and in others papular.

(e) *Mode of action*

It is not known how DDS acts. It is difficult to explain the results obtained only by its bacteriostatic activity. As I have mentioned, there is no relationship between the blood levels attained and the amount required *in vitro* to inhibit *N. brasiliensis*. In addition, the experiment in which guinea pigs were infected with *N. brasiliensis*, and treated with doses of 300, 200 and 100 mg/kg during an average of 2 months, was inconclusive.

3. The Sulphonamides

In view of the mediocre or negative results with DDS in cases with extensive lesions, and marked bone or lung involvement, González-Ochoa and his colleagues have continued examining drugs and methods of treatment that might be better and more reliable.

Strauss, Kligman & Pillsbury (1951) pointed out the usefulness of the sulphonamides in the treatment of nocardiosis. Halde & Newstrand (1955) carried out extensive research on the susceptibility of 23 strains of *N. asteroides* and four strains of *N. brasiliensis* to sulphadiazine, sulphathiazole, sulphisoxazole, sulphamethazine and sulphamerazine. Sulphamerazine appeared to be the most effective of the sulphonamide drugs in inhibiting the growth of *N. asteroides in vitro*. It produces complete inhibition at dilutions usually 10-, and occasionally, 100 fold less than the other sulphonamides. In general, a strain which was sensitive to one drug at a low concentration was also comparatively sensitive to the other sulphonamides. The other strains were by comparison sensitive only to high concentrations of the drugs.

Guzmán-Toledano (1962), Beirana & Hernández (1963) and Ferreira-López (1963) published papers referring to the action of 4,aminobenzenesulphanilamido-5,6-dimethoxy pyrimidine (Fanasil) in the treatment of actinomycotic mycetoma caused by *N. brasiliensis*. The drug's chemical structure which conditions its slow elimination and low toxicity is very well known.

In a paper presented at the VI International Congress of Chemotherapy González-Ochoa, Stark & Vazquez-Ibarra (1967) presented the results on the sensitivity of *N. brasiliensis* towards Fanasil. The *in vitro* studies showed inhibition that varied within a range of 1 : 4000 to 1 : 20 000 according to the method, culture media, temperature and strain used. These preliminary clinical experiments seemed to indicate the superiority of Fanasil to DDS, and the authors have continued to prefer this drug for the treatment of that mycetoma.

However, the oral administration of Fanasil does have disadvantages; the clinical response can be low, an increase in dosage may not cause a proportional increase in blood concentration, and marked vascular involvement of the injured tissue can occur. For these reasons, and because most cases of mycetoma occur in the lower extremities, González-Ochoa and his colleagues have experimented with the intra-arterial administration of the drug in large doses.

(a) *Dosage*

The observations from the oral treatment were based on 60 cases: 28 without or with very light bone involvement, and 32 with severe bone involvement or lung invasion. the drug was administered at dosages much higher than those usually employed for bacterial infections, in periods that varied between 2 and 4 years, and under four different schemes of treatment: (1) 2 g/weekly, (2) 0·50 g/daily, (3) 1 g/daily and (4) 1·50 g/daily. This means that patients received respectively 2, 3·50, 7 and 10·50 g/weekly. The selection of the treatment scheme was made in accordance with the weight of patient, the extent of the lesions, but basically depended on the situation in the bones and lung. The dosage was higher when there was a lack of response to the smaller doses.

For the intra-arterial administration 12 severe cases, considered as candidates for amputation due to the severity and extent of their lesions, as well as to the degree of bone involvement, were selected. The drug was dissolved in 1000 ml of normal saline and administered by slow perfusion over the space of 18 to 20 h via the femoral artery of the affected limb. The dosage varied from 4 to 8 g depending on the weight and age of the patient. The number of perfusions varied from 1 to 3, and these were repeated at intervals of 1–2 months. Oral treatment was continued after the injections, when the Fanasil blood levels fell to approximately 10 mg % (within an average of 10 days). The oral dose varied from 0·50 g to 2 g daily the dosage being adjusted in each case so as to ensure a blood level about 20–25 mg %.

(b) *Clinical results*

Given the diversity of clinical manifestations in mycetoma (duration of the disease, extension and intensity of soft tissue, bone and lung lesions, age and general status of patients) it is difficult to make an accurate evaluation of the activity of a drug, as a group of patients is never homogeneous. Of the 60 patients treated orally for 2–4 years, it can be seen (Table 1) that there was clinical cure in

Table 1

Clinical results in 60 patients treated orally with Fanasil for 2 to 4 years

Status of the patients	Clinical cure	Evident improvement	Failure
Without or with slight bone involvement 28	18 (64%)	10 (36%)	
With marked bone involvement or lung invasion 32	—	26 (81%)	6 (19%)
60	18 (30%)	36 (60%)	6 (10%)

18 patients (30%) although two of them relapsed and a marked improvement in 36 patients (60%). Six patients (10%) were considered failures. With some exceptions these results, could be related to the severity of the disease. Better results were obtained when the higher doses of the drug were administered.

Twelve patients were treated with the arterial perfusion. Most of them showed evident improvement 3 weeks after the perfusion. A clinical cure was obtained in five cases after some months, and three of these with extensive lesions and serious bone involvement, were cured in 4 months. In one of the latter cases the results were spectacular, given the severity of the lesions in the soft tissues (see Figs 3, 4, 5 and 6). This patient remains asymptomatic after 9 years of treatment.

(c) *Blood level*

In the cases treated orally the average blood levels obtained with the four different schemes of treatment logically varied according to the amount of drug used, (scheme (1) 11 mg %, (2) 12 mg %, (3) 20 mg % and (4) 25 mg %). Nevertheless in some cases a lack of relationship between the dosage and the blood level was observed, so were individual variations of the blood levels with the same dosage, and in many cases there seemed to exist a level of acceptance to the drug that was only surpassed with large doses.

The blood levels obtained at the end of the arterial perfusion with 4, 6 and 8 g, were 24 mg %, 30 mg % and 40 mg % respectively; but exceptions were observed as in a case treated with 6 g that reached a blood level of 80 mg %.

(d) *Side effects*

No important renal lesions have been observed in patients treated orally. In 70% of the cases treated no manifestations of renal impairment were observed. Albumin traces and hyaline casts were found in the urine of 20% of the patients but they disappeared spontaneously without treatment or withdrawal of the drug. In 10% of the patients the urine contained granular casts and red blood cells and blood chemistry analyses found a moderate elevation of N.P.N. and bilirubin. However, a discontinuation of the treatment or lowering of the dose for some weeks caused these abnormalities to disappear and treatment was continued without trouble. In four cases, a slight leucopenia was observed, but 1–2 months after treatment was discontinued the white cell count returned to normal.

In the patients treated by intra-arterial infusions signs of renal impairment were observed in almost half of them regardless of the dosage administered. Some received 4 g and showed abnormalities while another with 8 g behaved normally. The collateral manifestations disappeared even after another infusion was given or when

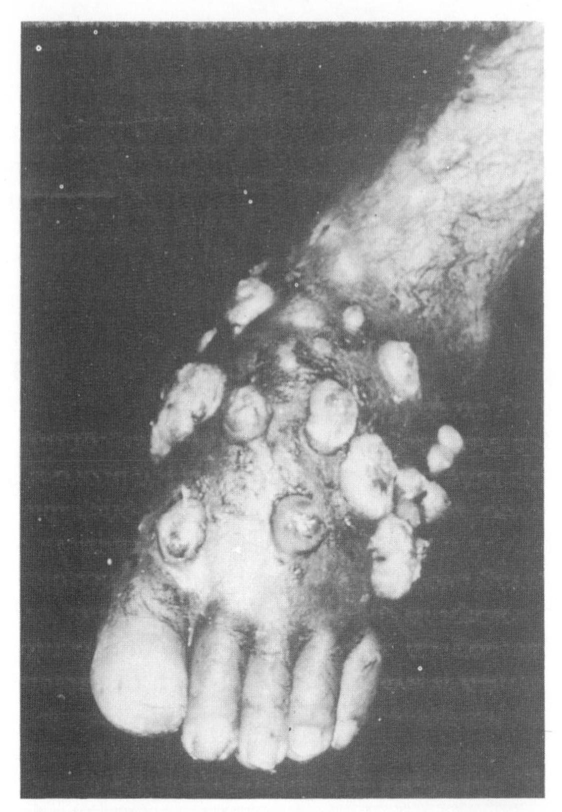

Fig. 3. Mycetoma before treatment with intra-arterial perfusion of Fanasil.

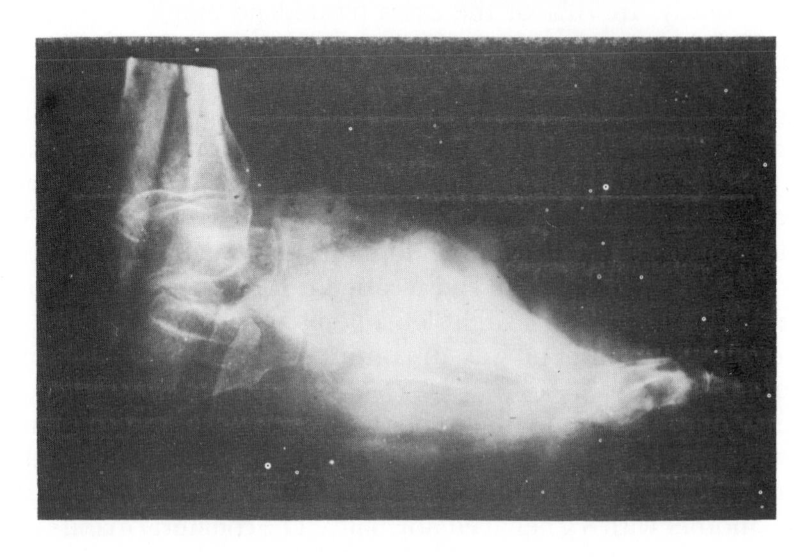

Fig. 4. Bone lesions of the case shown in Fig. 3.

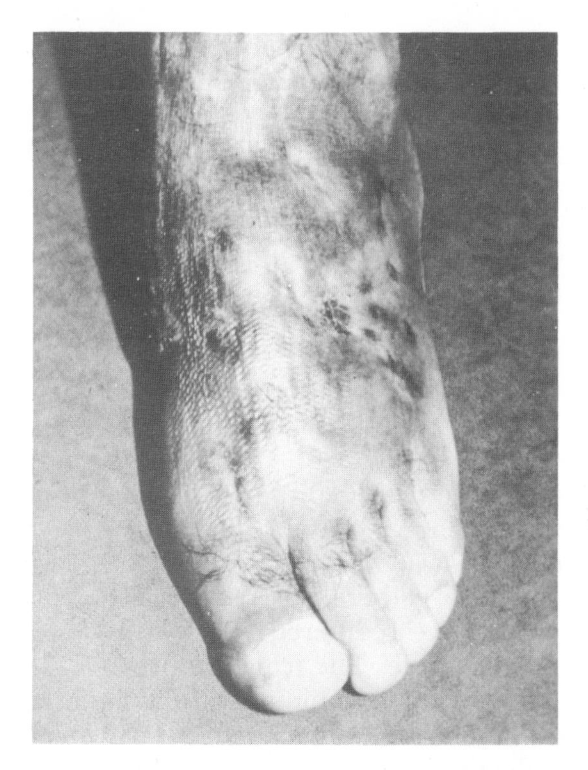

Fig. 5. The case shown in Fig. 3 4 months after treatment with intra-arterial perfusion.

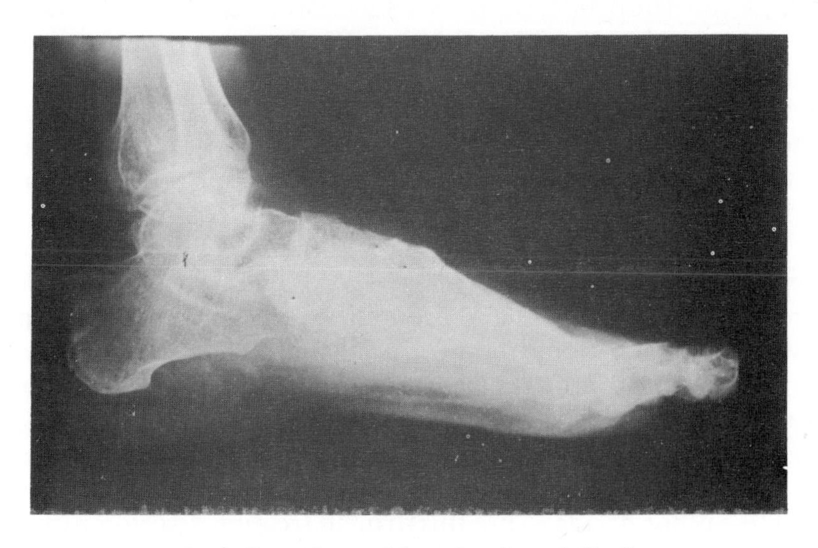

Fig. 6. X-ray picture of the patient shown in Fig. 5.

the drug was administered by mouth. These side effects were observed 1–2 weeks after infusion (mean 10 days), and disappeared in 2–3 weeks. All of the cases with signs of renal impairment had traces of albumin, hyaline casts and biliary pigments in the urine. In half of the cases showing the abnormalities slight elevations of blood uric acid and creatinine were observed, as were granular casts and erythrocytes in the urine. In a few cases a high blood urea was noted in addition to the abnormalities mentioned.

4. Treatment with trimethoprim plus sulphamethoxazole

In 1966 studies were begun in which trimethoprim was used on laboratory animals and in the clinic. The results showed synergistic clinical effectiveness when this agent was associated with the sulpha drugs. That is to say, there was reciprocal potentiation of the trimethoprim and to the different sulphonamides used, particularly sulphamethoxazole. The trimethroprim, or 2,4 diamino-5-(3,4,5 trimethoxybenzyl pyrimidine), is a bacteriostatic substance with antifolic action. The sulphamethoxazole corresponds to 5-methyl-3-sulphanilamidoisoxazole. The association of these two drugs (Bactrim) interrupts two successive stages in a metabolic process that is essential to the life of susceptible organisms. The mixture of the drugs inhibits both the enzyme responsible for synthesizing dihydrofolic acid (by virtue of the sulphamethoxazole) and that responsible for the reduction of this substance to tetrahydrofolic acid (by virtue of the trimethoprim). Both inhibitions block the formation of DNA and RNA in the microorganisms, thereby not only making it impossible for them to reproduce but also causing their death. These successive actions are strongly synergistic, and their effectiveness increases 10 times over that which would be produced by inhibiting each enzyme separately. Bactrim's action covers a broad range of Gram-positive and Gram-negative bacteria and actinomycetes, including the taxa *Salmonella, Neisseria, Escherichia coli, Proteus, Nocardia brasiliensis, N. asteroides, N. caviae, Actinomadura pelletieri* and *A. madurae.*

The author decide to use this drug in treating mycetoma caused by actinomycetes. The preliminary results of the therapeutic trial

(González-Ochoa & Tamayo, 1969), were very encouraging. In 14 cases of actinomycotic mycetoma caused by *N. brasiliensis* (Table 2) clinical cures were effected after between 2 and 7 months of treatment, in six patients (43%) and marked improvement was achieved in the remaining.

Table 2
Results of treatment of actinomycotic mycetoma, due to N. brasiliensis, *with Bactrim for 2–7 months*

Number of cases	Clinical cure	Marked improvement	Failure
14	6 cases (43%)	8 cases (57%)	0 cases (0%)

Continuing the observations on the treatment for *N. brasiliensis* mycetoma, González-Ochoa (1970), on the occasion of the International Symposium on Mycoses organized by the Pan American Health Organization in Washington, D.C., presented the results of further studies where the number of patients had been increased from 14 to 20 and the duration of treatment to 16 months. The results obtained were very similar to those achieved initially (Table 3) since for the 14 cases covered by the initial study, plus the 6 cases treated subsequently, a clinical cure was obtained in 50% of the patients, marked improvement in 45%, and in one case it was necessary to suspend treatment due to the occurrence of leucopenia in a patient who had been taking the drug for 2 months.

Table 3
Results of treatment of actinomycotic mycetoma, due to N. brasiliensis, *with Bactrim for 2–16 months*

Number if cases	Clinical cure	Marked improvement	Failure
20	10 cases (50%)	9 cases (45%)	1 case (5%)

In subsequent studies the number of patients and the duration of the treatment were increased, and larger doses of the drug were used in cases where a relapse occurred after a clinical cure had been obtained. *In vitro* studies on the susceptibility of various strains of *Nocardia* to Bactrim were also carried out.

(a) In vitro *studies*

The nocardiae studied included four strains of *N. asteroides*, five of *N. brasiliensis* and four of *N. caviae*.

In the method employed dilutions of the drug at concentrations of 1,2,4 and 6 μg/ml were prepared. Sabouraud's agar with 2% w/v glucose was used as the medium. The latter was contained in tubes, and after it had melted the inoculum and drug were added. The inoculum used consisted of two drops of a suspension of triturated actinomycetes equivalent to number 1 capacity on McFarland's nephelometer. After mixing the inoculum and the drug with the medium, the tubes were tilted until the medium solidified, and then incubated at 28°C for 6 days.

The results showed that the *N. asteroides* strains were the most susceptible (Table 4) since inhibition was observed for three of the four strains at a concentration of 4 μg/ml, while the fourth strain showed only slight growth at this concentration.

Table 4
Activity of trimethoprim plus sulphamethoxazole (*Bactrim*) *against* N. asteroides in vitro

Strains	μg/ml 1	2	4	6	Controls
4408	+	±	−	−	+++
4414	++	±	−	−	+++
4415	+	±	±	−	+++
4440	+	±	−	−	+++

+ = growth.
− = inhibition.

N. brasiliensis was in second place with regards to its sensitivity as there was total inhibition of all five strains at a concentration of 6 μg/ml (Table 5).

Table 5
Activity of trimethoprim plus sulphamethoxazole (Bactrim) against N. brasiliensis in vitro

Strains	μg/ml 1	2	4	6	Controls
4220	+++	++	–	–	+++
4220 a	+++	+++	++	–	+++
4220 b	+++	+++	+	–	+++
4243	+++	++	–	–	+++
4244	–	–	–	–	+++

+ = growth.
– = inhibition.

Finally, of the four strains of *N. caviae* which were studied (Table 6) three were partially inhibited at 6 μg/ml, and one was resistant at that concentration.

Table 6
Activity of trimethoprim plus sulphamethoxazole (Bactrim) against N. caviae in vitro

Strains	μg/ml 1	2	4	6	Controls
518	+++	+	±	±	+++
4403	+++	+++	++	+	+++
4404	+++	+++	+	±	+++
4420	+++	+++	+++	+++	+++

+ = growth.
– = inhibition.

(b) *Dosage*

The drug, which is available in the form of tablets containing 80 mg of trimethoprim and 400 mg of sulphamethoxazole, was administered, except in the case of three patients, in accordance with the following two treatment schedules. Schedule I, called for one tablet every 12 h from the start of the treatment until clinical cure was achieved; schedule II, provided for one tablet every 24 h beginning at the time clinical cure was effected in order to prevent a relapse. In general, the second schedule was continued for 1–4 years depending on the bone involvement.

The three exceptions were a 24-year-old adult, a 10-year-old-boy and a 21-year-old adult, for whom as will be seen later, the dosage had to be increased.

(c) *Clinical results*

As the duration of treatment was extended the number of clinical cures increased at the expense of cases showing marked improvement (Table 7). Of the 39 patients treated for periods of 2 months to 4 years, 67% achieved a clinical cure and 5% showed a marked recovery.

Table 7

Results of treatment of actinomycotic mycetoma, due to N. brasiliensis, *with Bactrim for 2 months to 4 years*

Number of cases	Clinical cure	Marked improvement	Failure	Relapsed after clinical cure	Deserted
39	26 cases (67%)	2 cases (5%)	2 cases (5%)	2 cases (5%)	7 cases (18%)

In two of the patients treated with higher doses something most unusual happened, namely that after showing obvious improvement they suffered an intense reactivation. The 10-year-old boy with mycetoma of the back who was successfully responding to a dose of one tablet a day on entering puberty suffered a sudden relapse, with the appearance of numerous lesions. The dose of Bactrim was therefore increased to four tablets a day; this patient weighed 36 kg. The second case was that of a 24-year-old adult with mycetoma on the left arm, for whom a clinical cure was achieved with a dose of two tablets a day over a 10-month period. The patient was released, but in spite of the recommendation that he continue to take one tablet a day, he suspended treatment and suffered a very severe relapse which could not be controlled with a dose of four tablets a day. As the patient was getting markedly worse, the arm was amputated but the mycetoma then recurred in the shoulder and chest, and penetrated into the lung, apparently through the blood stream, before the lesions in the shoulder and chest wall were noticed. The dose of Bactrim was therefore increased to six tablets a day; this patient weighed 67 kg. In the case of mycetoma of the thigh, which was amputated due to resistance to DDS, the mycetoma recurred in the pelvis. This patient was cured after taking four tablets of Bactrim a day for 3 years. The first two cases have been under treatment for a year, without a marked improvement.

The remaining patients, with actinomycotic mycetoma, who achieved a clinical cure with two tablets a day continued to take one tablet daily for 1 or 2 years depending on whether there was any lesion of the bone or not. They have remained cured, and in some cases almost 5 years have passed since the drug was suspended.

The number of patients with nocardiosis caused by *N. asteroides* who have been treated with Bactrim is much smaller. Three of the four cases studied were located in the lungs of old people, and had been confused with tuberculosis. The final case was that of a young man with a subcutaneous abscess of the scalp. The doses administered were the same as those used in the schedules adopted for the treatment of actinomycotic mycetoma casued by *N. brasiliensis*. The clinical and radiological results were excellent both in the case of clearing the pulmonary lesions in the three adults and healing the abscess of the scalp. However, in all four cases the primary ailment, leukemia and Hodgkin's disease in the adults and diabetes mellitus in the young man, which was responsible for the patients being susceptible to *Nocardia*, caused the death of the patients.

It has been pointed out that erythromycin and ampicillin show some action *vis-a-vis N. asteroides* but not *vis-a-vis N. brasiliensis* and *N. caviae*. However, in infections caused by *N. asteroides*, most authors prefer sulphonamides, while we prefer Bactrim.

As nocardiosis is, in the great majority of cases, an opportunistic mycosis, it is fundamental to correct the pathological or iatrogenic situation determining its appearance and, in a complementary way, to treat the infection caused by *N. asteroides*.

(d) *Side effects*

The drug is well tolerated at the doses used in the two treatment schedules, and in spite of the prolonged periods of administration, only one case of leucopenia was observed after 2 months of taking one tablet of Bactrim every 12 h. In this case, when the drug was renewed 2 months later after white cells levels had returned to normal, the drug was tolerated perfectly.

In a case of mycetoma, in which a thigh was amputated due to resistance to treatment with DDS, the mycetoma recurred in the pelvis, but four tablets a day of Bactrim administered for 3 years resulted in a clinical cure. In this case the drug was suspended when toxic hepatitis occurred, but the clinical cure has been maintained for the last 2 years and the hepatic symptomatology disappeared 3 months after suspending the treatment.

5. Conclusions

Patients with *Nocardia brasiliensis, N. caviae* and *N. asteroides* mycetoma of soft tissue or with slight invasion of the bone, systematically achieve a clinical cure with a short course of treatment lasting from 6 to 8 months, at a dose of two tablets of Bactrim per day. Treatment for a further 2 years at a dose of one tablet a day practically eliminates relapses. In patients with extensive mycetoma and moderate invasion of the bone, a clinical cure is obtained in the majority of cases after 2 to 4 years treatment following the same schedule. Cases with extensive, deep invasion of the bone show marked improvement but do not achieve a clinical cure at least within the 5 years period. In three extremely serious cases which could not be treated surgically, the dose was increased to four and six tablets a day, with good results, and the drug was well tolerated.

When the duration of treatment is increased, there is also an increase in the number of cures. Thus, when the drug was administered for 7 months, clinical cures were obtained in 43% of the cases. When the treatment period was increased to 16 months, the percentage of clinical cures rose to 50%; and with a further increase to 4 years clinical cures rose to 67%.

Our experience with nocardiosis is scanty but Bactrim is the preferred treatment, following the schedules, proposed for the mycetoma caused by *N. brasiliensis,* with which the control of the infection by *N. asteroides* is achieved. However, the pathological cause determining the development of nocardiosis may result in the death of the patient.

The drug is well tolerated at the doses mentioned, in spite of prolonged administration, since only one case of leucopenia was observed after 2 months of taking one tablet of Bactrim every 12 h.

even in this case when the drug was renewed 2 months later, after the white cells levels had returned to normal, it was tolerated perfectly. Toxic hepatitis occurred in a case after 3 years treatment with four tablets a day, but disappeared 3 months after the drug was suspended.

6. Acknowledgements

This work was supported, in part, by a grant from F. Hoffman-La Roche & Co. Ltd. Basel, Switzerland. The 4-sulfanilamido-5,6-dimethoxy pyrimidine, and the trimethoprim plus sulphamethoxazole, used in this research were supplied by the same Institution under the trade names of "Fanasil" and "Bactrim", respectively.

7. References

Beirana, L. & Hernández, J. D. (1963). Ensayo Clínico del Ro-4-4393 (Sulfa para 8 días). *Mem. 11 Congr. Mex. Dermatatol., Guadalajara, Jal. Mexico.* p. 558.

Blanchard, R. (1896). Parasites végétaux a l'exclusion des bactéries. In *Traité de Pathologie Genérale,* vol. 2, Ed. Bouchard. Paris: G. Masson.

Buttle, G. A. H., Stephenson, D., Smith, S., Dewing, T. & Foster, G. E. (1937). Treatment of streptococcal infections in mice with 4 : 4′ diaminodiphenylsulphone. *Lancet* **1,** 1331.

Eppinger, H. (1891). Über eine neue pathogene *Cladothrix* und eine durch sie nervorgerufene Pseudotuberculosis (Cladothrichia). *Beitr. path. Anat.* **9,** 287.

Erikson, D. (1935). The pathogenic aerobic organisms of the actinomyces group. *Med. Res. Coun. (Gt. Brit.) Spec. Rep. Ser.* **203,** 5.

Faget, G. H., Pogge, R. C., Johansen, F. A., Dinan, J. F., Prejean, B. M. & Eccles, C. G. (1943). The promin treatment of leprosy. A progress report. *Publ. Hlth Rep., Wash.* **58,** 1729.

Ferreira-López, C. (1963). Tratamento da Blastomicose Sul-Americana e de Nocardiose com a Sulfamida Ro 4-4393. *Actas Finales V. Congr. Iber. Lat.-Am. Dermatol., Buenos Aires, Argentina.* p. 211.

Fourneau, E., Tréfouël, J., Nitti, F., Bovet, D. & Tréfouël, J. (1937). Action antistreptococcique des dérivés sulfurés organiques. *C.r. hebd. Séanc. Acad. Sci., Paris* **204,** 1763.

Fromm & Whittman. (1908). Quoted in Encyclopedie medico-chirugicale. Dermatologie, cahier 24e, 2040 A^{50}, 1953.

González-Ochoa, A. (1953). El micetoma tóracopulmonar por *Actinomyces bovis,* y *Nocardia brasiliensis. Gac med. Mexico* **83,** 109.

González-Onchoa, A. (1970). The prevention and treatment of subcutaneous mycoses. *Proc. Intern. Symp. Mycoses, PAHO, Washington, D.C.*

González-Ochoa, A., Martínez-Baéz, M., Lavalle, P. & Castro-Estrada, S. (1952). Diamino-difenil-sulfona, dihidroestreptomicina y ácido para-amino salicílico, en el fratamiento de la lepra lepromatosa. *Boln Of. sanit. pan-am.* **33,** 306.

González-Ochoa, A., Shiels, J. & Vázquez, P. (1952). Acción de la 4,4'-diamino-difenil-sulfona frente a *Nocardia brasiliensis. Gac. méd. Mexico* **83,** 345.

González-Ochoa, A., Stark, B. & Vázquez-Ibarra, R. (1967). Fanasil in the actinomycote mycetoma caused by *Nocardia brasiliensis.* Oral and intra-arterial administration. *Proc. V. Intern. Contr. Chemotherapy, Vienna.* p. 407.

González-Ochoa, A. & Tamayo, L. (1969). Tratamiento del micetoma actinomicético por *N. brasiliensis* con Ro 6-2580/11. Medicina, Mex. **49,** 473.

Gordon, R. E. & Mihm, J. M. (1962). Identification of *Nocardia caviae* (Erikson) nov. comb. *Ann. N.Y. Acad. Sci.* **98,** 628.

Guzmán-Toledano, R. (1962). Micetomas, su Tratamiento Moderno con Sulfadrogas de Elininacion Lenta. *Tesis Rec. Univ. Nal. Aut. Mexico.*

Halde, C. & Newstrand, D. (1955). The sensitivity of pathogenic actinomycetes to various sulfonamide and sulfone compounds. In *Therapy of Fungal Diseases.* Eds T. H. Stenberg & V. D. Newcomer. Boston: Little, Brown & Co.

Lindenberg, A. (1909). Un nouveau mycetome. *Arch. Parasitol.* **13,** 265.

Pinoy, E. (1913). Actinomycoses et mycetomes. *Annls Inst. Pasteur, Paris* **11,** 929.

Rist, N. (1939). Action du p-aminodiphénylsulfamide et de la p-diaminodiphenylsulfone sur la culture des bacilles tuberculeux des mammiferes et des oiseaus. *C.r. Séanc. Soc. Biol.* **130,** 972.

Strauss, R. E., Kligman, A. M. & Pillsbury, D. M. (1951). The chemotherapy of actinomycosis and nocardiosis. *Am. Rev. Tuberc. pulm. Dis.* **63,** 441.

18. Metabolism in Nocardiae and Related Bacteria

I. TÁRNOK

Institut für Experimentelle Biologie und Medizin, Forschungin-stitut Borstel, 2061 Borstel, Federal Republic of Germany

Contents

1. Introduction

The author dealing with the metabolism of nocardioform bacteria has great difficulties due to the poor taxonomy of these microorganisms. several strains labelled "*Nocardia*" are probably mis-identified and belong to other genera. Since taxonomists have different opinions about the composition of the genus *Nocardia*, this paper deals not only with *Nocardia sensu stricto* but also with strains belonging to the related genera *Actinomadura, Oerskovia* and to the *rhodochrous* complex. The review, therefore, will undoubtably contain data about *Nocardia sensu lato.*

A review on metabolism cannot be limited to the specification of enzyme-substrate interactions for it is also necessary to mention analytical data about lipids, deoxyribonucleic acids (DNA) and other cell components.

Some readers may be disappointed at the fragmentary data on topics such as vitamin and protein metabolism but in such areas little information is available. The author, however, hopes that the incomplete nature of the review will stimulate interest in the neglected areas of nocardial metabolism.

2. Nutrition and Growth

(a) *General aspects of the nutrition of nocardioform organisms*

Members of the genus *Nocardia* and related genera (*Oerskovia Actinomadura* and rhodochrous-like bacteria) are heterotrophic aerobic organisms (Bergey, 1974). The majority can grow on media supplemented with organic substrates such as peptone, amino acids, egg white, egg yolk and milk. Synthetic media are used for special purposes, such as in the search for growth factors, isolation of auxotrophic mutants, determination of oligocarbophilic properties, and in fermentation studies.

(i) *Autotrophy in* Nocardia

Nocardia autotrophica (Takamiya & Tubaki, 1956) and *Nocardia opaca* strain 1b (Aggag & Schlegel, 1973) are markedly autotrophic though the latter can also grow heterotrophically. Both organisms belong to the rare Gram-positive hydrogen bacteria, which can grow in an atmosphere consisting of hydrogen, carbon dioxide and oxygen, and in media which contain a nitrogen source and the necessary salts. Heterotrophically grown *N. opaca* does not contain hydrogenase (hydrogen-(NAD^+)*-oxidoreductase) but when transferred to autotrophic conditions, a nearly linear increase in hydrogenase activity occurs (Probst & Schlegel, 1973).

Carbon dioxide can also be readily utilized by *Nocardia petroleophila* (Hirsch, 1958); the energy necessary for its assimilation coming from the oxidation of alkanes. Finally CO can serve as a carbon source for some nocardiae (Hirsch, 1965; Jochens, 1974); several strains can utilize 80% CO presented to them in the vapour phase. The observed gas uptake suggests the oxidation of CO to CO_2:

$$2\,CO + O_2 \rightarrow 2\,CO_2$$

* Nicotinamide—adenine dinucleotide.

Oligocarbophily, defined as the ability to adsorb and incorporate organic substances from the atmosphere (Beijerinck & Van Delden, 1903), has been described for *Nocardia corallina* and *N. hydrocarbonoxydans* (Hirsch, 1964; Nolof & Hirsch, 1962). These strains are able to utilize traces of pyridine and other organic compounds.

(b) *Carbon and nitrogen sources*

A great number of organic substances can be used as carbon and carbon+nitrogen sources. Detailed studies were carried out by Watanabe, Kumagaya & Murooka (1963), Dasani & De Sa (1965), Goodfellow (1971) and others to investigate the availability of different compounds for growth.

(i) *Alkanes*

Alkanes are good carbon sources. It seems to be an advantage for nocardiae to use alkanes with relatively longer chains for they can be better metabolized than those with shorter chains. Paraffin, a fraction of alkanes, was used as sole carbon source in earlier experiments but was substituted later by well defined hydrocarbons (Erikson, 1949; Schneidau & Schaffer, 1957; Nolof, 1962; Watanabe, Kumagaya & Murooka, 1963).

(ii) *Carbohydrates and related compounds*

Extensive studies have been carried out by Goodfellow (1971) concerning the utilization of carbohydrates and related derivatives. *Nocardia asteroides, N. caviae, N. brasiliensis, N. pelletieri, Oerskovia tubata* and *Mycobacterium rhodochrous* strains utilize glucose as a sole carbon source; fructose, however, seems to be less effective in enhancing growth. L-arabinose, D-xylose, L-rhamnose, D-galactose, L-mannose, cellobiose, lactose, sucrose, maltose, trehalose, melezitose, raffinose, glycogen and some polyols were differently suitable as carbon sources.

(iii) *Organic acids*

Sodium acetate, propionate and butyrate are good carbon sources for strains of the taxa mentioned above. Some dicarboxylic acids, citrate, benzoate and hippurate can be used with variable success for growth (Goodfellow, 1971).

(iv) *Nitrogenous derivatives and amino acids*

Inorganic compounds such as NH_4^+, NO_2^- and NO_3^- can be utilized as sole nitrogen source e.g. for *Nocardia asteroides* and *N. brasiliensis* (Hotchkiss & Simons, 1959). Acetamide can serve as sole carbon and nitrogen source for *N. asteroides, N. brasiliensis, Actinomadura madurae* and *Mycobacterium rhodochrous* (Goodfellow, 1971). Bönicke & Juhász (1965) found that *m*-nitrophenol can be metabolized by *Nocardia pellegrino, N. rubra, N. corallina* and *M. rhodochrous* and by some *N. asteroides* strains.

L-glutamic acid, L-aspartic acid, L-lysine, L-leucine, L-isoleucine and L-valine support good growth in many *Nocardia* species (Uesaka & McClung, 1961).

(c) *Vitamin and growth factor requirements*

Nocardia corallina and *N. rubra* require thiamine for growth (Martin & Batt, 1957*a,b*; Adams, 1970). Indeed 1 μg thiamine/ml can markedly enhance the growth of *Nocardia asteroides, N. caviae, N. alba, N. coeliaca, N. salmonicolor* and *N. lutea* strains (Tárnok & Wenzel, unpublished data, Table 1). To date nocardiae have not been found to need specific growth factors.

(d) *Generation time*

Although the generation times for nocardiae are not generally known, some data are available. Aggag & Schlegel (1973) estimated the generation time of the hydrogen bacterium *Nocardia opaca* to be about 7·5 h. It can be assumed that under optimal growth conditions, the generation time for the majority of the *Nocardia* could be 3–7 h; under unfavorable culture conditions, however, it can be prolonged to one week or longer. Hirsch (1964) measured the increase of the amount of organic materials in *Nocardia petroleophila* and *N. paraffinae* cultures under oligocarbophilic condition and also without traces of petroleum in the vapor phase; a 2–6 fold increase of the organic materials has been observed after 42–437 days.

(e) *Temperature and pH ranges for optimal growth*

Some *Nocardia* are able to grow at rather extreme temperatures such as 10° or 50°C but the best temperature range seems to be

Table 1
Influence of thiamine on the growth of some nocardioform bacteria

Strain		Medium	
		Minimal medium	Minimal medium +1 μg/ml thiamine
N. polychromogenes	SN5001	++	++
N. pellegrino	SN5101	+	++
N. rubra	SN5201	+	++
N. corallina	SN5303	++	++
N. asteroides	SN5401	+	++
N. brasiliensis	SN5501	ϕ	ϕ
N. caviae	SN5601	ϕ	+
N. farcinica	SN5701	+	+
N. rubropertincta	SN5800	++	++
N. opaca	SN5900	++	++
N. minima	SN6000	++	++
N. alba	SN6100	+	+
N. paraffinae	SN6200	ϕ	ϕ
N. blackwellii	SN6300	++	++
N. globerula	SN6400	ϕ	ϕ
N. caprae	SN6500	+	++
N. coeliaca	SN6600	++	++
N. salmonicolor	SN6700	ϕ	++
N. lutea	SN6800	ϕ	++

Interpretation of the results:
++ Confluent growth.
+ Weak growth, only single colonies.
ϕ No growth.

between 27°–37°C. Similarly, growth has been obtained in media with relatively low or high pH values (pH 5 or 10) but the best results have been observed on media with pH 6–9 (Goodfellow, 1971; Schneidau & Shaffer, 1957).

3. Lipid Composition and Metabolism

(a) *Lipid composition*

(i) *Saturated, straight-chain fatty acids*

These compounds are common in nocardiae and in *rhodochrous* strains; their relative amounts, however, can vary between broad limits depending upon the strain and cultural conditions. *Nocardia asteroides, Actinomadura pelletieri, N. brasiliensis, N. pellegrino, N. rugosa, N. rubropertincta* and *N. lurida* contain relatively large amounts of palmitic acid and, with the exception of *N. rugosa,* stearic acid. *Nocardia asteroides* and *N. rugosa* contain low amounts of margaric acid (Bordet & Michel, 1963; Ballio & Barcellona, 1968; Farshtchi & McClung, 1970). Palmitic and oleic acids can be isolated from the free lipid part of *Nocardia kirovani* (Vacheron *et al.*, 1972).

(ii) *Branched-chain fatty acids*

The amounts of the branched-chain fatty acids are low in nocardiae e.g. in *Nocardia asteroides, N. brasiliensis, N. rugosa* and *N. lurida. Nocardia asteroides* contains C_{14-15} acids (Farshtchi & McClung, 1970). 10-methylstearic acid (tuberculostearic acid) can be found in *N. asteroides* and *N. brasiliensis* (10–30% of the whole fatty acids; Bordet & Michel, 1963).

(iii) *Unsaturated fatty acids*

Nocardia asteroides and *N. brasiliensis* are rich in unsaturated fatty acids; the main component is oleic acid (up to 30% of the whole fatty acids; Bordet & Michel, 1963).

(iv) *Poly-unsaturated fatty acids*

These compounds have been detected as glycerol esters in *Nocardia kirovani* (Vacheron *et al.*, 1972).

(v) *Variability in the fatty acid composition*

The fatty acid composition can be influenced qualitatively and quantitatively depending upon the cultural conditions (Farshtchi & McClung, 1970). If C_{10-20} *n*-alkanes are used as carbon source, the chain-length of the major bacterial fatty acids reflects the length of that of the starting material (Davis, 1964).

(vi) *α-Branched, β-hydroxy, unsaturated, high-molecular weight fatty acids (mycolic acids)*

This type of fatty acids is characteristic for mycobacteria, nocardiae, *rhodochrous* strains and some corynebacteria (Goodfellow *et al.*, 1973) but is absent in *Oerskovia* and *Actinomadura* strains. The C_{58-66} acids are common in true nocardiae, and are generally termed nocardomycolic acids. Nocardomycolic acids have been found in *Nocardia asteroides* (Bordet *et al.*, 1965; Etémadi & Lederer, 1965), *N. corallina* (Batt, Hodges & Robertson, 1971), *N. brasiliensis*, *N.* (*Mycobacterium*) *pellegrino* and *N. asteroides* (Lanéelle, Asselineau & Castelnuovo, 1965), *N. kirovani* (Vacheron *et al.*, 1972), *N. rhodochrous*, *N. caviae*, *N. calcarea* (Bordet *et al.*, 1965; Goodfellow *et al.*, 1973; Minnikin, Patel & Goodfellow, 1974) *N. opaca* (Etémadi, Markovits & Pinte, 1966) and other species. The general structure of nocardomycolic acids is shown in Fig. 1.

A nocardomycolic acid-D-arabinose monoester has been found in the bound lipid fraction of *Nocardia brasiliensis*; the free lipids of this strain contain a glycolipid fraction consisting of nocardomycolic acids, arabinose, galactose and amino acids (Lanéelle & Asselineau, 1970).

$$CH_3-(CH_2)_7-CH=CH-(C_{24}H_{45}\pm 3CH_2)-\underset{\substack{|\\OH}}{CH}-\underset{\substack{|\\(CH_2)_n\\|\\CH_3}}{CH}-COOH$$

$n = 11, 13$

$$CH_3-(CH_2)_7-CH=CH-(CH_2)_x-CH=CH-(CH_2)_y-CH=CH-(CH_2)_z-\overset{\substack{OH\\|}}{CH}-\underset{\substack{|\\(CH_2)_n\\|\\CH_3}}{CH}-COOH$$

$x+y+z = 15, 17, 19, 21$

$n = 11, 13$

$C_{46}H_{86}O_3$
$C_{48}H_{90}O_3$
$C_{50}H_{94}O_3$
$C_{52}H_{98}O_3$

$$CH_3-(CH_2)_7-CH=CH-(CH_2)_x-CH=CH-(CH_2)_y-\overset{\substack{OH\\|}}{CH}-\underset{\substack{|\\(CH_2)_n\\|\\CH_3}}{CH}-COOH$$

$x+y = 25, 27, 29$

$n = 11, 13$

$C_{54}H_{104}O_3$
$C_{56}H_{108}O_3$
$C_{58}H_{112}O_3$

Fig. 1. Structure of nocardomycolic acids.

"Cord factors" (characteristic for mycobacteria) (Bloch, 1950) have been isolated from strains of *Nocardia rhodochrous* and *N. asteroides* and have been identified as diesters of trehalose with nocardomycolic acids (Ioneda, Lederer & Rozanis, 1970).

(vii) *Stability of the nocardomycolic acid composition*

In contrast to the low-molecular weight fatty acid composition, the nocardomycolic acid composition of nocardiae seems to be constant and characteristic for the strains (Goodfellow *et al.*, 1973). Thus, the determination of the nocardomycolic acids is decisive in the taxonomy of nocardiae.

The analysis of the nocardomycolic acids became easier after the observation that pyrolysis of nocardomycolic acids yields a mixture composed of predominantly palmitic and stearic acids (Lanéelle, Asselineau & Castelnuovo, 1965). Lechevalier, Horan & Lechevalier (1971) carried out the pyrolysis of the nocardomycolic acids and their esters in the gas chromatograph and obtained C_{14-18} fatty acid esters whereas mycobacterial mycolates yielded C_{22-26} esters.

(viii) *High-molecular weight alcohols and ketones in* Nocardia

These compounds, called nocardols (Michel & Lederer, 1962; Bordet & Michel, 1966*a*) and nocardones (Bordet & Michel, 1966*a*; Lanéelle, 1966) have been isolated from nocardiae. Nocardols can be found in the cells as mixtures of mono-, di- and tri-unsaturated monoalcohols. The major compounds in *Nocardia asteroides* are nocardenol, nocardodienol and nocardotrienol (Bordet & Michel, 1966*b*). A new alcohol, termed 16-hentriacontanol was found in *Nocardia asteroides* (Bordet & Michel, 1964). Nocardones occur as mixtures of mono- and di-unsaturated ketones in *Nocardia brasiliensis* (Lanéelle, 1966).

(ix) *Polar lipids in nocardiae*

Different types of polar lipids have been found in *Nocardia coeliaca* and *N. polychromogenes* (Khuller & Brennan, 1972). Mannophosphoinositides from the dimannosyl type are among the most obvious phospholipids in *N. coeliaca.* Glucose containing phospholipids and glycolipids are also encountered (Khuller & Brennan, 1972). A peptidolipid termed peptidolipin NA, isolated from *Nocardia asteroides* consists of seven different amino acids and 3-hydroxyeicosanic acid which has never been found previously in natural products (Guinand & Michel, 1966). The polar lipids of nocardiae were also studied by Pommier & Michel (1973).

(b) *Metabolism of fatty acids*

(i) *Low-molecular weight fatty acids*

Little is known about the details of the biosynthesis of the low-molecular weight fatty acids in nocardioform bacteria. It seems certain that the methyl group of the tuberculostearic acid in *Nocardia asteroides* is derived from methionine (Farshtchi & McClung, 1970) in keeping with the previous work of Lennarz, Scheuerbrandt & Bloch (1962) on the biosynthesis of this compound in *Mycobacterium phlei.* The assumption that fatty acids, and also alcohols, can be derived by enzymatical oxidation of the terminal C-atom is supported by the observation that cetyl palmitate was isolated when an unidentified *Nocardia* strain was grown on *n*-hexadecane (Raymond & Davis, 1960). Monocarboxylic acids such as propionate are oxidized in *Nocardia corallina* cultures; for the propionate breakdown, the pathway

$$\text{propionate} \rightarrow \text{succinate} \rightarrow \text{pyruvate}$$

has been proposed (Martin & Batt, 1957*b*).

(ii) *Metabolism of the nocardomycolic acids*

Etémadi & Lederer (1965) and Bordet & Michel (1969) postulated a pathway for nocardomycolic acid, nocardol and nocardone biosynthesis (Fig. 2). One molecule of β-hydroxy, α-branched C_{32} acid is formed by a condensation reaction between two molecules of palmitic acid and by hydrogen addition. After an ω-oxidation step and a head-tail condensation with a fatty acid molecule, the elongation of the chain and the formation of unsaturated linkages is obtained. Part of this mechanism was proposed by Kanemasa & Goldman (1965) for mycobacteria. Finally, decarboxylation, reduction and dehydrogenation steps lead to an unsaturated acid. On a similar pathway, high-molecular weight alcohols and ketones could also be synthesized.

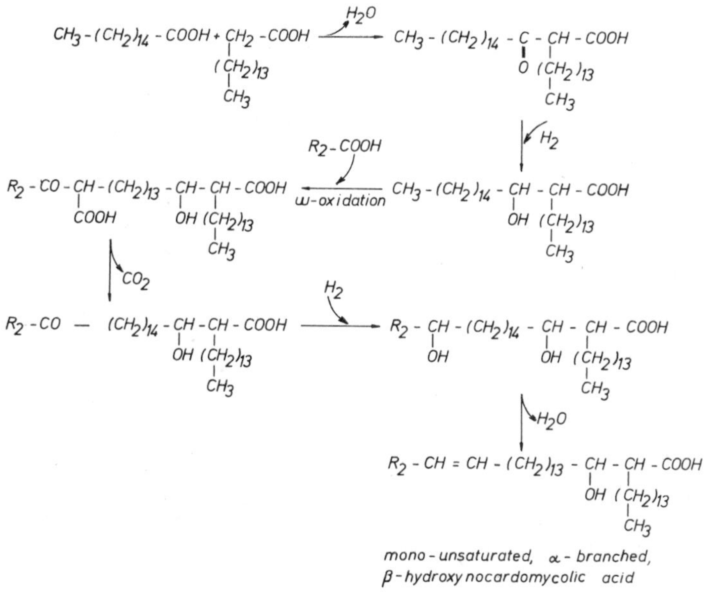

Fig. 2. Biosynthesis of nocardomycolic acids.

4. Metabolism of Steroids and Sterols

So far steroids and sterols have not been found in nocardiae: however, the capability of these organisms to transform substrates with a steroid skeleton is unusually high. It is conceivable that the conversions are carried out by unspecific enzymes. Regardless of this possibility it is necessary to give a short summary on the steroid transformations since they also belong to the metabolic capacities of the nocardiae. The data are numerous and it is impossible to give a complete review here; readers are referred to the excellent paper of Raymond & Jamison (1971).

The first papers about steroid metabolism were published by Turfitt (1948) and Krámli & Horváth (1948, 1949) who reported the conversion of cholesterol to 4-dehydroetiocholanic acid by a *Nocardia* sp. and to 7-hydroxycholesterol by *Proactinomyces roseus*, respectively.

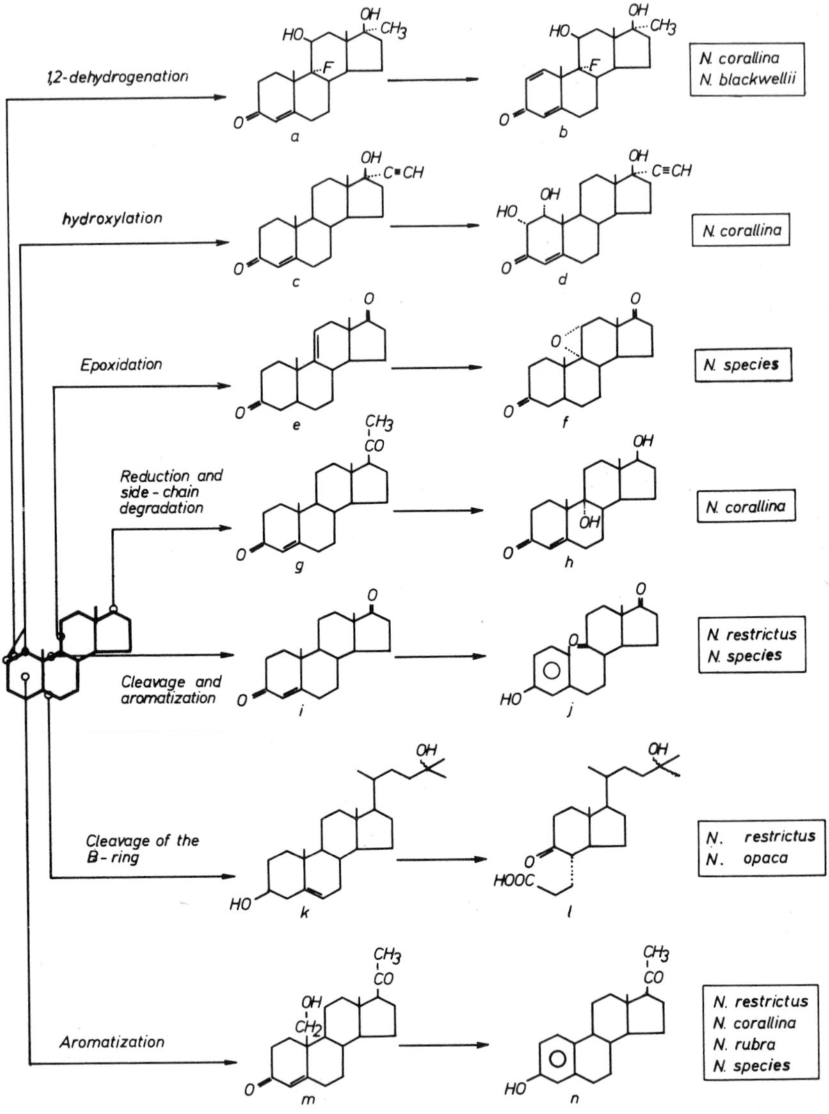

Fig. 3. Transformation of steroids by *Nocardia*. Designation of the steroids: (a) 9α-fluoro-11β,17β-dihydroxy-17α-methylandrost-4-en-3-one, (b) 9α-fluoro-11β,17β-dihydroxy-17α-methylandrost-1,4-diene-3-one; (c) 17α-ethynyl-17β-hydroxy-androst-4-en-3-one; (d) 17α-ethynyl-1α,2α,17β-trihydroxyandrost-4-en-3-one; (e) androst-9(11)-ene-3,17-dione; (f) androst-9α,11α-oxido-3,17-dione; (g) progesterone; (h) 9α-hydroxytestosterone; (i) androst-4-en-3,17-dione; (j) 9,10-seco-3-hydroxyandrosta-1,3,5(10)-triene-9,17-dione; (k) 25ε-hydroxycholesterol; (l) 7aβ-methyl-1β[1'-methyl-5'-hydroxyisohexyl]-5-oxo-3aα-hexahydro-4-indanpropionic acid; (m) 19-hydroxyprogesterone; (n) 3-hydroxy-19-norpregna-1,3,5(10)-trien-20-one.

(a) *Transformation of steroids by nocardiae* (*Fig. 3*)

(i) *Dehydrogenation*

The steroid nucleus can be dehydrogenated in the 1,2-position by *Nocardia corallina* and *N. blackwellii* (Sax *et al.*, 1965; Stoudt *et al.*, 1958).

(ii) *Hydroxylation*

Hydroxylation has been obtained in the 6α-position by *Nocardia restrictus* (Lee & Sih, 1964), in the 9α-position by cell-free extracts of this organism (Chang & Sih, 1964), in the 1α, and 2α-position by *N. corallina* (Sax *et al.*, 1965), in the 9α-position by a *Nocardia* sp. (Dodson & Muir, 1958*a*).

(iii) *Epoxydation*

Introduction of an epoxy group into the 9α, 11α-position can be carried out by a non-identified *Nocardia* sp. (Sih, 1962).

(iv) *Reduction, side-chain degradation and hydroxylation*

Nocardia corallina can reduce the C_{20} keto group, introduce an OH-group into the 9α-position and simultaneously, shorten the side-chain by splitting off the C_{21} methyl group (British Patent, 1959; U.S. Patent, 1963). Germain *et al.* (1972) described a soluble reductase in *Nocardia corallina* extracts.

(v) *Ring cleavage and aromatization*

The B-ring of the steroid nucleus can be split off and the A-ring aromatized in *Nocardia restrictus* and a *Nocardia* sp. (Sih & Wang, 1963; Wang & Sih, 1963; Gibson *et al.*, 1966; Coombe *et al.*, 1966; Tjhing-Lok *et al.*, 1972).

(vi) *The cleavage of the B-ring and the elimination of the A-ring*

An indanpropionic acid derivative was obtained from hydroxycholesterol after several enzymatical steps if *Nocardia restrictus* and *Nocardia opaca* were used as test organisms (Sih *et al.*, 1968). During the degradation of cholesterol, C_{22} acids occur (Sih, Wang & Tai, 1968).

(vii) *Aromatization of the A-ring*

The A-ring can be aromatized by some *Nocardia* e.g. *N. restrictus* (Brodie, Possanza, & Townsley, 1968). *Nocardia corallina, N. erythropolis* and *N. asteroides* (Kluepfel & Vézina, 1970; Sehgal & Vézina, 1970; Singh, Marshall & Vézina, 1970) transform androsta-1,2,7-triene-3,17-dione into equilin and equilenin. The aromatization of the A-ring can also be carried out by an unidentified *Nocardia* sp. (Dodson & Muir, 1958*b*).

(viii) *Further transformations*

Data about the degradation of tomatidine (Belič & Sočič, 1970), progesterone and its two C_{20} epimeric alcohols (Strijewski *et al.*, 1972) and of derivatives with an androstene nucleus (Büki, Ambrus & Szabo, 1969), have also been reported.

As intermediate in the degradation of androst-4-ene-3,17-dione, an indanpropionic acid derivative, was obtained which could be metabolized further by *Nocardia restrictus* and *N. opaca* (Kondo, Stein & Sih, 1969).

Nocardia restrictus converts 19-hydroxycholest-4-en-3-one to estrone and also to acidic products: 3-hydroxy-19-norbisnorchola-1,3,5(10)-trien-22-oic acid and 3-hydroxy-19-norbisnorchola-1,3,5(10),17(20)-tetraen-22-oic acid (Sih, Wang & Tai, 1967). *Nocardia restrictus* is able to hydrolyze steroid esters. The esterase has been purified; it can hydrolyze 15α, 6α, 2β and other acetoxy groups (Sih, Laval & Rahim, 1963).

Nocardia pellegrino converts testosterone into androstenedione (Tárnok & Wenzel, pers. comm.).

(b) *Steroids as carbon source*

Some *Nocardia* and *rhodochrous* strains are able to grow on testosterone as a sole carbon source (Goodfellow, 1971). Testosterone can also be mixed into a basal medium; clear areas around colonies are due to the utilization and/or degradation of testosterone (Goodfellow, pers. comm.). The method can be used for the differentiation of nocardiae.

5. Metabolism of Alkanes, Cycloaryl and Cycloalkyl Derivatives and some of their Oxidation Products

(a) *Methodological aspects*

Although conventional techniques can be used for metabolic studies (e.g. incorporation of the substrate into the medium and isolation of the fermentation products), a new method termed "co-oxidation" has been successfully employed. Leadbetter & Foster (1959) observed that nongrowth hydrocarbons ("co-substrates") can be oxidized when metabolizable hydrocarbons were also present in the medium. This technique is not limited to hydrocarbon cosubstrates and has found wide application in the transformation of unusual non-metabolizable products and also for large-scale production of some derivatives (Abbott & Gledhill, 1971; Raymond, Jamison & Hudson, 1971; Raymond & Jamison, 1971).

(b) *Biosynthesis of dicarboxylic acids*

Purified cell-free extracts of *Nocardia corallina* are able to carboxylate acetyl-, butyl- or propionyl-CoA; the products of this carboxylation reaction are malonyl- and methylmalonyl-CoA (Baugh *et al.*, 1962). Biotin is necessary for the reaction.

(c) *Tricarboxylic acid cycle in* Nocardia

There is evidence for the presence of the tricarboxylic acid cycle in *Nocardia corallina* (Brown & Clark, 1961). Theoretically, on this pathway, all organic substrates yielding the degradation product acetyl-CoA can be oxidized to CO_2 and water; it is not known, however, what kind of additional mechanisms exist for the complete oxidation of a compound in other nocardioform bacteria.

(d) *Metabolism of alkanes*

(i) *Dehydrogenation*

Dodecane and hexadecane enter the biosynthetic pathway in *Nocardia opaca* after a previous dehydrogenation step (Webley & DeKock, 1952). Accumulation of unsaturated intermediates of the alkane degradation has been observed e.g. 1-hexadecene from hexadecane in a culture of a *Nocardia* sp. (Wagner, Zahn & Buhring, 1967), and mixed-type monoalkenes from hexa- and octadecane in a *Nocardia salmonicolor* strain (Abbott & Casida, 1968).

(ii) *Oxidation*

Treccani, Canonica & de Girolamo (1955) obtained the oxidation of C_{5-28} alkanes by a *Nocardia* sp. P_2; the similarity of this oxidation mechanism to that involved in a common fatty acid metabolism was pointed out. Mixtures of different hydrocarbons (Essovarsol 145/200 Ev) can be oxidized to organic acids by *Nocardia petroleophila* (Seeler, 1962); low-molecular weight fatty acids, succinic acid, caproic and caprylic acids and alcohols have been isolated from the medium. According to Seeler, the first step is a terminal oxidation of the substrates. In iron-, zinc- and manganese-deficient *Nocardia opaca* cells, the oxidation of organic compounds (e.g. dodecane) was strongly inhibited (Webley, Duff & Anderson, 1962).

From the fatty acids and alcohols, obtained after the appropriate oxidation of alkanes, esters can be formed by *Nocardia* strains (see also (b)(i); Raymond & Davis, 1960).

(e) *Metabolism of aromatic derivatives*

Although a great number of aromatic compounds can be utilized, transformed or degraded by nocardiae, the chemistry of the transformations can be summarized to be ring substitution, side-group transformation and ring cleavage. Aryl-substituted alkanes which can be considered to be either substituted paraffins or substituted aromatic compounds and can be readily metabolized but phenyl-substitution of an alkane seems to decrease the availability for growth (McKenna, 1966).

(i) *Hydroxylation of the ring*

From *p*-xylene, 3,6-dimethylpyrocatechol was obtained in a *Nocardia corallina* culture by the co-oxidation technique; *n*-hexadecane was used as the metabolizable substrate (Jamison, Raymond & Hudson, 1969). *Nocardia corallina, N. minima* and *N. salmonicolor* are able to introduce two hydroxyl groups into *p*-xylene (*n*-hexadecane was used as substrate; Raymond, Jamison & Hudson, 1967; Fig. 4). Benzene yields catechol and its degradation products if *Mycobacterium rhodochrous* is used as the test organ-

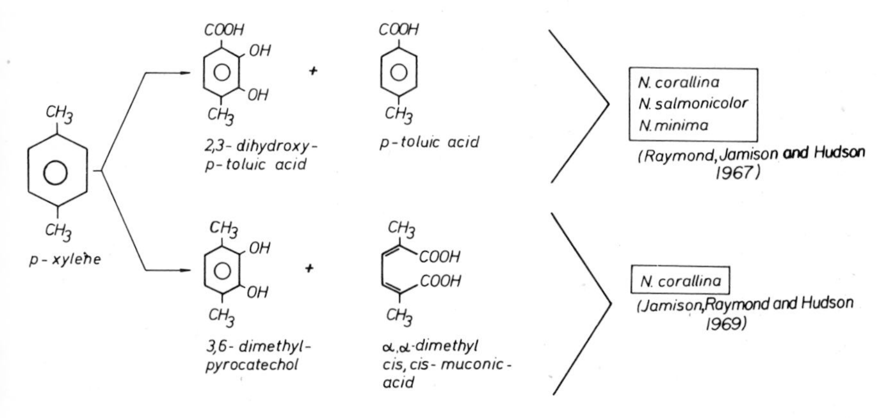

Fig. 4. Oxidation of the side-chain of *p*-xylene and introduction of hydroxyl groups into the ring by *Nocardia*.

ism (Marr & Stone, 1961). *Nocardia corallina* oxidizes benzene to phenol (Kleinzeller & Fencl, 1952). Toluene can be dihydroxylated and degraded to methylmuconic acid by *Nocardia corallina* (Jamison, Raymond & Hudson, 1969). *m*-Xylene yields 2,3-dihydroxy-*m*-toluic acid in *n*-hexadecane-grown *N. corallina, N. salmonicolor* and *N. minima* cultures (Hosler & Eltz, 1969); these strains can oxidize *o*-xylene to 2,3-dihydroxy-*o*-toluic acid. 1-Phenylalkanes can be hydroxylated by *Nocardia salmonicolor* (Sari-Aslani, Harper & Higgins, 1972, 1974). It was assumed (Sari-Aslani, Harper & Higgins, 1972) that the catabolism of 1-phenyldodecane is initiated by a side-chain β-oxidation to 4-phenylbutyrate; a β- and a subsequent ω-oxidation yields phenylacetic acid.

(ii) *Enzymatical modification of the side-group*

Oxidation of the side-group is often accompanied by the introduction of an OH-group into the ring. In *Nocardia salmonicolor, N. corallina* and *N. minima* cultures, 2,3-dihydrobenzoic acid was found deriving from the substrate toluene (Hosler & Eltz, 1969); the first two species were able to synthesize toluic acid, 2,3-dihydroxy-*p*-xylene and 3,6-dimethylpyrocatechol from *p*-xylene (Raymond, Jamison & Hudson, 1967; Fig. 4). The oxidation of *m*-xylene to 2,3-dihydroxy-*m*-toluic acid was also obtained. Toluene and *o*- and *m*-xylene yield benzoic acid and *o*- and *m*-toluic acids in *N. corallina, N. salmonicolor, N. minima* and

Mycobacterium rhodochrous cultures (Jamison, Raymond & Hudson, 1969; Raymond, Jamison & Hudson, 1967).

The side-chain of alkyl-substituted aromatics can be shortened after an initial oxidation of the terminal C-atom to a carboxyl group; the acid can be degraded to acids with shorter chain lengths (probably by β-oxidation). *n*-Nonylbenzene and *n*-dodecylbenzene yielded phenylpropionic acid and phenylacetic acid in a *Nocardia salmonicolor* culture (Davis & Raymond, 1961). It has been shown that in the presence of a metabolizable compound (e.g. hexadecane), cyclic hydrocarbons with alkyl substituents can be oxidized at the terminal C-atom to the corresponding acid (ethylbenzene to phenylacetic acid etc.).

3-Phenyldodecane, 3-phenylpentane, 4-phenylheptane and 2-phenylbutane can be converted to 1-phenylbutyrate, 2-phenylvalerate and 4-phenylhexanoate by two unidentified *Nocardia* sp. (Baggi *et al.*, 1972).

Extensive studies were carried out by Webley, Duff & Anderson (1962) on the breakdown of phenyl-substituted fatty acids. A β-oxidation mechanism was found to be operating; it should be mentioned that the ring remained intact under the employed experimental conditions.

(iii) *Opening of the ring*

The β-adipate pathway. Catechol and protocatechuic acid play a central role in the degradation of aromatics (Stanier & Ornston, 1973). By the action of oxygenases, their ring can be opened by intradiol or extradiol cleavage (*ortho* and *meta* pathway). The *ortho* pathway yields from catechol *cis,cis*-muconate, β-ketoadipate, succinate and acetyl-CoA, and the *meta* pathway α-hydroxy-muconic acid semialdehyde, formate, acetaldehyde and pyruvate. Protocatechuic acid can be degraded on the *ortho* pathway to β-carboxy-*cis,cis*-muconate and β-ketoadipate; the *meta* pathway yields α-hydroxy-ω-carboxy-muconic semialdehyde, formate and pyruvate. The catechol and protocatechuate pathways are shown in Fig. 5.

Wieland, Criss & Haccius (1958) reported the accumulation of *trans, trans*-muconate in the culture medium of a non-identified *Nocardia* strain when benzene was used as sole carbon source; it was assumed that the first step of the benzene degradation was it's

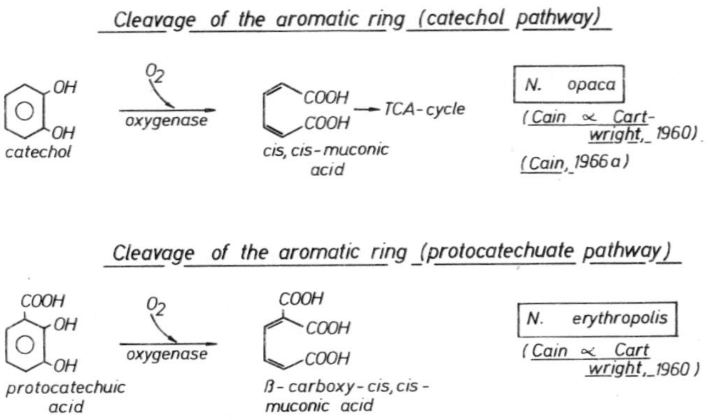

Fig. 5. Breakdown of catechol and protocatechuic acid by *Nocardia*.

direct oxidation to pyrocatechol. This benzene-oxidizing strain was identified as *Nocardia corallina* (Haccius & Helfrich, 1958). Benzene-oxidase isolated from a *Nocardia* sp. converted benzene to *β*-adipate (Treccani & Bianchi, 1959). *Nocardia opaca* and *N. erythropolis* can degrade catechol by the *ortho* pathway (Cain & Cartwright, 1960) and *o*-nitrobenzoate-grown *N. opaca* metabolizes catechol to *cis,cis*-muconate (Cain, 1966*a*, Fig. 5 and Fig. 6). Wodzinski & Johnson (1968) reported on a non-identified *Nocardia* sp. which grew on benzene as sole carbon source. Treccani *et al.* (1968) found in the cell-free extracts of a napthalene-grown *Nocardia* strain R the enzyme catechol-2,3-oxygenase.

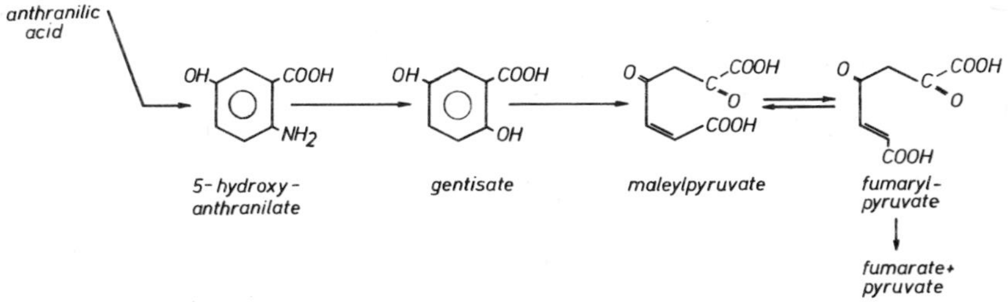

Fig. 6. Catechol and gentisate pathway in *Nocardia opaca*.

Tanzil & Bönicke (1968*b*) used catechol derivatives for the differentiation of nocardiae. The degradation patterns seemed to be characteristic for the species; no data were given, however, about the degradation mechanism.

Schumann (1974) studied the occurrence of some enzymes involved in catechol degradation (catechol: oxygen 1,2-oxidoreductase and protocatechuate-3,4-oxygenase) using different *Nocardia* strains as test organisms (Table 2). As suggested by Schumann, the catechol degradation follows the β-adipate pathway in *Nocardia salmonicolor*, an observation in agreement with those of Jamison, Raymond & Hudson (1969) and Holsler & Eltz (1969) regarding the oxidation of *p*-xylene to 3,6-dimethylpyrocatechol (Fig. 4) and α,α'-dimethyl-*cis,cis*-muconic acid by *N. corallina* (co-oxidation technique).

Table 2

Occurrence of catechol-1,2-oxygenase and protocatechuate oxygenase in Nocardia *and* rhodochrous *strains* (*According to K. P. Schumann, 1974*)

Nocardia and *Actinomadura*	catechol-1,2-oxygenase	protocatechuate oxygenase
pellegrino	+	+
rubra	+	+
opaca	+	+
alba	+	+
A. madurae	+	+
globerula	+	+
coeliaca	+	+
salmonicolor	+	+
lutea	+	+
A. pelletieri	+	+
asteroides	+	−
corallina	+	−
rubropertincta	+	−
minima	+	−
paraffinae	+	−
blackwellii	+	−
gibsonii	+	−
polychromogenes	−	+
brasiliensis	−	+
vaccinii	−	+

Schumann also observed the degradation of protocatechuic acid and protocatechualdehyde. Generally, both compounds can be metabolized but in *Nocardia asteroides* SN5401 only the aldehyde cleavage was observed, and Schumann postulated tentatively a protocatechualdehyde pathway different from the usual protocatechuic acid degradation.

Protocatechuate-3,4-oxygenase, 3-carboxymuconate cycloisomerase, 4-carboxymuconolactone decarboxylase and 3-oxoadipate enol lactone hydrolase (protocatechuate-branch-enzymes) can be induced by 3-oxo-adipate in *Nocardia opaca* (Rann & Cain, 1969, 1973). Identical regulatory systems for these enzymes were found in *N. globerula, N. caviae, N. erythropolis, N. corallina* and *N. rubra* strains.

Halogen substitution of the ring (e.g. halogenated benzoic acid) inhibited the growth of the bacteria but the compound could not be split by the test strain *Nocardia erythropolis* (Smith, Traster & Cain 1968). It was observed, however, that benzoate-grown *N. erythropolis* cells were able to oxidize 2-fluoro-*p*-nitrobenzoate to fluoroacetate (Cain, Traster & Darrah, 1968).

The gentisate pathway. Gentisic acid can be degraded to maleylpyruvic acid which may be decomposed further to pyruvate and fumarate. Cain (1968) described a *Nocardia opaca* strain which could degrade anthranilic acid on both the β-adipate and the gentisate pathway. Anthranilate can be hydrolized and decarboxylysed yielding the intermediate catechol, or hydroxylated to 5-hydroxyanthranilate. The latter can be degraded to gentisate and this again to pyruvate and fumarate. According to Cain (1968), the gentisate pathway does not seem to be the major route (Fig. 6).

Sari-Aslani, Harper & Higgins (1972, 1974) observed the catabolism of phenylalkanes *via* homogentisate and maleylacetoacetate; as end products, acetoacetic acid and fumarate were isolated.

(f) *Metabolism of cycloalkyl derivatives*

(i) *Enzymatical modification of the side-group*

Cyclohexane acetic acid was isolated from a *Nocardia salmonicolor* culture when *n*-butylcyclohexane and *n*-octadecane were present in the medium; the accumulation of acids in the culture fluid was obtained if the cells were grown on mineral oil (Davis & Raymond, 1961).

(ii) *Substitution of the ring*

From bicyclohexyl, 4,4′-dihydroxybicyclohexyl was obtained in a *Nocardia* culture (Fonkin, Herr & Murray, 1966). Ooyama & Foster (1965) observed the formation of cyclohexanone from cyclohexane by a *Nocardia* sp. Cyclohexanol is metabolized in *Nocardia globerula* to 7-hydroxy-1-oxa-2-oxocycloheptane by a cyclohexanone oxygenase which has been isolated and purified (Norris & Trudgill, 1971, 1972, 1973); the enzyme exists in two electrophoretically different forms. A *Nocardia* sp. can oxidize cyclohexanone derivatives *via* 2-hydroxycyclohexane-1-one (Murray, Scheikowski & MacRae, 1974); this pathway seems to be different from that described by Norris & Trudgill. Cyclohexane-1,2-diols (*trans*- and *cis*-forms) are degraded to adipate in *Nocardia globerula* (Donoghue & Trudgill, 1973); 2-hydroxycyclohexanone and 7-hydroxy-1-oxa-2-oxo-cycloheptane were identified as intermediates.

(g) *Metabolism of polynuclear hydrocarbons*

(i) *Hydroxylation of the ring*

1,2-dihydroxynaphthalene was isolated from the culture medium of a *Nocardia* strain R (Walker & Wiltshire, 1953; Treccani, Canonica & de Girolamo, 1955).

(ii) *Oxidation of the side-group*

This oxidation step is often accompanied by the introduction of hydroxyl groups into the ring. *Nocardia corallina* oxidizes 1,2-dimethylnaphthalene and 2,3-dimethylnaphthalene to 1-methyl-3-naphthoic acid and 3-methyl-2-naphthoic acid, respectively (Raymond, Jamison & Hudson, 1969). 1-(1-naphthyl)-undecane is oxidized by *Nocardia opaca* to 3-(1-naphthyl)propionic acid and 3-(1-naphthyl)-acrylic acid (Webley, Duff & Anderson, 1962).

(iii) *Opening of the ring(s)*

Nocardia coeliaca and a *Nocardia* sp. degrade naphthalene to 4-(2-hydroxyphenyl)-2-ketobutyric acid (Raymond, Jamison & Hudson, 1971). Different ketobutyric acid derivatives can be obtained if substituted naphthalenes (1-methyl-, 1,2-dimethyl-, 1,3-dimethyl or 2-ethylnaphthalene) are used; the co-oxidation technique has been successfully employed (Sun Oil Co., 1969). Naphthalene cleavage (due probably to an oxygenase action) has been observed in *Nocardia coeliaca* and a *Nocardia* sp. (Raymond, Jamison & Hudson, 1971). Figure 7 gives some examples for these transformation steps.

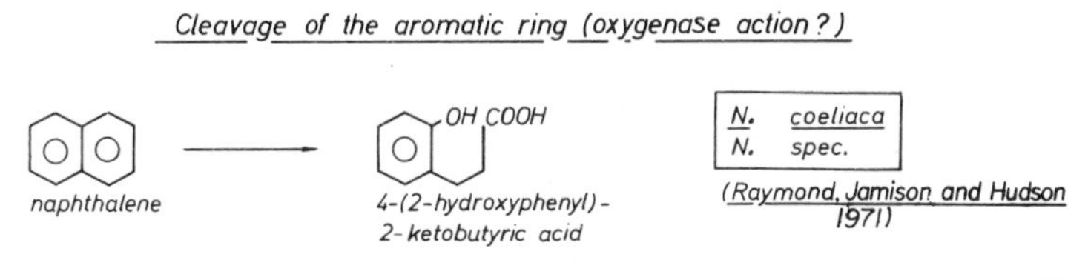

Fig. 7. Breakdown of naphthalene (probably by oxygenase action) by nocardiae.

Salicylate accumulation has been observed in a non-identified *Nocardia* culture grown on naphthalene (Treccani, 1953); catechol could also be isolated as an intermediate. A *Nocardia* sp. can split naphthalene to *o*-hydroxybenzalpyruvic acid (Wegner, 1969).

6. Metabolism of Carbohydrates and Related Compounds

(a) *General remarks*

Nocardioform bacteria utilize a wide variety of carbohydrates; this property is used for differentiation (Goodfellow, 1971). Usually, acid production from sugars under aerobic and anaerobic conditions and the growth of strains on the carbohydrate in question are the first steps in sugar metabolism studies. Using this simple system, however, no data can be obtained about the details of the actual degradation pathways.

The endogenous respiration and oxidative assimilation in *Nocardia corallina* strain S were studied by Midwinter & Batt (1960), and the respiratory quotients calculated (also for glucose as substrate). Carbohydrate respiration was studied in *Nocardia rubra* and the effect of glucose on the utilization of some sugars has been determined; in the presence of glucose, the usually non-metabolizable sugars can be rapidly oxidized. This phenomenon seems to be similar to the "co-oxidation" mechanism (McClung, Salser & Santoro, 1960).

(b) *The Embden–Meyerhof–Parnas pathway*

The presence of this pathway was suggested for *Nocardia opaca* and *N. corallina* (Duff & Webley, 1959; Brown & Reda, 1967). The metabolism of glucose in *N. corallina* shows quantitative changes associated with the period for the impermeability for glucose.

(c) *The hexosemonophosphate shunt*

Glucose 6-phosphate and gluconate 6-phosphate dehydrogenases were demonstrated in cell-free extracts of *Nocardia corallina* (Brown & Clark, 1966*a*). The specific activity of glucose 6-phosphate dehydrogenase seems to be correlated with the life cycle of this organism.

(d) *The Entner–Doudoroff pathway*

Probst & Schlegel (1973) demonstrated the presence of the key enzymes of the Entner–Doudoroff pathway (6-phosphogluconate dehydrase and 2-keto-3-deoxy-6-phosphogluconate aldolase) in the hydrogen bacterium *Nocardia opaca* strain 1b. Pang & Bradley (1974*a*) failed to detect the latter enzyme in *Nocardia erythropolis* and the Entner–Doudoroff pathway seems to be absent in strains of this species.

(e) *Uptake of carbohydrates and related compounds*

Cérbon & Ortigoza-Ferado (1968) demonstrated the phosphate dependence of the monosaccharide transport in *Nocardia asteroides* and *N. brasiliensis*; for the function of the transport mechanism phosphate or arsenate are required. The observation that phosphate can be replaced by arsenate indicates the possible involvement of phosphorylated compounds (phosphoinositides?) in the active transport mechanism.

The uptake of glycerol by *Nocardia asteroides* has been extensively studied by Calmes & Deal (1972); at high glycerol concentrations (more than 80 μM) transport could be due to free diffusion while at concentrations less than 80 μM a specific active transport system seemed to be in action.

Evidence was given for the occurrence of regulatory mechanisms for glucose, glycerol and mannitol uptake in *Nocardia erythropolis* (Pang & Bradley, 1974*a,b*). Glucose and glycerol uptake are constitutive in this strain, while mannitol uptake is inducible. The induced uptake of mannitol and the constitutive uptake of glucose and glycerol are energy dependent and carrier mediated.

(f) *Metabolism of some carbohydrate derivatives*

Glycosides and high-molecular weight polysaccharides such as amygdaline, aesculin, arbutin, dextrin, starch and glycogen can be degraded by some *Nocardia* and *rhodochrous* strains (Goodfellow, 1971). Xylan was degraded only by *Oerskovia turbata.* Amygdaline, arbutin, dextrin and starch enhanced growth in *Nocardia asteroides, N. caviae, N. brasiliensis, Actinomadura madurae, A. pelletieri* and *Oerskovia turbata* to different extents.

7. Metabolism of Inorganic Nitrogen Compounds and Organic Nitrogen Containing Derivatives

(a) *Nitrogen fixation*

Metcalf & Brown (1957) reported the fixation of nitrogen in *Nocardia cellulans* (probably *Oerskovia*) and *Nocardia calcarea* strains. Hill & Postgate (1969), however, were not able to reproduce their results, and nitrogen fixation in nocardiae remains an open question but generally, they are considered to belong to the non-fixing organisms.

(b) *Nitrate assimilation and nitrification*

Although there are insufficient data to separate nitrate assimilation (utilization of nitrate as nitrogen source by the involvement of nitrate reductase) from denitrification (nitrate as electron acceptor for the cytochromes), observations in regard to the nitrate, nitrite and ammonia metabolism of the *Nocardia* can give a hint to the participation of these organisms in the recycling of nitrogen in the soil. Both nitrification and denitrification steps can be carried out by members of this genus.

Alexander, Marshall & Hirsch (1960) found three out of 20 nocardiae isolates which were nitrifiers and converted different nitrogenous compounds into (probably) nitrite. The opposite reaction e.g. the reduction of nitrate or nitrite is also well known. Nitrate reductase is common in nocardiae (Goodfellow, 1971), and nitrite reductase can also be detected. Although ammonia should be found as a consequence of this latter reaction usually this is not so probably due to the rapid uptake of ammonia by the cells. Our preliminary experiments show that ammonia can indeed be rapidly taken up by nocardiae, and that this is unaffected by the presence of different organic compounds in the reaction mixture (Tárnok, unpublished data).

A nitro-reductase has been isolated by Villanueva (1964*a,b*) from a *Nocardia* strain V. The activity of the preparations could be successfully determined using *p*-dinitrophenol as substrate.

(c) *Metabolism of alkylamines*

Diamineoxydases (diamine: O_2-transhydrogenases) occur in *Nocardia farcinica, N. polychromogenes* and *N. pellegrino*. Using different diamines, characteristic degradation patterns for some selected species were obtained and the test seems to have diagnostic value (Tanzil & Bönicke, 1968*a*).

(d) *Metabolism of acid amides*

Acid amides such as acetamide, malonamide, succinamide, pyrazinamide, benzamide, isonicotinamide, nicotinamide, asparagine, and urea can be hydrolyzed by a great number of mycobacteria, and the determination of their amidase patterns has been successfully used for their differentiation (Bönicke, 1961). Schneidau (1963) used 20 different amides for the differentation of nocardiae but obtained great variability between strains belonging to the same species. Bönicke & Juhász (1965) also reported the inhomogeneity of *Nocardia* strains in regard to their amidase patterns. According to our observations (Tárnok, unpublished data, Table 3) the amidase degradation by *Nocardia* is characteristic for these strains and as a rule good reproducibility is obtained; the observed discrepancies are probably due to the poor taxonomy of this genus.

In many cases, acid amides can also serve as sole carbon and nitrogen sources. This seems to be the case for urea which can be taken up in the presence of pyruvate or glycerol by *Nocardia*

Table 3
Hydrolysis of acid amides by Nocardia

Species (number of the strains tested)	1	2	3	4	5	6	7	8	9	10
N. asteroides (9)	±	±	+	∓	±	±	−	∓	−	∓
N. pellegrino (12)	+	−	−	−	∓	∓	−	−	−	−
N. rubra (8)	∓	−	+	−	−	∓	−	∓	−	−
N. corallina (3)	−	−	+	−	−	−	−	∓	−	−
N. brasiliensis (1)	−	−	+	−	+	+	−	−	−	+
N. caviae (5)	−	−	±	−	+	+	−	∓	−	−
N. farcinica (7)	+	−	+	−	∓	∓	−	+	∓	−
N. minima (5)	±	∓	±	−	±	∓	−	∓	−	−

Incubation time: 24 h at 37°C.
List of the amides: 1. acetamide; 2. benzamide; 3. urea; 4. isonicotinamide; 5. nicotinamide; 6. pyracinamide; 7. salicylamide; 8. allantoin; 9. succinamide; 10. malonamide.
Interpretation of the results:
+ Strong hydrolysis.
− No hydrolysis.
± The majority of the tested strains are positive.
∓ The majority of the tested strains are negative.

pellegrino. The uptake depends on the pyruvate or glycerol concentration which suggests a requirement for energy. The involvement of urease can be excluded since this enzyme is not detectable in *N. pellegrino* (Tárnok, 1971).

(e) *Metabolism of aromatic compounds with nitrogenous substituents*

(i) *Nitro-group containing derivatives*

Cartwright & Cain (1959) reported the reduction of *o*-, *m*- and *p*-nitrobenzoate, *m*-dinitrobenzene, 2,4-dinitrobenzoate, nitrobenzene, 2,4- and 2,5-dinitrophenols, nitrofurazone, chloramphenicol and some other compounds to the corresponding amino derivatives by *Nocardia erythropolis* and *N. opaca* strains. The nitrate reductase of these strains and also of the *Nocardia* strain *V* (Villanueva, 1964*a,b*) need NADPH*; all these enzymes are non-specific. A *Nocardia alba* strain decomposed some nitrophenol derivatives and nitrobenzoate to metabolizable compounds; oxyhydrochinon occurs as an intermediate (Germanier & Wuhrmann, 1963) entering the β-ketoadipate pathway.

Bönicke & Juhász (1965) observed the ability of some nocardiae to metabolize *m*-nitrophenol: strains of *Nocardia asteroides, N. rubra* and *N. pellegrino* showed growth on this substrate. The test was originally suggested for the differentiation of mycobacteria using *p*- and *m*-nitrophenols (Tacquet, 1962).

(ii) *Amino group containing derivatives*

The anthranilate metabolism in *Nocardia opaca* was studied extensively by Cain (1958; 1968). Anthranilate could be transformed to catechol by decarboxylation and hydrolysis or also to 5-hydroxyanthranilate.

(f) *Metabolism of heterocyclic derivatives*

(i) *Metabolism of pyridine*

Pyridine can be utilized as a carbon and nitrogen source by *Nocardia* strain *Z1* (Watson & Cain, 1972); C_5 compounds could be isolated suggesting that pyridine metabolism in this strain occurs by a ring fission adjacent to the nitrogen atom together with a ring

*Nicotinamide adenine dinucleotide phosphate.

reduction. Houghton & Cain (1972) reported the formation of dihydroxypyridine from pyridine in *Nocardia* strain *Z1*.

(ii) *Metabolism of pyrimidine derivatives*

Batt (1961) suggested the presence of an inducible system for pyrimidine catabolism in *Nocardia corallina*; both uracil and thymine oxidases were inducible. Batt & Woods (1961) observed the degradation of barbituric and methylbarbituric acid by *Nocardia corallina*; according to these findings the oxidative pyrimidine degradation pathway (*via* barbituric acid) seems to be present. It should be mentioned that according to our experience, strains designated as *Nocardia corallina* (Culture Collection of the Institute Borstel) are not able to oxidize thymine or uracil to the corresponding derivatives.

Nocardia brasiliensis, N. farcinica and *N. rangoonensis* degrade cytosine, uracil, dihydrouracil and ureidopropionic acid indicating the occurrence of the known reductive degradation pathway; in *N. globerula* and *N. vaccinii*, cytosine was deaminated to uracil but the latter could not be decomposed (Fig. 8).

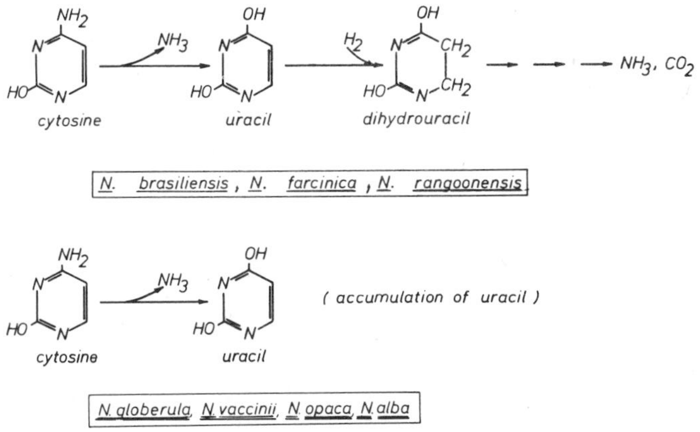

Fig. 8. Breakdown of cytosine and uracil by nocardiae.

Nocardia opaca, N. alba and *N. rubra* degrade thymine but neither dihydrothymine nor methylbarbituric acid can be metabolized to ammonia and CO_2; consequently, for these strains the oxidative pathway (*via* methylbarbituric acid) can be excluded.

In experiments according to Warburg, a strong thymine oxidation was measured; however, the intermediates of this oxidation are not known (Fig. 9). *Nocardia farcinica* degrades thymine on the reductive (*via* dihydrothymine) pathway.

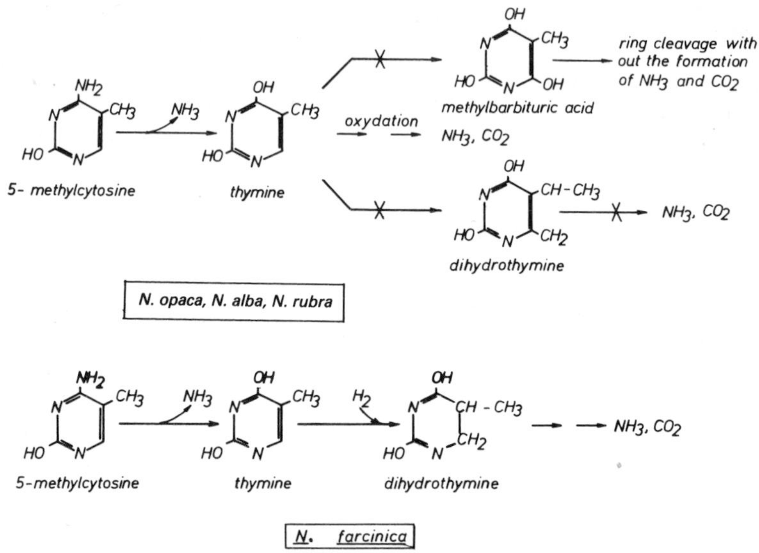

Fig. 9. Breakdown of 5-methylcytosine and thymine by nocardiae.

Röhrscheidt, Tárnok & Tárnok (1970) carried out systematic experiments in regard to the pyrimidine catabolism in nocardiae; members of the species *Nocardia rubra* and *N. brasiliensis* were able to degrade thymine and uracil.

(iii) *Metabolism of purine derivatives*

According to Röhrscheidt, Tarnok & Tarnok (1970), *Nocardia corallina, N. brasiliensis* and *N. asteroides* were positive in the purine degradation test but the species were not homogeneous in this regard.

The degradation pathway of adenine (Fig. 10) was found to be identical with the pathway postulated for other microorganisms. However, in *Nocardia asteroides* strain SN5421, degradation of xanthine was not obtained but both hypoxanthine and uric acid were slowly metabolized; a rather unusual hypoxanthine metabolism can be tentatively suggested (Tárnok *et al.*, unpublished

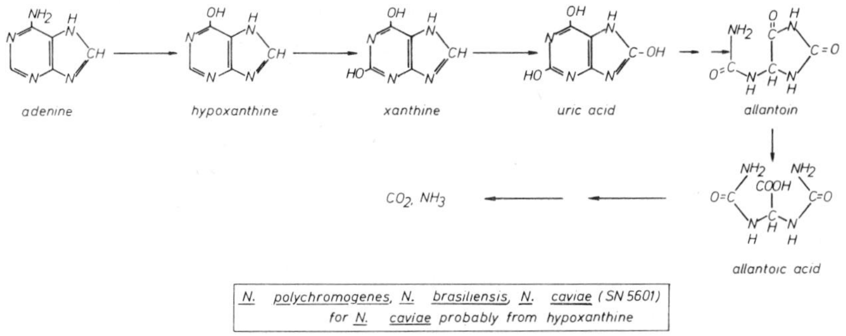
adenine
hypoxanthine
xanthine
uric acid
allantoin
allantoic acid
CO_2, NH_3
N. polychromogenes, N. brasiliensis, N. caviae (SN 5601)
for N. caviae probably from hypoxanthine

Fig. 10. Metabolism of purine derivatives by nocardiae.

data). Table 4 gives a summary of the purine and pyrimidine degradation patterns of the nocardiae.

Occurrence of cyclic purine derivative in nocardiae. Cyclic adenosine monophosphate (Ado-3′ : 5′-P) was found in *Nocardia erythropolis* and its role in metabolism and enzyme induction discussed (Ide, 1971).

8. Composition and Metabolism of Nocardial Deoxyribonucleic Acids (DNA)

(a) *Base composition of DNA*

Generally, the overall base composition of nocardiae and nocardioform bacteria is estimated to be between 62 to 77·3 mol % guanine + cytosine (GC) (Belozersky & Spirin, 1960; Jones & Bradley, 1964; Frontali, Hill & Silvestri, 1965; Bradley, 1966; Hill, 1966; Prauser, 1966; Tewfik & Bradley, 1967; Yamaguchi, 1967; Tárnok, Röhrscheidt & Bönicke (1967) reported a relatively low GC mol % (57·5–60·0%) for some *Nocardia*, and Adams (1970) found a low GC mol % (62·5%) for *N. erythropolis* and *N. canicruria*.

The DNA content of nocardiae has been compared by Tsukamura & Mizuno (1969) and found to be different from strain to strain whereas the DNA-content of mycobacteria was more uniform.

(b) *Degradation of DNA by nocardial enzymes*

Besides these analytical data very little is known about the metabolism of nucleic acids in nocardiae. However, the determination of the deoxyribonuclease (DNase) activity has been found to be

Table 4

Degradation of pyrimidines and purines by Nocardia *and* rhodochrous *strains*

Species	Substrate 1	2	3	4	5	6	7	8	9	10	11	12	13
N. polychromogenes	—	—	—	—	—	—	—	—	—	w	w	w	w
N. pellegrino	—	—	—	—	—	—	—	—	—	w	—	—	—
N. rubra	—	—	—	—	s	—	—	—	—	w	—	—	—
N. corallina	w	—	—	—	—	—	—	w	—	—	—	—	—
N. asteroides	—	—	—	—	—	—	—	—	—	—	var	var	—
N. brasiliensis	w	var	w	—	var	—	—	w	—	var	var	var	var
N. caviae	—	w	w	—	—	w	—	—	—	—	var	var	var
N. "farcinica"	s	s	s	—	s	s	—	s	—	—	—	—	w
N. rubropertincta	—	—	—	—	—	—	—	—	—	—	w	w	—
N. opaca	s	—	—	—	s	—	—	s	—	—	—	—	—
N. minima	—	—	—	—	—	—	—	—	—	—	—	—	—
N. alba	s	—	—	—	s	—	—	w	—	—	—	—	—
N. paraffinae	—	w	w	—	—	w	—	—	—	—	w	w	w
N. blackwellii	—	—	—	—	—	—	—	—	—	—	—	—	—
N. globerula	s	—	—	—	—	—	—	s	—	—	—	—	—
N. caprae	—	—	—	—	—	—	—	—	—	—	—	—	—
N. coeliaca	—	—	—	—	—	—	—	—	—	w	—	w	w
N. salmonicolor	—	—	—	—	—	—	—	—	—	w	—	—	—
N. lutea	—	—	—	—	—	—	—	—	—	—	—	—	—
N. vaccinii	w	—	—	—	—	—	—	w	—	—	—	—	—

The degradation was determined according to Röhrscheidt, Tárnok & Tárnok (1970).
Substrate: 1. Cytosine; 2. Uracil; 3. Dihydrouracil; 4. Barbituric acid; 5. Thymine; Dihydrothymine; 7. 5-Methylbarbituric acid; 8. 5-Methylcytosine; 9. 5-Hydroxymethyluracil; 10. Adenine; 11. Hypoxanthine; 12. Xanthine; 13. Uric acid.
interpretation of results:
— No ammonia detected during the reaction.
w 10–20% out of the maximal amount of ammonia liberated during the reaction.
s 21–100% out of the maximal amount of ammonia liberated during the reaction.
var Different results obtained in different strains.

useful for diagnostical purposes (Goodfellow, 1971), strains of *Nocardia asteroides* and *N. caviae* being DNase positive. According to our observations, DNase is relatively common in nocardiae (Tárnok & Wenzel, unpublished data) (Table 5), but contrary to Goodfellow, we found *Nocardia asteroides* and *N. caviae* to be only very weakly positive.

Table 5
Deoxyribonuclease (DNase) activity in nocardioform bacteria

Taxa and number of strains tested	DNase activity
Nocardia polychromogenes (1)	strong
N. pellegrino (12)	negative
N. rubra (8)	weak
N. corallina (3)	weak
N. asteroides (9)	very weak
N. brasiliensis (1)	weak
N. caviae (5)	weak
N. farcinica (7)	negative
N. rubropertincta (1)	strong
N. opaca	strong
N. minima (5)	negative
N. globerula (1)	negative
N. caprae (1)	strong
N. coeliaca (1)	negative
N. salmonicolor (1)	negative
N. lutea (1)	negative
Actinomadura pelletieri (1)	strong
A. madurae (1)	weak
N. gardneri (1)	strong
N. gibsonii(1)	strong
N. rangoonensis (1)	strong

Incubation time: 3 days at 37°C; DNA-agar plate method.

9. Metabolism of Proteins, Amino Acids and Related Compounds

Only a few data are available about these topics.

(a) *Proteolytic activity of nocardiae*

Proteolytic activities seem to be characteristic of some species (Bergey, 1974) and attempts have been made to use this property for diagnosis (Goodfellow, 1971). Casein, gelatin and keratin can be hydrolyzed by *Nocardia asteroides, Actinomadura madurae* and other nocardioform strains but to date proteolytic enzymes involved have not been isolated and purified.

(b) *Peptidases*

Peptide bonds as present in hippurate can be hydrolyzed by some *Nocardia* (Gordon, 1966; Gordon & Horan, 1968); it seems possible, however, that these *Nocardia* are mislabelled and belong to the species *Streptomyces griseus*.

(c) *Amino acid metabolism*

Amino acids not only enhance growth of nocardiae (Uesaka & McClung, 1961) but they undergo different transformation steps.

Röhrscheidt (pers. comm.) measured the histidine ammonialyase (histidase) activity in *Nocardia asteroides* strain 5401 and postulated the following pathway for histidine degradation:

histidine → urocanic acid → (imidazolone propionic acid) →
(formamidino glutamic acid) → formamide + glutamic acid.

Urocanic acid and glutamic acid could be detected in the reaction mixture. The compounds in parenthesis are the probable intermediates but were not identified with certainty.

An aspartic acid decarboxylase was isolated from *Nocardia globerula* (Crawford, 1958). The enzyme is specific for aspartic acid decarboxylizing it to α-alanine.

Kajinami & Smith (1971) isolated a glutamine synthetase (glutamate ammonia ligase) from *Nocardia corallina*. The purified enzyme preparation formed glutamine from ammonia and glutamic acid in the presence of ATP.

10. Pigments of *Nocardia* and Related Genera

(a) *Occurrence of pigments with different chemical structures*

(i) *Carotenoid pigments*

El'Registan *et al.* (1965) reported the occurrences of carotenoids in *Proactinomyces* (*Nocardia*) *asteroides*. According to the observation of Röhrscheidt & Tárnok (1972*a*) pigments belonging probably to the carotenes and related compounds can be detected in nocardiae; α-, β- and γ-carotenes, however, seem to be absent. Vacheron, Arpin & Michel (1970) isolated a glucose-xanthophyll ester from *Nocardia kirovani*; the structure of the pigment resembles the so-called *Mycobacterium phleixanthophyll.*

The red pigment of *Nocardia corallina* previously suggested to be a carotenoid (Webb, 1956) is a nitrogen-containing compound with a non-carotenoid character (Brown & Clark, 1966*b*).

(ii) *Pigments with chinoid structures*

Aiso *et al.* (1954) isolated a red pigment termed nocardorubin from a non-identified *Nocardia* sp. and considered it to be a chinoid.

(iii) *Phenazines*

Different phenazines have been isolated from *Actinomadura dassonvillei* and related species. 1,6-Phenazinediol-5,10-dioxide, 1,6-phenazinediol-5-oxide and 1,6-phenazinediol were found in *A. dassonvillei* together with 1-phenazinol-10-oxide and 2-amino-1-carboxy-3H-phenoxazin-3-one (Gerber, 1966, 1969).

(iv) *Prodiginines*

Pigments with a methoxy-tripyrrole skeleton, related to prodigiosin were obtained from *Actinomadura madurae* and *A. pelletieri* (Gerber, 1969, 1970). The occurrence of phenazines and prodigiosin-like pigments is without taxonomical value; for details, the reader is referred to the review of Lechevalier, Lechevalier & Gerber (1971).

(b) *Evidence for the influence of light on pigment synthesis*

(i) *Light-inducible pigment synthesis*

It has been observed that a short exposure to light increases the pigment synthesis in *Nocardia corallina* and *N. rubropertincta* (Röhrscheidt & Tárnok, 1972*b*). In the light-exposed cells, some pigments present only in traces in bacteria grown in the dark were extracted and separated by t.l.c.

(ii) *Inhibition of the pigment synthesis by exposure of the cultures to light*

Under the influence of even a short light exposure *Actinomadura madurae* strain EB2301 shows a strong inhibition of its pigment synthesis. The corresponding dark-grown bacteria synthesize a relatively great number of pigments. This phenomenon has been termed "negative photochromogeneity" (Röhrscheidt & Tárnok, 1972*b*).

(c) *Evidence for the different types of binding of pigments in the orange and yellow strains of* Nocardia pellegrino

It has been observed that both the deep-orange pigmented wild-type bacteria and also the yellow mutant contain qualitatively and nearly quantitatively the same pigment components. As a consequence, the different pigmentation of the wild-type cells and of the mutant can not be explained by differences in the biosynthesis of the single compounds. The cell-bound pigments of the yellow mutant showed an indicator character: a color change from yellow to deep orange occurs after treatment of the bacteria with alkali. By acidification of the previously alkali-treated cells, the original color has been obtained. In contrast the wild-type bacteria showed no change of pigmentation in the presence of acid or alkali. It was found using t.l.c. that two out of the three pigment components isolated from the yellow and red bacteria also showed the indicator character. These two components contain one and two hydroxyl groups which can be acetylated; acetylation prevents the color change of the extracted components in alkali or acid. We assume that the yellow or red color of the cells can be due mostly to hydroxyl-containing components being differently bound in the wild-type and mutant cells; in the yellow cells, the OH-groups are accessible to pH-changes while in the wild-type cells they are protected against the action of alkali or acid. It would seem that this observation is the first one giving a hint for an unchanged biosynthesis of the pigments in a phenotypically modified mutant and for the probably different type of pigment binding in the mutant cells (Wenzel & Tárnok, in prep.).

11. Antibiotics in *Nocardia*

(a) *Nocardorubin*

This compound, which has a chinoid structure, was isolated from a non-identified *Nocardia* sp; it shows a marked inhibitory activity against several Gram-positive microorganisms (Aiso *et al.*, 1954).

(b) *Formycins*

The biosynthesis of a group of antibiotics termed formycins has been studied in *Nocardia interforma* (Hori *et al.*, 1964; Koyama & Umezawa, 1965; Koyama *et al.*, 1966; Sawa *et al.*, 1968). Formycin A (7-amino-3-(β-D-ribofuranosyl)pyrazolo-[4,3-d]-pyrimidine) and formycin B (3-(β-ribofuranosyl)pyrazolo-4,3-d-6(H)-7-pyrimidone) show inhibitory effects against *Xanthomonas oryzae*, *Mycobacterium tuberculosis* and other microorganisms. The terminal steps of the formycin biosynthesis are as follows:

formycin B-monophosphate → formycin A-monophosphate →
formycin A → formycin B → oxoformycin B

12. Vitamin Metabolism in *Nocardia*

(a) *Vitamin B_{12} and related compounds*

Vitamin B_{12} biosynthesis has been extensively studied in *Nocardia rugosa* strains and mutants. Barchielli *et al.* (1957) isolated a guanine-containing analogue of vitamin B_{12} from the fermentation broth of a *Nocardia* strain. A substance of the vitamin B_{12} group, identified as guanosine diphosphate factor B has also been described for a *N. rugosa* strain (Barchielli *et al.*, 1960). There is evidence that vitamin B_{12} production and porphyrin synthesis are in inverse correlation in *Nocardia rugosa* (Bardi *et al.*, 1958). In their studies on *N. rugosa* mutants, Di Marco & Spalla (1961) proposed that the biosynthetic pathway for these strains was *via* cobyrinic acid hexamide, cobynamide (factor B), factor B phosphate, guanosine-diphosphate-factor B and B_{12}. An intermediate in the biosynthesis of vitamin B_{12} was isolated from *N. rugosa* and identified as a corrinoid conjugate (Migliacci & Rusconi, 1961). Di Marco *et al.* (1962) observed that for the B_{12} biosynthesis in *N. rugosa* strains 635 and 959 a substance called V_1 was necessary which could be synthesized by another *N. rugosa* strain (mutant 466).

(b) *Thiamine and folic acid*

Some data are available about the biosynthesis of these vitamins. Quantitative determination of both of these compounds was carried out in *Nocardia rhodnii* (Harington, 1960). At present, no information is available for the genetic block in the thiamine synthesis in thiamine-requiring strains, but pyrithiamine or 4-amino-5-aminomethyl-2-methylpyrimidine are able to replace thiamine in thiamine-requiring *Nocardia corallina* strain S suggesting that the thiazole moiety of the vitamin is available in the cells (Martin & Batt, 1957*a*).

13. Metabolism of Compounds with Little or Still Unknown Physiological Importance

(a) *Esterases*

(i) *Arylsulphatases*

Enzymes hydrolyzing sulphate esters seem to be present in *Nocardia*; Wayne, Juarez & Nichols (1958) obtained positive results with *N. asteroides, N. brasiliensis* and *Actinomadura* strains. However, the species were not homogeneous in regard to their arylsulphatese activities and the method does not seem to be suitable for diagnostic purposes.

(ii) *Tween esterases*

Tweens 20, 40, 60 and 80 are hydrolyzed by *Nocardia asteroides, N. caviae, N. brasiliensis* and *rhodochrous* strains (Goodfellow, 1971). Tanzil & Hidajat (1970) tested some *Nocardia* strains for their Tween-80 hydrolyzing activity and found *N. brasiliensis, N. caviae* to be positive. Their method is suitable for diagnostic purposes.

(iii) *Phosphatase*

Banerjee & Nandi (1966) studied the inhibition of the phosphatase excretion by *Actinomadura madurae* cells; glycine inhibited the phosphatase diffusion completely.

(b) *Penicillin inactivating enzymes penicillinase and penicillin acylase*

A non-identified *Nocardia* sp. produces both types of these penicillin-inactivating enzymes (Nyiri, 1963); details on the kinetics were also given.

(c) *Alkaloid synthesis in nocardiae*

Relatively high amounts of ergothioneine and hercynine were produced by a *Nocardia asteroides* strain (Genghof, 1970). It is not yet known whether histidine can be used as a precursor for the biosynthesis.

(d) *Degradation of pesticides*

Given their structures, the breakdown of the following pesticides should be discussed in the section "Metabolism of alkanes . . . " but in regard to their agricultural importance, it seems to be worthwhile to describe it in a separated paragraph. Chacko, Lockwood & Zabik (1966) reported the inactivation of DDT (1,1,1-trichloro-2,2-bis(*p*-chlorophenyl)-ethane) and PCNB (pentachloronitrobenzene) by a *Nocardia* sp. to 1,1-dichloro-2,2-bis(*p*-chlorophenyl)-ethane (DDD) and pentachloroaniline, respectively. Namdeo (1972) observed the biodegradation of paraquat (1,1′-dimethyl-4,4′bipyridylium dichloride) probably to 1-methyl-4,4′-bipyridylium ion and 1-methyl-4-carboxypyridylium ion by a non-identified *Nocardia.*

14. Concluding Remarks: Taxonomical Aspects of the Metabolical Studies

Regarding their known metabolic activities, nocardiae and rhodochrous-like organisms seem to be similar to many other aerobic bacteria; this statement has validity for the utilization of organic compounds and nitrogen containing derivatives, lipid metabolism and other metabolic steps. At present it appears that *Nocardia* differs from other genera only in the capability for nocardomycolic acid synthesis; the determination of this end product is a powerful tool for taxonomists (see chapter 7).

Given their common metabolic activities, *Nocardia* are to some extent a connecting link between *Mycobacterium, Corynebacterium* and *rhodochrous*-like organisms sometimes showing similar overlapping activities with them. It is conceivable, however, that some metabolic activities show periodical changes in correlation with the different phases of growth thereby allowing *Nocardia* to be distinguished from related organisms; relevant studies are, however, extremely rare.

15. References

Abbott, B. J. & Casida, L. E., Jr. (1968). Oxidation of alkanes to internal monoalkenes by a *Nocardia*. *J. Bact.* **96,** 925.

Abbott, B. J. & Gledhill, W. E. (1971). The extracellular accumulation of metabolic products by hydrocarbon-degrading microorganisms. *Adv. appl. Microbiol.* **14,** 249.

Adams, J. N. (1970). Recombination and segregation of resistance and auxotrophic characters in crosses in *Nocardia* species. (*Mycobacterium rhodochrous* [Overbeck] Gordon & Mihm). In *Variability of Mycobacteria under Experimental and Clinical Conditions.* Eds E. Freerksen, R. Bönicke & J. H. Thumim. p. 164. Berlin–Göttingen–Heidelberg: Springer-Verlag.

Aggag, M. & Schlegel, H. G. (1973). Studies on a gram-positive hydrogen bacterium, *Nocardia opaca* strain 1b. I. Description and physiological characterization. *Arch. Mikrobiol.* **88,** 299.

Aiso, K., Arai, T., Shidara, I. & Ogi, K. (1954). Nocardorubin, an antibiotic produced by *Nocardia* sp. *J. Antibiot., Tokyo* **7,** 1.

Alexander, M., Marshall, K. C. & Hirsch, P. (1960). Autotrophy and heterotrophy in nitrification. *Transactions* **II,** 586.

Baggi, G., Catelani, D., Galli, E. & Treccani, V. (1972). The microbial degradation of phenylalkanes, 2-phenylbutane, 3-phenylpentane, 3-phenyldodecane and 4-phenylheptane. *Biochem. J.* **126,** 1091.

Ballio, A. & Barcellona, S. (1968). Relations chimiques et immunologiques chez les actinomycétales. I. Les acides gras de 43 souches d'actinomycétes aérobies. *Annls Inst. Pasteur, Paris* **114,** 121.

Banerjee, A. K. & Nandi, P. (1966). Studies on extracellular phosphatase in actinomycetes. *Indian J. exp. Biol.* **4,** 188.

Barchielli, R., Boretti, G., Di Marco, A., Julita, P., Migliacci, A., Minghetti, A. & Spalla, C. (1960). Isolation and structure of a new factor of the vitamin B_{12} group: guanosine diphosphate factor B. *Biochem. J.* **74,** 382.

Barchielli, R., Boretti, G., Julita, P., Migliacci, A. & Mingehetti, A. (1957). A natural guanine-containing analogue of vitamin B_{12}. *Biochim. biophys. Acta* **25,** 452.

Bardi, U., Boretti, G., Di Marco, A., Julita, P., Marnati, M. & Spalla, C. (1958). Produzione di vitamina B_{12} e di porfirine in *Nocardia rugosa. G. Microbiol.* **6,** 81.

Batt, R. D. (1961). Induction of enzymes for pyrimidine catabolism in *Nocardia corallina. J. Bact.* **81,** 59.

Batt, R. D., Hodges, R. & Robertson, J. G. (1971). Gas chromatography and mass spectrometry of the trimethylsilyl ether methyl ester derivatives of long chain hydroxy acids from *Nocardia corallina. Biochim. biophys. Acta* **239,** 368.

Batt, R. D. & Woods, D. D. (1961). Degradation of pyrimidines by *Nocardia corallina. J. gen. Microbiol.* **24,** 207.

Baugh, C. L., Bates, D. S., Claus, G. W. & Werkman, C. H. (1962). Carboxylation of fatty acids by extracts of *Nocardia corallina. Iowa St. J. Sci.* **37,** 23.

Beijerinck, M. W. & Van Delden, A. (1903). Über eine farblose Bacterie, deren C-Bedarf aus der atmosphärischen Luft herrührt. *Zentbl. Bakt. ParasitKde Abt.* **11,** 1033.

Beliĉ, I. & Soĉiĉ, H. (1970). Microbiological dehydrogenation of tomatidine. *Experimentia* **27,** 626.

Belozersky, A. N. & Spirin, A. S. (1960). In *The Nucleic Acids*. Eds E. Chargaff, J. N. Davidson. **3,** 147. New York & London: Academic Press.

Bergey's Manual of Determinative Bacteriology (1974). Eds R. E. Buchanan & N. E. Gibbons. Baltimore: Williams and Wilkins.

Bloch, H. (1950). Studies on the virulence of tubercle bacilli. Isolation and biological properties of a constituent of virulent organisms. *J. exp. Med.* **91,** 197.

Bönicke, R. (1961). Die Bedeutung der Acylamidasen für die Identifizierung und Differenzierung der verschiedenen Arten der Gattung *Mycobacterium*. In *Jahresbericht Borstel*. Ed. E. Freerksen. **5,** 7. Berlin–Göttingen–Heidelberg: Springer-Verlag.

Bönicke, R. & Juhász, S. E. (1965). Morphogenetische und biochemische Kriterien zur Unterscheidung der Gattungen *Nocardia* und *Mycobacterium*. *Ann. Inst. Super. Sanitá* **1,** 629.

Bordet, C. & Michel, G. (1963). Etude des acides gras isolés de plusieurs espèces de *Nocardia*. *Biochim. biophys. Acta* **70,** 613.

Bordet, C. & Michel, G. (1964). Isolement d'un nouvel alcool, le 16-hentriacontanol a partir des lipides de *Nocardia brasiliensis*. *Bull. Soc. Chim. biol.* **46,** 1101.

Bordet, C. & Michel, G. (1966*a*). Isolement et structure des nocardones, cétones à haut poids moléculaire de *Nocardia asteroides*. *C.r. hebd. Séanc. Acad. Sci., Paris* **262,** 1810.

Bordet, C. & Michel, G. (1966*b*). Les nocardols, alcool à haut poids moléculaire de *Nocardia asteroides*. *C.r. hebd. Séanc. Acad. Sci., Paris* **262,** 1294.

Bordet, C. & Michel, G. (1969). Structure et biogenèse des lipides a haut poids moléculaire de *Nocardia asteroides*. *Bull. Soc. Chim. biol.* **51,** 527.

Bordet, C., Etémadi, A-H., Michel, G. & Lederer, E. (1965). Structure des acides nocardiques de *Nocardia asteroides*. *Bull. Soc. chim. Fr.* 234.

Bradley, S. G. (1966). Genetics in applied microbiology. *Adv. appl. Microbiol.* **8,** 29.

British Patent (1959). No. 862,701. (Cited in: Raymond, R. L. & Jamison, V. W. (1971).)

Brodie, H., Possanza, G. & Townsley, J. D. (1968). Studies on the mechanism of estrogen biosynthesis. V. Stereochemical comparison of aromatization in placental and microbiological systems. *Biochim. biophys. Acta* **152,** 770.

Brown, O. & Clark, J. B. (1961). The tricarboxylic acid cycle in *Nocardia corallina*. *Proc. Okla. Acad. Sci.* **41,** 94.

Brown, O. & Clark, J. B. (1966*a*). Evidence for the pentose cycle in *Nocardia corallina*. *Proc. Soc. exp. Biol. Med.* **122,** 887.

Brown, O. & Clark, J. B. (1966*b*). Pigment in *Nocardia corallina*. *J. Bact.* **92,** 1844.

Brown, O. & Reda, S. (1967). Enzyme and permeability changes during morphogenesis of *Nocardia corallina*. *J. gen. Microbiol.* **47,** 199.

Büki, K. G., Ambrus, G. & Szabó, A. (1969). Microbiological decomposition of 17α-methyl-17β-hydroxy steroids with androstane nucleus. *Acta microbiol. hung.* **16,** 253.

Cain, R. B. (1958). The microbial metabolism of nitro-aromatic compounds. *J. gen. Microbiol.* **19,** 1.

Cain, R. B. (1966*a*). Induction of an anthranilate oxidation system during the metabolism of ortho-nitrobenzoate by certain bacteria. *J. gen. Microbiol.* **42,** 197.

Cain, R. B. (1966*b*). Utilization of anthranilate and nitrobenzoic acids by *Nocardia opaca* and a *Flavobacterium. J. gen. Microbiol.* **42,** 219.

Cain, R. B. (1968). Anthranilic acid metabolism by microorganisms. Formation of 5-hydroxyanthranilate as an intermediate in anthranilate metabolism by *Nocardia opaca. Antonie van Leeuwenhoek* **34,** 417.

Cain, R. B. & Cartwright, N. J. (1960). On the properties of some aromatic ring-opening enzymes of species of the genus *Nocardia. Biochim. biophys. Acta* **37,** 197.

Cain, R. B., Tranter, E. K. & Darrah, J. A. (1968). The utilization of some halogenated aromatic acids by *Nocardia.* Oxidation and metabolism. *Biochem. J., Tokyo* **106,** 211.

Calmes, R. & Deal, S. J. (1972). Glycerol transport by *Nocardia asteroides. Can. J. Microbiol.* **18,** 1703.

Cartwright, N. J. & Cain, R. B. (1959). Bacterial degradation of nitrobenzoic acids. 2. Reduction of the nitro group. *Biochem. J.* **73,** 305.

Cérbon, J. & Ortigoza-Ferado, J. (1968). Phosphate dependence of monosaccharide transport in *Nocardia. J. Bact.* **95,** 350.

Chacko, C. I., Lockwood, J. L. & Zabik, M. (1966). Chlorinated hydrocarbon pesticides: degradation by microbes. *Science, N.Y.* **154,** 893.

Chang, F. N. & Sih, C. J. (1964). Mechanisms of steroid oxidation by microorganisms. VII. Properties of the 9α-hydroxylase. *Biochemistry, N.Y.* **3,** 1551.

Coombe, R. G., Tsong, Y. Y., Hamilton, P. B. & Sih, C. J. (1966). Mechanisms of steroid oxidation by microorganisms. X. Oxidative cleavage of estrone. *J. biol. Chem.* **241,** 1587.

Crawford, L. V. (1958). Studies on the aspartate-decarboxylase of *Nocardia globerula. Biochem. J.* **68,** 221.

Dasani, U. P. & de Sa, J. D. H. (1965). Utilization of carbon and nitrogen compounds by Indian *Nocardia* species isolated from soil and other sources. *Indian J. exp. Biol.* **3,** 130.

Davis, J. B. (1964). Microbial incorporation of fatty acids derived from *n*-alkanes into glycerides and waxes. *Appl. Microbiol.* **12,** 210.

Davis, J. B. & Raymond, R. L. (1961). Oxidation of alkyl-substituted cyclic hydrocarbons by a *Nocardia* during growth on *n*-alkanes. *Appl. Microbiol.* **9,** 383.

Di Marco, A., Marnati, M. P., Migliacci, A., Rusconi, A. & Spalla, C. (1962). Requirement of some *Nocardia rugosa* mutant strains for the biosynthesis of vitamin B_{12}. *Int. Z. Vitam Forsch* **31,** 350.

Di Marco, A. & Spalla, C. (1961). Mutazioni e auxotrofia nello studio della biosintesi de vitamina B_{12} in *Nocardia rugosa. G. Microbiol.* **9,** 237.

Dodson, R. M. & Muir, R. D. (1958*a*). Microbiological transformation. II. The microbiological aromatization of steroids. *J. Am. chem. Soc.* **80,** 5004.

Dodson, R. M. & Muir, R. D. (1958*b*). Microbiological transformations. III. The hydroxylation of steroids at C-9. *J. Am. chem. Soc.* **80,** 6148.

Donoghue, N. A. & Trudgill, P. W. (1973). The metabolism of cyclohexane-1,2-diols by *Nocardia globerula* CL1. *Biochem. Soc. Transactions* **I,** 1287.

Duff, R. B. & Webley, D. M. (1959). Production of pentose intermediates during growth of *Nocardia opaca* and other saprophytic soil nocardias and mycobacteria. *Biochem. biophys. Acta* **34,** 398.

El-Registan, G. I., Kirillova, N. F. & Krasilnikov, N. A. (1965). Carotenoid pigments in *Nocardia asteroides* (*Proactinomyces asteroides*). *Izv. Akad. Nauk. SSSR. Ser. Biol.* **30,** 128.

Erikson, D. (1949). Differentiation of the vegetative and sporogeneous phases of the actinomycetes. 4. The partially acid-fast proactinomycetes. *J. gen. Microbiol.* **3,** 361.

Etémadi, A-H. & Lederer, E. (1965). Sur la biogénèse des acides nocardiques de *Nocardia asteroides. Bull. Soc. Chim. biol.* **47,** 107.

Etémadi, A. H., Markovits, J. & Pinte, F. (1966). Sur les différents types d'acides nocardiques isolés de *Nocardia opaca. C.r. hebd. Séanc. Acad. Sci., Paris* **263,** 835.

Farshtchi, D. & McClung, N. M. (1970). Effect of substrate on fatty acid production in *Nocardia asteroides. Can. J. Microbiol.* **16,** 213.

Fonkin, G. S., Herr, M. E. & Murray, H. C. (1966). U.S. Patent 3,281,330. (Cited in: Abbott, B. J. & Gledhill, W. E. (1971).)

Frontali, C., Hill, L. R. & Silvestri, L. G. (1965). The base composition of deoxyribonucleic acids of *Streptomyces. J. gen. Microbiol.* **38,** 243.

Genghof, D. S. (1970). Biosynthesis of ergothioneine and hercynine by fungi and Actinomycetales. *J. Bact.* **103,** 475.

Gerber, N. N. (1966). Phenazines and phenoxazinones from some novel Nocardiaceae. *Biochemistry, N.Y.* **5,** 3824.

Gerber, N. N. (1969). New microbial phenazines. (1). *J. Heterocycl. Chem.* **6,** 297.

Gerber, N. N. (1970). A novel, cyclic, tripyrrole pigment from *Actinomadura* (*Nocardia*) *madurae. Tetrahedron Lett.* **11,** 809.

Germain, P., Lefebvre, G., Bena, B. & Gay, R. (1972). Étude *in vitro* des caractéristiques de la 5α-$\Delta 4$ stéroide réductase de *Nocardia corallina. C.r. Séanc. Soc. Biol.* **166,** 1123.

Germanier, R. & Wuhrmann, K. (1963). Über den aeroben mikrobiellen Abbau aromatischer Nitroverbindungen. *Pathologia Microbiol.* **26,** 569.

Gibson, D. T., Wang, K. C., Sih, C. J. & Whitlock, H., Jr. (1966). Mechanisms of steroid oxidation by microorganisms. IX. On the mechanism of ring A cleavage in the degradation of 9,10-seco steroids by microorganisms. *J. biol. Chem.* **241,** 551.

Goodfellow, M. (1971). Numerical taxonomy of some nocardioform bacteria. *J. gen. Microbiol.* **69,** 33.

Goodfellow, M., Minnikin, D. E., Patel, P. V. & Mordarska, H. (1973). Free nocardomycolic acids in the classification of nocardias and strains of the "*rhodochrous*" complex. *J. gen. Microbiol.* **74,** 185.

Gordon, R. E. (1966). Some strains in search of a genus—*Corynebacterium, Mycobacterium, Nocardia* or what? *J. gen. Microbiol.* **43,** 329.

Gordon, R. E. & Horan, A. C. (1968). A piecemeal description of *Streptomyces griseus* (Krainsky) Waksman and Henrici. *J. gen. Microbiol.* **50,** 223.

Guinand, M. & Michel, G. (1966). Structure d'un peptidolipide isole de *Nocardia asteroides*, la peptidolipine NA. *Biochim. biophys. Acta* **125,** 75.

Haccius, B. & Helfrich, O. (1958). Untersuchungen zur mikrobiellen Benzoloxydation. 2. Mitt. Beschreibung und systematische Stellung benzolabbauender Mikroorganismen. *Arch. Mikrobiol.* **28,** 394.

Harington, J. S. (1960). Synthesis of thiamine and folic acid by *Nocardia rhodnii*, the micro-symbiont of *Rhodnius prolixus*. *Nature, Lond.* **188,** 1027.

Hill, L. R. (1966). An index to deoxyribonucleic acid base compositions of bacterial species. *J. gen. Microbiol.* **44,** 419.

Hill, S. & Postgate, H. R. (1969). Failure of putative nitrogen fixing bacteria to fix nitrogen. *J. gen. Microbiol.* **58,** 277.

Hirsch, P. (1958). Stoffwechselphysiologische Untersuchungen an *Nocardia petroleophila* n. sp. *Arch. Mikrobiol.* **29,** 368.

Hirsch, P. (1964). Oligocarbophilie (Wachstum auf Kosten von Luftverunreinigungen) bei Mycobakterien und einigen ihnen nahestehenden Actinomyceten. In *Stoffwechsel und Stoffwechselprodukte von Mycobakterien*. Eds E. Freerksen & R. Bönicke. p. 70. Stuttgart: Gustav Fischer Verlag.

Hirsch, P. (1965). Bacterial oxidation and utilization of carbon monoxide. *Bact. Proc.* 108.

Hori, M., Ito, E., Takita, T., Koyama, G., Takeuchi, T. & Umezawa, H. (1964). A new antibiotic, Formycin. *J. Antibiot., Tokyo* **17,** 96.

Hosler, P. & Eltz, R. W. (1969). In *Fermentation Advances*. Ed. D. Pearlman, p. 789. New York: Academic Press.

Hotchkiss, M. & Simons, M. (1959). Utilization of inorganic nitrogen by *Nocardia asteroides* and *Nocardia brasiliensis*. *Bact. Proc.* **1959,** 42.

Houghton, C. & Cain, R. B. (1972). Microbial metabolism of the pyridine ring. Formation of pyridinediols (dihydroxypyridines) as intermediates in the degradation of pyridine compounds by micro-organisms. *Biochem. J.* **130,** 879.

Ide, M. (1971). Occurrence of cyclic AMP in *Nocardia erythropolis*. *Archs. Biochem. Biophys.* **144,** 262.

Ioneda, T., Lederer, E. & Rozanis, J. (1970). Sur la structure des diesters de tréhalose ("cord factors") produits par *Nocardia asteroides* et *Nocardia rhodochrous*. *Chem. Phys. Lipids* **4,** 375.

Jamison, V. W., Raymond, R. L. & Hudson, J. O. (1969). Microbial hydrocarbon co-oxidation. III. Isolation and characterization of an α,α'-dimethyl-cis,cis-muconic acid-producing strain of *Nocardia corallina*. *Appl. Microbiol.* **17,** 853.

Jochens, G. (1974). Ph.D. Thesis. Personal communication from Peter Hirsch.

Jones, L. A. & Bradley, S. G. (1964). Phenetic classification of actinomycetes. *Devs ind. Microbiol.* **5,** 267.

Kajinami, S. & Smith, E. C. (1971). Glutamine synthetase from *Nocardia corallina*. *J. Biochem., Tokyo* **69,** 429.

Kanemasa, Y. & Goldman, D. S. (1965). Direct incorporation of octanoate into long-chain fatty acids by soluble enzymes of *Mycobacterium tuberculosis*. *Biochem. biophys. Acta* **98,** 476.

Khuller, G. K. & Brennan, P. J. (1972). The polar lipids of some species of *Nocardia*. *J, gen. Microbiol.* **73,** 409.

Kleinzeller, A. & Fencl, Z. (1952). *Chem. Listy* **46,** 300. (Cited in: Abbott, B. J. & Gledhill, W. E. (1971).)

Kluepfel, D. & Vézina, D. (1970). Microbial aromatization of androsta-1,4,7-triene-3,17-dione. *Appl. Microbiol.* **20,** 515.

Kondo, E., Stein, B. & Sih, C. J. (1969). Microbial metabolism of tetra- and hexahydroindanpropionic acid derivatives. *Biochim. biophys. Acta* **176,** 135.

Koyama, G., Maeda, K., Umezawa, H. & Iitaka, Y. (1966). The structural studies of formycin and formycin B. *Tetrahedron Lett.* **6,** 597.

Koyama, G. & Umezawa, H. (1965). Formycin B and its relation to formycin. *J. Antibiot., Tokyo* **18,** 175.

Krámli, A. & Horváth, J. (1948). Microbiological oxidation of sterols. *Nature, Lond.* **162,** 619.

Krámli, A. & Horváth, J. (1949). Microbiological oxidation of sterols. *Nature, Lond.* **163,** 219.

Lanéelle, M-A. (1966). La relation entre les cétones a haut poids moléculaire isolés de *Nocardia brasiliensis* et les acides nocardomycoliques. *C.r. hebd. Séanc. Acad. Sci., Paris* **263,** 560.

Lanéelle, M-A. & Asselineau, J. (1970). Caracterisation de glycolipides dans une souche de *Nocardia brasiliensis. FEBS Letters* **7,** 64.

Lanéelle, M-A., Asselineau, J. & Castelnuovo, G. (1965). Etudes sur les mycobactéries et les nocardiae. IV. Composition des lipides de *Mycobacterium rhodochrous, M. pellegrino,* sp., et de quelques souches de nocardiae. *Annls Inst. Pasteur, Paris* **108,** 69.

Leadbetter, E. & Foster, J. (1959). Oxidation products formed from gaseous alkanes by the bacterium *Pseudomonas methanica. Archs Biochem. Biophys.* **82,** 491.

Lechevalier, M. P., Horan, A. C. & Lechevalier, H. (1971). Lipid composition in the classification of *Nocardia* and mycobacteria. *J. Bact.* **105,** 313.

Lechevalier, H. A., Lechevalier, M. P. & Gerber, N. N. (1971). Chemical composition as a criterion in the classification of actinomycetes. *Adv. appl. Microbiol.* **14,** 47.

Lee, S. S. & Sih, C. J. (1964). Mechanisms of steroid oxidation by microorganisms. VI. Metabolism of 3β-hydroxy-5α,6α-oxidoandrostan-17-one. *Biochemistry, N.Y.* **3,** 1267.

Lennarz, W. J., Scheuerbrandt, G. & Bloch, K. (1962). The biosynthesis of oleic- and 10-methylstearic acids in *Mycobacterium phlei, J. biol. Chem.* **327,** 664.

Marr, E. K. & Stone, R. W. (1961). Bacterial oxidation of benzenes. *J. Bact.* **81,** 425.

Martin, J. K. & Batt, R. D. (1957*a*). Studies on the nutrition of *Nocardia corallina. J. Bact.* **74,** 225.

Martin, J. K. & Batt, R. D. (1957*b*). Oxidation of propionic acid by *Nocardia corallina. J. Bact.* **74,** 359.

McClung, N. M., Salser, J. S. & Santoro, T. (1960). Growth studies of *Nocardia* species. I. Respiration of carbohydrates by *Nocardia rubra. Mycologia* **52,** 845.

McKenna, E. J. (1966). Ph.D. Thesis, University of Iowa, Ames, Iowa. (Cited in: Raymond, R. L. & Jamison, V. W. (1971).)

Metcalf, G. & Brown, M. E. (1957). Nitrogen fixation by a new species of *Nocardia. J. gen. Microbiol.* **17,** 567.

Michel, G. & Lederer, E. (1962). Isolement et constitution chimique des nocardols de *Nocardia asteroides*. *Bull. Soc. chim. Fr.* 651.

Midwinter, G. G. & Batt, R. D. (1960). Endogenous respiration and oxidative assimilation in *Nocardia corallina*. *J. Bact.* **79,** 9.

Migliacci, A. & Rusconi, A. (1961). An intermediate in the biosynthesis of vitamin B_{12} present in *Nocardia rugosa* as a corrinoid conjugate. *Biochim. biophys. Acta* **50,** 370.

Minnikin, D. E., Patel, P. V. & Goodfellow, M. (1974). Mycolic acids of representative strains of *Nocardia* and the "*rhodochrous*" complex. *FEBS Letters* **39,** 322.

Murray, J. R., Scheikowski, T. A. & MacRae, I. C. (1974). Utilization of cyclohexanone and related substances by a *Nocardia* sp. *Antonie van Leeuwenhoek* **40,** 17.

Namdeo, K. N. (1972). Biodegradation of paraquat dichloride. *Indian J. exp. Biol.* **10,** 133.

Nolof, G. (1962). Beitrage zur Kenntnis des Stoffwechsels von *Nocardia hydrocarbonoxydans* n. spec. *Arch. Mikrobiol.* **44,** 278.

Nolof, G. & Hirsch, P. (1962). *Nocardia hydrocarbonoxydans* n. spec., ein oligocarbophiler Actinomycet. *Arch. Mikrobiol.* **44,** 266.

Norris, D. B. & Trudgill, P. W. (1971). The metabolism of cyclohexanol by *Nocardia globerula* CL1. *Biochem. J.* **121,** 363.

Norris, D. B. & Trudgill, P. W. (1972). The purification and properties of cyclohexanone oxygenase from *Nocardia globerula* CL1. *Biochem. J.* **130,** 30.

Norris, D. B. & Trudgill, P. W. (1973). Multiple forms of cyclohexanone oxygenase from *Nocardia globerula* CL1. *Biochem. Soc. Transactions* **1,** 1252.

Nyiri, L. (1963). Etude des propriétés des enzymes pénicillinase et pénicilline acylase à l'occasion de leur coexistence. *Acta microbiol. hung.* **10,** 261.

Ooyama, J. & Foster, J. W. (1965). Bacterial oxidation of cycloparaffinic hydrocarbons. *Antonie van Leeuwenhoek*, **31,** 45.

Pang, R. H. L. & Bradley, S. G. (1974*a*). Regulation of catabolism in *Nocardia erythropolis*. *Mycologia* **66,** 48.

Pang, R. H. L. & Bradley, S. G. (1974*b*). Regulation of substrate uptake in *Nocardia erythropolis*. *J. Bact.* **118,** 400.

Pommier, M-T. & Michel, G. (1973). Phospholipid and acid composition of *Nocardia* and nocardoid bacteria as criteria of classification. *Biochem. Syst.* **1,** 3.

Prauser, H. (1966). New and rare actinomycetes and their DNA base composition. *Publ. Fac. Sci. Univ. Brno* **475,** 268.

Probst, I. & Schlegel, H. G. (1973). Studies on a gram-positive hydrogen bacterium, *Nocardia opaca* strain 1b. II. Enzyme formation and regulation under the influence of hydrogen or fructose as growth substrates. *Arch. Mikrobiol.* **88,** 319.

Rann, D. L. & Cain, R. B. (1969). The regulation of the enzymes of aromatic ring fission in an actinomycete. *Biochem. J.* **114,** 77.

Rann, D. L. & Cain, R. B. (1973). Regulation of the enzymes of aromatic ring fission in the genus *Nocardia*. *Biochem. Soc. Transactions* **1,** 658.

Raymond, R. L. & Davis, J. B. (1960). n-Alkane utilization and lipid formation by a *Nocardia*. *Appl. Microbiol.* **8,** 329.

Raymond, R. L. & Jamison, V. W. (1971). Biochemical activities of *Nocardia*. *Adv. appl. Microbiol.* **14,** 93.

Raymond, R. L., Jamison, V. W. & Hudson, J. O. (1967). Microbial hydrocarbon co-oxidation. I. Oxidation of mono- and dicyclic hydrocarbons by soil isolates of the genus *Nocardia*. *Appl. Microbiol.* **15,** 357.

Raymond, R. L., Jamison, V. W. & Hudson, J. O. (1969). Microbial hydrocarbon co-oxidation. II. Use of ion-exchange resins. *Appl. Microbiol.* **17,** 512.

Raymond, R. L., Jamison, V. W. & Hudson, J. O. (1971). Hydrocarbon co-oxidation in microbial systems. *Lipids* **6,** 453.

Röhrscheidt, E. & Tárnok, I. (1972*a*). Investigation on *Nocardia* pigments. Chromatographic properties of the pigments and their significance for differentiation of pigmented *Nocardia* strains. *Zentbl. Bakt. ParasitKde Abt. I.* **221,** 221.

Röhrscheidt, E. & Tárnok, I. (1972*b*). Studies on the pigmented strains of the genus *Nocardia*. Influence of light on the pigmented synthesis. Photochromogenic, negative-photochromogenic and scotochromogenic *Nocardia* species. *Zentbl. Bakt. ParasitKde Abt. I.* **222,** 510.

Röhrscheidt, E., Tárnok, Zs. & Tárnok, I. (1970). Purin- und pyrimidinabbauende Enzyme in Mycobakterien und Nocardien. *Zentbl. Bakt. ParasitKde Abt. I.* **215,** 550.

Sari-Aslani, F. S., Harper, D. B. & Higgins, I. J. (1972). Catabolism of 1-phenylalkanes by *Nocardia salmonicolor* (NCIB. 9701). *Biochem. J.* **130,** 19.

Sari-Aslani, F. S., Harper, D. B. & Higgins, I. J. (1974). Microbial degradation of hydrocarbons. Catabolism of 1-phenylalkanes by *Nocardia salmonicolor*. *Biochem. J.* **140,** 31.

Sawa, T., Fukagawa, Y., Homma, I., Wahashiro, T., Takeuchi, T., Hori, M. & Komai, T. (1968). Metabolic conversion of formycin B to formycin A and to oxoformycin B in *Nocardia interforma*. *J. Antibiot., Tokyo* **21,** 334.

Sax, K. J., Holmlund, C. E., Feldman, L. I., Evans, R. H., Jr., Blank, R. H., Shay, A. J., Schultz, J. S. & Dann. M. (1965). Microbiological formation of 1,2-dihydroxysteroids. *Steroids* **5,** 345.

Schneidau, J. D., Jr. & Shaffer, M. F. (1957). Studies on *Nocardia* and other Actinomycetales. I. Cultural studies. *Am. Rev. Tuberc. pulm. Dis.* **76,** 770.

Schneidau, J. D., Jr. (1963). The amidase activity of certain species of *Nocardia* and *Mycobacterium*. *Am. Rev. resp. Dis.* **84,** 563.

Schumann, K. P. (1974). Wege des enzymatischen Abbaus von Catechol und seinen Derivaten durch Nocardien. M.D. Thesis, University of Kiel, Federal Republic of Germany.

Seeler, G. (1962). Untersuchungen über den oxydativen Abbau von Alkanen durch *Nocardia petroleophila*. *Arch. Mikrobiol.* **43,** 213.

Sehgal, S. N. & Vézina, C. (1970). Microbial aromatization of steroids into equilin. *Appl. Microbiol.* **20,** 875.

Sih, C. J. (1962). Microbiological epoxidation of steroids. *J. Bact.* **84,** 382.

Sih, C. J., Laval, J. & Rahim, M. A. (1963). Purification and properties of steroid esterase from *Nocardia restrictus*. *J. biol. Chem.* **238,** 566.

Sih, C. J., Tai, H. H., Tsong, Y. Y., Lee, S. S. & Coombe, R. G. (1968). Mechanisms of steroid oxidation by microorganisms. XIV. Pathway of cholesterol side-chain degradation. *Biochemistry, N.Y.* **7,** 808.

Sih, C. J., Wang, K. C. & Tai, H. H. (1967). C_{22} acid intermediates in the microbiological cleavage of the cholesterol side chain. *J. Am. chem. Soc.* **89,** 1956.

Sih, C. J. & Wang, K. C. (1963). Mechanisms of steroid oxidation by microorganisms. II. Isolation and characterization of $3a\alpha$-H-4α-[3'-propionic acid]-$7a\beta$-methylhexahydro-1,5-indanedione. *J. Am. chem. Soc.* **85,** 2135.

Sih, C. J., Wang, K. C. & Tai, H. H. (1968). Mechanisms of steroid oxidation by microorganisms. XIII. C_{22} acid intermediates in the degradation of the cholesterol side chain. *Biochemistry, N.Y.* **7,** 796.

Singh, K., Marshall, D. J. & Vézina, C. (1970). Microbial transformation of 19-hydroxypregnanes. *Appl. Microbiol.* **20,** 23.

Smith, A., Tranter, E. K. & Cain, R. B. (1968). The utilization of some halogenated aromatic acids by *Nocardia*. Effects on growth and enzyme induction. *Biochem. J.* **106,** 203.

Stanier, R. Y. & Ornston, L. N. (1973). The β-ketoadipate pathway. *Adv. Microbial Physiol.* **9,** 89.

Stoudt, T. H., McAleer, W. J., Kozlowski, M. A. & Marlatt, V. (1958). The microbial dehydrogenation of some pregnanes and allopregnanes to 1,4-pregnadienes. *Archs Biochem. Biophys.* **74,** 280.

Strijewski, A., Tan, T., Bozler, G., Zahn, W. & Wagner, F. (1972). Mikrobiologischer Abbau von 5-pregnen-3β,20α-diol, 5-pregnen-3β,20β-diol und Progesteron. *Hoppe–Seyler's Z. Physiol. chem.* **353,** 1440.

Sun Oil Co. (1969). German Patent 1,810,026. (Cited in: Abbott, B. J. & Gledhill, W. E. (1971).)

Tacquet, A. (1962). Techniques récentes d'identification des mycobactéries. *Annls Soc. belge Méd. trop.* **4,** 383.

Takamiya, A. & Tubaki, K. (1956). A new form of *Streptomyces* capable of growing autotrophically. *Arch. Mikrobiol.* **25,** 58.

Tanzil, H. O. K. & Bönicke, R. (1968*a*). Über das Vorkommen von Diaminoxydasen im genus *Nocardia. Zentbl. Bakt. ParasitKde Abt. I.* **209,** 112.

Tanzil, H. O. K. & Bönicke, R. (1968*b*). The "o-diphenol-row", a new method for the differentiation of *Nocardia*: A preliminary report. *Tubercle, Lond.* **49,** 413.

Tanzil, H. O. K. & Hidajat, S. (1970). The Tween-80 hydrolyzing activity of certain species of *Nocardia. Tubercle, Lond.* **51,** 452.

Tárnok, I. (1971). Pyruvatabhängige Harnstoffaufnahme in *Nocardia pellegrini. Zentbl. Bakt. ParasitKde Abt. I.* **218,** 335.

Tárnok, I., Röhrscheidt, E. & Bönicke, R. (1967). Basenzusammensetzung der Desoxyribonukleinsäure (DNS) von Mycobakterien und verwandten Mikroorganismen. *Rass. Patol. Appar. resp.* **XVII,** 4.

Tewfik, E. M. & Bradley, S. G. (1967). Characterization of deoxyribonucleic acids from streptomycetes and *Nocardia. J. Bact.* **94,** 1994.

Tjhing-Lok Tan, Strijewski, A. & Wagner, F. (1972). Degradation of 5-pregnene-3β,20β-diol by a *Nocardia* sp. Some factors influencing the production of pregnendiol-secoacid. *Arch. Mikrobiol.* **87,** 249.

Treccani, V. (1953). *Ann. Microbiol.* **5,** 232. (Cited in: Raymond, R. L. & Jamison, V. W. (1971).)

Treccani, V., Canonica, L. & de Girolamo, M. G. (1955). *Ann. Microbiol.* **6,** 183. (Cited in: Raymond, R. L. & Jamison, V. W. (1971).)

Treccani, V. & Bianchi, B. (1959). *Atti Congr. Naz. Soc. Ital. Microbiol.*, 10 *Bologna.* p. 1.

Treccani, V., Galli, E., Catelani, D. & Sorlini, C. (1968). Induction of 1,2- and 2,3-diphenol oxygenases in *Pseudomonas desmolyticum. Z. allg. Mikrobiol.* **8,** 65.

Tsukamura, M. & Mizuno, S. (1969). DNA content of mycobacteria and nocardias. *Kekkaku* **44,** 155.

Turfitt, G. E. (1948). The microbiological degradation of steroids. 4. Fission of the steroid molecule. *Biochem. J.* **42,** 376.

Uesaka, I. & McClung, N. M. (1961). The effect of amino acids on growth and acid-fastness of *Nocardia* sp. isolates. *Bact. Proc.* **1961,** 102.

U.S. Patent (1963). No. 3,087,864. (Cited in: Raymond, R. L. & Jamison, V. W. (1971).)

Vacheron, M-J., Arpin, N. & Michel, G. (1970). Isolement d'esters de phleixanthophylle de *Nocardia kirovani. C.r. hebd. Séanc. Acad. Sci., Paris* **271,** 881.

Vacheron, M-J., Guinand, M., Michel, G. & Ghuysen, J-M. (1972). Structural investigation on cell walls of *Nocardia* sp. The wall lipid and peptidoglycan moieties of *Nocardia kirovani. Eur. J. Biochem.* **29,** 156.

Villanueva, J. R. (1964*a*). Nitro-reductase from a *Nocardia* sp. *Antonie van Leeuwenhoek* **30,** 17.

Villanueva, J. R. (1964*b*). The purification of a nitro-reductase form *Nocardia* V. *J. biol. Chem.* **239,** 773.

Wagner, F., Zahn, W. & Buhring, U. (1967). 1. Hexadecene, an intermediate in the microbial oxidation of n-hexadecene *in vivo* and *in vitro. Angew. Chem.* **6,** 359.

Walker, N. & Wiltshire, G. H. (1953). The breakdown of naphthaline by a soil bacterium. *J. gen. Microbiol.* **8,** 273.

Wang, K. C. & Sih, C. J. (1963). Mechanisms of steroid oxidations by microorganisms. IV. Seco intermediates. *Biochemistry, N.Y.* **2,** 1238.

Watanabe, S., Kumagaya, A. & Murooka, H. (1963). Biochemical and cytological studies on the growth of *Nocardia* sp. Part I. *J. gen. Microbiol.* **9,** 363.

Watson, G. K. & Cain, R. B. (1972). Metabolism of pyridine by soil bacteria. *Biochem. J.* **127,** 44.

Wayne, L. G., Juarez, W. J. & Nichols, E. G. (1958). Arylsulfatase activity in aerobic actinomycetes. *J. Bact.* **75,** 367.

Webb, R. (1956). Ph.D. Thesis, Univ. of Oklahoma, Norman. (Cited in: Brown, O. R. & Clark, J. B. (1966).)

Webley, D. M. & De Kock, P. C. (1952). The metabolism of some saturated aliphatic hydrocarbons, alcohols and fatty acids by *Proactinomyces opacus* Jensen (*Nocardia opaca* Waksman & Henrici). *Biochem. J.* **51,** 371.

Webley, D. M., Duff, R. B. & Anderson, G. (1962). The metabolism of iron-, zinc- and manganese-deficient *Nocardia opaca. J. gen. Microbiol.* **29,** 179.

Wegner, G. H. (1969). *Bact. Proc.* p. 62. (Cited in: Abbott, B. J. & Gledhill, W. E. (1971).)

Wieland, Th., Griss, G. & Haccius, B. (1958). Untersuchungen zur mikrobiellen Benzoloxydation. I. Nachweis und Chemismus des Benzolabbaus. *Arch. Mikrobiol.* **28,** 383.

Wodzinski, R. S. & Johnson, M. J. (1968). Yields of bacterial cells from hydrocarbons. *Appl. Microbiol.* **16,** 1886.

Yamaguchi, T. (1967). Similarity in DNA of various morphologically distinct actinomycetes. *J. gen. appl. Microbiol., Tokyo* **13,** 63.

Subject Index

Index of Scientific Names